NUMERICAL METHODS FOR WAVE PROPAGATION

FLUID MECHANICS AND ITS APPLICATIONS
Volume 47

Series Editor: R. MOREAU
MADYLAM
Ecole Nationale Supérieure d'Hydraulique de Grenoble
Boîte Postale 95
38402 Saint Martin d'Hères Cedex, France

Aims and Scope of the Series

The purpose of this series is to focus on subjects in which fluid mechanics plays a fundamental role.

As well as the more traditional applications of aeronautics, hydraulics, heat and mass transfer etc., books will be published dealing with topics which are currently in a state of rapid development, such as turbulence, suspensions and multiphase fluids, super and hypersonic flows and numerical modelling techniques.

It is a widely held view that it is the interdisciplinary subjects that will receive intense scientific attention, bringing them to the forefront of technological advancement. Fluids have the ability to transport matter and its properties as well as transmit force, therefore fluid mechanics is a subject that is particulary open to cross fertilisation with other sciences and disciplines of engineering. The subject of fluid mechanics will be highly relevant in domains such as chemical, metallurgical, biological and ecological engineering. This series is particularly open to such new multidisciplinary domains.

The median level of presentation is the first year graduate student. Some texts are monographs defining the current state of a field; others are accessible to final year undergraduates; but essentially the emphasis is on readability and clarity.

For a list of related mechanics titles, see final pages.

Numerical Methods for Wave Propagation

Selected Contributions from the Workshop held in Manchester, U.K., Containing the Harten Memorial Lecture

Edited by

E. F. TORO

Department of Computing and Mathematics,
Manchester Metropolitan University,
Manchester, U.K.

and

J. F. CLARKE

Department of Computing and Mathematics,
Manchester Metropolitan University,
Manchester, U.K.

SPRINGER-SCIENCE+BUSINESS MEDIA, B.V.

A C.I.P. Catalogue record for this book is available from the Library of Congress.

DOI 10.1007/978-94-015-9137-9

Printed on acid-free paper

Contents

PREFACE

In May 1995 a meeting took place at the Manchester Metropolitan University, UK, with the title *International Workshop on Numerical Methods for Wave Propagation Phenomena.* The Workshop, which was attended by 60 scientists from 13 countries, was preceded by a short course entitled *High–Resolution Numerical Methods for Wave Propagation Phenomena.* The course participants could then join the Workshop and listen to discussions of the latest work in the field led by experts responsible for such developments. The present volume contains written versions of their contributions from the majority of the speakers at the Workshop.

Professor Amiram Harten, but for his untimely death at the age of 50 years, would have been one of the speakers at the Workshop. His remarkable contributions to Numerical Analysis of Conservation Laws are commemorated in this volume, which includes the text of the *First Harten Memorial Lecture*, delivered by Professor P. L. Roe from the University of Michigan in Ann Arbour, USA.

In dealing with wave propagation, the Workshop addressed a topic that lies at the heart of a huge array of scientific, and therefore technological, problems. Specifically, the audience was to hear on topics that ranged from astrophysics to acoustics, from civil engineering to radar, from weather prediction to fluids containing bubbles; the materials involved in these studies covered the whole range from gases, through liquids, to solids, often with these phases interacting in essential ways. The common thread that links all of this work lies in the fact that matter is capable of transmitting information about changes in its state, at some *source* location, from that location to another place; as a consequence the local state of the material at the *target* location, often very far from the *source*, is changed and all without having to move the bulk material itself from *source* to *target*.

Connections between the origins of a disturbance and the latter's destination, as well as being potentially distant, can also be subtle in character... and it is easy to believe that numerical studies of these connections will need complementary subtleties of approach at the level of designing numerical methods. For example, it is one of the prominent features of many wave motions that, though they may be born out of continuous data, they may, nonetheless, progress to a condition for which the local data at some subsequent time exhibits a *sharp gradient*. Such states can be quite well modelled as propagating discontinuities... most famously of course as the shock wave, or *Rankine–Hugoniot* shock, that has become familiar from its association with very high–speed flight in the atmosphere as well as from some (usually unwanted) explosive event. In view of the considerable period of time during which propagating finite–amplitude discontinuities have been known and accepted as excellent models of physical reality (seminal papers by Rankine and Hugoniot are dated 1870 and 1889, respectively, and Riemann's work on nonlinear wave propagation was done in the 1860s), it is surprising to realise that it is only in quite recent times that the logical incorporation of such discontinuous solutions into the formulation of wave problems and their numerical solution has taken place through the medium of *integral*, as opposed to *differential*, forms of *conservation laws*,

Unprecedented advances in the design, and application, of numerical methods for partial differential/integral equations governing wave propagation phenomena has taken place in the last two decades, and one of the principal motivations of the Workshop was to divulge some of these developments by bringing the relevant ideas together in one place. The presentations ranged from design and analysis of new numerical methods to ambitious applications of mature schemes to problems of scientific and technological interest. A variety of approaches were represented at the Workshop, ranging from explicit schemes to implicit and from central differencing to upwinding. In spite of the advances reported, it is, however, not sensible to relax one's vigilance when it comes to mathematical formulation and numerical solution of wave propagation problems. The continuing need for careful thought about the way in which the *continuous* problem is transformed into a *discrete* problem, so the latter will generate physically meaningful solutions, is a subtext that underlies the work that is presented here.

We thank all the participants to the Workshop, and particularly we thank all the speakers and contributors to this volume. We also thank Kluwer Academic Publishers, and especially Dr. Karel Neverdeen, for their professional support and for making the publication of this volume a reality. Thanks are also due to Ms Tracy McKenna, Dr S J Billett, Dr D M Ingram and Ms Wei Hu for their assistance in the running of the short course, the Workshop and in the preparation of this volume.

Eleuterio F. Toro and John F. Clarke (the editors)

Manchester, UK, January 1998

THE HARTEN MEMORIAL LECTURE-NEW APPLICATIONS OF UPWINDING

PHILIP ROE
W.M.Keck Foundation Laboratory
for Computational Fluid Dynamics
Department of Aerospace Engineering
University of Michigan
Ann Arbor, Michigan 48109-2118, USA

1. Introduction

High resolution upwind schemes based on nonlinear limiting and the solution of Riemann problems have matured into a standard technology that has become invaluable in many scientific and industrial contexts. Pivotal contributions due to the late Ami Harten include the development of the Total-Variation-Diminishing (TVD) concept [29], approximate Riemann solvers [35], and the achievement of very high accuracy by means of Essentially-Non-Oscillatory (ENO) schemes [36, 34]. He also wrote a number of excellent survey papers [30, 31] that brought the underlying concepts into the common culture of scientific computation, and by his energy, enthusiasm, and extensive travelling forged enduring links between the mathematical and engineering communities. Because of these links, high resolution schemes combine today great practical utility with a high degree of theoretical sophistication. This workshop testifies to both aspects.

The initial development of upwind schemes was inspired and fuelled by the need to provide accurate solutions of the Euler equations governing compressible inviscid flow to the aerospace and armament industries, but was often sustained, in the case of Ami Harten and others, by the traditional academic values attached to rigor, curiosity, and aesthetics. It has been fascinating to watch how work that was sometimes criticised as 'academic' and impractical has led to procedures that are now widely employed.

I shall not, in this paper, try to survey Ami Harten's own contribution. I do not really believe that I would be the best person to do that because I

E.F. Toro and J.F. Clarke (eds.), Numerical Methods for Wave Propagation, 1–31.

sometimes disagreed with him on quite fundamental issues. For him, everything came down to mathematics and in particular a rather personal view of information theory; the basic question was, 'In order to describe and compute this flow, what information do we really need to have?' For some time before his death he had been working on what he called multiresolution methods, which combined ideas from multigrid theory and wavelet decompositions. I believe that he had worked out his ideas with enough coherence that others will be able to carry them forward (for example [11]) but it is a matter for great regret that we will never know what his own final thoughts would have been.

My own stress has always been at the interface of mathematics and physics; seeking those mathematical formulations of a problem that best reveal the underlying physics and the numerical pocesses that most faithfully preserve them. I recall with pride the arguments that this difference led to, because from Ami it was a high compliment that he regarded your ideas as worth disagreeing with. Indeed, he regarded argument with a worthy opponent as the most fruitful means of working toward clarity, and he was a formidable debater. So it is my own view of high resolution methods that will be presented here, sadly unopposed.

Because these methods are, from any viewpoint, firmly based in the strongly descriptive mathematics of hyperbolic partial differential equations, they are applicable not only to the Euler equations, but to any other set of equations sharing the same structure. The extension to flows of gases with non-ideal equations of state appears in several places; many of the earlier references are summarised in [80]. The paper in these proceedings by Marti presents results for relativistic gas dynamics, and documents the superiority of high-resolution upwind methods over more empirical techniques in this very challenging field. See also [21, 73]. A nice account of the difficulties arising from application to multicomponent flows with chemistry has been given, together with a promising method for solving them, in [77]. Alternative, quasi-conservative approaches are to be found in [1, 42] and a highly impressive application in [68]. Application to ionised flows is discussed in [16], to suspensions of liquid droplets in [72], and to shocks travelling through multiple condensed phases in [54]. A treatment of nonlinear elastic waves in solids appears in [78]. Although discontinuous behaviour is most often thought of in conjunction with compressible flows, there are of course many essentially incompressible flows where a density jump is either imposed in the initial conditions (interface problems) or develops in the course of the flow due to chemical reactions. Such problems are treated in, for example [2, 9, 44, 55]. Application to Maxwell's equations is relatively straightforward; a method that includes dispersion effects is given by [56].

Several of these papers describe computations in which the natural ability of upwind schemes to resolve discontinuities is further enhanced by local mesh refinement in strongly-varying regions of the solution [2, 54, 68, 78]. Upwind schemes appear to be especially well suited to this procedure because the use of wave decompositions greatly reduces spurious reflections at mesh interfaces [10, 63, 66, 67]. Extensions have been made to flows in which the geometry, and even the topology, of the bodies involved changes with time [8]. This list of applications is not intended to be comprehensive; the production of a complete bibliography would be a formidable task, but one worth doing.

The first part of this paper starts with a brief review of the upwind methodology, although complete beginners should perhaps precede this by consulting Leveque's textbook [46]. I also include some observations that relate to the efficient coding of large systems. I will then describe other recent applications, in which I have had some personal involvement, of 'traditional' upwind methods, based like those mentioned above on the one-dimensional physics of Riemann problems. Specifically, I consider the equations of extended thermodynamics and of magnetohydrodynamics. In both cases the system of unknowns is considerably larger than we are accustomed to when dealing with the Euler equations. The first step, of course, is to determine the eigenstructure of any new system. Although the immediate objective may simply be to make a characteristic decomposition of the equations, leading to a properly upwinded numerical flux function, the exercise inevitably reveals much of interest about the physics of the system, and sometimes this imposes a distinctive twist to the way the method must be applied in that particular case.

In the second part of this paper, some non-traditional characteristic-based methods will be described, that are applicable to linear problems and are designed to minimise, in fact to eliminate, numerical dissipation. They are intended for use in predicting the propagation of acoustic, electromagnetic, or elastic waves over many wavelengths.

2. Classical Upwind Schemes

2.1. FINITE-VOLUME METHODS

The most usual formulation is to divide the computational domain into non-overlapping control volumes called cells, and to take as unknowns the values of the conserved variables (mass, momentum, energy, whatever) within each cell. In the simplest, first-order, version of the scheme the conserved variables are assumed to be uniformly distributed within each cell, and one then asks the question, if the initial data were so distributed, how would the solution evolve? At least for small times, the answer is that each pair of cells with a common interface would exchange plane waves across that interface,

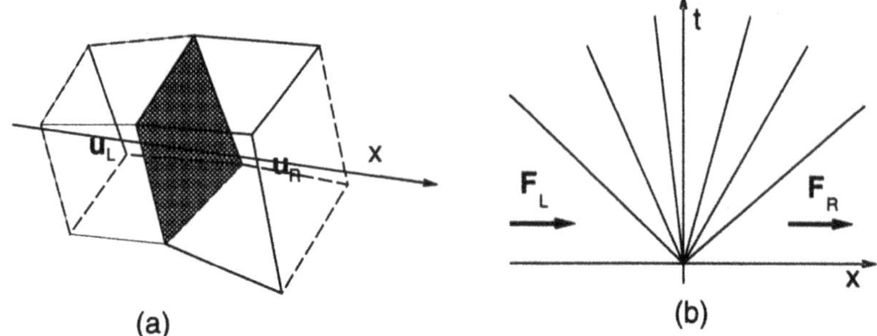

Figure 1. (a) Two finite volumes with a common interface. (b) Time evolution of the solution along line normal to the interface, if the initial data is piecewise constant.

and the resulting flux of conserved quantities would cause a certain amount of mass, say, to be lost from one cell and gained by the other. Determining the strengths of these waves requires of course that one solves the Riemann problem with the states in these two cells as the 'left' and 'right' data. Specifically, if the governing equations are written in conservation form as

$$\mathbf{u}_t + \mathbf{F}_x + \mathbf{G}_y + \mathbf{H}_z = 0, \tag{1}$$

with x chosen normal to the interface, we solve the one-dimensional problem

$$\mathbf{u}_t + \mathbf{F}_x = 0, \tag{2}$$

with initial data $\mathbf{u} = \mathbf{u}_L$ for $x < 0$ and $\mathbf{u} = \mathbf{u}_R$ for $x > 0$, where $\mathbf{u}_L, \mathbf{u}_R$ are the states in the two cells. The flux function $\mathbf{F}$ then has to be evaluated at $x = 0$.

It usually proves economical, and does not degrade the accuracy, to solve the Riemann problem approximately, through some some form of linearisation. Thus, one writes

$$\mathbf{F}_R - \mathbf{F}_L = \Delta\mathbf{F} = \sum_j \Delta\mathbf{F}_j = \sum_j \alpha_j \lambda_j \mathbf{r}_j, \tag{3}$$

where ΔF_j is the change of flux across the jth wave, of which α_j is the amplitude, λ_j is the speed, and $\mathbf{r}_j$ is a right eigenvector of the matrix $A = \partial\mathbf{F}/\partial\mathbf{u}$. To evaluate the flux at $x = 0$, one can either begin with $\mathbf{F}_L$ and add the changes due to each backward-moving wave, or else begin with $\mathbf{F}_R$ and subtract the changes due to each forward-moving wave. The resulting flux formulae are, respectively,

$$\mathbf{F}^* = \mathbf{F}_L + \sum_{j:\lambda_j<0} \alpha_j \lambda_j \mathbf{r}_j, \tag{4}$$

$$\mathbf{F}^* = \mathbf{F}_R - \sum_{j:\lambda_j>0} \alpha_j \lambda_j \mathbf{r}_j. \tag{5}$$

An elegant alternative is to average the two expressions, as first noted by Harten et al [35]

$$\mathbf{F}^* = \frac{1}{2}[\mathbf{F}_L + \mathbf{F}_R] - \frac{1}{2}\sum_j \alpha_j |\lambda_j| \mathbf{r}_j. \quad (6)$$

Theoretically, this form is attractive because it displays the flux as the mean (central-difference) flux, plus a 'numerical dissipation'. Reduction to this standard form enables comparison with other methods. Practically, the form (6) avoids the use of logical tests (which slow down a vector machine), replacing them with the vectorisable absolute value function. The formula (6) appears in numerous texts and reports.

However, for large systems of equations, it may not actually be the best practical choice.The sum in (6) is over all j, whereas one of the formulae (4,5) will involve fewer than half the total number of waves. Whenever vectorisation is not an issue, the formula (4) should be used if the local particle speed is positive, and (5) if it is negative. When the MHD code described in Section 2.3 was modified in this way, the run time was cut by half. There, the ability to reduce a sum over eight waves to a sum over at most three clearly repays the small logical overhead. However, the trick is also well worthwhile for the three-dimensional Euler equations, where no more than one wave out of five need ever be evaluated. This is true even for multiphase flows, if the wave structure consists of two acoustic waves, together with a contact discontinuity separating flows of different composition. It needs to be stressed that this modification to the flux function will not, to machine accuracy, alter the numerical results.

Higher-order accuracy is commonly achieved by 'reconstructing' the data within each cell, at the beginning of the time step, as a local polynomial function. The reader is referred to the original papers of van Leer [47] and Harten [36], and also to more recent interpretations [7, 6, 39].

2.2. EXTENDED THERMODYNAMICS

The application here is to problems involving dilute gases, whose molecules are so sparse that a description in terms of a continuum substance in thermodynamic equilibrium would be misleading. A relevant parameter is the *Knudsen number* defined by $\mathrm{Kn} = \lambda/L$ where λ is the mean free path and L is a typical body dimension. For gases of everyday experience this number is small, but for high altitude flight λ may be large, and for microminiature devices L may be small. In the limit of large Kn, the only satisfactory description is via Kinetic Theory, and a numerical simulation must either track individual particles (direct simulation) or else use the Boltzmann equation to compute the probability $F(\vec{x}, t, \vec{u})$ that a particle found at $(\vec{x}, t)$ will have velocity $\vec{u}$. There is an intermediate regime where Kn is of order unity which is also of practical importance; typically, the heat transfer to a reentering spacecraft is greatest then [45]. The attempt to compute in this regime by using 'direct simulation' methods proves extremely expensive because of the excessive number of particles required and the long times over which their behaviour must be averaged. The Bolzmann equation is also expensive because the solutions have to be found in a seven-dimensional space. One may attempt to apply continuum concepts in this regime, but the gas can not be regarded as in thermodynamic equilibrium. For example, there is no 'equation of state' connecting pressure with density or internal energy. Indeed, pressure ceases to be a well defined scalar quantity, and must be replaced with a stress tensor.

One way to incorporate nonequilbrium effects into a continuum model is to add higher order derivatives to the Euler equations; second-order terms yield the Navier-Stokes equations, third-order the Burnett equations, fourth-order the Super-Burnett equations. Computationally, this is not an attractive route, because the higher derivatives are hard to evaluate on non-smooth grids, and the stable timesteps for an explicit method are very restrictive. Indeed, the mathematical models often prove to be ill-posed [81], containing unstable modes of behaviour that may be unimportant analytically but create havoc with computer codes. (It is possible that some of these ill-posedness issues have been recently removed [5].)

An alternative that keeps only first-order derivatives is to increase the number of unknowns by introducing the moments of the velocity distribution. Let c_i be the ith component of the random velocity (i.e. the particle velocity relative to the mean flow) , and let the notation $<>$ denote integration over the entire velocity space. Then we have, if m is the mass of one particle,

$$m < F > = \rho \tag{7}$$

$$m < c_i F > = 0 \tag{8}$$

$$m < c_i c_j F > = P_{ij} \tag{9}$$

$$m < c_i c_j c_k F > = Q_{ijk} \tag{10}$$

$$m < c_i c_j c_k c_l F > = R_{ijkl} \tag{11}$$

and so on. Here, P_{ij} is the pressure tensor referred to above, Q_{ijk} is a third-order tensor of heat fluxes, and R_{ijkl} is a fourth-order tensor describing fluxes of heat fluxes. The quantities P, Q, R, etc., are treated as additional, independent, unknowns. In principle, an infinite number of these moments would be required to describe completely an arbitrary distribution of velocities. However, a totally random, unorganised distribution is rather unlikely, and the hope is that some reduced description will be adequate.

One way to proceed is to assume some functional form for the distribution function F as it depends on the random velocities. For example, assuming

$$F(\vec{c}) \propto \exp(-|\vec{c}|^2/2RT),$$

which is the *Gaussian* distribution characteristic of a flow that has had ample time for its molecules to reach a state of equilibrium, reduces Boltzmann's equation to the Euler equations. There is a long history of attempts to find the correct generalisation of this idea. The choice

$$F(\vec{c}) \propto \exp(-|\vec{c}|^2/2RT)\mathcal{P}(\vec{c})$$

is the basis for Grad's approach [27], where $\mathcal{P}(\vec{c})$ is a polynomial correction factor. Substituting assumed distribution functions into moments of the Boltzmann equation gives evolution equations for the moment quantities.

Many problems crop up with such an approach. The time evolution of a particular moment function gets expressed in terms of space derivatives of moments one order higher, often leading to an unending regression that must be broken by means of some closure assumption. The function F, which is supposed to be a probability, may fail to be positive in some regions of velocity space. This may not be fatal, since we are only using F to predict a finite number of moments, but there may not in fact be any positive F that yields these particular moments. The predicted moments are then said to be *unrealisable* and the mathematical model is self-contradictory. Another possibility is that the evolution equations may turn out to have complex wavespeeds, so that an initial-value problem is no longer well-posed. These forms of failure cast grave doubt on the significance of any results obtained.

Recently, Levermore[46] has constructed an heirarchy of moment approximations that by design avoid all of these difficulties. He assumes a distribution function of the form

$$F(\vec{c}) \propto \exp(-\mathcal{P}(\vec{c}))$$

where $\mathcal{P}(\vec{c})$ is a member of a family of polynomials having certain properties of symmetry and completeness. There is an heirarchy of such polynomials, leading to sets of equations with $5, 10, 14, 21, 35\ldots$ members. All such sets are guaranteed to be closed, hyperbolic, and to possess an entropy function, and hence lead to well-posed initial-value problems.

The 5-moment approximation merely recovers the Euler equations, but the 10-moment approximation, somewhat equivalent to neglecting heat transfer in the Navier-Stokes equations, and actually dating back to Maxwell

[52] has an elegant and fascinating structure. The distribution function is

$$F(\vec{c}) \propto \exp\left(-\frac{1}{2}\rho P_{ij}^{-1} c_i c_j\right) \tag{12}$$

where P_{ij} is the pressure tensor. The equations can be written as

$$\mathbf{V}_t + A\mathbf{V}_x + B\mathbf{V}_y + C\mathbf{V}_z = \mathbf{S}/\tau,$$

where $\mathbf{V} = (\rho, u_x, u_y, u_z, P_{xx}, P_{xy}, P_{xz}, P_{yy}, P_{yz}, P_{zz})^T$ is the vector of primitive variables, $\mathbf{S}$ is a source vector deriving from the collision terms in the Boltzmann equation, τ is a relaxation time equal to μ/p, where μ is the viscosity and p the static pressure, and the matrix A, for example, is given for a monatomic gas by

$$A = \begin{pmatrix} u_x & \rho & 0 & 0 & 0 & 0 & 0 & 0 & 0 & 0 \\ 0 & u_x & 0 & 0 & 1/\rho & 0 & 0 & 0 & 0 & 0 \\ 0 & 0 & u_x & 0 & 0 & 1/\rho & 0 & 0 & 0 & 0 \\ 0 & 0 & 0 & u_x & 0 & 0 & 1/\rho & 0 & 0 & 0 \\ 0 & 3P_{xx} & 0 & 0 & u_x & 0 & 0 & 0 & 0 & 0 \\ 0 & 2P_{xy} & P_{xx} & 0 & 0 & u_x & 0 & 0 & 0 & 0 \\ 0 & 2P_{xz} & 0 & P_{xx} & 0 & 0 & u_x & 0 & 0 & 0 \\ 0 & P_{yy} & 2P_{xy} & 0 & 0 & 0 & 0 & u_x & 0 & 0 \\ 0 & P_{yz} & P_{xz} & P_{xy} & 0 & 0 & 0 & 0 & u_x & 0 \\ 0 & P_{zz} & 0 & 2P_{xz} & 0 & 0 & 0 & 0 & 0 & u_x \end{pmatrix}.$$

The source term is

$$\mathbf{S} = \left(0, 0, 0, 0, P_{xx} - \frac{\mathrm{tr}(P)}{3}, P_{xy}, P_{xz}, P_{yy} - \frac{\mathrm{tr}(P)}{3}, P_{yz,}, P_{zz} - \frac{\mathrm{tr}(P)}{3}\right)^T.$$

The eigenvalues of the matrix A are

$$\begin{aligned} \lambda_1 &= u_x - \sqrt{3}c_{xx}, \\ \lambda_{2,3} &= u_x - c_{xx}, \\ \lambda_{4,5,6,7} &= u_x, \\ \lambda_{8,9} &= u_x + c_{xx}, \\ \lambda_{10} &= u_x + \sqrt{3}c_{xx}, \end{aligned}$$

where $c_{xx}^2 = P_{xx}/\rho$, so that the wavespeeds are all real provided that the normal stress in the direction of propagation remains positive. Thus, the wavespeeds are real for all directions if the pressure tensor is positive definite.

The matrix R of right eigenvectors is

$$\begin{pmatrix}
\frac{1}{3} & 0 & 0 & 1 & 0 & 0 & 0 & 0 & 0 & \frac{1}{3} \\
-\frac{c_{xx}}{\sqrt{3}\rho} & 0 & 0 & 0 & 0 & 0 & 0 & 0 & 0 & \frac{c_{xx}}{\sqrt{3}\rho} \\
-\frac{c_{xy}^2}{\sqrt{3}\rho c_{xx}} & -\frac{c_{xx}}{\rho} & 0 & 0 & 0 & 0 & 0 & 0 & \frac{c_{xx}}{\rho} & \frac{c_{xy}^2}{\sqrt{3}\rho c_{xx}} \\
-\frac{c_{xz}^2}{\sqrt{3}\rho c_{xx}} & 0 & -\frac{c_{xx}}{\rho} & 0 & 0 & 0 & 0 & \frac{c_{xx}}{\rho} & 0 & \frac{c_{xz}^2}{\sqrt{3}\rho c_{xx}} \\
c_{xx}^2 & 0 & 0 & 0 & 0 & 0 & 0 & 0 & 0 & c_{xx}^2 \\
c_{xy}^2 & c_{xx}^2 & 0 & 0 & 0 & 0 & 0 & 0 & c_{xx}^2 & c_{xy}^2 \\
c_{xz}^2 & 0 & c_{xx}^2 & 0 & 0 & 0 & 0 & c_{xx}^2 & 0 & c_{xz}^2 \\
\frac{c_{xx}^2 c_{yy}^2 + 2c_{xy}^4}{3c_{xx}^2} & 2c_{xy}^2 & 0 & 0 & c_{xx}^2 & 0 & 0 & 0 & 2c_{xy}^2 & \frac{c_{xx}^2 c_{yy}^2 + 2c_{xy}^4}{3c_{xx}^2} \\
\frac{c_{xx}^2 c_{yz}^2 + 2c_{xy}^2 c_{xz}^2}{3c_{xx}^2} & c_{xz}^2 & c_{xy}^2 & 0 & 0 & c_{xx}^2 & 0 & c_{xy}^2 & c_{xz}^2 & \frac{c_{xx}^2 c_{yz}^2 + 2c_{xy}^2 c_{xz}^2}{3c_{xx}^2} \\
\frac{c_{xx}^2 c_{zz}^2 + 2c_{xz}^4}{3c_{xx}^2} & 0 & 2c_{xz}^2 & 0 & 0 & 0 & c_{xx}^2 & 2c_{xz}^2 & 0 & \frac{c_{xx}^2 c_{zz}^2 + 2c_{xz}^4}{3c_{xx}^2}
\end{pmatrix}$$

These eigenvectors are simple, and have clear physical meaning. The first and tenth represent acoustic waves, and are genuinely nonlinear (so care may be needed to distinguish computationally between shocks and rarefactions). The second, third, eighth, and ninth affect transverse velocities, the fourth is an entropy wave, and the remainder transport transverse shear forces. Because the eigenvectors are sparse, the computational work involved in setting up an upwind flux function is not excessive, particularly if the asymmetric formulae mentioned in Section 2.1 are employed. In fact, it appears from preliminary studies in [13] (also J. A. F. Hittinger, private communication) that calculations based on the 10-moment model may be less expensive than those based on the Navier-Stokes equations because of much more rapid convergence, especially at high Mach numbers. The rapid convergence appears to derive from replacing a diffusive process with a wave process. In particular, an explicit numerical scheme can take the $\mathcal{O}(\Delta t)$ timesteps associated with hyperbolic problems rather than the $\mathcal{O}(\Delta t^2)$ timesteps associated with parabolic problems.

The numerical process can follow rather closely the standard techniques of high-order Godunov methods, using limited reconstruction within cells followed by the solution of approximate (linearised) Riemann problems. The non-standard aspect is the need to incorporate the source terms from the right-hand side, and these create problems only if they are large compared with the convective terms. This depends chiefly on the *stiffness parameter* $k = \Delta t/\tau$ where Δt is the timestep permitted for the source-free problem by the CFL rule, and τ is the relaxation time characterising the return to equilibrium. Accuracy may be lost when this parameter is large. The problem is then described as *stiff*, a situation that occurs in this context when computing flows close to equilibrium. A similar difficulty occurs when attempting to incorporate rapid chemical reactions into the Euler equations.

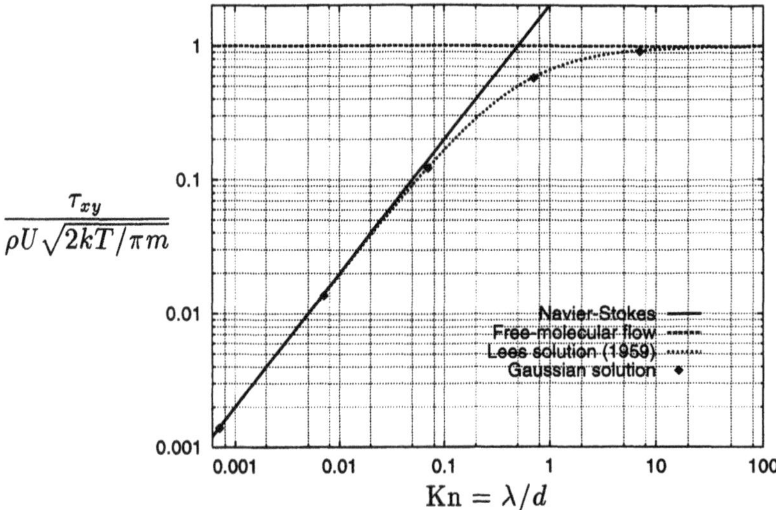

Figure 2. Nondimensional wall shear stress for Couette flow according to the 10-moment approximation and other theories.

Several authors [3, 40, 41, 62] have recently discussed methods for dealing with stiffness of these kinds; it is their concensus that the simple device of *operator splitting* [74] can be greatly improved on.

It is not yet clear to what extent the 10-moment approximation can serve as an adequate model for physical situations, or whether it is merely a stepping-stone toward more complete descriptions. There may however be situations where shear stresses are much more important than heat flow. Possibly this is true of laminar flow in pipes and channels at extremely low Reynolds numbers and moderate Knudsen numbers, such as are encountered in the technology of Micro Electro-Mechanical (MEM) devices. Figs 2 and 3 display the results of a preliminary study by Dr Clinton Groth on the Couette flow between two sliding plates. For a discussion of this problem see ([79], pp 424-433). This particular example is not computationally demanding, but is offered as showing promise that at least some aspects of rarefied transitional flows are well-modelled by a computationally tractable mathematical model.

For more details of the 10-moment model [13, 49] may be consulted. Application to a hypersonic blunt-body flow has been made in [14], again by means of a rather standard Godunov-type approach. Analysis of a 35-moment model is presented in [28], and although space does not permit discussion of it here, the eigenstructure is again surprisingly elegant and tidy. This surely bodes well both for the realism of the mathematical model and also for its computational effciency. For a different approach to rarified

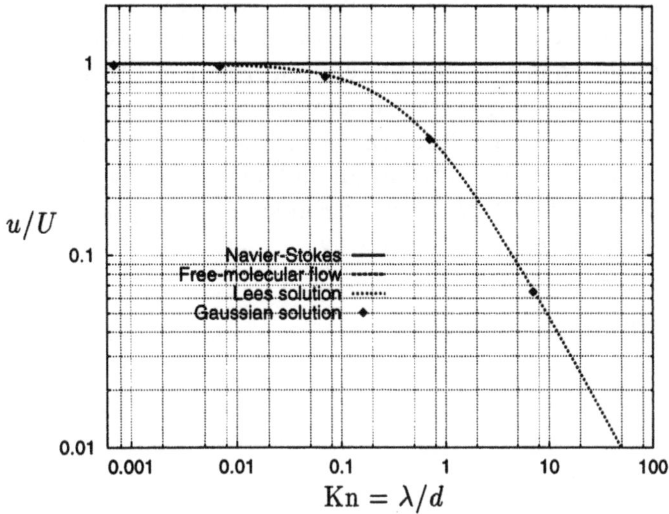

Figure 3. Nondimensional slip velocity for Couette flow according to the 10-moment approximation and other theories.

flows, but again making use of high-resolution upwind schemes, see [85].

2.3. IDEAL MAGNETOHYDRODYNAMICS (MHD)

Here we consider a fluid that is in thermodynamic equilibrium, and satisfies a well-defined equation of state. The complication comes from the fluid being electrically conducting, and interacting with a magnetic field. The equations of motion are commonly written in conservation form as [38]

$$\frac{\partial}{\partial t}\begin{pmatrix}\rho\\ \rho\mathbf{u}\\ \mathbf{B}\\ E\end{pmatrix}+\nabla\cdot\begin{pmatrix}\rho\mathbf{u}\\ \rho\mathbf{u}\mathbf{u}+\mathbf{I}(p+\frac{\mathbf{B}\cdot\mathbf{B}}{2})-\mathbf{B}\mathbf{B}\\ \mathbf{u}\mathbf{B}-\mathbf{B}\mathbf{u}\\ (E+p+\frac{\mathbf{B}\cdot\mathbf{B}}{2})\mathbf{u}-\mathbf{B}(\mathbf{u}\cdot\mathbf{B})\end{pmatrix}=0, \tag{13}$$

where ρ is the fluid density, p is the pressure, $\mathbf{u}$ is the velocity, $\mathbf{B}$ is the magnetic field, $\mathbf{I}$ is a 3×3 identity matrix, and E is the energy,

$$E=\frac{p}{\gamma-1}+\rho\frac{\mathbf{u}\cdot\mathbf{u}}{2}+\frac{\mathbf{B}\cdot\mathbf{B}}{2}.$$

These equations have to be solved subject to the constraint that

$$\nabla\cdot\mathbf{B}=0,$$

which expresses the non-existence of magnetic monopoles.

For one-dimensional waves causing no variation in the y or z directions, the divergence contraint can be simplified to a statement that B_x is constant, and the equations can be reduced from eight to seven. The eigenstructure of the resulting system is fairly complex, and has only recently [12, 71, 86] been given in a form that is numerically useful (older analysis gives expressions that are singular or indeterminate in many cases). We have in fact, in terms of primitive variables $(\rho, u_x, u_y, u_z, B_y, B_z, p)^T$, the matrix R of right eigenvectors given by

$$\begin{pmatrix} \alpha_f \rho & 0 & \alpha_s \rho & 1 & \alpha_s \rho & 0 & \alpha_f \rho \\ -\alpha_f c_f & 0 & -\alpha_s c_s & 0 & \alpha_s c_s & 0 & \alpha_f c_f \\ \alpha_s c_s \beta_y S & -\beta_z & -\alpha_f c_f \beta_y S & 0 & \alpha_f c_f \beta_y S & \beta_z & -\alpha_s c_s \beta_y S \\ \alpha_s c_s \beta_z S & \beta_y & -\alpha_f c_f \beta_z S & 0 & \alpha_f c_f \beta_z S & -\beta_y & -\alpha_s c_s \beta_z S \\ \alpha_s \sqrt{4\pi\rho} a \beta_y & -\sqrt{4\pi\rho}\beta_z S & -\alpha_f \sqrt{4\pi\rho} a \beta_y & 0 & -\alpha_f \sqrt{4\pi\rho} a \beta_y & -\sqrt{4\pi\rho}\beta_z S & \alpha_s \sqrt{4\pi\rho} a \beta_y \\ \alpha_s \sqrt{4\pi\rho} a \beta_z & \sqrt{4\pi\rho}\beta_y S & -\alpha_f \sqrt{4\pi\rho} a \beta_z & 0 & -\alpha_f \sqrt{4\pi\rho} a \beta_z & \sqrt{4\pi\rho}\beta_y S & \alpha_s \sqrt{4\pi\rho} a \beta_z \\ \alpha_f \rho a^2 & 0 & \alpha_s \rho a^2 & 0 & \alpha_s \rho a^2 & 0 & \alpha_f \rho a^2 \end{pmatrix}. \tag{14}$$

The first and seventh waves are the fast magnetoacoustic waves with speeds $u_x \pm c_f$, the second and sixth are Alfvèn waves with speeds $u_x \pm B_x/\sqrt{4\pi\rho}$, the third and fifth are slow magnetoacoustic waves with speeds $u_x \pm c_s$, and the fourth is an entropy wave with speed u_x. Here $c_{f,s}$ are the positive roots, with $c_f \geq c_s$, of the equation

$$c^4 - (a^2 + b^2)c^2 + a^2 b_x^2 = 0,$$

and we define

$$\alpha_f = \sqrt{\frac{a^2 - c_s^2}{c_f^2 - c_s^2}}, \qquad \alpha_s = \sqrt{\frac{c_f^2 - a^2}{c_f^2 - c_s^2}}.$$

Other definitions are

$$b_{x,y,z} = B_{x,y,z}/\sqrt{4\pi\rho}, \qquad \beta_{y,z} = \frac{b_{y,z}}{\sqrt{b_y^2 + b_z^2}}, \qquad b^2 = b_x^2 + b_y^2 + b_z^2,$$

$$a^2 = \gamma p/\rho. \qquad S = \operatorname{sgn} B_x.$$

The only singular case remaining is not algebraic but has a physical significance; it occurs when the transverse magnetic field vanishes ($b_y = b_z = 0$), and the Alfvèn speed coincides with the sound speed ($a^2 = b_x^2$). In that case, the fast, slow, and Alfvèn speeds all coincide. More seriously the eigenvectors are no longer distinct (the parameters $\alpha_{f,s}$ become indeterminate) and the solution of a Riemann problem becomes ill-posed; the strengths of the various waves are not continuously dependent on the left and right states.

However, it can be shown [71] that the linearised flux formulae (4, 5, or 6) are in fact well-posed, and it seems that for computational purposes the singularity can be ignored. Of course this merely says that a stable code can be constructed; it does not guarantee that the delicate physics of these *resonant waves* will be correctly reproduced.

An interesting aspect of the nonlinear MHD Riemann problem is that when the direction of the magnetic field changes sign the Alfvèn waves may merge with either or both of the magnetoacoustic waves to form *intermediate shocks*, which are non-standard shocks into which either more, or fewer characteristics converge than usual. Although their ultimate stability is questionable, it is felt that they may persist for long times [82, 83]. They are, however, predicted by codes using orthodox Riemann solvers that do not specially account for them [4, 12, 57].

When an MHD problem is solved in more than one dimension, the divergence-free condition cannot be satisfied in any simplified form, and so the full system of eight equations needs to be considered. When the 8×8 version of the matrix A is considered, it is found to have a zero eigenvalue, combined with a highly unpleasant eigenvector. The stationary eigenvalue is a very suspicious feature; it does not respect Gallilean invariance, and it will not be damped numerically. This analysis, even though prompted merely by coding issues, actually suggests that in (14) we may not be dealing with the correct governing equations.

A recent analysis by Marcel Vinokur (NASA Ames, private communication) derives the MHD equations from Ampère's Law and Faraday's Law, without at any point assuming that div$\mathbf{B} = 0$, although it is easily shown that if a divergence-free field is imposed as initial data, the field will remain divergence-free. He finds a *non-conservative* system that can be written as

$$\frac{\partial}{\partial t}\begin{pmatrix} \rho \\ \rho\mathbf{u} \\ \mathbf{B} \\ E \end{pmatrix} + \nabla\cdot\begin{pmatrix} \rho\mathbf{u} \\ \rho\mathbf{uu} + \mathbf{I}(p + \frac{\mathbf{B\cdot B}}{2}) - \mathbf{BB} \\ \mathbf{uB} - \mathbf{Bu} \\ (E + p + \frac{\mathbf{B\cdot B}}{2})\mathbf{u} - \mathbf{B}(\mathbf{u}\cdot\mathbf{B}) \end{pmatrix} = \begin{pmatrix} 0 \\ \mathbf{B} \\ \mathbf{u} \\ \mathbf{u}\cdot\mathbf{B} \end{pmatrix} \nabla\cdot\mathbf{B} \tag{15}$$

In any exact solution of these equations that derives from divergence-free initial conditions the terms on the right vanish identically. Also, the wave structure of (15) remains identical to that of (14) in every respect but one. If a nonzero magnetic field divergence is introduced into the initial conditions (as it might be by numerical processes unless some very special precautions are taken) the conservative form predicts that the divergence will not change with time, whereas the nonconservative form predicts that it will convect as a passive scalar.

Interestingly, Powell [65] had already derived the same equations by a purely computational argument. He placed an arbitrary multiple of div$\mathbf{B}$

on the RHS of (15) and showed that only one such modification led to 'nice' computational properties while also preserving the known one-dimensional wave structure.

There is room for a certain amount of debate about the nature of these discoveries. We have two alternative mathematical models that for divergence-free initial conditions are completely equivalent and yet which have very different computational behaviour. The nonconservative model allows that div$\mathbf{B}$ may not be identically zero, and also proposes waves across which the normal component of $\mathbf{B}$ jumps discontinuously. Naturally no-one claims that Nature allows such behaviour. However, we do have a *mathematically consistent* set of equations that proves much easier to discretise than the traditional (div$\mathbf{B} \equiv \mathbf{0}$) description. The eighth wave is merely a device that allows small nonzero div$\mathbf{B}$ to be harmlessly convected instead of remaining where it was generated. In Vinokur's opinion, the extra wave is needed numerically to allow for the inconsistency of a finite-volume scheme in representing both the fluid variables and the magnetic field variables as volume averages, whereas the field variables make more sense as surface averages. In fact, Dai and Woodward [17] resolve the issue by defining their magnetic fields at interfaces and then interpolating them to get data for their Riemann problems. This stresses once again that Computational Physics is sometimes the physics of a computational model as much as the computation of a physical model.

The modified system (15) indeed turns out to be a very satisfactory basis for multidimensional MHD calculations. One application that has been carried out is to the bow shock wave of a comet approaching the sun [24]. Comets consist mostly of rocks held together by ice. As they come closer to the sun the ice begins to sublime into water vapour, subsequently dissociating and then ionising. The charged particles then interact with other charged particles streaming outward from the sun (the solar wind), so that a shockwave eventually forms, whose stand-off distance is of order 10^6 km, compared with a typical comet radius of 10 km.

An adaptive grid is called for to resolve this disparity of scale, and it proved quite simple to modify an adaptive Cartesian-grid code originally developed for the Euler equations by de Zeeuw and Powell [19]. Merely some variables had to be renamed, some array sizes redeclared, and a new flux function substituted. The hoped-for universality of the upwind method was eventually achieved, but only after some fairly subtle preparation. (Essentially the same code was used to produce the results for the 10-moment equations in the previous subsection.)

The calculation is matched to the known size and speed of comet Halley. The comet is assumed to move at an angle of 45° to the solar wind, so that the situation is fully three-dimensional, but steady. (An earlier calculation

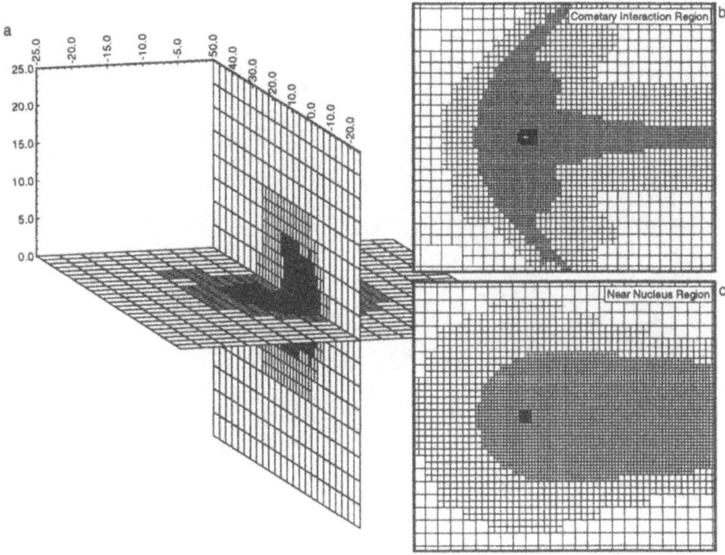

Figure 4. View of the computational box for the comet calculation.

assumed axial symmetry [26].) The calculation is done within a box measuring 50 × 50 × 75 Gm (1Gm=10^9 meters). Fig 4(left) shows this box, and some of the coarse cells used to define the outer part of the domain, together with two cuts through the grid. Within these cuts we can see how cells in inner regions of the grid are subdivided to yield smaller cells. There are in fact sixteen levels of refinement and the cells at each level have half the linear dimensions of their 'parents'. Fig 4(top right) shows a section of the horizontal (equatorial) plane measuring 2.8 × 2.4 Gm. This occupies rather less than one of the coarsest cells, which are 3.125 Gm on each side. Enlarging the inmost region of this figure by a factor of about 80 leads to Fig 4(bottom right), depicting a region roughly 37,000 × 32,000 km. Even this is much larger than the finest cells, whose sides measure only 48 km. The comet itself is not resolved, having a radius of about 10 km. The flow is driven by the 'outgassing' water vapour, of which the comet is effectively a point source. The total number of mesh points required to represent this flow is about 486,000. To fill the whole box with the finest cells would have required 1.7 ×10^9 points, which is an impressive advertisement for the power of adaptive grids.

Fig 5 corresponds to a region slightly larger than that of Fig 4(top right). Pale lines show the direction of the magnetic field and the grey -scale background has an intensity proportional to the field strength. Clustering of the grid reveals the location of the bow shock, whose calculated position is also marked by a pale line. Fig 6 corresponds to the region of Fig 4(bottom

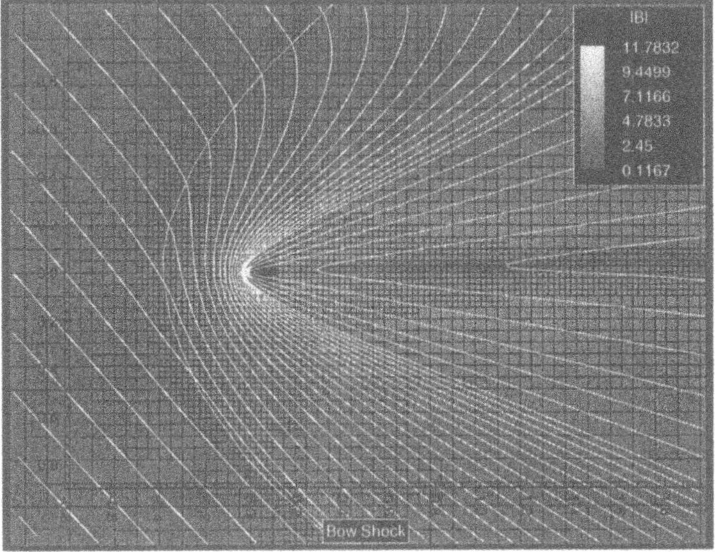

Figure 5. Magnetic field lines and magnitude superposed on the computational grid. The region shown is roughly 3 × 2 Gm and contains the forward portion of the bow shock.

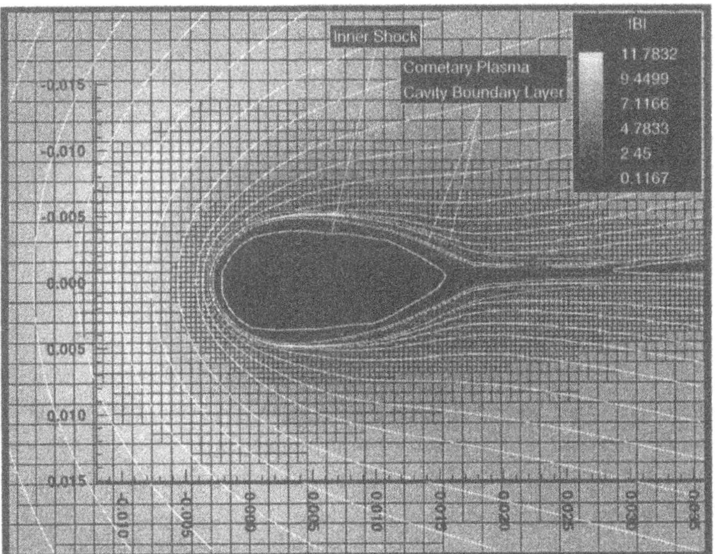

Figure 6. Magnetic field lines and magnitude superposed on the computational grid. The region shown is roughly 50,000 km × 30,000 km and contains the inner shock and diamagnetic cavity.

right). An inner shock appears, which the field lines do not cross, and within which the magnetic field is virtually zero. More images, in color and accompanied by greatly extended discussion, are to be found in [24]. This paper also contains convincing comparison with experimental data gathered from the fly-by missions of space probes Vega1, Vega2, Suisei, and Giotto.

An even more ambitious project is underway to model the interaction of the solar wind with the interstellar medium, a phenomenon that extends to several hundred times the radius of the earth from the sun. A preliminary account appears in [25]. More comprehensive results (Timur Linde, private communication) show impressive detail over a region covering some 3,000AU $\times$ 2,400AU $\times$ 2,400AU, and reproducing many known features. A paper is in preparation.

3. Upwind Leapfrog Methods

These intriguing schemes were first described by Iserles [37], who set them purely in the context of the linear advection equation in one dimension,

$$u_t + au_x = 0. \tag{16}$$

Recall that leapfrog schemes are distinguished by being reversible in time, and hence free from numerical dissipation. One way in which to achieve time-reversibility is to make the stencil symmetric, as in Figure 7a, corresponding to the classical leapfrog scheme. Iserles realised, however, that a weaker condition of point symmetry, as in Figs 7 (b,c,d) is sufficient, and obtained some interesting theorems relating the geometry of the stencil to the properties of the scheme. For three-level schemes he considered stencils comprising $k+1$ mesh points at level n having spatial indices $(j, j+1, j+2, \ldots j+k)$, with additional points at $j+p$ at level $n+1$ and $j+k-p$ at level $n-1$. He then showed that the only stable schemes were those with $p = k/2$ or $p = (k-1)/2$ (if k is even), or with $p = (k+1)/2$ (if k is odd). He showed also that the group velocity of schemes in the first two classes is sometimes negative, but never for the third class.

We will concentrate first on the very simple scheme that can be developed on the stencil of Figure 7b. It is, in one dimension

$$u_j^{n+1} - u_{j-1}^{n-1} = (1-2\nu)(u_j^n - u_{j-1}^n).$$

If we define the numerical phase speed to be a_N, and the phase speed error of a discrete solution $u_j^n = \exp i(n\phi - j\theta)$ to be

$$\varepsilon_p = \frac{a_N - a}{\frac{\Delta x}{\Delta t}},$$

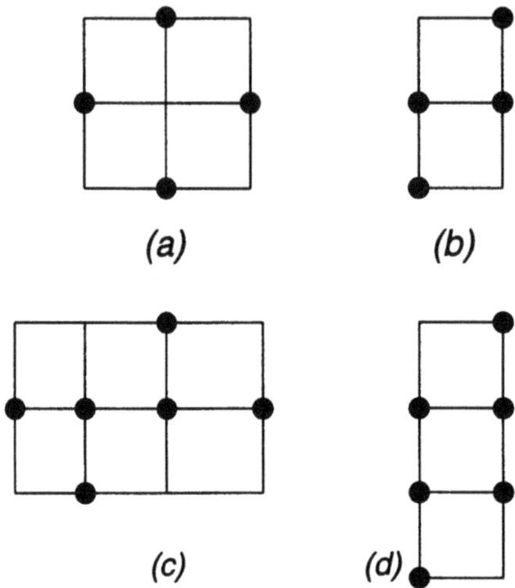

Figure 7. Four symmetric stencils (x, j horizontal, t, n vertical) for the discretisation of (16).

where $a_N = (\phi\Delta x)/(\theta\Delta t)$, then we find for small θ that

$$\varepsilon_p \simeq \frac{\theta^2}{12}\nu(1-\nu)(1-2\nu).$$

The scheme is accurate to within one percent if six or more grid points are available to resolve each wavelength, and to one-tenth of a percent if 18 points are used. It may be observed that the scheme is exact (for linear advection) if $\nu = 0.5$, reducing simply to $u_j^{n+1} = u_{j-1}^{n-1}$.

The extension of this scheme to higher dimensions involves some rethinking of the basic upwind strategy. Paradoxically, it has proved easier to do this for systems of wave equations than for scalar problems. As a prototype, we may consider the equations of linear acoustics,

$$\begin{aligned} p_t + \rho_0 a_0^2(u_x + v_y) &= 0, \\ u_t + (1/\rho_0)p_x &= 0, \\ v_t + (1/\rho_0)p_y &= 0. \end{aligned} \tag{17}$$

From these, we may construct, by linear combination, the bicharacteristic equation,

$$(\partial_t + a_0(\cos\theta\partial_x + \sin\theta\partial_y))(p + \rho_0 a_0(u\cos\theta + v\sin\theta)) = \rho_0 a_0(\sin\theta\partial_x - \cos\theta\partial_y)(v\cos\theta - u\sin\theta). \tag{18}$$

A bicharacteristic equation is one containing only derivatives that lie within a subspace of dimension at least one lower than that of the problem domain; here all derivatives are interior to a plane whose normal in (t, x, y)

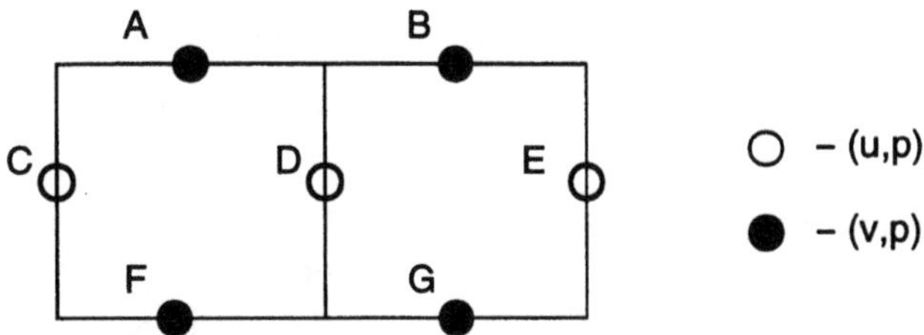

Figure 8. Staggered storage for the secord-order bicharacteristic method applied to the update of u and p.

is $(a_0, -\cos\theta, -\sin\theta)$. Such a plane is tangent to the cone $a_0^2 t^2 = x^2 + y^2$. There are an infinite number of bicharacteristic equations, in contrast to the one-dimensional situation where there are exactly as many characteristic equations as unknowns.

Clearly not more than three of the bicharacteristics equations can be linearly independent, since they are all derived from the three original equations 17. It is also clear that any equation whose physical significance derives from relating interior derivatives in a plane should be discretised, so far as possible, with data lying in that plane. However, the number of bicharacteristic planes that can be appropriately discretised relates to the symmetry of the grid. On a regular grid in two space dimensions this number is four, rather than three. A resolution of this conflict can be found in the use of staggered storage [70].

The arrangement is shown in Figure 8. As with familiar methods for incompressible flow, each cell edge is used to store the velocity component normal to itself. However, instead of storing the pressure at the center of each cell, it is collocated with both of the velocity components. Using the data of Fig 5 we discretise the bicharacteristic equations corresponding to wave propagation in the directions $\theta = 0$ and $\theta = \pi$. For example, when $\theta = 0$, and the wave is travelling in the positive x-direction, (18) becomes

$$(\partial_t + a_0\partial_x)(p + \rho_0 a_0 u) = -\rho_0 a_0 \partial_y v.$$

The time derivative is given by the average of a forward difference at D and a backward difference at C, x-derivatives are found from CD and y-derivatives from AF, giving a prediction for $p + \rho_0 a_0 u$ at D. To predict $p - \rho_0 a_0 u$ at D, we use the equation for waves travelling in the negative x-direction, with $\theta = \pi$. Time derivatives are averaged from a forward difference at D and a backward difference at E, x-derivatives are found from DE and y-derivatives from BG. The quantities p and u can now both be found at D. The scheme is rotated through $\pi/2$ to update v and the values of p collocated with it at the points marked with solid symbols. This

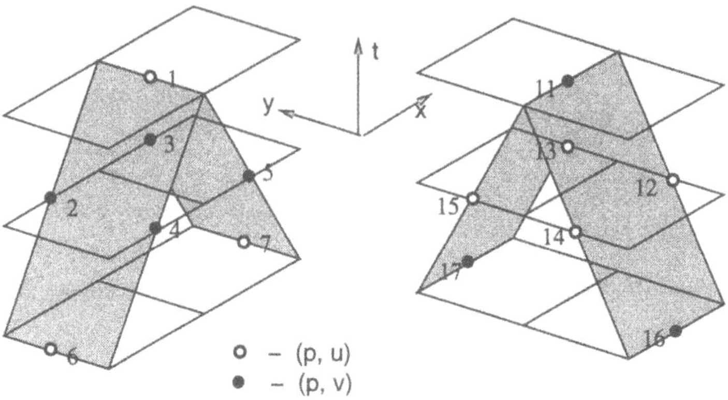

Figure 9. Bicharacteristic stencils in space-time for the special case when $\nu = 0.5$.

scheme has been successfully applied to the acoustic equations [75, 76], and modified for Maxwell's equations [59, 60] and the elastic wave equations [70].

As in one dimension, the scheme simplifies drastically when the Courant number is set to 0.5. In that case, all terms derived from points not lying in the bicharacteristic planes cancel, so that we have a 'pure' bicharacteristic method. Fig 9(left) shows the update for a point lying on an x-interface and marked with an open symbol. The points 1246 define a (conventional!) leapfrog discretisation of the bicharacteristic equation for a wave moving with speed $+a$ along the x-axis. Similarly the points 1357 discretise the wave moving in the opposite direction. The difference method becomes, for (u, p) points such as 1,

$$p_1 = \frac{1}{2}(p_6 + p_7) - (u_7 - u_6) - \frac{1}{2}(v_2 + v_3 - v_4 - v_5), \tag{19}$$

$$u_1 = \frac{1}{2}(u_6 + u_7) - \frac{1}{2}(p_2 + p_3 - p_4 - p_5). \tag{20}$$

At (v, p) points such as 11 (Fig 9(right)), we have

$$p_{11} = \frac{1}{2}(p_{16} + p_{17}) - (v_{17} - v_{16}) - \frac{1}{2}(u_{12} + u_{13} - u_{14} - u_{15}), \tag{21}$$

$$v_{11} = \frac{1}{2}(v_{16} + v_{17}) - \frac{1}{2}(p_{12} + p_{13} - p_{14} - p_{15}). \tag{22}$$

When the Courant number is not equal to 0.5, the additional points may be thought of as interpolating the data into the characteristic plane.

Although the method described above has proved very reliable and has excellent accuracy in comparison with most second-order schemes, even greater accuracy is needed to propagate waves over large distances with a faithful reproduction of the interference patterns, so we have designed a fourth-order scheme along similar lines. We begin again in one dimension with the scheme

$$\begin{aligned} u_{j+1}^{n+1} - u_j^{n-1} &= -\frac{1}{2}(1+\nu)(2-\nu)(1-2\nu)\left(u_{j+1}^n - u_j^n\right) \\ &\quad +\frac{1}{6}\nu(1-\nu)(1-2\nu)\left(u_{j+2}^n - u_{j-1}^n\right), \end{aligned} \tag{23}$$

which was also noted by Iserles [37]. The dispersion relationship can be arranged to read

$$\sin(\phi - \theta/2) = (2\nu - 1)\sin\theta/2\left[1 + \frac{2}{3}\nu(1-\nu)\sin^2\theta/2\right].$$

Stability requires that the RHS remains in the interval $[-1, 1]$ and it is easily shown that this will be so for $0 \leq \nu \leq 1$. The phase error is given for small θ by

$$\varepsilon_p \simeq \frac{\theta^4}{120}\nu(2-\nu)(1-\nu^2)(1-2\nu).$$

An error better than 1% is achieved for resolution with three points per wavelength, and better than 0.1% with as few as five points per wavelength.

To achieve accurate interference patterns over several tens of wavelengths, it seems that the more ambitious target of 0.1% is required, and then the fourth-order method uses less than a third of the storage of the second order method, and will take one-ninth of the nodal updates to reach a specified final time. If the same advantages can be realised in three dimensions, the factors will be 1/27 and 1/81. Of course the number of operations required to realise each update will also increase substantially, but for problems on sufficiently large domains the more accurate formula will always prove to be more economical. These higher-order schemes are not so easy to obtain, however, in higher dimensions. The basic procedure is simple enough; we first obtain the modified equations of the second-order method (a task much simplified by the use of symbolic manipulation packages). These contain error terms proportional to the third-order spatial derivatives of the solution. Adding finite-difference corrections to these errors creates a fourth-order scheme, if care is taken to retain the time-reversible form of the discrete operators. This is guaranteed if the terms added to each bicharacteristic equation are formed symmetrically with respect to the centroid of the original stencil. Fig 10 shows the minimal extension of the stencil needed to accomodate these correction terms.

There is, however, no guarantee that the resulting scheme is stable at all frequencies. It usually turns out to be neutrally stable for almost all

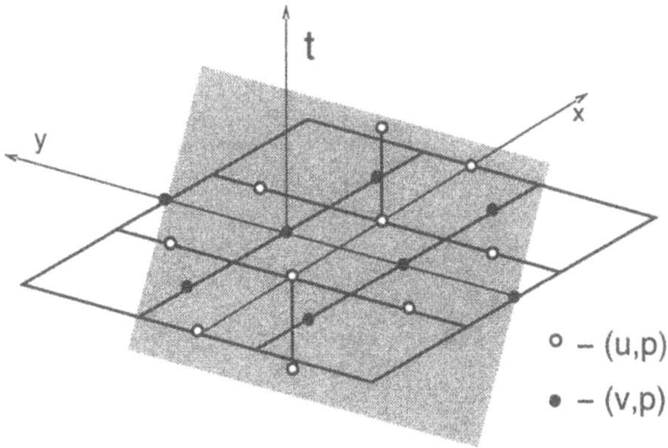

Figure 10. Stencil for incorporating fourth-order corrections into a particular bicharacteristic equation.

frequencies but with occasional bifurcations where a pair of neutral roots split into one stable and one unstable root. These bifurcations have to be found by numerical searching. They are often triggered by adding to the basic scheme such additional effects as source terms, nonuniform grids, and three-dimensionality. Usually, however, the instability turns out to be extremely mild, and often has a recognisable physical character that makes it easier to remove. Commonly, there is for the acoustic equations a spurious production of vorticity, which should remain zero according to the analytical solution. Adding numerical damping to remove something that is anyway below the level of truncation error does not change the formal order of accuracy. Similarly, we have sometimes found it necessary to add damping to couple the pressures stored at different locations, but this also does not affect the order of accuracy.

Although these corrections have not yet been made completely systematic it has proved possible to apply the high-order schemes to some demanding test problems. Results are given for a problem proposed as part of the Second Workshop on Computational Aeroacoustics, held in Talahassee, Florida, in November, 1996. The problem is two-dimensional, and involves the diffraction of sound by a rigid circular cylinder of radius $r = 0.5$. Centered at $r = 4.0$, $\theta = 0$ is an initial pressure disturbance

$$p(x, y, 0) = \rho a_0^2 \exp\left[-\ln 2\left(\frac{(x-4)^2 + y^2}{(0.2)^2}\right)\right]$$

This represents a fairly point-like disturbance; its amplitude falls to 0.5 at about a radius of $\frac{1}{6}$ from its center.

This problem has been solved with both the second- and fourth-order schemes on a polar grid extending to $r = 10$. It is a drawback of polar

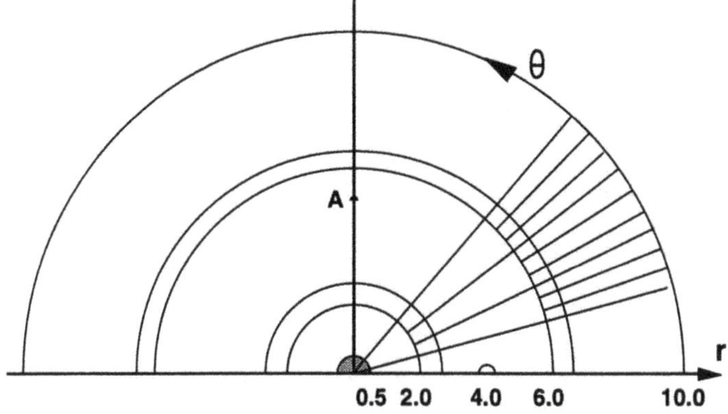

Figure 11. Schematic of grid structure for acoustic diffraction problem.

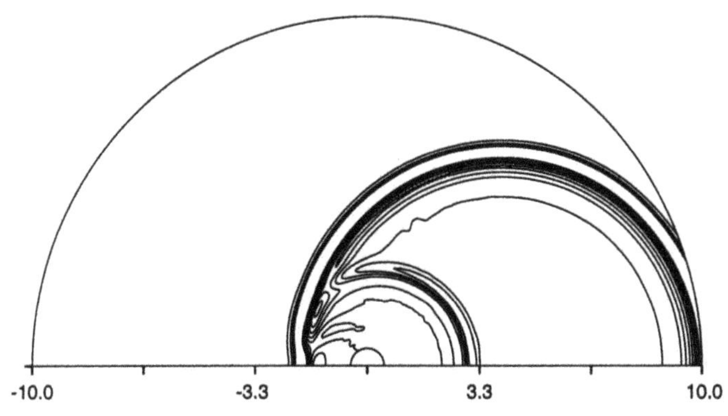

Figure 12. Pressure contours at t=6.0

grids that if they are constructed straightforwardly with constant intervals $\Delta r, \Delta\theta$ then the aspect ratio of the cells changes by a factor $r_{\max}/r_{\min}$, which is here equal to 20. Such a strong variation introduces anisotropy into the wave propagation. Therefore the present grid uses three subgrids as shown in Fig 3. In each subgrid the value of $\Delta\theta$ is one third of the value in the subgrid closer in. Missing information is supplied by polynomial interpolation on the coarser grid.

The solution in Fig 11 shows pressure contours at time $t = 6.0$ and there are no visible disturbances from the grid interfaces. Ability to handle abrupt grid refinement without causing wave reflection is one of the customary merits of upwind methods. Figure 13 shows the time history of the pressure at the point A ($r = 5.0,\ \theta = 90^\circ$) in Figure 3, according to the second-order

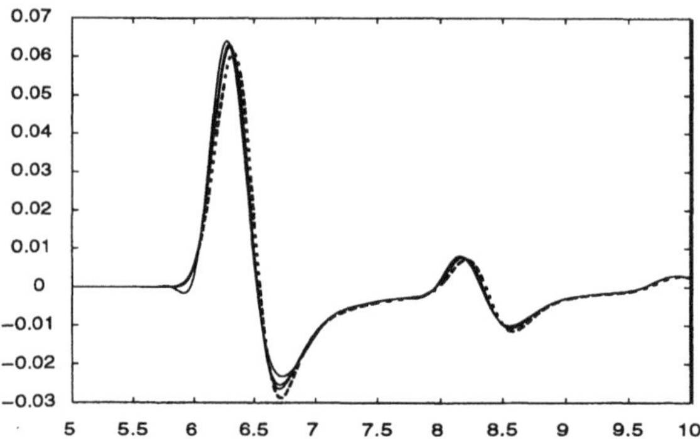

Figure 13. Pressure-time history for point A according to the second-order scheme. The mesh sizes are $N = 24, 32, 40$. The dashed line shows an 'exact' solution.

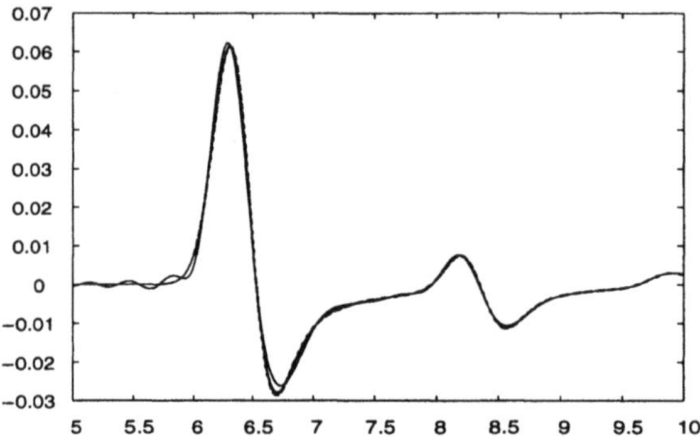

Figure 14. Pressure-time history for point A according to the fourth-order scheme. The mesh sizes are $N = 12, 18, 24. The dashed line shows an `exact' solution.$

method. Results are shown for several different grid sizes, which are defined by the number N of mesh intervals across a diameter of the cylinder. Here $N = 24, 32, 40$. Also, the numerical results are compared with an 'exact' solution kindly provided by Prof C. K. W. Tam, produced by numerical integration of the Green's function. Corresponding results from the fourth-order method are shown in Fig 14 on coarser grids with $N = 12, 18, 24$.

The increased accuracy from the fourth-order scheme can be seen clearly in Fig 15 where a grid convergence study is shown for the pressure at point A at a time $t = 6.7$ corresponding to arrival of the first pressure minimum. Both sets of results appear to be converging at the appropriate rate to a result very close to that given by the analytical expression. Since the first pressure minimum is due to a wave that has not reflected from the cylinder, the exercise is repeated in Fig 16 for the second minimum at $t = 8.65$. Again there is excellent agrement with the 'exact' solution. The saving of a factor

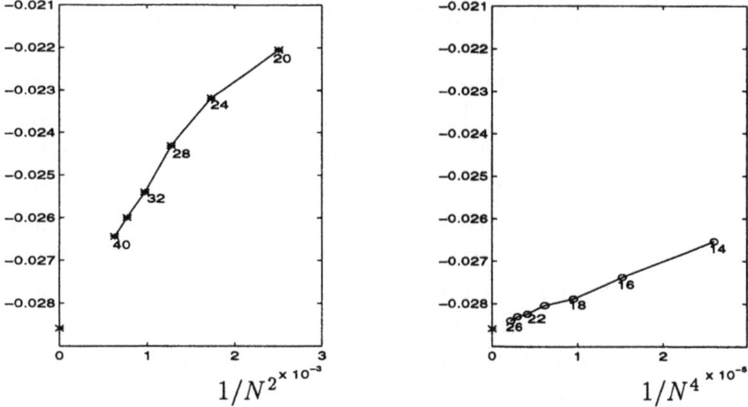

Figure 15. Grid convergence study for point A at $t = 6.7$; arrival of the first pressure minimum; (left) second-order scheme, (right) fourth-order scheme.

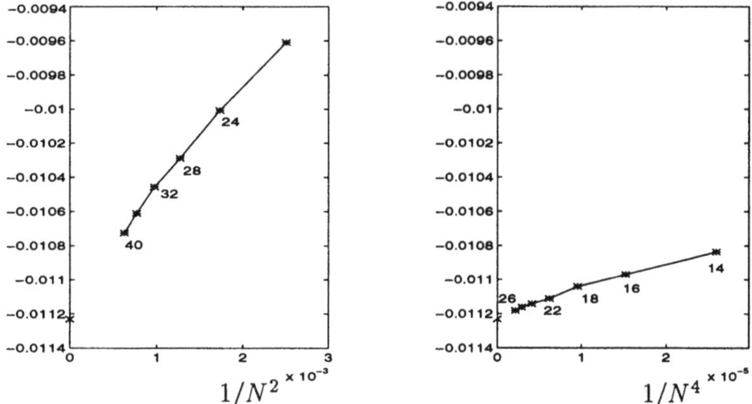

Figure 16. Grid convergence study for point A at $t = 8.65$; arrival of the second pressure minimum; (left) second-order scheme, (right) fourth-order scheme.

of three in mesh density appear to carry over from the one-dimensional analysis. Further implementational details of the method and additional results will appear in [43]

Since even this linear two-dimensional model problem proves difficult to compute, one may be sceptical about the possibility of 'realistic' computations of noise generation and scattering. We need to note that there is little chance of making economies through adaptive mesh refinement, because in this context the solution is equally 'busy' at all scales. Time will show whether Ami Harten's ideas of 'multiresolution' can be employed effectively here. Certainly such calculations are going to remain extremely expensive for a long while to come. We will need methods that are of higher-order accuracy than the ones currently in practical use, and they will have to be capable of dealing with flow equations more complex than the linear acoustic model used here. On the other hand, the range of wavelengths over which the disturbances need to be propagated may not be all that much greater than we have here, because there are semi-analytic 'far-field' theories that can provide the final stages of wave propagation from an integral performed [22] over a 'Kirchoff surface' surrounding the strongly perturbed region. Probably a practical computation will employ some form of domain decomposition, with separate mathematical models and algorithms deployed in the strongly nonlinear, weakly nonlinear, and linear asymptotic parts of the flow. It is hard to predict what the eventual components of such a code will be, and all I wish to claim here is that long-range, virtually dissipationless schemes can be constructed on upwind principles. If they survive the introduction of additional complexities they may have much to offer in the form of robustness, compactness, and the natural handling of boundary conditions that upwind schemes usually provide.

4. Discussion

The strong basis of upwind Euler codes in the theory of hyperbolic conservation laws always held the promise of extension to other equation sets. This promise was one of the sustaining faiths of many of the pioneers such as Ami Harten. If it has been rather slow in coming to fruition it is because these larger equation sets relate to complex physical situations of genuine intellectual difficulty.

Writing an upwind code for an unfamiliar set of equations, or even for a familiar set on the first occasion, forces one to place the governing equations under intense scrutiny and to think carefully about the small-scale physics of the problem. There are several benefits to be expected from undertaking this obligation. One of these may be the discovery of new and significant facts about the mathematical model itself. Another may be the discovery of a requirement to be more precise about certain details of the numerical procedure. Initially, these discoveries will slow down the process of obtaining results, but in the long run they will almost certainly lead to the development of more robust and accurate codes.

Robustness in the face of strong nonlinearity is indeed the chief virtue associated with upwind schemes since their earliest days. It does not however guarantee robustness in all circumstances. It has been recognised for a long time [20] that many upwind schemes break down for problems featuring widely varying ranges of pressure or density. High-energy flows that expand to near vacuum conditions are a good example. The property that pressure and density remain positive throughout a calculation is rather easier to achieve with particle-type codes than with continuum-based codes, but particle codes have difficulty recognising the proper wave behaviour. There have been several attempts recently to merge the two approaches by combining the robustness of particle methods with respect to low density with the robustness of upwind schemes with respect to nonlinear waves [15, 23, 51, 58, 69, 84, 85]. Given the existence of a first-order method that preserves positive pressure and density in one dimension, one can create schemes that have similar properties on unstructured multidimensional grids with high-order reconstruction [50, 64]. Although these 'provably robust' schemes have been devised so far only for the Euler equations they will undoubtedly be extended soon to more complex equation sets.

Thus, although I described the upwind methodology in my Introduction as 'mature', it is by no means static, and continues to support a prolific literature both at the fundamental and applied levels. There is moreover the possibility, since one-dimensional physics is incomplete physics, that schemes of even greater sophistication may prove worthwhile, and in some contexts perhaps necessary. To give but one example, flows in which strong vorticity interacts with shocks are poorly described, at the qualitative level, by Riemann problems. The contribution by Baines to these procedings gives an indication of how things may develop (see also [18, 53, 61]) although these multidimensional methods are as yet fragile creatures and we do not yet know how tough they can become.

Similar considerations apply to the bicharacteristic schemes discussed in Section 3. Because they aim at neutral stability they are very vulnerable to apparently small details of implementation. Overcoming these difficulties has so far only been possible by resorting to the distasteful expedient of artificial smoothing. It may be, however, that once these details are properly understood and correctly implemented the sensitivity will vanish and a new generation of methods that are both powerful and delicate will arise. Their fixed and relatively compact stencils may prove to be coding advantages in comparison with some rival techniques such a ENO schemes.

If the newer ideas can demonstrate robustness and versatility, then methods of the older kind described in Section 2 may come to seem merely one of the lower life-forms on the evolutionary ladder of Computational Science. But, like the dinosaurs, they will have enjoyed a long stretch of domination within their own environmental niche. For the present, Riemann-problem based methods provide a technique able to solve a variety of problems, including many that prove extremely hard for other methods. Although the most ambitious applications do require quite substantial analysis of the governing equations, my belief is that effective computation without analysis is probably impossible. Analysing even the one-dimensional structure of the equations to be solved shows dividends that abundantly repay the effort.

5. Acknowledgements

I thank the following students and colleagues who have contributed to the work described above; Necdet Aslan, Dinshaw Balsara, Shawn Brown, Darren De Zeeuw, Tamas Gombosi, Clinton Groth, Cheolwan Kim, Bram van Leer, Timur Linde, Rob Lowrie, Rho-Shin Myong, Suichi Nakazawa, Brian Nguyen, Ken Powell, Jeff Thomas, Marcel Vinokur, and Pete Washabaugh.

References

1. R. Abgrall, How to prevent pressure oscillations in multicomponent calculations: a quasi-conservative approach. *J. Comput. Phys.* **125**, pp 150-160, 1996.
2. A. S. Almgren, J. B. Bell, P. Colella, T. Marthaler, A cell-centred Cartesian grid projection method for the incompressible Euler equations in complex geometries AIAA paper 95-1743-CP in *Proceedings 12th AIAA CFD Conference (San Diego, June, 1995)*, AIAA, 1995.
3. M. Arora, Explicit characteristic-based high-resolution algorithms for hyperbolic conservation laws with stiff source terms. *Ph. D. Thesis*, Department of Aerospace Engineering, University of Michigan, 1996.
4. N. Aslan, Solutions of one-dimensional MHD equations by a fluctuation approach, *Num. Meth. in Fluids*, **22**, pp 569-580, 1996.
5. R. Balakrishnan, R. Agarwal, Entropy-consistent formulation and numerical simulation of the BGK-Burnett equations using a kinetic wave/particle flux splitting scheme, 15th Int Conf Num Meth Fluid Dyn, Monterey, July, 1996. Springer, to appear.
6. T. J. Barth, Aspects of unstructured grids and finite-volume solvers for the Euler and Navier-Stokes equations, in Special Course on Unstructured Grid Methods for Advection Dominated Flows, AGARD Rept 787, 1992.
7. T. J. Barth, P. O. Fredrickson, High-order solution of the Euler equations on unstructured grids using quadratic reconstruction, AIAA paper 90-0013, January 1990.
8. S. Bayyuk, K. G. Powell, B. van Leer, Computation of unsteady Euler flows with moving boundaries and fluid-structure interaction, 15th Int Conf Num Meth Fluid Dyn, Monterey, July, 1996.
9. J. B. Bell, P. Colella, H. M. Glaz, A second-order projection method for the incompressible Navier-Stokes equations, *J. Comput. Phys.* **85**, pp 257-283 , 1989.
10. M. J. Berger, P. Colella, Adaptive mesh refinement for hyperbolic differential equations, *J. Comput. Phys.* **53**, p 484, 1989.
11. B. L. Bihari, Multiresolution schemes for conservation laws with viscosity, *J. Comput. Phys.* **123**, pp 207-225, 1996.
12. M. Brio, C. C. Wu, An upwind differencing scheme for the equations of ideal magnetohydrodynamics, *J. Comput. Phys.* **75**, p 400, 1988.
13. S. L. Brown, P. L. Roe, C. P. Groth, Numerical solution of a 10-moment model for nonequilibrium gasdynamics, AIAA paper, 95-1677-CP, 1995.
14. P. Charrier, B. Dubroca, J-L. Feugas, Etude numerique des modèles au moment de D. Levermore, Rep 96-014, Mathematiques Appliquées, Univ de Bordeaux, 1996.
15. F. Coquel, M-S. Liou, Stable and low diffusive hybrid upwind splitting methods, in First European CFD Conference, Brussels, 1992.
16. F. Coquel, C. Marmignon, A Roe-type linearization for the Euler equations for weakly ionized multi-component and multi-temperature gas, AIAA paper 95-1675, in AIAA CP 956, Proc 12th AIAA CFD Conference, (San Diego 1995).
17. W. Dai, P. R. Woodward, On the divergence-free condition and conservation laws in numerical simulations for supersonic magnetohydrodynamical flows, Minnesota Supercomputer Research Institute Report UMSI 96/205, 1996.

18. H. Deconinck, P. L. Roe, R. Struijs, A multidimensional generalisation of Roe's flux difference splitter for the Euler equations, *Computers and Fluids*, **22**, p 215-222, 1993.
19. D. De Zeeuw, K. G. Powell, An adaptively refined Cartesian mesh solver for the Euler equations, *J. Comput. Phys.* **104**, pp 56-68, 1992.
20. B. Einfeldt, C. D. Munz, P. L. Roe, B. Sjögreen, Godunov-type methods near low densities, *J. Comput. Phys.* **92**, pp 273-295, 1991.
21. F. Eulderink, G. Mellema, General relativistic hydrodynamics with a Roe solver, *Astronomy and Astrophysics Supplement Series* , **110** pp 587-623, 1995.
22. F. Farassat, M. K. Myers, Extension of Kirchoffs formula to radiation from moving sources, *J. Sound Vibr.*, **123**, pp 451-460, 1988.
23. M. Fey, Decomposition of the multidimensional Euler equations into advection equations, SAM Research Report 95-14, ETH Zürich, 1995.
24. T. I. Gombosi, D. De Zeeuw, R. M. Häberli, K. G. Powell, Three-dimensional multiscale MHD model of cometary plasma environments, *J. Geophysical Res.- Space Phys*, **101** No. A7, p 15,233, 1996.
25. T. I. Gombosi, D. De Zeeuw, T. Y. Linde, K. G. Powell, Solar wind interaction with comets, lessons for modelling the heliosphere. in, *Cosmic Winds and the Heliosphere*, eds. J. R. Jokipii, C. P. Sonnett, M. S. Giampapa, University of Arizona Press, to appear.
26. T. I. Gombosi, K. G. Powell, D. De Zeeuw, Axisymmetric modelling of cometary mass loading on an adaptively refined grid; MHD results, *J. Geophysical Res.- Space Phys*, **99** No. A11, p 21,525, 1994.
27. H. Grad, On the kinetic theory of gases, *Comm. Pure Appl. Math.* **2**, pp331-407, 1952.
28. C. P. T. Groth, P. L. Roe, T. I. Gombosi, S. L. Brown, On the nonstationary wave structure of a 35-moment closure for rarefied gas dynamics, AIAA paper 95-2312-CP, 1995.
29. A. Harten, High resolution schemes for hyperbolic conservation laws, *J. Comput. Phys.*, **49**, pp 357-372, 1983.
30. A. Harten, On the nonlinearity of modern shock-capturing schemes, in *Wave Motion, Theory, Modelling and Computation*, proceedings of a conference in honour of the 60th birthday of Peter Lax, eds A.J. Chorin, A. Majda, Mathematical Sciences Research Institute,pp 147-201, pub. Springer Verlag, 1987.
31. A. Harten, Recent developments in shock-capturing schemes, Proceedings of the International Congress of Mathematicians, Kyoto, Japan, 1990. Also ICASE Report 91-8, 1991.
32. A. Harten, Adaptive multiresolution schemes for shock computations, *J. Comput. Phys.*, **115**, pp 319-338, 1994.
33. A. Harten, Multiresolution representation and numerical algorithms: a brief review, Proc ICASE/LaRC Workshop on Parallel Numerical Algorithms, Kluwer/Academic Press, 1996.
34. A. Harten, B. Engquist, S. Osher, S. Chakravarthy, Uniformly high-order accurate non-oscillatory schemes III, *J. Comput. Phys.*, **71**, pp 231-303, 1987.
35. A. Harten, P. D. Lax, B. van Leer, On upstream differencing and Goduov-type schemes for hyperbolic conservation laws, *SIAM Review*, **25**, pp 35-61, 1983.
36. A. Harten, S.Osher, Uniformly high-order accurate non-oscillatory schemes I, *SIAM J. Num Anal.* **24**, p 279, 1987.
37. A. Iserles, Generalised leapfrog schemes, *IMA Jnl. Num. Anal.* vol 6, 1986.
38. A. Jeffrey, T. Tanuiti, *Non-Linear Wave Propagation*, Academic Press, 1966.
39. G-S. Jiang, C-W. Shu, Efficient implementation of weighted ENO schemes, *J. Comput. Phys.*, **126**, pp 202-228, 1996.
40. S. Jin, Runge-Kutta methods for hyperbolic conservation laws with stiff relaxation terms, *J. Comput. Phys.*, **112**, p 31-43, 1994.
41. S. Jin, C. D. Levermore, Numerical schemes for hyperbolic conservation laws with

stiff relaxation terms. *J. Comput. Phys.*, to appear.
42. S. Karni, Multicomponent flow calculations by a consistent primitive algorithm, *J. Comput. Phys.* **122**, p 51, 1995.
43. C. Kim, P. L. Roe, Accurate schemes for advection and aeroacoustics, paper accepted for 13th AIAA CFD Conference, Snowmass, Colorado, 1997.
44. R. Klein, Semi-implicit extension of a Godunov-type scheme based on low Mach number asymptotics I: one-dimensional flow, *J. Comput. Phys.*, **121**, pp 213-237, 1995.
45. G. Koppenwallner, Low Reynold number influence on aerodynamic performance of hypersonic lifting vehicles, pp 11.1-11.14 of AGARD CP-428, 1987.
46. R. J. Leveque, *Numerical Methods for Conservation Laws*, Birkhauser, 1992.
47. B. van Leer, Towards the ultimate conservative differencing scheme V. A second-order sequel to Godunov's method. *J. Comput. Phys.*, **32**, pp 101-136, 1979.
48. C. D. Levermore, Moment closure heirarchies for kinetic theory, *J. Statistical Physics*, **83**, pp 1021-1065, 1996.
49. C. D. Levermore, The Gaussian closure for gas dynamics, *SIAM J. App. Math.*, to appear.
50. T. Linde, P. L. Roe, On multidimensional positively-conservative high-resolution schemes, *Challenges and Barriers in CFD*, ICASE/Larc Workshop, August 1996, Kluwer, to appear.
51. M-S. Liou, C. Steffen, A new flux-splitting scheme, *J. Comput. Phys.*, **107**, pp 23-39, 1993.
52. J. C. Maxwell, On the dynamical theory of gases, *Phil. Trans. Roy. Soc. London* **157** pp 49-88, 1867; also in *The Scientific Papers of James Clerk Maxwell*, **2**, pp 26-78, Dover, New York, 1965.
53. L.M. Mesaros, P. L. Roe, Multidimensional fluctuation splitting schemes based on decomposition methods, AIAA 95-1699, 12th AIAA CFD Conference, AIAA CP-956.
54. G. H. Miller, E. G. Puckett, A high-order Godunov method for multiple condensed phases, *J. Comput. Phys.*, **128**, p 134-164, 1996.
55. M. L. Minion, On the stability of Godunov-projection methods for incompressible flow, *J. Comput. Phys.*, **123**, p 435-449, 1996.
56. A. H. Mohammadian, V. Shankar, W. Hall, Computation of electromagnetic scattering and radiation using a time-domain finite-volume disretization procedure, *Computer Physics Communications*, **68**, pp 175-196, 1991.
57. R-S. Myong, Theoretical and computational investigations of nonlinear waves in magnetohydrodynamics, Ph. D. Thesis, Aerospace Engineering, University of Michigan, 1996.
58. B. T. Nadiga, D. I. Pullin, A method for near-equilibrium discrete-velocity gas flows, *J. Comput. Phys.*, **112**, p 162-172, 1994.
59. B. T. Nguyen, Investigation of three finite-difference time-domain methods for multidimensional acoustics and electromagnetics, *Ph.D. Thesis*, Aerospace Engineering, University of Michigan, 1995.
60. B. Nguyen, P. L. Roe, Application of an upwind leapfrog method for electromagnetics, in *10th Annual Review of Progress in Applied Computational Electromagnetics*, vol 1, pp 446-458, 1994.
61. H.Paillière, H. Deconinck, P. L. Roe, Conservative upwind residual-distribution schemes based on the steady characteristics of the Euler equations, AIAA 95-1700, 12th AIAA CFD Conference, AIAA CP-956.
62. R. Pember, Numerical methods for hyperbolic conservation laws with stiff relaxation, II higher-order Godunov methods, *SIAM J. Sci. Stat. Comput.*, **14**, no 4, p 824, 1993.
63. R. Pember, J. B. Bell, P. Colella, W. Y. Crutchfield, M. L. Welcome, An adaptive Cartesian grid method for unsteady compressible flow in irregular regions, *J. Comput. Phys.*, **120**, p 278-304, 1995.

64. B. Perthame, C-W. Shu, On positive preserving finite-volume schemes for compressible Euler equations, *Numerische Mathematik*, **73**, pp119-130, 1996.
65. K. G. Powell, An approximate Riemann solver for magnetohydrodynamics (that works in more than one dimension) ICASE Report 94-24, 1994.
66. K. G. Powell, P. L. Roe, J. J. Quirk, Adaptive-mesh algorithms for computational fluid dynamics, in *Algorithmic Trends in Computational fluid dynamics*, eds M.Y Hussaini, A. Kumar, M.D. Salas, Springer, 1993.
67. J. J. Quirk, A parallel adaptive mesh refinement algorithm for computational shock hydrodynamics, *Applied Numerical Mathematics*, to appear.
68. J. J. Quirk, S. Karni, On the dynamics of a shock-bubble interaction, ICASE Report 94-75, also *Journal of Fluid Mechanics*, **318**, pp 129-164, 1996.
69. R. Radespeil, N. Kroll, Accurate flux-vector splitting for shocks and shear layers, *J. Comput. Phys.*, **121**, p 66-78, 1995.
70. P. L. Roe, Linear bicharacteristic schemes without dissipation, *SIAM J. Sci. Comp.*, to appear.
71. P. L. Roe, D. S. Balsara, Notes on the eigensystem of magnetohydrodynamics, *SIAM J. App. Math.***56**, pp 57-67 , 1996.
72. L. Sansaulieu, Finite-volume approximation of two-phase fluid flows based on an approximate Roe-type Riemann solver, *J. Comput. Phys.*, **121**, pp 1-28, 1995.
73. V. Schneider, U. Katscher, D. H. Rischke, B. Waldhauser, J. A. Maruhn, C-D. Munz, New algorithms for ultrarelativistic hydrodynamics, *J. Comput. Phys.* **105**, p 92-107, 1993.
74. G. Strang, On the construction and comparison of difference schemes. *SINUM* **5**, pp 506-517, 1968.
75. J. P. Thomas, An investigation of the upwind leapfrog method for scalar advection and acoustic/aeroacoustic wave propagation. *Ph.D. Thesis*, Aerospace Engineering, University of Michigan, 1995.
76. J. P. Thomas, P. L. Roe, Development of non-dissipative numerical schemes for computational aeroacoustics, AIAA paper 93-3382-CP, Orlando, 1993, AIAA CP 933, 1993.
77. V. T. Ton, Improved shock-capturing methods for multicomponent and reacting flows, *J. Comput. Phys.* **128**, pp 237-253, 1996.
78. J. R. Trangenstein, Adaptive mesh refinement for wave propagation in nonlinear solids, *SIAM J. Sci. Comput.*, **16**, p 809, 1995.
79. W. G. Vincenti, C. H. Kruger, *Introduction to Physical Gas Dynamics*, John Wiley, 1965.
80. M. Vinokur, J-L. Montagné, Generalized flux-vector splitting and Roe average for an equilibrium real gas. *J. Comput. Phys.*, **89**, pp 276-300, 1990.
81. W. T. Welder, D. R. Chapman, R. W. MacCormack, Evaluation of various forms of Burnett equations, AIAA paper 93-3094,Orlando, 1993.
82. C. C. Wu, The MHD intermediate shock intermediate shock interaction with an intermediate wave, are intermediate shocks physical? *J. Geophysical Res.*, **93**, pp 987-999, 1988.
83. C. C. Wu, Formation, structure, and stability of MHD intermediate shocks, *J. Geophysical Res.*, **95**, pp 8149-8175, 1990.
84. K. Xu, L. Martinelli, A. Jameson, Gas-kinetic finite volume methods, flux-vector splitting, and artificial diffusion, *J. Comput. Phys.* **120**, pp 48-65, 1995.
85. J. Y. Yang, J. C. Huang, Rarefied flow computations using nonlinear model Boltzmann methods, *J. Comput. Phys.* **120**, pp 323-339, 1995.
86. A. Zachary, A. Malagoli, P. Colella, A higher-order Godunov method for multidimensional ideal magnetohydrodynamics, *SIAM.J. Sci. Comp.*, **15**, pp 263-284, 1994.

MULTIDIMENSIONAL UPWINDING WITH GRID ADAPTATION

M.J. BAINES AND M.E. HUBBARD
ICFD, The University of Reading,
Department of Mathematics,
Whitknights, P.O. Box 220,
Reading, RG6 6AF.

1. Introduction

Over the past ten years multidimensional upwinding techniques have been developed with the intention of superseding traditional conservative upwind finite volume methods which rely on the solution of one-dimensional Riemann problems. The new methods attempt a more genuinely multidimensional approach to the solution of the Euler equations by considering a piecewise linear continuous representation of the flow with the data stored at the nodes of the grid. The schemes are then constructed from three separate stages: the decomposition of the system of equations into simple (usually scalar) components, the construction of a consistent, conservative discrete form of the equations and the subsequent solution of the decomposed system using scalar fluctuation distribution schemes. A detailed description of each of these stages can be found in [1, 2, 3].

As we shall show, the quality of the solution of a system of differential equations can be improved by means of grid adaptation. For example, multidimensional upwind schemes will capture shocks within two or three cells when they are aligned with the grid [3] and adaptation can be used to take advantage of this. On unstructured grids this can be accomplished by refinement, which reduces the size of the cells, and by edge swapping, which realigns the grid. However, both selective refinement and edge alignment can, to a large extent, be achieved by a third option, grid movement, which has the added advantage of avoiding the expensive process of changing the number of nodes or the connectivity of the grid.

In this paper one of the most recent and successful of the multidimensional upwind algorithms [1, 3] is described. Following this, a very simple

E.F. Toro and J.F. Clarke (eds.), Numerical Methods for Wave Propagation, 33–54.

and cheap algorithm for moving nodes is presented, which improves the accuracy of two-dimensional steady state solutions of the Euler equations on unstructured triangular grids.

2. Multidimensional upwinding

2.1. DECOMPOSITION OF THE EULER EQUATIONS

The two-dimensional Euler equations in conservative form are written

$$\underline{\mathbf{U}}_t + \underline{\mathbf{F}}_x + \underline{\mathbf{G}}_y = \underline{\mathbf{0}} \,, \tag{1}$$

where

$$\underline{\mathbf{U}} = \begin{pmatrix} \rho \\ \rho u \\ \rho v \\ e \end{pmatrix} , \quad \underline{\mathbf{F}} = \begin{pmatrix} \rho u \\ \rho u^2 + p \\ \rho u v \\ u(p+e) \end{pmatrix} , \quad \underline{\mathbf{G}} = \begin{pmatrix} \rho v \\ \rho u v \\ \rho v^2 + p \\ v(p+e) \end{pmatrix} , \tag{2}$$

are the vectors of conserved variables and the corresponding conservative fluxes, respectively, in which ρ is density, u and v are the x- and y-velocities, p is pressure, and e is total energy, related to the other variables by an equation of state which, for a perfect gas, is

$$e = \frac{p}{\gamma - 1} + \frac{1}{2}\rho(u^2 + v^2) \,. \tag{3}$$

The decomposition stage of the algorithm dictates how the flux balance within each triangle of the grid, namely

$$\underline{\mathbf{\Phi}}_U = -\int\!\!\int_\triangle \left(\underline{\mathbf{F}}_x + \underline{\mathbf{G}}_y\right) \, dx \, dy \,, \tag{4}$$

is divided into simpler components. In this paper each of these components will be scalar (although in some cases transforming the equations results in a simple elliptic subsystem which could be considered without further decomposition). The decomposed flux balance takes the form of a sum of scalar contributions,

$$\underline{\mathbf{\Phi}}_U = \sum_{k=1}^{N} \phi^k \underline{\mathbf{r}}_U^k \,, \tag{5}$$

where

$$\phi^k = -S_\triangle \left(\vec{\lambda}^k \cdot \vec{\nabla} W^k + q^k\right) \tag{6}$$

is the general form of the fluctuation due to the kth component of the decomposition, in which $S_\triangle$ is the area of the cell. $\underline{\mathbf{r}}_U^k$ is the vector which maps this flux balance contribution back to the conservative variables and N is the number of components (or waves) in the model.

Unlike in one dimension where a unique decomposition is available, many different wave models have been proposed for the two-dimensional Euler equations. These can be divided into groups: some decompose the flux balance into contributions corresponding to simple wave solutions of the Euler equations [4, 5], while others use a similarity transformation of the system of equations into 'characteristic' variables [6]. However, the most successful models which have been produced are based on the decomposition of a preconditioned form of the Euler equations [3, 7, 8].

These 'preconditioned' decompositions are derived by considering the system (1) in the streamwise variables, ξ and η, and in terms of the symmetrising variables $\underline{\mathbf{Q}}$, defined by

$$\partial \underline{\mathbf{Q}} = \begin{pmatrix} \frac{\partial p}{\rho a} \\ \partial q \\ q \partial \theta \\ \partial p - a^2 \partial \rho \end{pmatrix} , \tag{7}$$

where a is the local speed of sound, $q = \sqrt{u^2 + v^2}$ is the flow speed and $\theta = \tan^{-1}\left(\frac{v}{u}\right)$ is the direction of the flow. When the symmetrised equations are preconditioned by the matrix $\mathbf{P}$ (see below) they can be written

$$\underline{\mathbf{Q}}_t + \mathbf{P}\left(\mathbf{A}_Q^s \underline{\mathbf{Q}}_\xi + \mathbf{B}_Q^s \underline{\mathbf{Q}}_\eta\right) = \underline{\mathbf{0}} , \tag{8}$$

where the new Jacobians are the symmetric matrices,

$$\mathbf{A}_Q^s = \begin{pmatrix} q & a & 0 & 0 \\ a & q & 0 & 0 \\ 0 & 0 & q & 0 \\ 0 & 0 & 0 & q \end{pmatrix} , \quad \mathbf{B}_Q^s = \begin{pmatrix} 0 & 0 & a & 0 \\ 0 & 0 & 0 & 0 \\ a & 0 & 0 & 0 \\ 0 & 0 & 0 & 0 \end{pmatrix} . \tag{9}$$

For the purposes of this work, the preconditioner $\mathbf{P}$ is chosen to be that of Deconinck and Paillère [7], a generalisation of the van Leer-Lee-Roe matrix [9], given by

$$\mathbf{P} = \frac{1}{q} \begin{pmatrix} \frac{\chi}{\beta_\epsilon^2} M^2 & -\frac{\chi}{\beta_\epsilon^2} M & 0 & 0 \\ -\frac{\chi}{\beta_\epsilon^2} M & \frac{\chi}{\beta_\epsilon^2} + 1 & 0 & 0 \\ 0 & 0 & \chi & 0 \\ 0 & 0 & 0 & 1 \end{pmatrix} , \tag{10}$$

in which β_ϵ and χ are

$$\beta_\epsilon = \sqrt{\max(\epsilon^2, |M^2 - 1|)} , \qquad \chi = \frac{\beta_\epsilon}{\max(M, 1)} . \tag{11}$$

The singularity which would otherwise occur at the sonic line is avoided by choosing the constant $\epsilon > 0$ (typically, $\epsilon = 0.05$).

The preconditioned system (8) can now be completely (in the case of supersonic flow) or partially decoupled by transforming it into a set of characteristic equations, becoming

$$\underline{\mathbf{W}}_t + \mathbf{A}_W^s \underline{\mathbf{W}}_\xi + \mathbf{B}_W^s \underline{\mathbf{W}}_\eta = \underline{\mathbf{0}} \,, \tag{12}$$

where $\underline{\mathbf{W}}$ is given by

$$\partial \underline{\mathbf{W}} = \begin{pmatrix} \beta_\epsilon \frac{\partial p}{\rho a} + Mq\partial\theta \\ \beta_\epsilon \frac{\partial p}{\rho a} - Mq\partial\theta \\ \frac{\partial p}{\rho a} + M\partial q \\ \partial p - a^2 \partial \rho \end{pmatrix} . \tag{13}$$

The flux Jacobians resulting from this set of variables now have the very simple form

$$\mathbf{A}_W^s = \begin{pmatrix} \chi\nu^+ & \chi\nu^- & 0 & 0 \\ \chi\nu^- & \chi\nu^+ & 0 & 0 \\ 0 & 0 & 1 & 0 \\ 0 & 0 & 0 & 1 \end{pmatrix}, \qquad \mathbf{B}_W^s = \begin{pmatrix} \frac{\chi}{\beta_\epsilon} & 0 & 0 & 0 \\ 0 & -\frac{\chi}{\beta_\epsilon} & 0 & 0 \\ 0 & 0 & 0 & 0 \\ 0 & 0 & 0 & 0 \end{pmatrix}, \tag{14}$$

where

$$\nu^+ = \frac{M^2 - 1 + \beta_\epsilon^2}{2\beta_\epsilon^2}, \qquad \nu^- = \frac{M^2 - 1 - \beta_\epsilon^2}{2\beta_\epsilon^2} . \tag{15}$$

Thus, the system can be written as four scalar equations of the form

$$W_t^k + \vec{\lambda}_s^k \cdot \vec{\nabla}_s W^k + q_s^k = 0 \,, \qquad k = 1, 2, 3, 4, \tag{16}$$

where the advection velocities in streamwise coordinates are given by

$$\vec{\lambda}_s^1 = \begin{pmatrix} \chi \\ \frac{\chi}{\beta_\epsilon} \end{pmatrix}_s, \qquad \vec{\lambda}_s^2 = \begin{pmatrix} \chi \\ -\frac{\chi}{\beta_\epsilon} \end{pmatrix}_s, \qquad \vec{\lambda}_s^3 = \vec{\lambda}_s^4 = \begin{pmatrix} 1 \\ 0 \end{pmatrix}_s, \tag{17}$$

and the coupling terms are

$$q_s^1 = \chi\nu^- W_\xi^2 \,, \quad q_s^2 = \chi\nu^- W_\eta^1 \,, \quad q_s^3 = q_s^4 = 0 \,. \tag{18}$$

These indicate that the decomposition is optimal since the third and fourth equations are always decoupled from the rest of the system, indicating the invariance of entropy and enthalpy along the streamlines, while for supersonic flow $\nu^- = 0$ and the system is completely decoupled. In subsonic flow it is possible to consider the first and second equations as an elliptic subsystem which can be decomposed no further [1] but for the purposes of this work they are considered independently, as follows.

Each scalar component is now treated separately. The fluctuations ϕ^k are cell-based quantities which can be distributed to the nodes of the grid to bring the system closer to equilibrium. The resulting contributions to the nodes can be transformed into updates to the conservative variables using the matrix

$$\mathbf{R}_U = \frac{\partial \underline{\mathbf{U}}}{\partial \underline{\mathbf{Q}}} \mathbf{P}^{-1} \frac{\partial \underline{\mathbf{Q}}}{\partial \underline{\mathbf{W}}} , \tag{19}$$

the columns of which are the vectors which provide the mapping in (6) of the characteristic system back on to the conservative system. $\mathbf{R}_U$ is non-singular, so its columns are linearly independent and this four-component decomposition is linearity preserving, *i.e.* a zero flux balance implies a zero contribution to each node, and should not destroy the higher order accuracy of the scalar distribution schemes. Note that, though the decomposition is itself continuous, the distribution is not since the acoustic subsystem will be treated differently for supersonic and subsonic flows.

2.2. CONSERVATIVE LINEARISATION

The above decomposition is based on an analysis of a quasilinear form of the Euler equations,

$$\underline{\mathbf{U}}_t + \mathbf{A}_U \underline{\mathbf{U}}_x + \mathbf{B}_U \underline{\mathbf{U}}_y = \underline{\mathbf{0}} , \tag{20}$$

where $\mathbf{A}_U = \frac{\partial \underline{\mathbf{F}}}{\partial \underline{\mathbf{U}}}$ and $\mathbf{B}_U = \frac{\partial \underline{\mathbf{G}}}{\partial \underline{\mathbf{U}}}$ are the conservative flux Jacobian matrices. In order to guarantee that any discretisation involving such a decomposition will give rise to a conservative scheme, a conservative linearisation of this system is necessary.

An appropriate and very neat linearisation may be obtained [10] by assuming that Roe's parameter vector variables,

$$\underline{\mathbf{Z}} = \sqrt{\rho} \begin{pmatrix} 1 \\ u \\ v \\ H \end{pmatrix} , \tag{21}$$

where $H = \frac{e+p}{\rho}$ is the total enthalpy, vary linearly within each cell. Under this assumption, each of the Jacobian matrices, $\frac{\partial \underline{\mathbf{U}}}{\partial \underline{\mathbf{Z}}}$, $\frac{\partial \underline{\mathbf{F}}}{\partial \underline{\mathbf{Z}}}$ and $\frac{\partial \underline{\mathbf{G}}}{\partial \underline{\mathbf{Z}}}$, depends linearly on the components of $\underline{\mathbf{Z}}$ and, because of this and the linear variation of $\underline{\mathbf{Z}}$, (4) can be integrated exactly using a one point quadrature, leading to a discrete form of the flux balance which is equal to the exact flux balance.

This ensures that, as long as the whole of each discrete flux balance is distributed to the nodes at each time-step, the sum of these contributions over the whole grid reduces to boundary contributions and the scheme is conservative.

The result is a consistent approximation to the flux balance of the form

$$\begin{aligned} \underline{\Phi}_U &= -S_\Delta \left(\widehat{\underline{\mathbf{F}}_x} + \widehat{\underline{\mathbf{G}}_y} \right) \\ &= -S_\Delta \left(\mathbf{A}_U(\hat{\underline{\mathbf{Z}}}) \widehat{\underline{\mathbf{U}}_x} + \mathbf{B}_U(\hat{\underline{\mathbf{Z}}}) \widehat{\underline{\mathbf{U}}_y} \right) , \end{aligned} \tag{22}$$

where the Jacobians are evaluated at the Roe-average state,

$$\hat{\underline{\mathbf{Z}}} = \frac{\underline{\mathbf{Z}}_1 + \underline{\mathbf{Z}}_2 + \underline{\mathbf{Z}}_3}{3} , \tag{23}$$

the mean of the values of $\underline{\mathbf{Z}}$ at the vertices of the cell, and the approximation to the gradient of the conservative variables, $\widehat{\vec{\nabla}\underline{\mathbf{U}}}$, is evaluated consistently from the discrete gradient of $\underline{\mathbf{Z}}$,

$$\widehat{\vec{\nabla}\underline{\mathbf{Z}}} = \frac{1}{2S_\Delta} \sum_{j=1}^{3} \underline{\mathbf{Z}}_j \vec{n}_j , \tag{24}$$

where $\vec{n}_j$ is the inward pointing normal to the edge opposite vertex j, scaled by the length of the edge. Importantly, the form of the linearisation implies that the analysis of the continuous system in the previous section will also hold at the discrete level without alteration, provided that all variables are evaluated at the average state $\hat{\underline{\mathbf{Z}}}$ and all discrete gradients are calculated consistently from $\widehat{\vec{\nabla}\underline{\mathbf{Z}}}$.

2.3. SCALAR FLUCTUATION DISTRIBUTION SCHEMES

Once the wave model has been used to decompose the system of equations into scalar components (5), the behaviour of these components (if the source terms are ignored for the moment) can be modelled by the scalar advection equation,

$$u_t + f_x + g_y = 0 \quad \text{or} \quad u_t + \vec{\lambda} \cdot \vec{\nabla} u = 0 , \tag{25}$$

where $\vec{\lambda} = (\frac{\partial f}{\partial u}, \frac{\partial g}{\partial u})^T$ defines the velocity of the advected variable u.

A scheme can be constructed for the solution of this equation by calculating the fluctuation,

$$\begin{aligned} \phi &= -\int\!\!\int_\Delta \vec{\lambda} \cdot \vec{\nabla} u \, dx \, dy \\ &= -S_\Delta \hat{\vec{\lambda}} \cdot \vec{\nabla} u , \end{aligned} \tag{26}$$

within each cell and then distributing it to the nodes of the grid, giving rise to a form of cell-vertex scheme. The integration, which can be done exactly because u is assumed to be linear within the cell, introduces the factor of $S_\triangle$, the area of the triangle, and a cell-averaged wave speed,

$$\widehat{\vec{\lambda}} = \frac{1}{S_\triangle} \int\int_\triangle \vec{\lambda} \, dx \, dy \, . \tag{27}$$

For simplicity and compactness, a cell is allowed to contribute its fluctuation only to its own vertices. Since summing the fluctuations over the whole domain reduces to a sum of boundary contributions, a conservative scheme is assured as long as the whole of each fluctuation is distributed.

If explicit forward Euler time-stepping is used, this leads to a scheme of the form

$$u_i^{n+1} = u_i^n + \frac{\Delta t}{S_i} \sum_{\cup \triangle_i} \alpha_i^j \phi_j \, , \tag{28}$$

where S_i is the area of the median dual cell for node i (one third of the total area of the triangles with a vertex at i), α_i^j is the distribution coefficient which indicates the proportion of the fluctuation ϕ_j to be sent from cell j to node i, and $\cup \triangle_i$ represents the set of cells adjacent to node i. Since each fluctuation is a linear function of the data, the scheme is of the form

$$u_i^{n+1} = \sum_k c_{ik} u_k^n \, . \tag{29}$$

If the coefficients c_{ik} are allowed to depend on the data, the scheme becomes nonlinear and can be designed to satisfy the following four criteria:

- Upwind - the fluctuation within a cell is only sent to the downstream vertices of that cell *i.e.* vertices opposite inflow edges for which $\vec{\lambda} \cdot \vec{n} > 0$, where $\vec{n}$ is the inward pointing normal to the edge.
- Positive - the coefficients c_{ik} are positive, so the scheme cannot produce new extrema in the solution at the new time-step, spurious oscillations will not appear in the solution and the scheme is stable for an appropriate time-step restriction.
- Linearity preservation - no update is sent to the nodes when a cell fluctuation is zero, so the scheme is second order accurate at the steady state on a regular mesh with a uniform choice of diagonals.
- Continuity - the contributions to the nodes, $\alpha_i^j \phi_j$, depend continuously on the data, avoiding limit cycling as convergence is approached.

A linear scheme cannot satisfy both the positivity and the linearity preservation properties simultaneously.

In the search for a scheme which satisfies all of the above properties it is initially advantageous to consider a linear, positive, upwind scheme, the N scheme. By the above definition of upwind, any triangle with only one downstream vertex will send the whole of its fluctuation to that node. Where it differs from other upwind schemes is in its treatment of cells with two inflow sides. Taking a single cell in isolation, with vertices i, j, k, of which i and j are downstream nodes, the N scheme can be written

$$\begin{aligned} S_i u_i^{n+1} &= S_i u_i^n - \Delta t k_i (u_i^n - u_k^n) \\ S_j u_j^{n+1} &= S_j u_j^n - \Delta t k_j (u_j^n - u_k^n) \\ S_k u_k^{n+1} &= S_k u_k^n , \end{aligned} \tag{30}$$

where $k_{i,j} = \frac{1}{2}\vec{\lambda} \cdot \vec{n}_{i,j}$ and contributions from other triangles are suppressed. By considering the complete nodal update (28), this scheme can be shown to be positive for a restriction on the time-step at a node i, given by

$$\Delta t \leq \frac{S_i}{\sum_{\cup \Delta_i} \max(0, k_i^j)} . \tag{31}$$

A linearity preserving scheme (which also retains the upwind and continuity properties) can be obtained from a positive upwind scheme [11] such as the N scheme by replacing the contributions, ϕ_i and ϕ_j, to the downstream nodes in the two-target case by 'limited' contributions,

$$\begin{aligned} \phi_i^* &= \phi_i - L(\phi_i, -\phi_j) \\ \phi_j^* &= \phi_j - L(\phi_j, -\phi_i) . \end{aligned} \tag{32}$$

where, in the case of the N scheme,

$$\phi_i = -k_i(u_i^n - u_k^n) , \qquad \phi_j = -k_j(u_j^n - u_k^n) . \tag{33}$$

$L(x, y)$ is any member of the family of symmetric limiter functions, although the minmod limiter,

$$L(x, y) = \frac{1}{2}(1 + \text{sgn}(xy)) \, \frac{1}{2}(\text{sgn}(x) + \text{sgn}(y)) \, \min(|x|, |y|) , \tag{34}$$

is the only one for which the 'limited' scheme remains positive. The resulting scheme, which is equivalent to the Positive Streamwise Invariant (PSI) scheme [2], satisfies all of the desired properties and will be used here to distribute all fluctuations without any associated source term.

When the flow is subsonic, not all of the components of the decomposition yield homogeneous advection equations and these are dealt with differently. In this case the two characteristic equations making up the acoustic subsystem are still solved with a scalar scheme, but one different to the PSI scheme described above.

The new scheme is derived by analogy with the finite element method and from consideration of a weak form of the linear advection equation, given by

$$\int\int_{\Omega} \tilde{\omega}_i u_t \, dx \, dy = -\int\int_{\cup\Delta_i} \tilde{\omega}_i \vec{\lambda} \cdot \vec{\nabla} u \, dx \, dy \,, \tag{35}$$

where the approximation to u in terms of the linear basis functions ω_i is

$$u(x,y) = \sum_{i=1}^{N_n} u_i \, \omega_i(x,y) \,. \tag{36}$$

Test functions $\tilde{\omega}_i$ are chosen [2] which add both linear and nonlinear dissipation terms to the standard linear 'tent' function, used as the basis function. When combined with mass-lumping and forward Euler time-stepping on the left hand side of (35), the result is a fluctuation distribution scheme for which the coefficients are given by

$$\alpha_i^j = \frac{1}{3} + \tau_1 \frac{k_i}{S_j} + \tau_2 \frac{(k_i)_{||}}{S_j} \,, \tag{37}$$

where

$$(k_i)_{||} = \frac{1}{2} \vec{\lambda}_{||} \cdot \vec{n}_i \,, \tag{38}$$

and

$$\tau_1 = C_1 \frac{h}{|\vec{\lambda}|} \,, \qquad \tau_2 = C_2 \frac{h}{|\vec{\lambda}_{||}|} \,. \tag{39}$$

C_1 and C_2 are both constants set to 0.5 [12], h is some measure of the size of the discretisation, taken to be the length of the longest edge of the triangle, and $\vec{\lambda}_{||}$ is the gradient dependent advection velocity, the projection of the advection velocity on to the solution gradient. This describes a nonlinear, mass-lumped, Streamline Upwind Petrov-Galerkin (SUPG) scheme, formulated in terms of fluctuation distributions. It is linearity preserving and continuous, but not generally positive nor, by the earlier definition, truly upwind.

3. Grid adaptation

The adaptation algorithm presented in this paper is a very simple form of node movement. It takes the form of an iteration where, at each step, nodes are moved to a weighted average of the positions of the centroids of the neighbouring triangles [13, 14]. The new nodal position can thus be written in terms of the old positions as

$$\vec{x}_i^{n+1} = \frac{\sum_{\cup\Delta_i} w_j \vec{x}_j^n}{\sum_{\cup\Delta_i} w_j} \,, \tag{40}$$

where the $\vec{x}_j$ are the positions of the centroids, w_j are the cell weights and the sums are over the cells adjacent to node i. The iteration (40) with constant, non-negative weights can be shown to converge. Moreover, provided that the weights themselves converge, convergence of the iteration can be shown for variable weights.

In the solution of the Euler equations, the weights w have been chosen to depend on local approximations to the first and second derivatives of the density of the flow, in the form

$$w = \frac{S_\Delta}{S_O}\left(1 + \alpha|\vec{\nabla}\rho|^2 + \beta(\vec{\nabla}^2\rho)^2\right)^{\frac{1}{2}} , \tag{41}$$

where α and β are arbitrary parameters, S_Δ is the current area of the triangle and S_O is the original area of that triangle. For the linear advection equation ρ is replaced by u, the advected variable. The choice of $\alpha = 1$ and $\beta = 0$ in (41) gives a simple generalisation of the weights which lead to arc length equidistribution in one dimension (assuming that the initial grid is equispaced so that S_O is constant throughout the domain).

Although there is no corresponding genuinely two-dimensional equidistribution property, the algorithm will still tend to move nodes towards regions where the weights are high. In the above case this means regions of high first and/or second derivatives, such as those found at shocks, but the weights can be modified so that nodes are attracted towards any detectable feature of the flow. Also, since the weights depend on derivatives of the flow, the degree of attraction to these features can be varied by scaling the grid - the smaller the grid size, the stronger the effect of the adaptation. In the present application, where the weights depend on a flow which will ultimately be steady, the gradients, and hence the w_i move towards steady state values, and at the limit the grid can be interpreted as having a local equidistribution property in the direction of the normal to ρ. The algorithm can also be easily generalised to three dimensions.

In one dimension mesh tangling can be avoided by ensuring that the chosen weights are always positive. In higher dimensions, though, particularly on the highly distorted grids which become common once the mesh is allowed to move, tangling occurs quite readily. Even with positive weights in (40), it is possible for a node at the vertex of a triangle to be overtaken by the opposite edge of that triangle, thus causing the cell to 'flip' and acquire a negative area.

This can be avoided by artificially limiting the distance which a node can move. A simple but rather restrictive limit is

$$(\Delta x_i)_{\max} = \min_{\cup\Delta_i}\left(\frac{S_j}{\max_{k=1,3} l_{jk}}\right) , \tag{42}$$

where S_j is the area of cell j and l_{jk} is the length of edge k of cell j. This expression is equivalent to half the smallest height of the surrounding triangles. A second restriction is also imposed which places a lower limit on the radius of the inscribed circle of each cell. This avoids extremely distorted meshes and the possibility of a prohibitively small limit on the time-step.

Using this strategy, a displacement can be found for all nodes, including boundary nodes, although the latter must be projected back on to the nearest point on the boundary and 'corner' nodes forced to remain fixed.

Once all the displacements have been found, the nodal positions are updated in a block. The solution is then obtained on the new grid using linear interpolation of the solution on the previous grid.

4. Solution strategy

The method by which node movement is combined with multidimensional upwinding to obtain steady state solutions to the two-dimensional Euler equations can be expressed in three stages:

1) Run the time-stepping algorithm on an initial, fixed grid until the solution appears steady (but long before convergence is achieved).
2) Run the time-stepping interspersed with the grid movement until the grid has adapted to the steady solution. In this work, each time-step is alternated with a single node movement iteration.
3) Fix the grid and run the time-stepping algorithm to convergence using the solution from step 2) as initial conditions.

The grid movement in step 2) can be initiated when the RMS of the residual over the grid drops below a certain level (typically a drop of 2 or 3 orders of magnitude from the initial residual), in effect when the flow has stopped changing.

It may well be possible that the combination of time-stepping and grid movement in stage 2) would lead to a converged solution if allowed to run indefinitely. However, it would be impractical to attempt this because the convergence of the overall scheme, depending as it does on two separate iterations, would be prohibitively slow. Also, this stage of the method is not, as it stands, conservative due to the interpolation step of the grid movement. However, since steady state solutions are sought, the grid can be frozen after a fixed number of time-steps (typically 500 for the Euler equations) after which the solution strategy returns solely to the conservative time-stepping scheme. Local time-stepping has been used throughout to accelerate convergence, particularly on the more distorted meshes.

5. Results

Adapted and unadapted steady state solutions will be presented here of both the scalar advection equation and the Euler equations. The first test case used is that of clockwise circular advection, $\vec{\lambda} = (y, -x)^T$ of a square wave profile within the domain $(x, y) \in [-1, 1] \times [0, 1]$. The mesh used is an isotropic triangulation alternating the direction of diagonals inserted into a regular quadrilateral grid with 65×33 nodes and of the form shown in Figure 1.

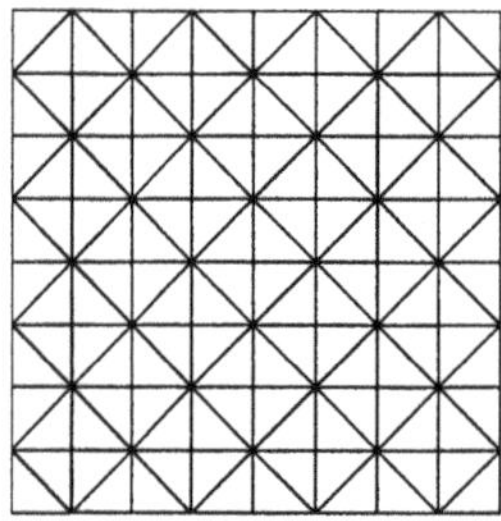

Figure 1. A section of the initial isotropic grid.

The boundary conditions for this test case are

$$\begin{aligned}
u(x, 0) &= 1 \quad \text{for} \quad -0.65 \leq x \leq -0.35 \\
u(x, 0) &= 0 \quad \text{for} \quad -1.0 \leq x < -0.65, \quad -0.35 < x \leq 0.0 \\
u(x, 1) &= 0 \quad \text{for} \quad 0.0 \leq x \leq 1.0 \\
u(0, y) &= 0 \quad \text{for} \quad 0.0 \leq y \leq 1.0\,.
\end{aligned}$$

On the rest of the boundary, where the flow is out of the domain, the solution is given the value predicted by the updates from the interior and initially the solution is set to zero in the whole of the domain's interior.

The steady state solution of this problem obtained using the PSI scheme (CFL = 0.8) is shown at the top of Figure 2. There are no spurious oscillations since the scheme is positive, and the linearity preservation property ensures that the discontinuities are captured reasonably sharply although a certain amount of numerical diffusion is unavoidable. The rest of Figure 2 shows both the adapted solution and the grid on which it has been obtained. As suggested in the previous section, the algorithm was run initially on the unadapted grid (for 200 iterations), then the grid movement was interleaved with the time-stepping for a further 200 iterations, after which the grid is fixed again and a converged solution obtained on the new grid. The CFL number can be kept at 0.8 throughout this procedure and the convergence of the algorithm on the adapted grid is not significantly

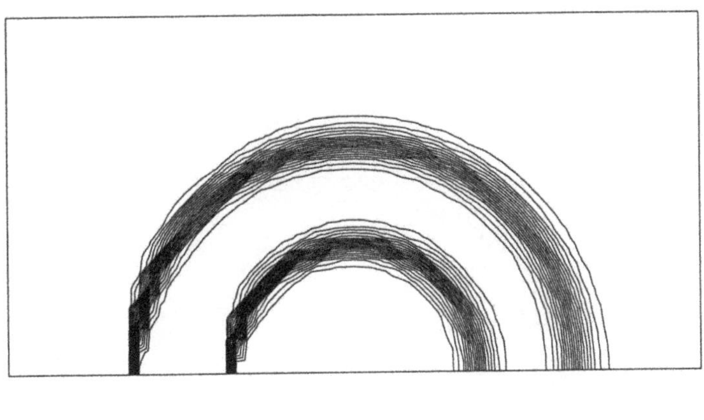

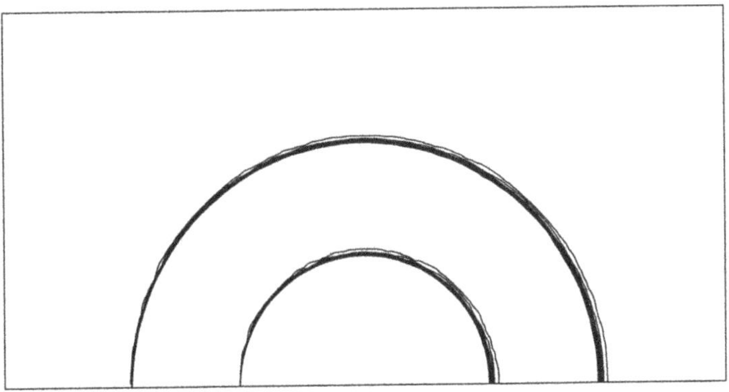

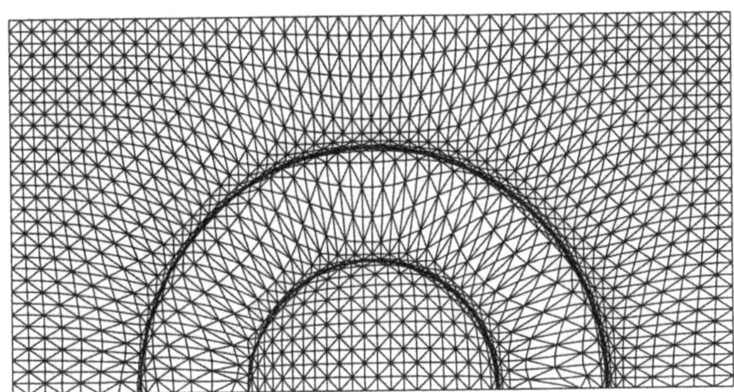

Figure 2. Circular advection of a square profile: solution on initial grid (top), solution on adapted grid with $\alpha = 1.0$, $\beta = 0.01$ (middle); adapted grid (bottom).

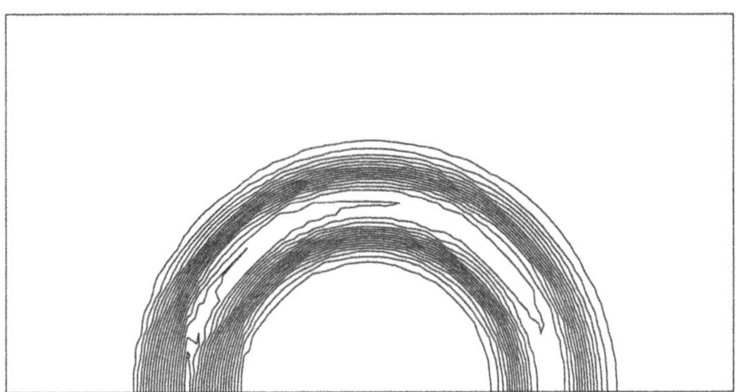

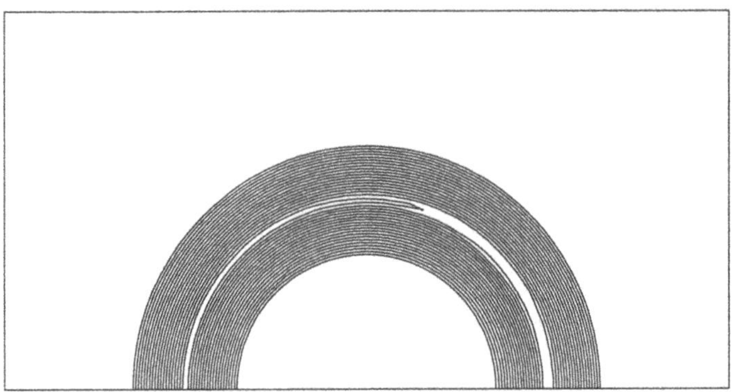

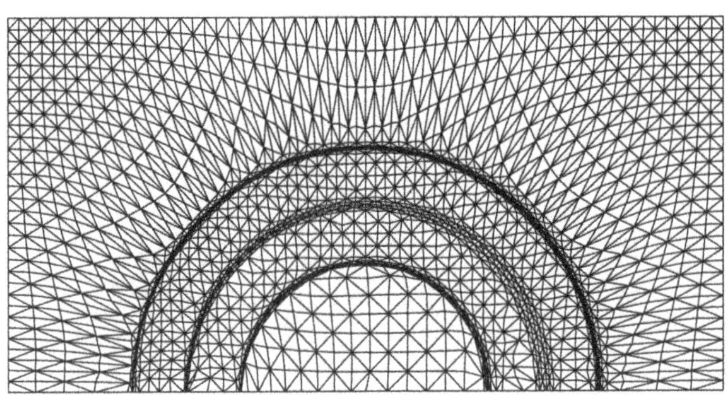

Figure 3. Circular advection of a triangular profile: solution on initial grid (top), solution on adapted grid with $\alpha = 0.1$, $\beta = 1.0$ (middle); adapted grid (bottom).

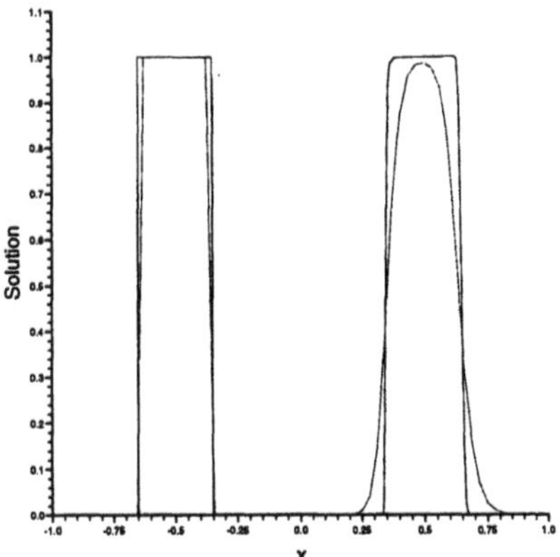

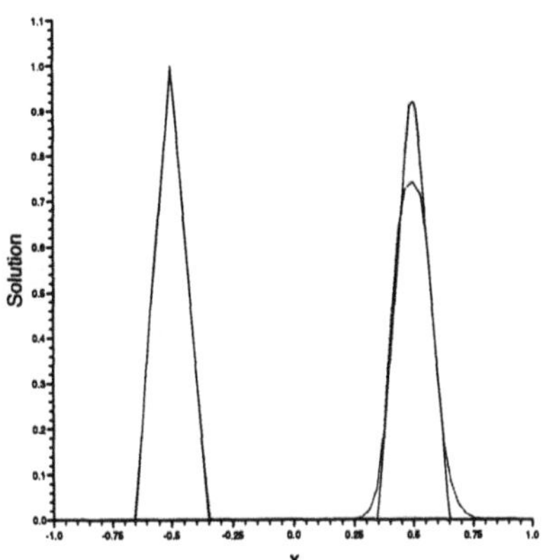

Figure 4. Solution on the boundary $y = 0$ for the advection of the square wave (left) and the triangular wave (right). Solid lines indicate the adapted solution, dotted lines the unadapted solution.

slower than on the initial, regular grid. However, the solution has improved enormously in quality; the discontinuities are captured extremely sharply and their positions are mirrored by the clustering of nodes in the grid. The grid movement parameters in (41) were chosen to be $\alpha = 1.0$ and $\beta = 0.01$, so emphasis was placed on adapting to the first derivative.

Figure 3 shows the steady state solution on both unadapted and adapted grids for a second circular advection test case, this time involving a triangular profile. The initial and boundary conditions are the same as before except that

$$u(x,0) = 1.0 - \frac{|x+0.5|}{0.15} \quad \text{for} \quad -0.65 \leq x \leq -0.35\,. \tag{43}$$

This time $\alpha = 0.1$ and $\beta = 1.0$ were chosen to give a greater significance to the second derivative, but the improvement in the quality of the solution is again remarkable.

Figure 4 shows the profiles of the adapted (solid line) and unadapted (dotted line) solutions on the boundary $y = 0$ for both test cases. When $x \leq 0$ this is an inflow boundary so the solution is exact, or as near as the grid will allow. The solution at outflow ($x > 0$) shows the improvement obtained by adapting the grid, and in both cases the shape and height of the solution has been maintained through its rotation.

However, it is necessary to be careful when choosing α and β. If $\beta = 0.0$, *i.e.* no second derivative contribution, the triangular profile becomes highly distorted as it rotates around the origin, Figure 5, and has almost become a

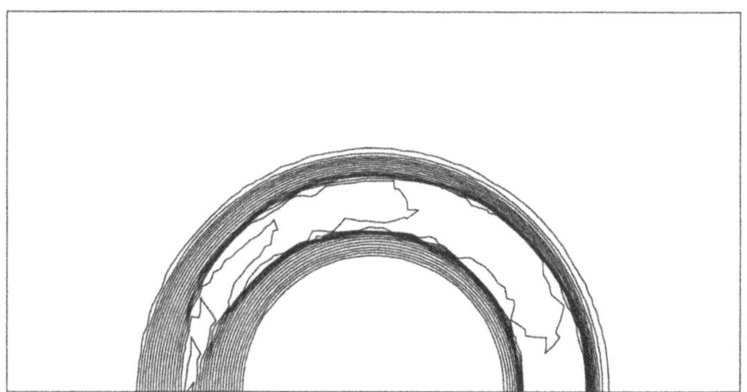

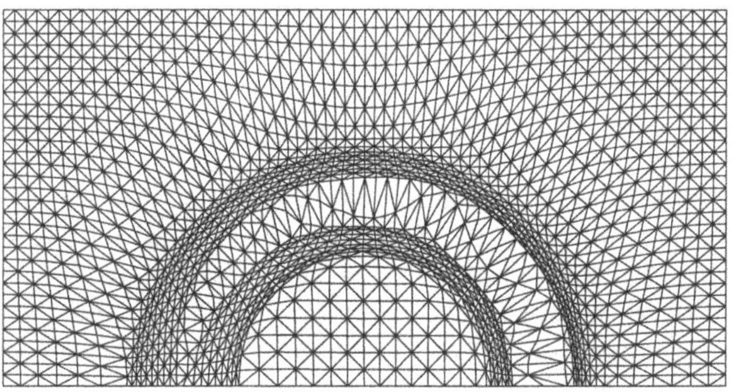

Figure 5. Circular advection of a triangular profile: solution on adapted grid with $\alpha = 1.0$, $\beta = 0.0$ (top); adapted grid (bottom).

square wave at outflow. This is because the flow varies linearly in the regions of high first derivative and can be modelled accurately using a coarse mesh, but changes rapidly at the base and peak of the triangle where the second derivative is high but the gradient is not. In fact, any solution in which capturing maxima or minima is important requires some contribution to the weights from the second derivative. Experience has shown that taking $\alpha = 1.0$ and $\beta = 0.1$ is a reasonably safe first choice for the weight parameters.

The first test case used for the Euler equations is that of flow through a walled channel of unit height with a circular arc bump on the lower surface which is 4% of the height of the channel. The freestream Mach number is $M_\infty = 1.4$. The time-stepping algorithm used decomposes the equations using the 'preconditioned' model described earlier and the PSI scheme on

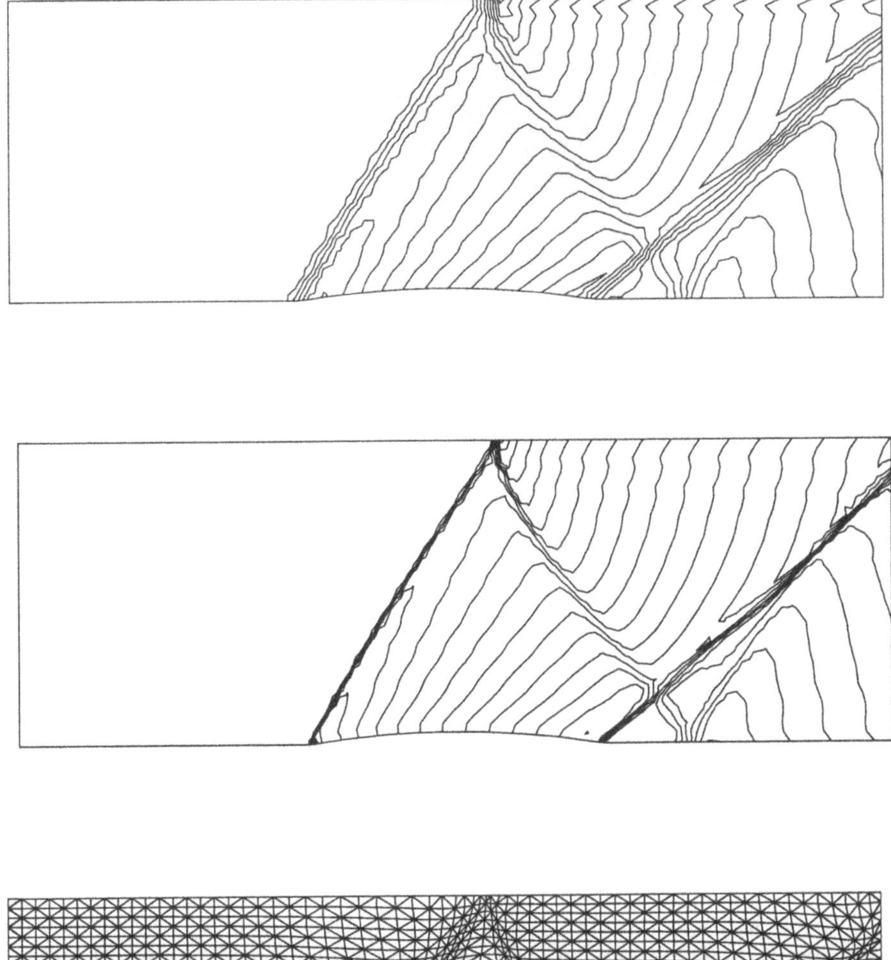

Figure 6. Inviscid flow through a channel of unit width with a 4% circular arc bump of unit length, $M_\infty = 1.4$: local Mach number contours of unadapted solution (top), adapted solution (middle); adapted grid (bottom).

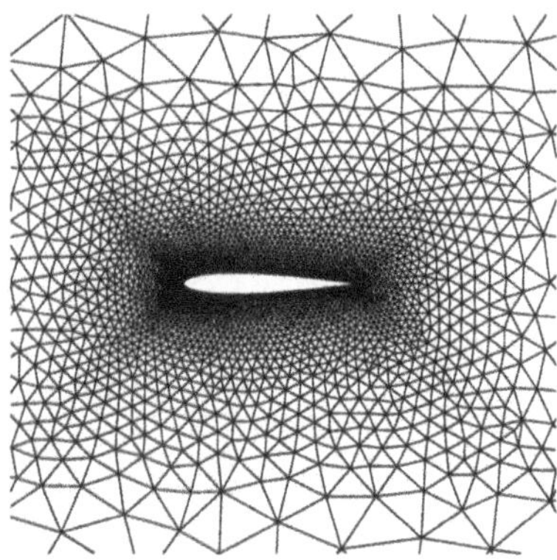

Figure 7. Initial grid for NACA0012 aerofoil.

the scalar components in all situations except for the acoustic subsystem in subsonic flow which is treated as two scalar equations and solved using the SUPG scheme. As a result, the CFL number used for these calculations is only 0.2. The boundaries are treated using a very simple characteristic-type boundary condition.

Figure 6 shows the local Mach number contours of the steady state solution obtained on a fixed isotropic grid, similar to that used for the scalar advection test cases, with 2145 nodes and 4096 cells. The adaptive algorithm is then used, with 5000 iterations being completed on this initial grid, 500 more being interleaved with the grid movement ($\alpha = 1.0$, $\beta = 0.1$) and then running to convergence on the adapted grid. The new grid and the solution obtained on it are both shown at the bottom of Figure 6 and again the improvement is marked, particularly in the capturing of the shocks. More surprisingly, perhaps, the oscillations which occur behind the shock reflection on the upper wall due to the non-positivity of the scheme in this small region of subsonic flow, are smoothed out by the adaptation. It is only where the shocks interact in a more complicated manner and nodes 'lock' that the improvement is less significant. Node locking can occur for two reasons, either because they are constrained to remain close to fixed objects such as boundaries or because they have an equal desire to move in more than one direction, *e.g.* within the triangle of shocks just behind the bump, and in either case some form of mesh refinement is needed if the solution is to be further improved.

Finally, two sets of results are given for inviscid flow around a NACA0012 aerofoil for which the initial grid is shown in Figure 7. No vorticity correction is included in the treatment of the far-field boundary so it has been

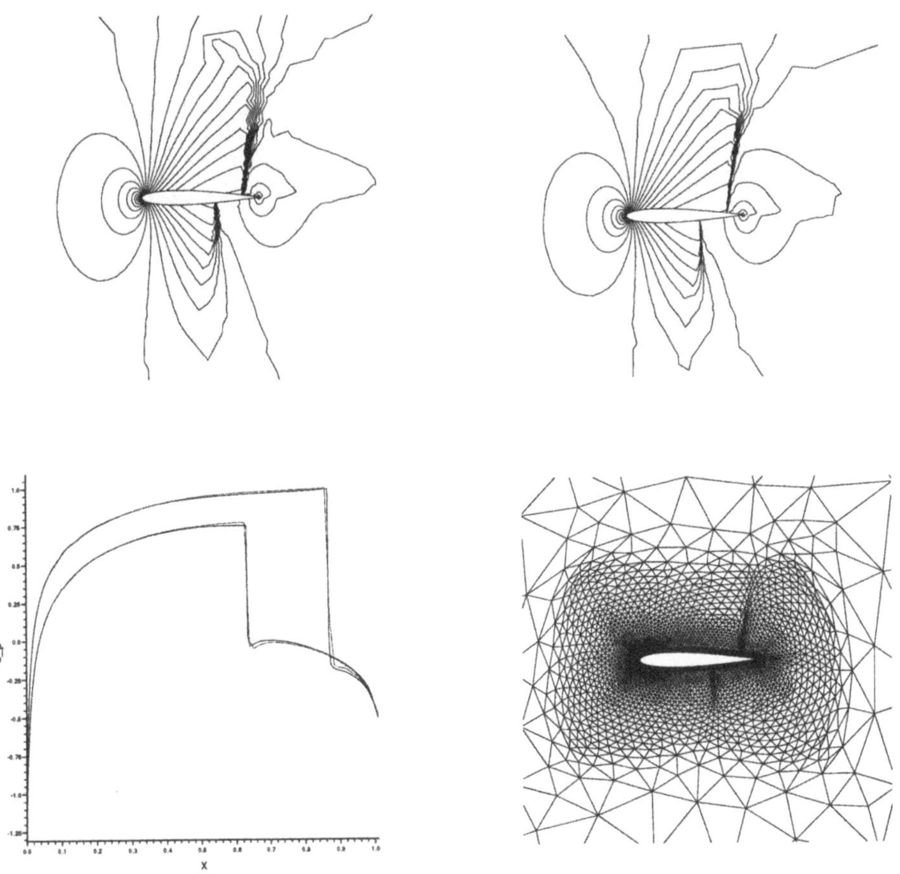

Figure 8. Inviscid flow over a NACA0012 aerofoil, $M_\infty = 0.85$, $\alpha = 1.0°$: local Mach number contours of unadapted solution, $c_l = 0.367$, $c_d = 0.050$ (top left), adapted solution, $c_l = 0.360$, $c_d = 0.049$ (top right); adapted grid (bottom right); comparison of c_p on aerofoil surface, unadapted - dotted line, adapted - solid line (bottom left).

placed at 30 chords distance from the aerofoil to reduce its effect on the solution.

The two standard test cases presented are

$$i)\quad M_\infty = 0.85, \alpha = 1.0°.$$
$$ii)\quad M_\infty = 0.8, \alpha = 1.25°.$$

Figures 8 and 9 show the local Mach number contours of the solutions obtained on the fixed and moved grids, together with the adapted grid and a graph comparing the pressure coefficient c_p on the surface of the aerofoil

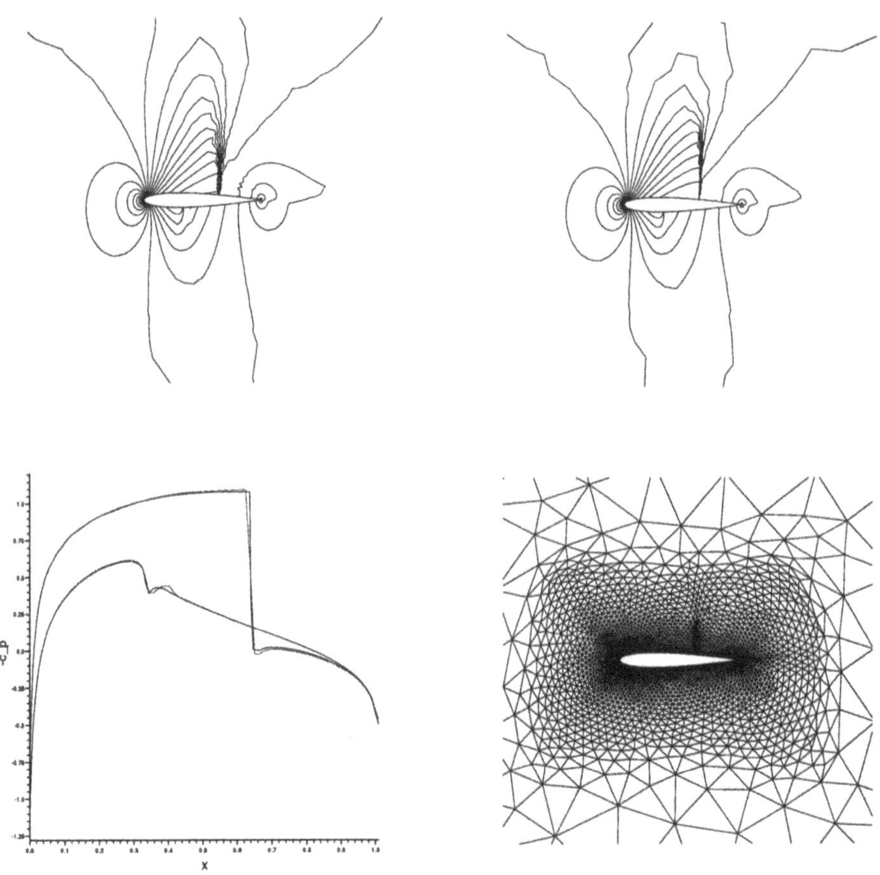

Figure 9. Inviscid flow over a NACA0012 aerofoil, $M_\infty = 0.8$, $\alpha = 1.25°$: local Mach number contours of unadapted solution, $c_l = 0.339$, $c_d = 0.015$ (top left), adapted solution, $c_l = 0.346$, $c_d = 0.016$ (top right); adapted grid (bottom right); comparison of c_p on aerofoil surface, unadapted - dotted line, adapted - solid line (bottom left).

on each grid. The solution algorithm was precisely that used to solve the channel flow, and the results again show how the capturing of shocks can be improved by the intelligent use of grid movement. It can also be seen that the oscillations behind the shocks have been reduced by moving the grid. In both cases the lift coefficient c_l is right in the middle of the range of values obtained from reference solutions using other codes [15], while the drag coefficients c_d are slightly lower. These are changed slightly by the adaptation but it is impossible to tell whether the new values are more accurate. Even so, the shocks have a significant effect on the lift and drag so the improvement in their capturing and the reduction in the oscillations behind them implies that c_l and c_d are likely to be more accurate on the adapted grids. In all of the above test cases, multidimensional has proved to be very robust even on highly distorted grids and, typically, the improved results obtained by using the adaptive algorithm take only 50% longer to produce than those on the fixed, regular grid.

6. Conclusions

Multidimensional upwinding techniques have been used to produce accurate steady state solutions to both the linear advection equation and the Euler equations in two dimensions in triangular grids, and a very simple and cheap node movement algorithm has been used to greatly improve the quality of these solutions. The strategy has proved to be robust, and steady state solutions have been obtained on all but the most highly distorted grids for a very small increase in the overall cost.

Even so, for many flows node movement is not enough to accurately model every feature and some form of grid refinement is necessary to obtain highly accurate solutions. This is particularly true where node locking prevents nodes from moving in the desired direction, a problem which can also be alleviated somewhat by the use of edge swapping. The best choices for the values of α and β, and the actual variable used to monitor the adaptation, are also unclear, and may vary depending on the flow. Finally, no measure of grid quality has been used here and indeed it is not clear what should be used since the improved solutions are obtained on grids which are much poorer in quality by any of the conventional criteria.

7. Acknowledgements

The second author would like to acknowledge the financial support of DRA Farnborough during the completion of this work.

References

1. H.Paillère, E.van der Weide and H.Deconinck. Multidimensional upwind methods for inviscid and viscous compressible flows. In *VKI LS 1995-02, Computational Fluid Dynamics*, 1995.
2. H.Deconinck, R.Struijs, G.Bourgois and P.L.Roe. High resolution shock capturing cell vertex advection schemes for unstructured grids. In *VKI LS 1994-05, Computational Fluid Dynamics*, 1994.
3. H.Paillère. Multidimensional upwind residual distribution schemes for the Euler and Navier-Stokes equations on unstructured grids. PhD Thesis, Université Libre de Bruxelles, 1995.
4. P.L.Roe. Discrete models for the numerical analysis of time-dependent multidimensional gas dynamics. In *J. Comp. Phys.*, 63:458-476, 1986.

5. M.A.Rudgyard. Multidimensional wave decompositions for the Euler equations. In *VKI LS 1993-04, Computational Fluid Dynamics*, 1993.
6. H.Deconinck, C.Hirsch and J.Peuteman. Characteristic decomposition methods for the multidimensional Euler equations. In *Lecture Notes in Physics*, 264:216-221, 1986.
7. H.Deconinck and H.Paillère. Multidimensional upwinding: preparing the future of computational methods for compressible flows? In *Proceedings of the 5th ICFD Conference*, 1995.
8. L.M.Mesaros and P.L.Roe. Multidimensional fluctuation splitting schemes based on decomposition methods. AIAA Paper 95-1699, 1995.
9. B.van Leer, W.-T.Lee and P.L.Roe. Characteristic time-stepping or local preconditioning of the Euler equations. AIAA Paper 91-1552-CP, 1991.
10. H.Deconinck, P.L.Roe and R.Struijs. A multi-dimensional generalization of Roe's flux difference splitter for the Euler equations. *Journal of Computers and Fluids*, 22:215-222, 1993.
11. D.Sidilkover and P.L.Roe. Unification of some advection schemes in two dimensions. ICASE Report, submitted to Math. Comp., 1995.
12. C.Johnson. Finite elements for flow problems. In *Unstructured grid methods for advection dominated flows*, AGARD-R-787, 1992.
13. G.Erlebacher and P.R.Eiseman. Adaptive triangular mesh generation. *AIAA Journal*, 25:1356-1364, 1987.
14. M.E.Hubbard and M.J.Baines. Multidimensional upwinding and grid adaptation. In *Proceedings of the 5th ICFD Conference*, 1995.
15. H.Viviand. AGARD Report, AGARD-AR-211, 1985.

WAVES PROPAGATION IN SATURATED RIGID POROUS MEDIA - NUMERICAL SIMULATION AND COMPARISON WITH EXPERIMENTS

G. BEN-DOR AND A. LEVY

Pearlstone Center for Aeronautical Engineering Studies
Department of Mechanical Engineering
Ben-Gurion University of the Negev, Beer Sheva, Israel

AND

S. SOREK

Pearlstone Center for Aeronautical Engineering Studies
Department of Mechanical Engineering
Ben-Gurion University of the Negev, Beer Sheva, Israel
Also affiliated with the Water Resource Research Center J. Blaustein Desert Research Institute, Sede Boker Campus, Israel.

Abstract. The head-on collision of planar shock waves with rigid porous materials was investigated experimentally and numerically. The experiments were conducted in a 75 mm x 75 mm shock tube. The rigid porous samples were made of silicon carbide (10 and 20 ppi) and alumina (30 and 40 ppi). The length of the models were in the range 40-100 mm. The shock wave Mach number range was 1.2-1.7. The pressure histories at various locations along the shock tube side-wall and at the end-wall were recorded in each experiment. The numerical study consisted of developing a 1-D TVD-based computer code capable of solving the macroscopic governing equations. The agreement between the experimental results and the numerical predictions was very good. Based on this agreement a parametric investigation using the computer code was carried out.

E.F. Toro and J.F. Clarke (eds.), Numerical Methods for Wave Propagation, 55–73.

1. Introduction

If a detailed analysis of the flow field which develops inside a porous medium which is struck head-on by a shock wave is required, then the process should be analyzed using the multi-phase approach. In this approach, the porous medium is considered as a multi-phase in which the various phases interact with each other. Unlike the mixing theory approach of Mazor et al. (1994) which practically leads to a single-phase model, a two-phase modeling approach was carried out.

Macroscopic mass, momentum and energy balance equations, for a saturated porous medium, were developed by Levy et al. (1995) by conducting a dimensional analysis on the macroscopic balance equations of Bear et al. (1992) and Sorek et al. (1992). The solid phase was assumed to be rigid and thermoelastic and the fluid phase was considered as a Newtonian air.

The modeling was based on conceptualizing the porous medium as a continuum composed of interacting compressible solid and multiple fluid phases. Macroscopic physical laws expressing mass, momentum and energy balances for the fluid phase and the solid matrix were formulated on the basis of Representative Elementary Volume (REV) concepts, as presented by Bear and Bachmat (1990). The generation of motion accounts for the domination of momentum inertial term thus obtaining the formulation of nonlinear waves.

A detailed derivation of the three-dimensional macroscopic governing equations describing the flow field in the porous medium can be found in Levy et al. (1995). Since the obtained set of equations was too complex to be solved analytically, it was decided to solve numerically the one-dimensional version as was developed and presented by Levy et al. (1995). It should be noted here that to the best of the authors' knowledge such a solution has not been conducted as yet. Neither had the compaction waves been simulated in a porous medium. As will be shown subsequently a novel 1-D TVD based computer code for solving the governing equations has been developed. Its predictions were compared to experimental results and very good to excellent agreement was evident. Many more details can be found in Levy (1995).

2. Present Study

In the present study, the one-dimensional version of the governing equations were solved. A detailed derivation of the three-dimensional macroscopic governing equations describing the flow field in a porous medium was developed by Levy et al. (1995). In the following only the assumptions used in the derivation of the governing equations and their final form are given:

2.1. THE ASSUMPTIONS

1. The fluid is ideal (i.e., $\mu_f = 0$ and $\lambda_f = 0$ where μ_f is the dynamic viscosity and λ_f is the thermal conductivity).
2. The fluid is a perfect gas.
3. The dispersive and diffusive mass fluxes of the fluid, and the dispersive mass flux of the solid, are much smaller than the corresponding advective ones and may, therefore, be neglected.
4. The dispersive flux of momentum of the fluid is much smaller than its advective one and may, therefore, be neglected.
5. The conductive and dispersive heat fluxes of the phases are negligibly small when compared to their advective heat flux.
6. The microscopic solid-fluid interfaces are material surfaces with respect to the mass of both phases.
7. The solid matrix is thermoelastic, and is assumed to undergo small deformations only.
8. The stress-strain relationship for the solid, at the microscopic level, and for the solid matrix, at the macroscopic level, have the same form.
9. The material of which the skeleton of the porous material is made is incompressible.
10. The specific heats of the fluid at constant volume, C_f, and of the solid at constant strain, C_s, are constant.
11. The heat transfer processes for the fluid and for the solid are reversible.
12. There are no external energy sources.
13. The energy associated with viscous dissipation is negligibly small.
14. The rate of heat transferred between the fluid and solid phases is negligibly small.

2.2. ONE-DIMENSIONAL GOVERNING EQUATIONS

In the present study, the one-dimensional form of the governing equations was solved. In a vector form these equations are written as follows

$$\frac{\partial \mathbf{U}}{\partial t} + \frac{\partial \mathbf{F}}{\partial x} = \mathbf{Q} \tag{1}$$

The variables vector, $\mathbf{U}$, is defined by

$$\mathbf{U} = [\, r_f, \quad r_s, \quad m_f, \quad m_s, \quad E_f, \quad E_s \,]^{\mathbf{T}} \tag{2}$$

where

$$\begin{matrix} r_f \equiv \phi\rho_f & m_f \equiv r_f V_f & E_f \equiv r_f e_f = r_f \left(C_f T_f + \dfrac{V_f^2}{2}\right) \\ r_s \equiv (1-\phi)\,\rho_s & m_s \equiv r_s V_s & E_s \equiv r_s e_s = r_s \left(C_s T_s + \dfrac{V_s^2}{2}\right) \end{matrix}, \tag{3}$$

ρ_f and ρ_s are the density of the fluid phase and the solid matrix, respectively, V_f and V_s denote their velocities, e_f and e_s denote their total energies per mass unit, T_f and T_s denote their temperatures and ϕ denotes the porosity.

The flux vector, $\mathbf{F}$, is

$$\mathbf{F} = \begin{bmatrix} m_f \\ m_s \\ \dfrac{m_f^2}{r_f} + T^*\phi P \\ \dfrac{m_s^2}{r_s} - \sigma_s' + (1-\phi T^*)\,P \\ \dfrac{m_f}{r_f}\,(E_f + T^*\phi P) \\ \dfrac{m_s}{r_s}\,(E_s - \sigma_s' + (1-\phi T^*)\,P) \end{bmatrix} \tag{4}$$

The source vector, $\mathbf{Q}$, is

$$\mathbf{Q} = \begin{bmatrix} 0 \\ 0 \\ T^*P\dfrac{\partial\phi}{\partial x} - \tilde{F}r_f\left|\dfrac{m_f}{r_f} - \dfrac{m_s}{r_s}\right|\left(\dfrac{m_f}{r_f} - \dfrac{m_s}{r_s}\right) \\ -T^*P\dfrac{\partial\phi}{\partial x} + \tilde{F}r_f\left|\dfrac{m_f}{r_f} - \dfrac{m_s}{r_s}\right|\left(\dfrac{m_f}{r_f} - \dfrac{m_s}{r_s}\right) \\ \dfrac{m_s}{r_s}\left(T^*P\dfrac{\partial\phi}{\partial x} - \tilde{F}r_f\left|\dfrac{m_f}{r_f} - \dfrac{m_s}{r_s}\right|\left(\dfrac{m_f}{r_f} - \dfrac{m_s}{r_s}\right)\right) \\ \dfrac{m_s}{r_s}\left(-T^*P\dfrac{\partial\phi}{\partial x} + \tilde{F}r_f\left|\dfrac{m_f}{r_f} - \dfrac{m_s}{r_s}\right|\left(\dfrac{m_f}{r_f} - \dfrac{m_s}{r_s}\right)\right) \end{bmatrix} \tag{5}$$

where P which denotes the pressure is prescribed by the equation of state,

$$\phi P = (\gamma - 1)\left[E_f - \frac{m_f^2}{2r_f}\right], \tag{6}$$

$\tilde{F}$ and T^* denote the Forchhiemer coefficient for an isotropic solid matrix and the tortuosity coefficient associated with the directional cosines at the solid-fluid interface, shape factor and hydraulic radius of the pores, respectively. σ_s' denotes the effective stress of a thermoelastic solid matrix as given by Bear et al. (1992). Its macroscopic constitutive relation is expressed as follows:

$$\sigma_s' = E_\varepsilon \varepsilon - E_T C_s (T_s - T_{s0}) \tag{7}$$

In the above equation $E_\varepsilon (\equiv \lambda_s'' + \mu_s')$ and $E_T (\equiv \eta / C_s)$ are the one dimensional macroscopic Lame coefficients for a thermoelastic solid (where μ_s' , λ_s'' and η denote the Lame's constant of a thermoelastic solid) and γ denotes the specific heat capacities ratio of the fluid phase. The porosity, ϕ, and the strain, ε, can be expressed as

$$\begin{aligned} \phi &= 1 - r_s/\rho_s \\ \varepsilon &= 1 - r_s/r_{s0} \end{aligned} \tag{8}$$

Hence, the effective stress, σ_s', can be rewritten to read

$$\sigma_s' = E_\varepsilon \left(\frac{r_{s0} - r_s}{r_{s0}} \right) - E_T \left(\frac{E_s}{r_s} - \frac{E_{s0}}{r_{s0}} - \frac{m_s^2}{2r_s^2} + \frac{m_{s0}^2}{2r_{s0}^2} \right) \tag{9}$$

where, the subscript s_0 denotes the solid properties at an initial state.

The above set of the governing equations consists of six partial differential equations and six unknowns namely; r_f, r_s, m_f, m_s, E_f and E_s . Consequently, in principle given the apropriate conditions, the set is complete and can be solved. However, due to the complexity of the equations, an analytical solution of the set is not feasable. Instead, a numerical solution has been performed. Details of the numerical method are given in the following section.

Note that in writing equation (1) we assumed that the gradients of the porosity were very small and therefor they might be written as source terms in the source vector, **Q**, given by (5). As a result the source vector, **Q**, contains derivative terms and thus affects the character of the equations (hyperbolic rather then elliptic-hyperbolic). This is very convenient for numerical purposes and is correct only when the gradients of the porosity are very small as is the case in this study. Applying this procedure, removes the ill-posedness associated with the elliptic-hyperbolic character of the equations which is usually manifested in highly oscillatory solutions as the mesh is refined.

3. The Numerical Method

An upwind TVD shock-capturing scheme, originally developed by Harten (1983), was extended to solve the problem of two-phase flow which describes waves propagation and interaction in saturated porous media. The scheme for solving equation (1) can be written in the following conservative form

$$\mathbf{U}_j^{n+1} = \mathbf{U}_j^n - \lambda\left(\bar{\mathbf{F}}_{j+1/2} - \bar{\mathbf{F}}_{j-1/2}\right) + \Delta t\mathbf{Q}_j \tag{10}$$

were the parameter λ is defined by

$$\lambda \equiv \Delta t/\Delta x \tag{11}$$

and the numerical flux, $\bar{\mathbf{F}}_{j+1/2}$, is evaluated by

$$\bar{\mathbf{F}}_{j+1/2} = \frac{1}{2}\left[\mathbf{F}\left(\mathbf{U}_j^n\right) + \mathbf{F}\left(\mathbf{U}_{j+1}^n\right) - \frac{1}{\lambda}\sum_{k=1}^{6}\beta_{j+1/2}^k \mathbf{R}_{j+1/2}^k\right] \tag{12}$$

$$\beta_{j+1/2}^k = \Psi^k\left(\nu_{j+1/2}^k + \gamma_{j+1/2}^k\right)\alpha_{j+1/2}^k - \left(g_j^k + g_{j+1}^k\right) \tag{13}$$

$$\nu_{j+1/2}^k = \lambda a^k\left(\mathbf{U}_{j+1/2}\right) \tag{14}$$

$$\gamma_{j+1/2}^k = \begin{cases} \left(g_{i+1}^k - g_i^k\right)/\alpha_{j+1/2}^k & \alpha_{j+1/2}^k \neq 0 \\ 0 & \alpha_{j+1/2}^k = 0 \end{cases} \tag{15}$$

$$g_i^k = \mathrm{sgn}\left(\tilde{\mathrm{g}}_{\mathrm{j}+1/2}^{\mathrm{k}}\right)\max\left[0, \min\left[\left|\tilde{\mathrm{g}}_{\mathrm{j}+1/2}^{\mathrm{k}}\right|, \tilde{\mathrm{g}}_{\mathrm{j}-1/2}^{\mathrm{k}}\mathrm{sgn}\left(\tilde{\mathrm{g}}_{\mathrm{j}+1/2}^{\mathrm{k}}\right)\right]\right] \tag{16}$$

$$\tilde{g}_{j+1/2}^k = \frac{1}{2}\left[\Psi^k\left(\nu_{j+1/2}^k\right) - \left(\nu_{j+1/2}^k\right)^2\right]\alpha_{j+1/2}^k \tag{17}$$

(18)

$$\Psi(x) = \begin{cases} x^2/4\xi + \xi & |x| < 2\xi \\ |x| & |x| \geq 2\xi \end{cases} \tag{19}$$

where

$$\xi = \begin{cases} 0.1 & \left(\partial a^k/\partial \mathbf{U}\right)\mathbf{R^k} \neq \mathbf{0} \\ 0 & \left(\partial a^k/\partial \mathbf{U}\right)\mathbf{R^k} = \mathbf{0} \end{cases}$$

The eigenvalues, a^k, of the Jacobian matrix $\mathbf{A}(\mathbf{U}) = \partial\mathbf{F}/\partial\mathbf{U}$ as obtained symbolically by using the application Mathematica were found to be

$$\begin{array}{lll} a^1 = V_s - a_s, & a^2 = V_f - a_f, & a^3 = V_s, \\ a^4 = V_f, & a^5 = V_f + a_f, & a^6 = V_s + a_s \end{array} \tag{20}$$

where a_f and a_s the equivalent speeds of sound of the fluid and the solid phases, respectively, can be expressed as

$$a_f^2 = \frac{\phi P\left(1 - T^* + \gamma T^*\right)T^*}{r_f} \tag{21}$$

and

$$
\begin{aligned}
a_s^2 &= \frac{E_\varepsilon}{r_{s0}} + \left(\frac{\rho_s}{(\rho_s - r_s)} + \frac{E_T\left(\rho_s(1-T^*)+T^* r_s\right)}{r_s^2}\right)\frac{\phi P}{(\rho_s - r_s)} \\
&- \frac{E_T E_\varepsilon}{r_s^2}\left(\frac{r_{s0}-r_s}{r_{s0}}\right) + \frac{E_T^2}{r_s^2}\left(\frac{E_s}{r_s} - \frac{E_{s0}}{r_{s0}} - \frac{m_s^2}{2r_s^2} + \frac{m_{s0}^2}{2r_{s0}^2}\right) \qquad (22)
\end{aligned}
$$

The corresponding right eigenvectors, $\mathbf{R^k}$, were found to be

$$
R^1 = \begin{bmatrix} 0 \\ 1 \\ V_s - a_s \\ 0 \\ 0 \\ H_s - V_s a_s \end{bmatrix}, \qquad
R^3 = \begin{bmatrix} 0 \\ 1 \\ 0 \\ V_s \\ 0 \\ H_s - \frac{r_s a_s^2}{E_T} \end{bmatrix},
$$

$$
R^2 = \begin{bmatrix} 1 \\ -\frac{a_f^2\left(\rho_s - (\rho_s - r_s)T^*\right)}{\left(a_s^2 - \vartheta_1^2\right)(\rho_s - r_s)T^*} \\ V_f - a_f \\ \frac{a_f^2\left(\rho_s - (\rho_s - r_s)T^*\right)(\vartheta_1 - V_s)}{\left(a_s^2 - \vartheta_1^2\right)(\rho_s - r_s)T^*} \\ H_f - V_f a_f \\ \frac{a_f^2\left(\rho_s - (\rho_s - r_s)T^*\right)(\vartheta_1 V_s - H_s)}{\left(a_s^2 - \vartheta_1^2\right)(\rho_s - r_s)T^*} \end{bmatrix}, \qquad
R^4 = \begin{bmatrix} 1 \\ 0 \\ V_f \\ 0 \\ V_f^2 \\ 0 \end{bmatrix},
$$

$$
R^5 = \begin{bmatrix} 1 \\ \frac{a_f^2\left((\rho_s - r_s)T^* - \rho_s\right)}{\left(a_s^2 - \vartheta_2^2\right)(\rho_s - r_s)T^*} \\ V_f + a_f \\ \frac{a_f^2\left(\rho_s(\rho_s - r_s)T^* - \rho_s\right)(\vartheta_2 + V_s)}{\left(a_s^2 - \vartheta_2^2\right)(\rho_s - r_s)T^*} \\ H_f + V_f a_f \\ \frac{a_f^2\left((\rho_s - r_s)T^* - \rho_s\right)(\vartheta_2 V_s + H_s)}{\left(a_s^2 - \vartheta_2^2\right)(\rho_s - r_s)T^*} \end{bmatrix}, \qquad
R^6 = \begin{bmatrix} 0 \\ 1 \\ V_s + a_s \\ 0 \\ 0 \\ H_s + V_s a_s \end{bmatrix}, \qquad (23)
$$

where

$$
H_f = \frac{E_f + T_f^* \phi P}{r_f} \qquad (24)
$$

and

$$
\begin{aligned}
H_s = \frac{E_s}{r_s} + \frac{(1-\phi T^*)P}{r_s} - E_\varepsilon\left(\frac{r_{s0}-r_s}{r_s r_{s0}}\right) \\
+ \frac{E_T}{r_s}\left(\frac{E_s}{r_s} - \frac{E_{s0}}{r_{s0}} - \frac{m_s^2}{2r_s^2} + \frac{m_{s0}^2}{2r_{s0}^2}\right) \qquad (25)
\end{aligned}
$$

are the enthalpies of the fluid and the solid phases, respectively. ϑ_1 and ϑ_2 are defined by

$$\vartheta_1 = a_f - V_f + V_s \tag{26}$$

and

$$\vartheta_2 = a_f + V_f - V_s \tag{27}$$

The parameters $\alpha^k_{j+1/2}$ were obtained by solving the following linear equations

$$\mathbf{U}_{j+1} - \mathbf{U}_j = \sum_{k=1}^{6} \alpha^k_{j+1/2} \mathbf{R}^k_{j+1/2} \tag{28}$$

to be

$$\begin{aligned}
\alpha^1_{j+1/2} &= (C_3 - C_4)/2 \\
&+ \frac{(\rho_s - (\rho_s - \hat{r}_s) T^*) \hat{a}_f^2 \left(\left(\hat{a}_s + \hat{V}_f - \hat{V}_s \right) C_1 - \hat{a}_f C_2 \right)}{2 (\rho_s - \hat{r}_s) T^* \hat{a}_s \left(\hat{a}_s - \hat{\vartheta}_1 \right) \left(\hat{a}_s + \hat{\vartheta}_2 \right)}, \\
\alpha^2_{j+1/2} &= (C_1 - C_2)/2, \\
\alpha^3_{j+1/2} &= [r_s] - C_3, \\
\alpha^4_{j+1/2} &= [r_f] - C_1, \qquad\qquad (29) \\
\alpha^5_{j+1/2} &= (C_1 + C_2)/2, \\
\alpha^6_{j+1/2} &= (C_3 + C_4)/2 \\
&+ \frac{(\rho_s - (\rho_s - \hat{r}_s) T^*) \hat{a}_f^2 \left(\left(\hat{a}_s - \hat{V}_f + \hat{V}_s \right) C_1 + \hat{a}_f C_2 \right)}{2 (\rho_s - \hat{r}_s) T^* \hat{a}_s \left(\hat{a}_s + \hat{\vartheta}_1 \right) \left(\hat{a}_s - \hat{\vartheta}_2 \right)}
\end{aligned} \tag{30}$$

where $[b] \equiv b_{i+1} - b_i$, $\hat{b}$ is an average property in the interval $[x_{i+1} - x_i]$ and the parameters C_k (for $k = 1, 2, 3, 4$) are as follows:

$$\begin{aligned}
C_1 &= \frac{(\gamma - 1) T^* \left([E_f] + \frac{1}{2} \hat{V}_f^2 [r_f] - \hat{V}_f [m_f] \right)}{\hat{a}_f^2 a}, \\
C_2 &= \frac{[m_f] - \hat{V}_f [r_f]}{\hat{a}_f i}, \\
C_3 &= \frac{E_T \left([E_s] + \left(\hat{V}_f^2 - \hat{H}_s + \frac{\hat{r}_s \hat{a}_s^2}{E_T} \right) [r_f] - \hat{V}_s [m_f] \right)}{\hat{r}_s \hat{a}_s^2}, \\
C_4 &= \frac{[m_s] - \hat{V}_s [r_s]}{\hat{a}_s}
\end{aligned} \tag{31}$$

The boundary condition on the shock-tube end-wall was simulated by using the image point method. The computational grid system was composed of 500 nodes. The computations were performed with DecStation 5000/260 and Indy R4400/150.

The performance of the presently developed TVD scheme was checked by simulating the well known 1-D shock tube problem. Figure 1 illustrates the analytical solution (solid lines) and the numerical predictions (open circles) for the fluid's density, velocity and pressure distributions. The initial conditions for this simulation were:

$$\mathbf{U}(x) = \begin{cases} \mathbf{U}_L, & x \leq 0 \\ \mathbf{U}_R, & x > 0 \end{cases}$$

$$\mathbf{U}_L = \begin{bmatrix} r_f \\ r_s \\ m_f \\ m_s \\ E_f \\ E_s \end{bmatrix}_L = \begin{bmatrix} 11.7683 \\ 0 \\ 0 \\ 0 \\ 2533125 \\ 0 \end{bmatrix}, \qquad \mathbf{U}_R = \begin{bmatrix} r_f \\ r_s \\ m_f \\ m_s \\ E_f \\ E_s \end{bmatrix}_R = \begin{bmatrix} 1.17683 \\ 0 \\ 0 \\ 0 \\ 253312.5 \\ 0 \end{bmatrix} \tag{32}$$

The comparison between the analytical solutions and the numerical predictions clearly indicated that the developed numerical code reproduced the 1-D shock tube problem excellently.

4. Experimental Study

The head-on collision of planar shock waves with rigid porous materials was investigated experimentally in order to validate the predictions of the physical model and the numerical code. The experiments were conducted in the 75 mm × 75 mm shock tube of the School of Mechanical Engineering of the University of Witwatersrand in Johannesburg, South Africa. The incident shock wave Mach number range was $1.2 \leq M_i \leq 1.7$, the initial pressures and temperatures throughout the experimental study were about 830 mbar and about 288 K, respectively.

The rigid porous samples were made of silicon carbide (SiC) and alumina (Al_2O_3). The SiC manufactured porous material had either 10 or 20 pores per inch and the Al_2O_3 manufactured porous material had either 30 or 40 pores per inch. The porosity of these porous materials as well as the initial length, L_o, of the samples which were cut from these materials are given in Table 1. The porosity, ϕ, was simply calculated from $\phi = 1 - \rho_b/\rho_s$ where ρ_b is the bulk density of the porous material and ρ_s is the material density of the skeleton of which the porous material is made. Note that the above expression is limited to gas saturated porous materials as was the case in the present study. The cross section dimensions of the samples

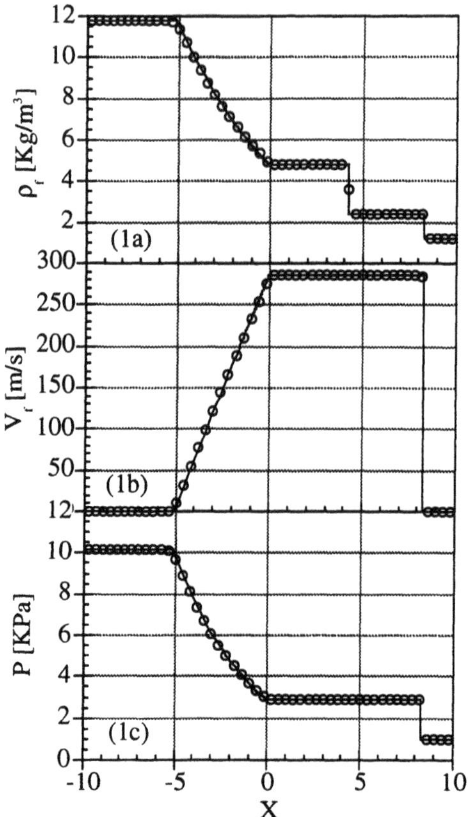

Figure 1. Comparison between the analytical solutions (solid lines) and the numerical predictions (open circles) of the classical 1-D shock tube problem. a) density, b) velocity, and c) pressure.

were 75 mm × 75 mm and hence as a result of the head-on collision with the shock wave they underwent a uni-axial strain (also known as tri-axial stress) compression. Twelve experiments were conducted with each sample. The samples were mounted at the end of the driven section of the shock tube in such a way that their back edges were supported by the shock-tube end-wall.

A schematic illustration of the experimental set-up used during the course of the present experimental study is shown in figure 2. Seven pressure transducer ports were used in each experiment. The pressure transducers in ports 1 and 2 were used to measure the incident shock wave velocity. Pressure transducers were mounted in ports 4, 5, 6, 7 and 8 for the samples having an initial length up to (inclusive) 83 mm (see Table 1). For the alumina sample with an initial length of 99 mm the pressure transducer of port

TABLE 1. The Experimental Samples

The Sample Material	Sample Type	Pores per Inch (ppi)	Porosity	Sample Length [mm]
Silicon Carbide	I	10	0.728±0.016	40, 60, 81
(SiC)	II	20	0.745±0.001	41, 62, 83
Alumina	III	30	0.814±0.010	48, 93
(Al_2O_3)	IV	40	0.821±0.007	50, 99

The structure of the porous materials was open cell.

4 was moved to port 3. This was done in order to have a measurement of the pressure changes caused by the shock wave which reflected backwards from the front edge of the porous material [for more details see Levy et al. (1993)]

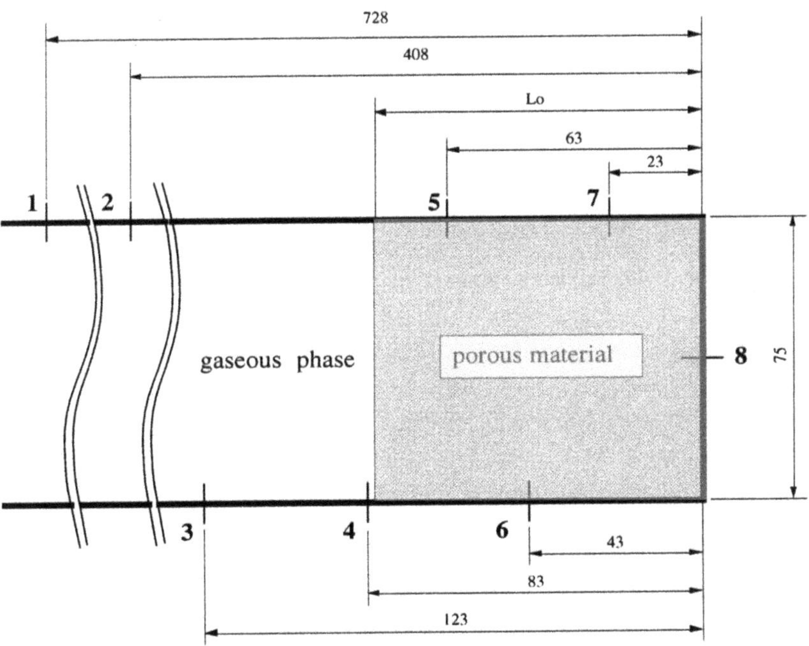

Figure 2. Schematic illustration of the experimental set-up and the location of the pressure transducers (numbered 1 to 8).

5. Comparison Between the Numerical Predictions and the Experimental Results

In order to validate the physical model and the numerical code, the governing equations were solved numerically for different samples and a variety of initial conditions and compared to experimental results. In order to solve the governing equations and compare the solution to the experimental results, the various parameters which appear in the physical model namely: the macroscopic Lame coefficients for a thermoelastic solid, E_ε and E_T, the Forchhiemer factor, $\tilde{F}$, the tortuosity, T^*, and the intrinsic density of the solid matrix, ρ_s, which appear in the physical model had to be estimated for each sample. Based on the properties of the materials of which the samples were made of and the fact that the upper limit of the elastic stress reduces when the porosity increases, in a $(1-\phi)^2$ manner, [see Gibson and Ashby (1988)] the order of magnitude of the macroscopic Lame coefficients, E_ε and E_T, were estimated to be identical for all the samples.

$$E_\varepsilon = 380 \times 10^7 \text{ Pa} \quad \text{and} \quad \mathrm{E_T} = 26.207 \text{ Kg}\big/\text{m}^3$$

The sensitivity of the predictions of the numerical code to these coefficients will be reported in a future study. The intrinsic density of the solid matrix for all the samples (i.e., silicon carbide and alumina) as provided by the manufacturer (FOSECO South Africa LTD.) was

$$\rho_s = 2000 \pm 60 \text{ Kg}\big/\text{m}^3$$

The tortuosity factors, T^*, for the various samples were found by estimating the ratio between the speed of sound of the air inside the porous medium, Eq. (21), and that in pure air. The Forchhiemer factors, $\tilde{F}$, for the various samples where found experimentally. In these experiments, each sample was mounted in a pipe and the pressure drop across it was measured as a function of the air flow rate. The pressure drop was found to be a parabolic function of the air velocity. This was done by assuming that the pressure drop depends linearly on the length of the sample. The values of the tortuosity and the Forchhiemer factors, as obtained experimentally for the various samples, are presented in Table 2.

In order to validate the physical model and the numerical code, Eq. (1) was solved numerically for different samples and initial conditions and compared to experimental results. Table 3 represents the initial conditions of 18 different experiments. The comparison between the predictions of the numerical simulations and the experimental results, for all the cases appearing in Table 3, are given in Levy (1995). In what follows, we will present two of these comparisons, the experiments of which are listed in Tabel 3 as test numbers 1 and 13.

TABLE 2. The Tortuosity and the Forchhiemer Factors for the Different Samples

The Sample Material	Sample Type	Pores per Inch (ppi)	$\tilde{F}$ [1/m]	T^*
Silicon Carbide	I	10	300	0.7
(SiC)	II	20	500	0.7
Alumina	III	30	900	0.75
(Al_2O_3)	IV	40	1800	0.75

The structure of the porous materials was open cell.

TABLE 3. The Initial Conditions for the Various Experiments

Test Number	Sample Material	Sample Length [mm]	Initial Pressure [KPa]	Initial Temperature [K]	Shock Mach Number-M_i
1		40	83.04	290.0	1.378
2	SiC	60	83.11	290.5	1.385
3		81	83.41	290.5	1.377
4	10 ppi	81	83.71	289.5	1.543
5		81	83.70	290.5	1.734
11		48	82.84	292.0	1.374
12	Al_2O_3	93	83.41	290.5	1.377
13	30 ppi	93	83.36	291.0	1.539
14		93	83.12	291.5	1.744
15		50	83.09	292.0	1.374
16	Al_2O_3	50	83.08	292.5	1.545
17	40 ppi	50	83.10	292.5	1.741
18		99	83.16	290.5	1.377

The structure of the porous materials was open cell.

Figures 3a to 3e and 4a to 4e represent typical experimental results and their numerical simulations for SiC and Al_2O_3 samples, respectively. P_1 is the pressure ahead of both the incident and the transmitted shock waves, P_2 is the theoretical pressure which should have been reached behind the incident shock wave, and P_5 is the theoretical pressure which should have been reached had the incident shock wave reflected head-on from a solid end-wall. The symbols represent the experimental results and the solid lines are the numerically predicted values. The pressure histories of the pure gas

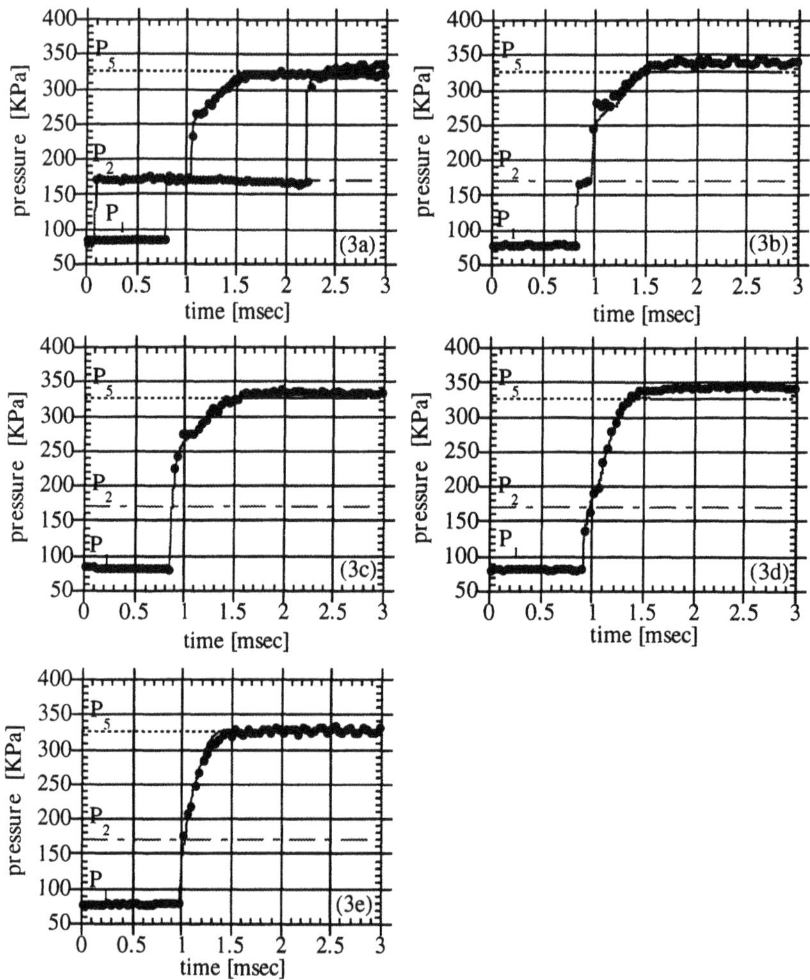

Figure 3. Typical experimental results and their numerical simulations with a 40 mm long sample made of SiC having 10 pores per inch (sample I in Table 1). The incident shock wave Mach number in this experiment was M_i=1.378 (test number 1 in Table 3). The pressure histories of the pure gas just ahead of the front edge of the porous material are shown in figures 3a and 3b. The pressure histories of the gas occupying the pores of the porous material along the shock tube side-wall and at its end-wall are shown in figures 3c to 3e, respectively. The symbols represent the experimental results and the solid lines are the numerically predicted values. P_1 is the pressure ahead of both the incident and the transmitted shock waves, P_2 is the theoretical pressure which should have been reached behind the incident shock wave, and P_5 is the theoretical pressure which would have been reached had the incident shock wave reflected head-on from a solid end-wall.

just ahead of the front edge of the porous material are shown in figures 3a, 3b for the SiC and 4a for the Al_2O_3. The pressure histories of the gas

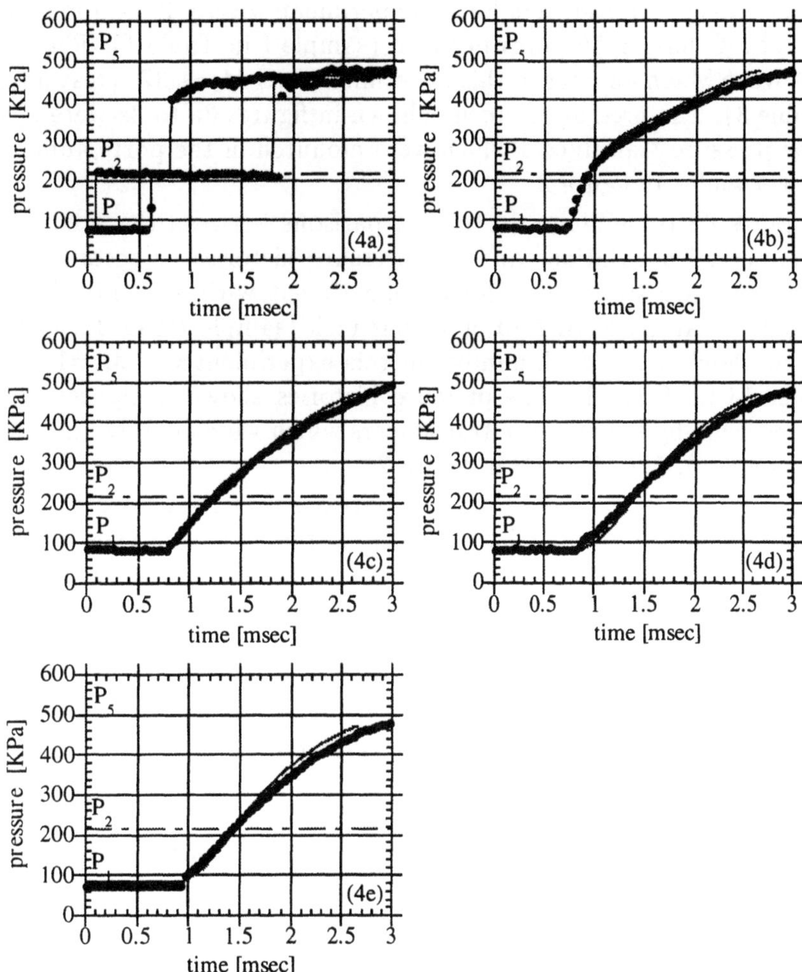

Figure 4. Typical experimental results and their numerical simulations with a 93 mm long sample made of Al_2O_3 having 30 pores per inch (sample III in Table 1). The incident shock wave Mach number in this experiment was M_i=1.539 (test number 13 in Table 3). The pressure histories of the pure gas just ahead of the front edge of the porous material are shown in figure 4a. The pressure histories of the gas occupying the pores of the porous material along the shock tube side-wall and at its end-wall are shown in figures 4b to 4e, respectively. The symbols represent the experimental results and the solid lines are the numerically predicted values. P_1, P_2, and P_5 are defined in the caption of figure 3.

occupying the pores of the porous material along the shock tube side-wall and at its end-wall are shown in figures 3c to 3e for the SiC samples and 4b to 4e for the Al_2O_3 samples.

Figures 3a to 3e illustrate the comparisons between the predictions of the numerical simulations and the pressure histories as recorded by the various pressure transducers for an experiment with a 40 mm long sample made of SiC having 10 pores per inch (sample I in Table 1). The incident shock wave Mach number in this experiment was M_i=1.378 (test number 1 in Table 3). The pressure histories shown in figures 3a to 3e were recorded by the pressure transducers which were mounted in the ports numbered 2, 4, 5, 6, 7 and 8 in figure 2.

Figures 4a to 4e illustrate the comparisons between the predictions of the numerical simulations and the pressure histories as recorded by the various pressure transducers for an experiment with a 93 mm long sample made of Al_2O_3 having 30 pores per inch (sample III in Table 1). The incident shock wave Mach number in this experiment was M_i=1.539 (test number 13 in Table 3). The pressure histories shown in figures 4a to 4e were recorded by the pressure transducers which were mounted in the ports numbered 2, 3, 5, 6, 7 and 8 in figure 2.

It is clearly evident from these comparisons that the agreement between the experimental results and the numerical predictions is very good. Note that there are disagreements between the numerical and the experimental positions of the reflected shock wave after it traveled to a relatively long distance from the front edges of the samples (see figures 3a and 4a). The agreement regarding the reflected shock wave position was better in the case of low incident shock wave Mach numbers and short samples. This may be caused by the way the porosity gradient terms were treated in the source vector, $\mathbf{Q}$.

Although figures 3a to 3e and 4a to 4e describe the results of only two typical experiments, one in the silicon-carbide (SiC) sample and one in an alumina (Al_2O_3) sample, similar good to excellent agreements were obtained in the comparisons with all the experiments which were conducted in the course of the present study. Details of all these comparisons are given in Levy (1995).

Based on this agreement the computer code was used to conduct a numerical parametric study of the flow fields which develop as a result of the head-on collision of planar shock waves with rigid porous media. Descriptions of the numerical predictions of the various flow field variables and their dependence on a variety of different parameters are given in Levy (1995).

6. The Riemann Problem in a Porous Medium

Finally, as a side-product of the developed code the solution of the Riemann problem inside a porous material was investigated. Since such a problem can not be set up in a conventional shock tube as is the case for pure gases, this problem is a theoretical one. The numerical solution of the Riemann

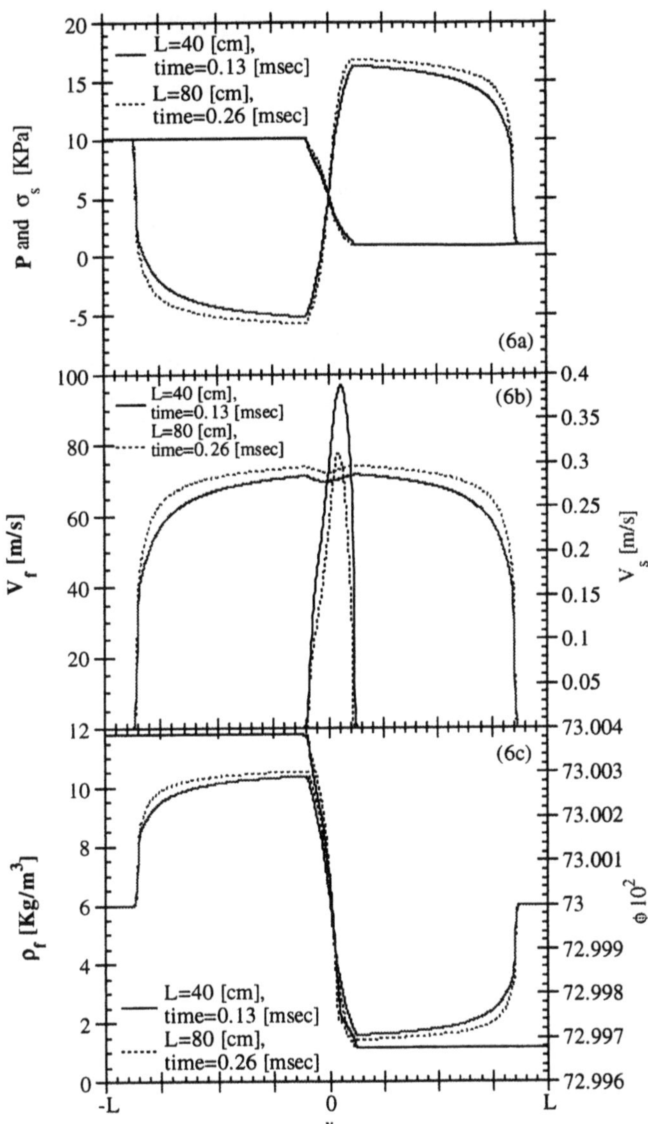

Figure 5. The numerical solution of the Riemann problem in a porous medium: a) pressure and solid normal stress, b) air and solid phase velocities, and c) air density and porosity.

problem inside the porous material is shown in figures 5a to 5c. The numerical solution revealed that the solution is not self-similar as is the case in pure gases. However, the solutions of the pressure, the solid phase normal stress, the density and the porosity of the fluid flow fields were found to be, from an engineering point of view, close enough to being self-similar.

Note that the solutions of the velocities field are not self-similar even in an engineering approximation.

7. Conclusions

The general macroscopic balance equations for a saturated rigid porous medium, were developed. Their one-dimensional version was solved numerically using a TVD-based numerical code which was developed during the course of this study. The numerical predictions were compared to experimental results and good to excellent agreement was evident for different porous materials and for a wide range of initial conditions. Based on this agreement the computer code was used to investigate numerically the flow fields which develop as a result of the head-on collision of planar shock waves with rigid porous media. Finally, the Riemann problem in a porous medium was also solved. To the best of the authors' knowledge, this is the first time that compaction waves are successfully simulated inside a porous medium.

Dedication This study which is a part of Dr. Avi Levy's Ph.D. research is dedicated to Professor Ami Harten. It was decided in the early stages of this study to develop a TVD-based computer code for simulating the phenomenon. For this reason Professor Ami Harten was asked to act as an external examiner. Ami was quite skeptical about the possibility of developing the code. However, as reported in this study, a TVD-based code was finally developed and the agreement between its numerical predictions and the experimental results was found to be very good to excellent. Unfortunately, Ami did not live to see the results.

References

Bear, J. and Bachmat, Y. (1990) *Introduction to Modeling of Transport Phenomena in Porous Media*, Kluwer Academic Publishers, Dordrecht, The Netherlands.

Bear, J., Sorek, S., Ben-Dor, G. and Mazor, G. (1992) Displacement Waves in Saturated Thermoelastic Porous Media. I. Basic Equations, Fluid Dynamic Research, Vol. 9, pp. 155–164.

Gibson, L.J. and Ashby, M.F. (1988) *Cellular Solids - Structure & Properties*, Program Press, Headington Hill Hall, Oxford , England.

Harten, A. (1983) High Resolution Schemes for Hyperbolic Conservation Laws, J. Comput. Phys., Vol. 49, pp. 357–393.

Levy, A. (1995) *Wave Propagation in a Saturated Porous Media*, Ph.D Thesis, Department of Mechanical Engineering, Ben-Gurion University of the Negev, Beer Sheva, Israel, (in Hebrew).

Levy, A., Ben-Dor, G., Skews, B.W. and Sorek, S. (1993) Head-On Collision of Normal Shock Waves with Rigid Porous Materials, Experiments in Fluids, Vol. 15, pp. 183–190.

Levy, A., Sorek, S., Ben-Dor, G. and Bear, J. (1995) Evolution of the Balance Equations in Saturated Thermoelastic Porous Media Following Abrupt Simultaneous Changes in Pressure and Temperature, Transport in Porous Media, to be published.

Mazor, G., Ben-Dor, G., Igra, O. and Sorek, S. (1994) Shock Wave Interaction with Cellular Materials. Part I: Analytical Investigation and Governing Equations, Shock Waves, Vol. 3, pp. 159–165.

Sorek, S., Bear, J., Ben-Dor, G. and Mazor, G. (1992) Shock Waves in Saturated Thermoelastic Porous Media, Transport in Porous Media, Vol. 9, pp. 3–13.

UNSPLIT WAF-TYPE SCHEMES FOR THREE DIMENSIONAL HYPERBOLIC CONSERVATION LAWS

S. J. BILLETT AND E. F. TORO
Department of Computing and Mathematics,
Manchester Metropolitan University,
Chester Street, Manchester, M1 5GD, U.K.

1. Introduction and Background

Riemann problem based methods form an important and popular class of numerical schemes for the solution of hyperbolic conservation laws today. The original method in this class was constructed by Godunov (1959). The method assumes that piecewise constant data (constant in each cell) is stored at each time level and, in the modern formulation of the method, makes use of the well-known conservative update formula introduced by Lax and Wendroff (1960). A numerical flux between adjacent cells is computed by first solving, exactly or approximately, the Riemann problem defined by the states in the two cells. The solution to a Riemann problem consists of a sequence of constant states separated by waves; the state that is upwind of all the waves is identified, and the intercell flux is taken as the flux on this state. This method is first order accurate and monotone, and has a Courant number stability limit of 1.

The physical derivation and robustness of Godunov's method has inspired many researchers to construct higher order schemes based on the solution of Riemann problems. Van Leer extended Godunov's scheme to second order accuracy with his Monotone Upwind Schemes for Conservation Laws (MUSCL schemes) (1977; 1979). He replaced the piecewise constant data used in Godunov's method with piecewise linear data. This step is commonly referred to as *reconstruction* of data or *sloping* of data. Other researchers have since produced other MUSCL-type schemes, for example Colella (1985) or Ben-Artzi and Falcovitz (1985). LeVeque (1985) has constructed a large timestep scheme that is stable for Courant numbers greater than 1. See also van Leer (1985) for the MUSCL-Hancock scheme. Riemann problem based schemes of higher than second order accuracy have also been

E.F. Toro and J.F. Clarke (eds.), Numerical Methods for Wave Propagation, 75–124.

developed; for example the Piecewise Parabolic Method (PPM) scheme of Colella and Woodward (1984). Details of Riemann solvers and Godunov-type methods can be found in the textbook by Toro (1997).

Of particular interest here is the Weighted Average Flux (WAF) method (Toro, 1989b; Toro, 1992; Toro, 1986), a second order accurate Riemann problem based method. Its origins lie in the Random Flux method (Toro, 1986; Toro *et. al.*, 1987). The intercell flux for the WAF method comes from an integral average of fluxes (or states) across the solution to the local Riemann problem based on *piecewise constant* data in cells. It has a stability restriction on the maximum Courant number of 1, and has been used in two dimensions via operator splitting in the thesis by Speares (1995). Since the original scheme was developed, the WAF approach has been extended in various ways. In one space dimension, it can be used with large timesteps (upto a maximum Courant number of 2) (Billett, 1994; Toro *et. al.*, 1993; Toro *et. al.*, 1997). This version of the scheme is a TVD Riemann-problem based extension of the Warming-Beam scheme (Warming *et. al.*, 1976), and has been used successfully in two space dimensions via operator splitting. Another extension has been to adapt the method for use on moving grids (Billett *et. al.*, 1995a), which has proved useful when combined with a CHIMERA approach to grid generation (Boden *et. al.*, 1997) to produce a scheme for solving two dimensional problems with moving boundaries. The WAF approach has also been extended for solving conservative systems in a way that is nonconservative away from shocks and in the vicinity of very weak shocks, but conservative at other shock waves, where conservation *is* needed (Toro, 1995a). The aim of this scheme is to avoid various defects of conservative schemes (Toro, 1995b; Karni, 1994), but the nonconservative part of this scheme could be developed to solve nonconservative hyperbolic systems.

The spurious oscillations associated with all (linear) schemes of higher than first order accuracy are normally eliminated in Riemann-problem based methods by locally enforcing Total Variation Diminishing (TVD) criteria (Harten, 1983; Sweby, 1984). This reduces the accuracy of schemes to first order in regions of rapidly varying gradient, but high order accuracy is maintained elsewhere. The robustness of the Riemann problem approach, high order accuracy away from regions of rapidly varying gradient and the use of TVD conditions to eliminate spurious oscillations have combined in to make the class of Riemann problem based schemes very popular in the field of computational physics. Advances in the efficiency of Riemann solvers (Harten *et. al.*, 1983; Gottlieb *et. al.*, 1988; Toro, 1997) over the last fifteen years can only have added to this popularity.

One dimensional Riemann problem based schemes were originally extended to more than one space dimension via operator splitting (Woodward *et. al.*, 1984). More recently, multidimensional unsplit schemes have been proposed by various authors. Colella's scheme (Colella, 1990) is a MUSCL type approach whereby the data on either side of a given cell interface is sloped both parallel and perpendicular to the interface. The resulting data immediately on either side of the interface is then used as the initial conditions for a Riemann problem perpendicular to the boundary. The intercell flux is taken as the Godunov flux based on this Riemann problem. Spurious oscillations are damped by applying slope limiters when reconstructing the data. Colella's scheme has been extended to three space dimensions by Saltzman (1994).

LeVeque's wave propagation method (LeVeque, 1988; LeVeque, 1997) has been developed for both conservative and nonconservative systems in two dimensions. In this method, a Riemann problem based on piecewise constant data is solved on each intercell boundary. Extra Riemann problems are then solved to determine the propagation of these waves transverse to the boundary. The speed, strength and direction of these waves are used to redistribute the variables at the next time level. For conservative systems, this redistribution is conservative. Spurious oscillations are damped by applying limiter functions to the strength of the waves.

Unsplit, multidimensional versions of the WAF method have been derived in (Billett *et. al.*, 1997b; Billett, 1994) and numerical results for the two dimensional Shallow Water equations were presented in these citations. Some numerical results for the three dimensional schemes were presented in (Billett *et. al.*, 1996). The schemes for nonlinear systems depend upon the solution to one dimensional grid-aligned Riemann problems, and in this sense can be thought of as belonging to the same class as the schemes of Colella (1990) and LeVeque (1988; 1997). There are two basic steps in the construction of an intercell flux for the multidimensional WAF schemes. Firstly, an intermediate state is found by solving, via a first order scheme, the system of equations reduced to the plane perpendicular to the boundary. This step accounts for propagation parallel to the cell boundaries, and helps stabilise the scheme. The intercell flux in then found by calculating from the intermediate state a one dimensional WAF flux perpendicular to the interface. Spurious oscillations are damped by the limiters used in the 1D flux calculation.

None of the three multidimensional schemes discussed in detail above is TVD, and there would be little point in developing TVD versions of these schemes since Goodman and LeVeque (Goodman *et. al.*, 1985) have shown that a strictly TVD scheme in more than one dimension is at most first order accurate. All the schemes successfully use one dimensional limiters to dampen spurious oscillations. A Total Variation Bounded (TVB) version of one of the 2D unsplit WAF schemes has been constructed for the linear advection equation in (Billett *et. al.*, 1995b), but has not yet been extended to nonlinear systems.

Unsplit schemes that do not rely on the solution to grid aligned Riemann problems commonly referred to as 'Multidimensional Upwind Schemes' are also under development; see e.g. (Roe, 1991; Roe, 1986; Morel, 1997; Baines, 1997). A good review of these schemes has been written by van Leer (1992).

The aim of this chapter is to discuss the class of three dimensional unsplit WAF schemes more fully than has been done in previous publications, and consider some details of their implementation. The schemes are extended to non-Cartesian geometries here by using body-fitted, curvilinear grids. Alternatively, one could use the Cartesian cut-cell approach (Quirk, 1994; Yang, 1996), although this is not discussed here. We present some numerical results for the Euler equations of Gas Dynamics. Results from an operator split version of the WAF scheme and an unsplit version of the MUSCL-Hancock scheme are also presented for comparison.

The rest of the chapter is organised as follows. In §2 we discuss the Euler equations, the Riemann problem and the WAF method in one space dimension, introduce some operator notation and discuss some operator splitting schemes in three space dimensions. In §3, we briefly review the derivation of unsplit WAF schemes in two space dimensions given in (Billett, 1994; Billett *et. al.*, 1996). In §4 we consider the construction of unsplit WAF schemes in three space dimensions and discuss their implementation. In §5 numerical results for the Euler equations of Gas Dynamics are presented. Results from an operator split WAF scheme and an unsplit version of the MUSCL-Hancock scheme (Quirk, 1994) are presented for comparison. The chapter is summarised in §6.

2. Equations of Interest and Numerical Preliminaries

2.1. THE EULER EQUATIONS OF GAS DYNAMICS

We consider the Euler Equations of Gas Dynamics

$$\mathbf{U}_t + \mathbf{F}(\mathbf{U})_x + \mathbf{G}(\mathbf{U})_y + \mathbf{H}(\mathbf{U})_z = 0 \quad (1)$$

where t denotes time x, y and z denote the three space dimensions, $\mathbf{U}$ is a vector of five conserved variables

$$\mathbf{U} = [\rho, \rho u, \rho v, \rho w, E]^T \quad (2)$$

and $\mathbf{F}$, $\mathbf{G}$ and $\mathbf{H}$ are vector-valued flux functions

$$\mathbf{F} = \left[\rho u, \rho u^2 + p, \rho uv, \rho uw, u(E+p)\right]^T \quad (3)$$

$$\mathbf{G} = \left[\rho v, \rho vu, \rho v^2 + p, \rho vw, v(E+p)\right]^T \quad (4)$$

$$\mathbf{H} = \left[\rho w, \rho wu, \rho wv, \rho w^2 + p, w(E+p)\right]^T \quad (5)$$

Here ρ is the density, u, v and w are velocity components in the x, y and z directions respectively, p is the pressure,

$$E = \frac{1}{2}\rho(u^2 + v^2 + w^2) + \rho e \tag{6}$$

is the total energy per unit volume and e is the specific internal energy. To close the system, we use the ideal equation of state

$$e = \frac{p}{(\gamma - 1)\rho} \tag{7}$$

where γ is the ratio of specific heats, assumed constant here.

We now define two important properties of the Euler equations, namely *rotational invariance* and *hyperbolicity*. Methods discussed in this chapter can be applied to any linear or nonlinear system of equations with these properties.

Definition 1 *Define two* component rotation matrices *as:*

$$\mathbf{R}_Y(\psi) = \begin{pmatrix} 1 & 0 & 0 & 0 & 0 \\ 0 & \cos\psi & 0 & \sin\psi & 0 \\ 0 & 0 & 1 & 0 & 0 \\ 0 & -\sin\psi & 0 & \cos\psi & 0 \\ 0 & 0 & 0 & 0 & 1 \end{pmatrix} \tag{8}$$

which rotates the momentum components of $\mathbf{U}$ *through an angle* ψ *about the* y *axis, and*

$$\mathbf{R}_Z(\theta) = \begin{pmatrix} 1 & 0 & 0 & 0 & 0 \\ 0 & \cos\theta & \sin\theta & 0 & 0 \\ 0 & -\sin\theta & \cos\theta & 0 & 0 \\ 0 & 0 & 0 & 1 & 0 \\ 0 & 0 & 0 & 0 & 1 \end{pmatrix} \tag{9}$$

which rotates the momentum components of $\mathbf{U}$ *through an angle* θ *about the* z *axis.*

Definition 2 *Define the* compound rotation matrix *as*

$$\mathbf{T}(\theta,\psi) = \mathbf{R}_Y(\psi)\mathbf{R}_Z(\theta) = \begin{pmatrix} 1 & 0 & 0 & 0 & 0 \\ 0 & \cos\psi\cos\theta & \cos\psi\sin\theta & \sin\psi & 0 \\ 0 & -\sin\theta & \cos\theta & 0 & 0 \\ 0 & -\sin\psi\cos\theta & -\sin\psi\sin\theta & \cos\psi & 0 \\ 0 & 0 & 0 & 0 & 1 \end{pmatrix} \tag{10}$$

which rotates first about the z *axis, then about the* y *axis.*

Theorem 1 *The Euler equations satisfy*

$$\cos\psi\ \cos\theta\ \mathbf{F} + \cos\psi \sin\theta\ \mathbf{G} + \sin\psi\ \mathbf{H} = \mathbf{T}^{-1}(\theta,\psi)\mathbf{F}(\mathbf{T}(\theta,\psi)\mathbf{U}) \quad (11)$$

for all vectors $\mathbf{U}$ *and angles* θ *and* ψ.

Proof 1 *Simple but tedious manipulation.*

Remark 1 *Property (11) is known as* rotational invariance, *and says that given a state* $\mathbf{W}$ *and two angles* θ *and* ψ, *the flux in the direction* $\mathbf{T}^{-1} \cdot (1,0,0,0,0)^T$ *can be computed by first rotating* $\mathbf{U}$ *through* (θ,ψ) *using* $\mathbf{T}$, *evaluating* $\mathbf{F}$ *and rotating back using* $\mathbf{T}^{-1}$. *This property is useful when applying numerical methods on non-Cartesian grids, as will be discussed in §2.4.*

Theorem 2 *The Euler equations have the property that the matrix*

$$\mathbf{J}(\mathbf{U},\theta,\psi) = \cos\psi\ \cos\theta \mathbf{A}(\mathbf{U}) + \cos\psi\ \sin\theta\ \mathbf{B}(\mathbf{U}) + \sin\psi\ \mathbf{C}(\mathbf{U}) \quad (12)$$

is diagonalisable $\forall\theta,\phi$. *Here* $\mathbf{A}$, $\mathbf{B}$ *and* $\mathbf{C}$ *are the Jacobians with respect to* $\mathbf{U}$ *of* $\mathbf{F}$, $\mathbf{G}$ *and* $\mathbf{H}$ *respectively.*

Proof 2 *The fact that* $\mathbf{A}$, $\mathbf{B}$ *and* $\mathbf{C}$ *are diagonalisable can be shown easily. (12) then follows trivially.*

Definition 3 *Property (12) is known as* hyperbolicity.

Remark 2 *The matrix* $\mathbf{J}$ *is just the Jacobian with respect to* $\mathbf{U}$ *of the left hand side of (11). Theorem 2 simply says that a complete set of real characteristic speeds exists in all directions.*

2.2. RIEMANN PROBLEMS

Definition 4 *The Riemann problem is an initial value problem consisting of a hyperbolic conservation law of the form (1) with initial data* $\mathbf{U}_0(x)$ *consisting of two constant states separated by an infinite plane, i.e:*

$$\mathbf{U}(x,y,z,0) = \mathbf{U}_0(x,y,z) = \begin{cases} \mathbf{U}_L & \text{if } \alpha x + \beta y + \gamma z < 0 \\ \mathbf{U}_R & \text{otherwise} \end{cases} \quad (13)$$

where α, β *and* γ *are real constants. We denote this Initial Value Problem* $\mathbf{RP}(\mathbf{U}_L,\mathbf{U}_R)$.

In practice, for a rotationally invariant system it is not necessary to solve the general problem (1),(13) for all possible data sets. It is sufficient to be able to solve the subclass of problems defined by the one dimensional hyperbolic conservation law

$$\mathbf{U}_t + \mathbf{F}(\mathbf{U})_x = 0 \quad (14)$$

with $\mathbf{U}$ and $\mathbf{F}$ as in (1) and initial data

$$\mathbf{U}(x,0) = \mathbf{U}_0(x) = \begin{cases} \mathbf{T}(\theta,\psi)\cdot\mathbf{U}_L & \text{if } x<0 \\ \\ \mathbf{T}(\theta,\psi)\cdot\mathbf{U}_R & \text{otherwise} \end{cases} \tag{15}$$

where θ and ψ are angles that specify the orientation of the plane $\alpha x + \beta y + \gamma z = 0$. See §2.6. If the solution to (15) is $\bar{\mathbf{U}}$, then the solution to (13) is $\mathbf{T}^{-1}(\theta,\psi)\cdot\bar{\mathbf{U}}$.

If $\mathbf{U}$ and $\mathbf{F}$ are taken as (2) and (3) respectively then system (14) is known as the *augmented one dimensional Euler equations*, because it consists of the one dimensional Euler equations augmented by momentum equations for the y and z directions. Henceforth, when referring to (14), we will assume it represents the augmented one dimensional Euler equations, although the discussion will extend readily to any hyperbolic system of that form.

System (14) is hyperbolic because $\mathbf{A}(\mathbf{U}) = \frac{\partial \mathbf{F}}{\partial \mathbf{U}}$ is diagonalisable. The eigenvalues are

$$e^{(1)} = u - a \qquad e^{(2)} = e^{(3)} = e^{(4)} = u \qquad e^{(5)} = u + a \tag{16}$$

where $a = \sqrt{\frac{\gamma p}{\rho}}$ is the sound speed. These eigenvalues correspond to five possible waves in the solution to the Riemann problem. $e^{(1)}$ and $e^{(5)}$ correspond to two nonlinear waves across which all primitive variables change. These can be either shock waves of rarefaction waves. The other three eigenvalues correspond to three coincident linear waves which always lie between the nonlinear waves. In out notation $e^{(2)}$ corresponds to a *contact wave*, across which only density changes, and $e^{(3)}$ and $e^{(4)}$ correspond to *shear waves*, across which only v and w change respectively. Fig. 1 shows a diagrammatic representation of one possible configuration of waves. Two constant states exist in the solution to the Riemann problem, between the nonlinear waves and the coincident linear waves. These are called the *left star state* and the *right star state*, and denoted $\mathbf{U}_L^\star$ and $\mathbf{U}_R^\star$ respectively. The pressure and velocity are equal in the two star states, and are denoted by $p^\star$ and $u^\star$ respectively. The density will, in general, change across the contact and thus be different in the two star states; the two values are denoted $\rho_L^\star$ and $\rho_R^\star$. The *star state variables* $p^\star$, $u^\star$, $\rho_L^\star$ and $\rho_R^\star$ can be found exactly from any exact Riemann solver for the one dimensional Euler equations in the literature; see, for example, (Gottlieb *et. al.*, 1988; Toro, 1989a; Toro, 1997). Alternatively, any one of a number of approximate Riemann solvers can be used, e.g (Roe, 1981; Osher *et. al.*, 1982; Toro, 1991; Toro, 1997). There also exist approximate Riemann solvers that give the fluxes $\mathbf{F}_L^\star$ and $\mathbf{F}_R^\star$ in the star states directly - see (Harten *et. al.*, 1983;

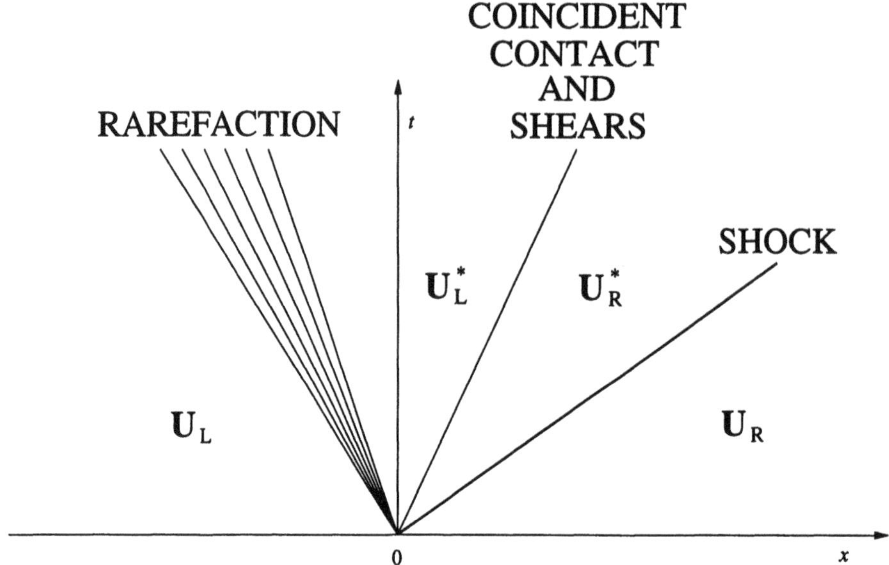

Figure 1. A possible wave configuration in the solution to the Riemann problem for the 1D augmented Euler equations.

Toro, 1994; Toro, 1997). The v and w components of velocity change across the shear waves, but not across the nonlinear waves, so in $\mathbf{U}_L^\star$, $v = v_L$ and $w = w_L$, and in $\mathbf{U}_R^\star$, $v = v_R$ and $w = w_R$. The actual speeds of the waves in the solution are the speeds from the Riemann solver for the one dimensional Euler equations. The solution to the Riemann Problem for the one dimensional augmented Euler equations is thus completely determined.

2.3. THE WEIGHTED AVERAGE FLUX METHOD FOR THE AUGMENTED 1D EULER EQUATIONS

We consider here the application of the Weighted Average Flux (WAF) method applied to the one dimensional augmented Euler equations (14), (2), (3). The method relies on the one dimensional conservative formula

$$\mathbf{U}_i^{n+1} = \mathbf{U}_i^n - \frac{\Delta t}{\Delta x}\left[\mathbf{F}_{i+\frac{1}{2}} - \mathbf{F}_{i-\frac{1}{2}}\right] \qquad (17)$$

where $\mathbf{U}_i^n$ is the integral average of the conserved variables in cell i at timelevel n, Δt is the timestep, Δx is the cell width and $\mathbf{F}_{i+\frac{1}{2}}$ is an intercell flux between cells i and $i+1$. The choice of flux $\mathbf{F}_{i+\frac{1}{2}}$ determines the scheme. For the WAF method it is taken as the integral average of fluxes across the solution to the local Riemann problem $\mathbf{RP}(\mathbf{U}_i^n, \mathbf{U}_{i+1}^n)$ at boundary $i + \frac{1}{2}$

at the half time step level

$$\mathbf{F}_{i+\frac{1}{2}} = \int_{-\frac{1}{2}\Delta x}^{\frac{1}{2}\Delta x} \mathbf{F}\left(\mathbf{U}^{\star}\left(\frac{x}{\frac{1}{2}\Delta t}\right)\right) dx \tag{18}$$

If all waves in the solution are represented as *single ray waves* this integral can be written as a weighted sum of the fluxes on the constant states in the solution

$$\mathbf{F}_{i+\frac{1}{2}} = \sum_{k=1}^{6} W_{i+\frac{1}{2}}^{(k)} \mathbf{F}_{i+\frac{1}{2}}^{(k)} \tag{19}$$

where $\mathbf{F}_{i+\frac{1}{2}}^{(k)}$ is the flux between waves $k-1$ and k and

$$W_{i+\frac{1}{2}}^{(k)} = \frac{1}{2}(\nu_{i+\frac{1}{2}}^{(k)} - \nu_{i-\frac{1}{2}}^{(k)}) \tag{20}$$

$$\nu_{i+\frac{1}{2}}^{(0)} = 1 \qquad \nu_{i+\frac{1}{2}}^{(6)} = -1$$

where $\nu_{i+\frac{1}{2}}^{(k)} = s_{i+\frac{1}{2}}^{(k)} \frac{\Delta t}{\Delta x}$ is the *Courant number* of wave k at boundary $i+\frac{1}{2}$ and $s_{i+\frac{1}{2}}^{(k)}$ is the speed of wave k at boundary $i+\frac{1}{2}$. Note that weights $W_{i+\frac{1}{2}}^{(3)} = W_{i+\frac{1}{2}}^{(4)} = 0$ since $\nu_{i+\frac{1}{2}}^{(2)} = \nu_{i+\frac{1}{2}}^{(3)} = \nu_{i+\frac{1}{2}}^{(4)}$. These weights become non-zero in the TVD version of the scheme discussed below. Fluxes $\mathbf{F}_{i+\frac{1}{2}}^{(3)}$ and $\mathbf{F}_{i+\frac{1}{2}}^{(4)}$ will be defined later. An alternative to (19) is to take

$$\mathbf{F}_{i+\frac{1}{2}} = \mathbf{F}\left(\sum_{k=1}^{6} \mathbf{W}_{i+\frac{1}{2}}^{(k)} \mathbf{U}_{i+\frac{1}{2}}^{(k)}\right) \tag{21}$$

where $\mathbf{U}_{i+\frac{1}{2}}^{(k)}$ is the state between waves $k-1$ and k. This formulation is computationally cheaper, since it requires less flux evaluations, and the resulting scheme is referred to as the *Weighted Average State* (WAS) *scheme.*

Scheme (17),(19),(20) reduces to the Lax-Wendroff scheme on the linear advection equation, and is thus second order accurate in space and time. It will thus suffer from spurious oscillations in regions of high gradient, as do all linear schemes of higher than first order accuracy. The oscillations are avoided by using a Total Variation Diminishing (TVD) version of the scheme, in which the weights (20) are replaced by

$$W_{i+\frac{1}{2}}^{(k)} = \frac{1}{2}(\phi_{i+\frac{1}{2}}^{(k)} - \phi_{i-\frac{1}{2}}^{(k)}) \tag{22}$$

$$\phi^{(0)}_{i+\frac{1}{2}} = 1 \qquad \phi^{(6)}_{i+\frac{1}{2}} = -1$$

The functions $\phi^{(k)}_{i+\frac{1}{2}}$ are *limiter functions* and must lie within data dependent bounds if the scheme is to be TVD. These bounds depend on the Courant number $\nu^{(k)}_{i+\frac{1}{2}}$ and a *flow parameter*

$$r^{(k)}_{i+\frac{1}{2}} = \frac{\Delta^{(k)}_{upw} q}{\Delta^{(k)}_{i+\frac{1}{2}} q} \tag{23}$$

where $\Delta^{(k)}_{upw} q$ is the jump in some variable q across the kth wave in the solution to the Riemann problem *upwind* of $i + \frac{1}{2}$, and $\Delta^{(k)}_{i+\frac{1}{2}} q$ is the jump in q across the kth wave in the solution to the Riemann problem at $i + \frac{1}{2}$. The bounds on $\phi^{(k)}_{i+\frac{1}{2}}$ were derived for the linear advection equation and extended to systems of equations by Toro (1989b; 1992). A popular limiter function is the $\phi^{(k)}_{i+\frac{1}{2}} = \phi^{VL}(r^{(k)}_{i+\frac{1}{2}}, \nu^{(k)}_{i+\frac{1}{2}})$ with

$$\phi^{VL}(r,\nu) = \begin{cases} 1 & r \leq 0 \\ 1 - 2(1- \mid \nu \mid) r/(1+r) & r > 0 \end{cases} \tag{24}$$

which is entirely equivalent to the VAN LEER limiter. Note that when limiter functions are used, weights $W^{(3)}_{i+\frac{1}{2}}$ and $W^{(4)}_{i+\frac{1}{2}}$ are, in general, non-zero since $\phi^{(2)}_{i+\frac{1}{2}}, \phi^{(3)}_{i+\frac{1}{2}}$ and $\phi^{(4)}_{i+\frac{1}{2}}$ depend on different flow parameters. Hence states $\mathbf{W}^{(3)}_{i+\frac{1}{2}}$ and $\mathbf{W}^{(4)}_{i+\frac{1}{2}}$ are needed to evaluate fluxes (19) or (21). These states can now be defined in a physically consistent way, as follows. We can write the limiter functions in the form

$$\phi^{(k)}_{i+\frac{1}{2}} = A^{(k)}_{i+\frac{1}{2}} \nu^{(k)}_{i+\frac{1}{2}} \tag{25}$$

where $A^{(k)}_{i+\frac{1}{2}}$ is a wave speed amplifier. These amplifiers were the original formulation of TVD limiter functions for the WAF scheme. See (Toro, 1989b; Toro, 1992). This gives an ordering on the three linear waves: larger limiter functions correspond to larger amplified wavespeeds. See Fig. 2 for an illustration of the case when $\phi^{(2)}_{i+\frac{1}{2}} < \phi^{(3)}_{i+\frac{1}{2}} < \phi^{(4)}_{i+\frac{1}{2}}$. Given that density jumps across the contact (represented by $\phi^{(2)}_{i+\frac{1}{2}}$), v jumps across one shear (represented by $\phi^{(3)}_{i+\frac{1}{2}}$) and w jumps across the other shear (represented by

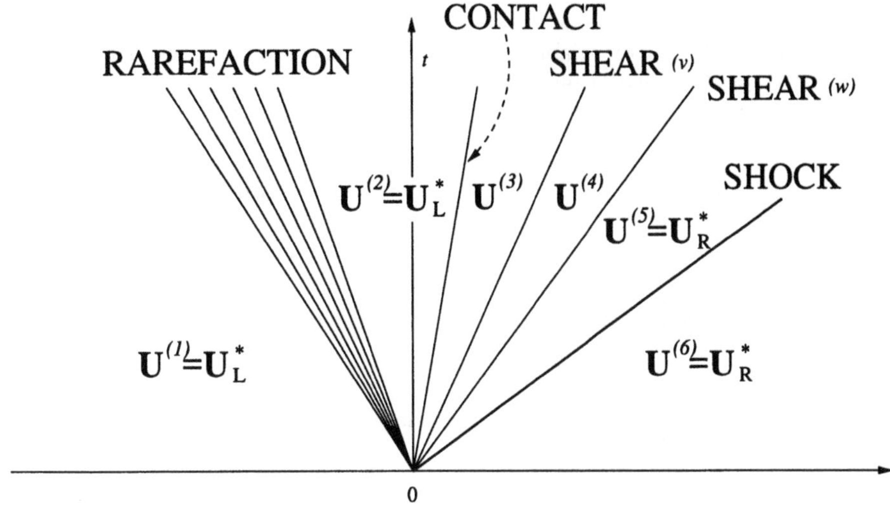

Figure 2. A possible wave configuration for the solution to the Riemann problem for the 1D augmented Euler equations, after limiting.

$\phi^{(4)}_{i+\frac{1}{2}}$), we take the states shown in Table 1. Note that the pressure is $p^\star$ and the u-component of velocity is $u^\star$ in both $\mathbf{W}^{(3)}_{i+\frac{1}{2}}$ and $\mathbf{W}^{(4)}_{i+\frac{1}{2}}$.

When the WAF scheme is fully limited, i.e $\phi^{(k)}_{i+\frac{1}{2}} \equiv 1\ \forall k$, it reduces to the Godunov method; when completely unlimited, i.e. $\phi^{(k)}_{i+\frac{1}{2}} \equiv \nu^{(k)}_{i+\frac{1}{2}}$, WAF is a nonlinear extension of the second order Lax-Wendroff scheme.

2.4. OPERATOR NOTATION

Many schemes discussed within this chapter can be efficiently described in operator notation. The operators used here are defined differently from those used in (Billett *et. al.*, 1997b). It is convenient to introduce the new operators here. Let

$$\mathbf{L}^q_{s_1...s_p,dt}(\mathbf{Q}_0, \mathbf{Q}_1, ..., \mathbf{Q}_p) \;:\; \Re^N \times \Re^{MX} \times \Re^{MY} \times \Re^{MZ} \rightarrow \Re^N \times \Re^{MX} \times \Re^{MY} \times \Re^{MZ} \quad (26)$$

denote an operator that applies a conservative update formula in p space dimensions in the 'plane' $s_1...s_p$, where $s_l = i, j$ or k, to the dataset $\mathbf{Q}_0$. Here N is the number of equations in the system being solved, and MX, MY and MZ are the number of cells in the i, j and k directions. The superscript q denotes the scheme used to calculate fluxes, with $q = G$ meaning Godunov's scheme and $q = W$ meaning the WAF scheme. The

TABLE 1. Primitive variables in states 3 and 4 corresponding to different orderings of the limiter functions

CASE	$\rho^{(3)}$	$v^{(3)}$	$w^{(3)}$	$\rho^{(4)}$	$v^{(4)}$	$w^{(4)}$
$\phi^{(2)}_{i+\frac{1}{2}} < \phi^{(3)}_{i+\frac{1}{2}} < \phi^{(4)}_{i+\frac{1}{2}}$	$\rho^\star_R$	v_L	w_L	$\rho^\star_R$	v_R	w_L
$\phi^{(2)}_{i+\frac{1}{2}} < \phi^{(4)}_{i+\frac{1}{2}} < \phi^{(3)}_{i+\frac{1}{2}}$	$\rho^\star_R$	v_L	w_L	$\rho^\star_R$	v_L	w_R
$\phi^{(3)}_{i+\frac{1}{2}} < \phi^{(2)}_{i+\frac{1}{2}} < \phi^{(4)}_{i+\frac{1}{2}}$	$\rho^\star_L$	v_R	w_L	$\rho^\star_R$	v_R	w_L
$\phi^{(3)}_{i+\frac{1}{2}} < \phi^{(4)}_{i+\frac{1}{2}} < \phi^{(2)}_{i+\frac{1}{2}}$	$\rho^\star_L$	v_R	w_L	$\rho^\star_L$	v_R	w_R
$\phi^{(4)}_{i+\frac{1}{2}} < \phi^{(2)}_{i+\frac{1}{2}} < \phi^{(3)}_{i+\frac{1}{2}}$	$\rho^\star_L$	v_L	w_R	$\rho^\star_R$	v_L	w_R
$\phi^{(4)}_{i+\frac{1}{2}} < \phi^{(3)}_{i+\frac{1}{2}} < \phi^{(2)}_{i+\frac{1}{2}}$	$\rho^\star_L$	v_L	w_R	$\rho^\star_L$	v_R	w_R

fluxes in direction s_l are computed as one dimensional fluxes perpendicular to boundaries from the dataset $\mathbf{Q}_j$ using scheme q for the 1D augmented equations. For example, $\mathbf{L}^W_{i,j,\Delta t}(\mathbf{U}^n, \mathbf{U}_i, \mathbf{U}_j)$ means update $\mathbf{U}^n$ via the two dimensional conservative formula

$$\mathbf{U}^n_{i,j,k} \rightarrow \mathbf{U}^n_{i,j,k} - \frac{\Delta t}{\Delta x}\left[\mathbf{F}_{i+\frac{1}{2},j,k} - \mathbf{F}_{i-\frac{1}{2},j,k}\right] - \frac{\Delta t}{\Delta y}\left[\mathbf{G}_{i,j+\frac{1}{2},k} - \mathbf{G}_{i,j-\frac{1}{2},k}\right] \quad (27)$$

where $\mathbf{F}_{i+\frac{1}{2},j,k}$ are WAF intercell fluxes in the i direction calculated from the augmented equations using data $\mathbf{U}_i$, and $\mathbf{G}_{i,j+\frac{1}{2},k}$ are WAF intercell fluxes in the j direction calculated from the augmented equations using data $\mathbf{U}_j$. Alternatively, $\mathbf{L}^G_{z,\frac{1}{2}\Delta t}(\mathbf{U}^n, \mathbf{U}_k)$ means update $\mathbf{U}^n$ via the one dimensional conservative formula

$$\mathbf{U}^n_{i,j,k} \rightarrow \mathbf{U}^n_{i,j,k} - \frac{\frac{1}{2}\Delta t}{\Delta z}\left[\mathbf{H}_{i,j,k+\frac{1}{2}} - \mathbf{H}_{i,j,k-\frac{1}{2}}\right] \tag{28}$$

where $\mathbf{H}_{i+\frac{1}{2}}$ are Godunov intercell fluxes in the k direction calculated from the augmented equations using data $\mathbf{U}_k$. Sometimes, we will use the operator with only one state defined, i.e

$$\mathbf{L}^q_{s_1\ldots s_p,dt}(\mathbf{Q}) : \Re^N \times \Re^{MX} \times \Re^{MY} \times \Re^{MZ} \rightarrow \Re^N \times \Re^{MX} \times \Re^{MY} \times \Re^{MZ} \tag{29}$$

In this case, all fluxes are computed on state $\mathbf{Q}$. Hence $\mathbf{L}^W_{x,y,\Delta t}(\mathbf{U}^n)$ means update $\mathbf{U}^n$ via the two dimensional conservative formula (27) with $\mathbf{F}_{i+\frac{1}{2},j,k}$ as WAF intercell fluxes in the i direction calculated from the augmented equations using data $\mathbf{Q}$ and $\mathbf{G}_{i,j+\frac{1}{2},k}$ are WAF intercell fluxes in the j direction calculated from the augmented equations using data $\mathbf{Q}$. Note that the operators (26) are unchanged under permutation of directions, e.g.

$$\mathbf{L}^q_{s_1,s_2,dt}(\mathbf{Q}_0, \mathbf{Q}_1, \mathbf{Q}_2) = \mathbf{L}^q_{s_2,s_1,dt}(\mathbf{Q}_0, \mathbf{Q}_2, \mathbf{Q}_1) \tag{30}$$

and have the property

$$\begin{aligned}\mathbf{L}^q_{s_1,s_2,dt}(\mathbf{Q}_0, \mathbf{Q}_1, \mathbf{Q}_2) &= \mathbf{L}^q_{s_2,dt}(\mathbf{L}^q_{s_1,dt}(\mathbf{Q}_0, \mathbf{Q}_1), \mathbf{Q}_2) \\ &= \mathbf{L}^q_{s_1,dt}(\mathbf{L}^q_{s_2,dt}(\mathbf{Q}_0, \mathbf{Q}_2), \mathbf{Q}_1)\end{aligned} \tag{31}$$

Property (31) is important for the memory efficient implementation of the unsplit schemes discussed later (see §4.4).

2.5. OPERATOR SPLIT SCHEMES

Perhaps the simplest schemes for solving multidimensional hyperbolic conservation laws are those based on schemes for one dimensional equations and the technique of *operator splitting* (Strang, 1968; Yanenko, 1971). Operator split schemes are those in which the augmented one dimensional equations in each direction are used separately to update a solution. For example, one operator split scheme in three dimensions can be written as

$$\mathbf{U}^{n+1} = \mathbf{L}^W_{k,\Delta t}(\mathbf{L}^W_{j,\Delta t}(\mathbf{L}^W_{i,\Delta t}(\mathbf{U}^n))) \tag{32}$$

Equation (32) means first apply the WAF scheme to the augmented 1D equations in the i-direction with data $\mathbf{U}^n$, *updating every cell in the domain.* This is called an i-sweep. Next, apply the WAF scheme to the augmented 1D equations in the j-direction with data resulting from the i-sweep. This is called a j-sweep. Finally, apply the WAF scheme to the augmented 1D equations in the k-direction with data resulting from the j-sweep. This is called a k-sweep. This splitting does seem the most obvious; however, using the approach of Strang (1968), it can be shown that although the three WAF operators (or any other second order operators) used in (32) are second order accurate in time in one space dimension, when combined in (32) the resulting scheme is only first order accurate in time. Shang (1995) has shown that the splitting

$$\mathbf{U}^{n+2} = \mathbf{L}^{S}_{i,\Delta t}(\mathbf{L}^{S}_{j,\Delta t}(\mathbf{L}^{S}_{k,\Delta t}(\mathbf{L}^{S}_{k,\Delta t}(\mathbf{L}^{S}_{j,\Delta t}(\mathbf{L}^{S}_{i,\Delta t}(\mathbf{U}^{n})))))) \tag{33}$$

is second order accurate every *second* timestep where S means any second order accurate scheme. See (Shang, 1995) for proof. Scheme (33) has optimal *Riemann solver count*: it requires the solution to just 3 Riemann problems per cell per timestep. It can be shown that the splitting

$$\mathbf{U}^{n+1} = \mathbf{L}^{S}_{i,\frac{1}{2}\Delta t}(\mathbf{L}^{S}_{j,\frac{1}{2}\Delta t}(\mathbf{L}^{S}_{k,\Delta t}(\mathbf{L}^{S}_{j,\frac{1}{2}\Delta t}(\mathbf{L}^{S}_{i,\frac{1}{2}\Delta t}(\mathbf{U}^{n}))))) \tag{34}$$

is second order at *every timestep*. The proof is similar to that found in (Shang, 1995) for Shang's splitting. Scheme (34) requires the solution to 5 Riemann problems per cell per timestep. We use splitting (34) later in this paper as a benchmark against which we compare results from our unsplit schemes.

The above operator split schemes combined with the WAF method for the one dimensional augmented Euler equations discussed in §2.3 provides a very efficient approach for solving the full three dimensional Euler equations. However, we have so far implicitly assumed a Cartesian grid; for most problems of physical importance the domain cannot be filled with a strictly Cartesian grid. There are two ways to overcome this problem and maintain logically cuboid cells, at least in most of the domain. The first is the Cartesian Cut Cell approach (Quirk, 1994; Yang, 1996), in which the vast majority of the computational domain is filled with a Cartesian grid. A narrow area around the boundary of the domain is filled with non-cuboid cells formed as the intersection of cuboid cells with the computational domain. The boundaries 'cut' the cells that intersect them. Small cells formed in this cutting process are merged with neighbouring cells. Any algorithm used to update the Cartesian cells that cover the vast majority of the domain has to be modified to integrate these cells. Some schemes are reduced to first order at the boundaries. An alternative to the cut cell approach is to use a *curvilinear grid*, i.e a logically cuboid, body fitted grid. All cells are logically cuboid, although sides may not meet at right angles. The modifications to the scheme described in §2.3 needed to implement it in an operator split fashion on curvilinear grids are described in the next section.

2.6. NON-CARTESIAN GRIDS

We now consider the calculation of intercell fluxes via Riemann problem based methods at intercell boundaries on non-Cartesian or *curvilinear* grids. Fluxes are required perpendicular to the cell sides. Before these can be found, we must first calculate for each boundary two angles θ and ϕ (see §2.1) that define the orientation of the boundary. Suppose the vector $\mathbf{n} = [n_1, n_2, n_3]^T$ is a *unit* vector normal to the boundary, and points *in the direction of increasing index* (i.e increasing i, j or k, depending on the boundary). This can be found easily from the cross product of the two vectors joining opposite corners of the cell side. If the four corners of a cell side are not coplanar, i.e the cell side is not a proper two dimensional quadrilateral, a 'normal' is computed in the same way, although in this case it is only an estimate of an 'average' normal. Once the normal has been found, the angles θ, ϕ of the cell side orientation are determined by

$$\mathbf{T}(\theta, \psi) \cdot \mathbf{n} = [1, 0, 0, 0, 0]^T \tag{35}$$

where $\mathbf{T}$ is given by (10). This can be rewritten as

$$\mathbf{T}^{-1}(\theta, \psi) \cdot [1, 0, 0, 0, 0]^T = \mathbf{n} \tag{36}$$

Using (10) in (36) and rearranging gives

$$\begin{aligned} \sin(\psi) &= n_3 \\ \cos(\psi) &= \sqrt{1 - \sin^2(\psi)} \\ \sin(\theta) &= \frac{n_2}{\cos(\psi)} \\ \cos(\theta) &= \frac{n_1}{\cos(\psi)} \end{aligned} \tag{37}$$

These sines and cosines should be computed for all boundaries at the start of the calculation and stored, as they are used a number of times in each timestep.

In order to calculate an intercell flux perpendicular to each boundary, we first determine the left and right data

$$\begin{aligned} \mathbf{U}_L &= [\rho_L, \rho_L u_L, \rho_L v_L, \rho_L w_L, E_L]^T \\ \mathbf{U}_R &= [\rho_R, \rho_R u_R, \rho_R v_R, \rho_R w_R, E_R]^T \end{aligned}$$

These states are transformed so as to obtain velocities perpendicular and parallel to the boundaries, i.e we compute

$$\begin{aligned} \bar{\mathbf{U}}_L &= \mathbf{T}(\theta,\psi)\mathbf{U}_L = [\rho_L, \rho_L\bar{u}_L, \rho_L\bar{v}_L, \rho_L\bar{w}_L, E_L]^T \\ \bar{\mathbf{U}}_R &= \mathbf{T}(\theta,\psi)\mathbf{U}_R = [\rho_R, \rho_R\bar{u}_R, \rho_R\bar{v}_R, \rho_R\bar{w}_R, E_R]^T \end{aligned}$$

where $\bar{u}_L$, $\bar{u}_R$ are the components of velocity perpendicular to the boundary, and $\bar{v}_L$, $\bar{v}_R$, $\bar{w}_L$ and $\bar{w}_R$ are parallel to the boundary. We then solve $\mathbf{RP}(\bar{\mathbf{U}}_L, \bar{\mathbf{U}}_R)$ and compute a flux $\mathbf{F}$ perpendicular to the boundary. The final intercell flux is

$$\mathbf{E} = \mathbf{T}^{-1}(\theta,\psi)\ \mathbf{F} \tag{38}$$

Note that this procedure for finding fluxes normal to boundaries relies on the rotational invariance of the system. See Theorem 1, equation (11).

When computing a WAF flux perpendicular to a boundary on a curvilinear grid, lengths that represent the cell widths on either side of the boundary are needed to compute Courant numbers $\nu^{(k)}_{i+\frac{1}{2}}$ perpendicular to the boundaries. These lengths are call *integration lengths*, and are discussed for the two dimensional case in (Speares, 1991). For each boundary, two integration lengths are needed, I_L, the *left integration length* and I_R, the *right integration length*. For the boundary between cells (i, j, k) and $(i+1, j, k)$ we compute these as

$$I_L = 2 \mid \mathbf{I}_L \cdot \mathbf{n} \mid \tag{39}$$

where $\mathbf{I}_L$ is the vector between the centre of cell (i, j, k) and the centre of the interface, and

$$I_R = 2 \mid \mathbf{I}_R \cdot \mathbf{n} \mid \tag{40}$$

where $\mathbf{I}_R$ is the vector between the centre of cell $(i+1, j, k)$ and the centre of the interface. Integration lengths for other types of boundaries are defined analogously.

Once fluxes have been computed for all boundaries, the solution is updated via the three-dimensional conservative formula for curvilinear grids:

$$\begin{aligned} \mathbf{U}^{n+1}_{i,j,k} &= \mathbf{U}^{n}_{i,j,k} - \frac{\Delta t}{Vi,j,k}\left(\mathbf{F}_{i+\frac{1}{2},j,k}\cdot A_{i+\frac{1}{2},j,k} - \mathbf{F}_{i-\frac{1}{2},j,k}\cdot A_{i-\frac{1}{2},j,k}\right) \\ &\quad - \frac{\Delta t}{Vi,j,k}\left(\mathbf{G}_{i,j+\frac{1}{2},k}\cdot A_{i,j+\frac{1}{2},k} - \mathbf{G}_{i,j-\frac{1}{2},k}\cdot A_{i,j-\frac{1}{2},k}\right) \\ &\quad - \frac{\Delta t}{Vi,j,k}\left(\mathbf{H}_{i,j,k+\frac{1}{2}}\cdot A_{i,j,k+\frac{1}{2}} - \mathbf{H}_{i,j,k-\frac{1}{2}}\cdot A_{i,j,k-\frac{1}{2}}\right) \end{aligned} \tag{41}$$

where Vi, j, k is the volume of cell (i, j, k) and $A_{i+\frac{1}{2},j,k}$ is the area of boundary $(i + \frac{1}{2}, j, k)$ (*not* to be confused with the 'amplifiers' discussed in the paragraph on limiters in §2.3). These volumes and areas need to be computed at the beginning of the calculation and stored for future use. The cell volumes are computed by dividing the cell into six tetrahedra, and computing the volume of each. The volume of a tetrahedron is given by

$$V = \frac{1}{6} \mid (\mathbf{s1} \times \mathbf{s2}) \cdot \mathbf{s3} \mid \tag{42}$$

where **s1**, **s2** and **s3** are vectors that lie along three edges of the tetrahedron that share a corner. The area of a cell side is taken as the absolute value of the cross product of the two vectors joining opposite corners of the cell side. This cross product is also used in the calculation of the orientation angles of the cell side, as discussed above.

Operator split schemes, which only update the solution in one grid direction at a time, use a pseudo one-dimensional conservative formulae in each grid direction. In the i-direction we use:

$$\mathbf{U}^{n+1}_{i,j,k} = \mathbf{U}^{n}_{i,j,k} - \frac{\Delta t}{V_{i,j,k}} \left(\mathbf{F}_{i+\frac{1}{2},j,k} \cdot A_{i+\frac{1}{2},j,k} - \mathbf{F}_{i-\frac{1}{2},j,k} \cdot A_{i-\frac{1}{2},j,k} \right) \tag{43}$$

Analogous formulae are used in the j and k directions. These one-dimensional formulae are also used in the 'predictor' stages of the unsplit multidimensional schemes discussed in §3 and §4.

3. Review of Unsplit Two Dimensional WAF-Type Schemes

The aim of this section is to briefly review the development of a two-dimensional unsplit WAF-type scheme discussed in (Billett, 1994; Billett *et. al.*, 1997b; Toro, 1997). We first construct two finite difference schemes for the two-dimensional linear advection equation (one first order, one second order) and then show how these can be combined and extended to give a scheme for systems of equations that is second order in most of the domain, but becomes first order in regions where gradients have rapid space variation in order to damp spurious oscillations.

3.1. UNSPLIT WAF-TYPE SCHEMES FOR THE TWO DIMENSIONAL LINEAR ADVECTION EQUATION

In two space dimensions the linear advection equation can be written as

$$u_t + f_x + g_y = 0 \tag{44}$$

where $f = au$ and $g = bu$ are flux components, a and b are components of the velocity vector (a, b) and (x, y) are two space coordinates. It has exact solution

$$u(x,y,t) = u^{(0)}(x - at, y - bt) \tag{45}$$

for given initial conditions $u^{(0)}(x,y)$.

Definition 5 *For a general grid (structured or unstructured), we can define a general* WAF *intercell flux* $\mathbf{e}_{AB}$ *across boundary* AB *as*

$$\mathbf{e}_{AB} = \frac{1}{t_2 - t_1}\frac{1}{V(I)}\int_{t_1}^{t_2}\int_I \mathbf{e}\cdot\mathbf{n}\, dx\, dy\, dt \tag{46}$$

where I *is a space integration range in* (x,y), $V(I)$ *is the area of* I, $\mathbf{e} = (au, bu)$ *is the general flux function, and* $\mathbf{n}$ *is the unit outward facing vector normal to the intercell boundary.*

We consider the simple case of regular Cartesian grids in this section. Let Δx, Δy be the mesh spacings, Δt be the timestep and assume that a and b are constant and positive. Consider the two dimensional conservation formula for Cartesian grids:

$$u_{i,j}^{n+1} = u_{i,j}^{n} - \frac{\Delta t}{\Delta x}\left[f_{i+\frac{1}{2},j} - f_{i-\frac{1}{2},j}\right] - \frac{\Delta t}{\Delta y}\left[g_{i,j+\frac{1}{2}} - g_{i,j-\frac{1}{2}}\right] \tag{47}$$

where $u_{i,j}^{n}$ is the integral average of u in cell (i,j) at time level n and $f_{i+\frac{1}{2},j}$ and $g_{i,j+\frac{1}{2}}$ are intercell fluxes in the x and y directions respectively. When applied to Cartesian grids, the formula (46) gives intercell fluxes in the x-direction as

$$f_{i+\frac{1}{2},j} = \frac{1}{t_2 - t_1}\frac{1}{V(I)}\int_{t_1}^{t_2}\int_I f(u^\star(x,y,t))dx\, dy\, dt \tag{48}$$

where $u^\star(x,y,t)$ is the solution to the initial value problem for (44) with initial data u^n at time level n. The fluxes in the y-direction follow by symmetry. Define the Courant numbers in the x and y directions as $\nu_x = a\frac{\Delta t}{\Delta x}$ and $\nu_y = b\frac{\Delta t}{\Delta y}$ respectively. We consider here only the case when the time limits are $t_1 = 0$ and $t_2 = \Delta t$, where Δt is the timestep chosen such that $\nu_x \leq 1$ and $\nu_y \leq 1$. For the flux $f_{i+\frac{1}{2},j}$ we take the space integration range as:

$$I = \left[-\frac{1}{2}\Delta x, \frac{1}{2}\Delta x\right] \times [0, \Delta y] \tag{49}$$

where, in local coordinates, $x = 0$ lie on the boundary, and $y = 0$, $y = \Delta y$ lie on the ends of the boundary. For the flux $g_{i,j+\frac{1}{2}}$ the obvious changes are made.

A useful scheme can be derived from (48) by using exact integration in time, the midpoint rule perpendicular to the boundary, and exact integration parallel to the boundary. The integral form of the intercell flux (48) simplifies in this case to:

$$f_{i+\frac{1}{2},j} = \frac{1}{\Delta t}\frac{1}{\Delta y}\int_0^{\Delta t}\int_{I\bigcap\{x=0\}} f(u^*(0,y,t))\, dy\, dt \tag{50}$$

where we integrate over the plane $x = 0$. This gives the intercell flux $f_{i+\frac{1}{2},j}$ as:

$$f_{i+\frac{1}{2},j} = \frac{1}{2}(2-\nu_y)f_{i,j} + \frac{1}{2}\nu_y f_{i,j-1} \tag{51}$$

The scheme resulting from this flux is a four point upwind scheme and is identical to the Corner Transport Upwind (CTU) scheme of Colella (1990) on (44). The following theorem has been proved in (Billett *et. al.*, 1997a).

Theorem 3 *Let p be some positive integer. Any scheme written in the form*

$$u_{i,j}^{n+1} = \sum_{\alpha,\beta} A_{\alpha,\beta} u_{i+\alpha,j+\beta}^n$$

for the solution of (44) is p^{th} order accurate in space and time if and only if

$$\sum_{\alpha,\beta} A_{\alpha,\beta} = (-\nu_x)^q(-\nu_y)^r \tag{52}$$

for all integer pairs (q,r) such that $q \geq 0$, $r \geq 0$ and $q+r \leq p$.

An interesting corollary follows from this Theorem.

Corollary 1 *The minimum stencil size for a pth order accurate scheme in two space dimensions is*

$$N = \frac{(p+2)!}{p!\ 2!}$$

Proof 3 *The proof of this requires two separate inductive proofs: the first to show from Theorem (3) that*

$$N = \sum_{i=1}^{p+1} i$$

and the second to show that this is equivalent to the formula in the corollary.

Thus a first order scheme in two space dimensions requires a minimum stencil of 3 points, a second order scheme requires 6 points and a third order scheme requires 10 points.

Application of conditions (52) to the CTU scheme shows that it is first order accurate in space and time. Colella proved this scheme to be stable provided

$$\max\{\nu_x, \nu_y\} \leq 1 \tag{53}$$

The second scheme that is of interest here is found by using the midpoint rule in time combined with exact integration in both space directions in (48),(49). The integral form of the intercell flux (48) simplifies in this case to:

$$f_{i+\frac{1}{2},j} = \frac{1}{\Delta x \Delta y} \int_I f(u^\star(x, y, \frac{1}{2}\Delta t)) \; dx \; dy \tag{54}$$

i.e we integrate over the plane $t = \frac{1}{2}\Delta t$. This gives intercell flux:

$$\begin{aligned} f_{i+\frac{1}{2},j} &= \frac{1}{4}(1+\nu_x)(2-\nu_y)f_{i,j} + \frac{1}{4}(1-\nu_x)(2-\nu_y)f_{i+1,j} \\ &+\frac{1}{4}(1+\nu_x)\nu_y f_{i,j-1} + \frac{1}{4}(1-\nu_x)\nu_y f_{i+1,j-1} \end{aligned} \tag{55}$$

The flux $g_{i,j+\frac{1}{2}}$ follows similarly. The resulting scheme has the eight point compact stencil. Application of conditions (52) shows that this scheme is second order accurate in space and time. When $\nu_x = 0$ or $\nu_y = 0$, the scheme reduces to the one dimensional Lax-Wendroff scheme. Numerical experiments in (Billett, 1994; Billett *et. al.*, 1997b) indicate that this scheme is stable provided (53) holds.

3.2. EXTENSION TO NONLINEAR SYSTEMS

The scheme for nonlinear systems proposed in (Billett *et. al.*, 1997b) was based on reinterpreting flux (55) as follows. It can be written as

$$f_{i+\frac{1}{2},j} = \frac{1}{2}(1+\nu_x)f\left(\bar{u}_{i,j}\right) + \frac{1}{2}(1-\nu_x)f\left(\bar{u}_{i+1,j}\right) \tag{56}$$

$$\bar{u}_{i,j} = u_{i,j} - \frac{\frac{1}{2}\Delta t}{\Delta y}\left[\bar{g}_{i,j+\frac{1}{2}} - \bar{g}_{i,j-\frac{1}{2}}\right] \tag{57}$$

where $\bar{g}_{i,j+\frac{1}{2}} = bu_{i,j}$. Equation (56) is just the Lax-Wendroff flux computed on the state $\bar{u}$, and $\bar{u}$ is found by applying a y-sweep of Godunov's method to the data u^n. An equivalent $g_{i,j+\frac{1}{2}}$ can be found in analogous form. Together they suggest a simple unsplit scheme, which can be written in operator notation for general curvilinear grids as:

$$u^{n+1} = \mathbf{L}^W_{i,j,\Delta t}(u^n, \mathbf{L}^G_{j,\frac{1}{2}\Delta t}(u^n), \mathbf{L}^G_{i,\frac{1}{2}\Delta t}(u^n)) \tag{58}$$

The use of limiter functions in the WAF operator $\mathbf{L}^W_{i,j,\Delta t}()$ in (58) does not make the scheme TVD but numerical experiments have shown that the limiter functions do control the spurious oscillations associated with the second order scheme. By the results of Goodman and LeVeque (Goodman *et. al.*, 1985), a TVD version of the scheme would be at most first order accurate, and thus too diffusive for practical applications. See (Billett *et. al.*, 1995b) for a Total Variation Bounded (TVB) version of the scheme for the linear advection equation. Note that when fully limited, the 1D WAF scheme reduces to the Godunov scheme, and (58) becomes

$$u^{n+1} = \mathbf{L}^G_{i,j,\Delta t}(u^n, \mathbf{L}^G_{j,\frac{1}{2}\Delta t}(u^n), \mathbf{L}^G_{i,\frac{1}{2}\Delta t}(u^n)) \tag{59}$$

which is easily shown to be the CTU scheme (47),(51) in operator form when applied to the linear advection equation on Cartesian grids.

Consider nonlinear hyperbolic systems of equations

$$\mathbf{U}_t + \mathbf{F}(\mathbf{U})_x + \mathbf{G}(\mathbf{U})_y = 0 \tag{60}$$

where $\mathbf{U}$ is a vector on N conserved variables and $\mathbf{F}$ and $\mathbf{G}$ are vector-valued flux functions. The schemes proposed in (Billett, 1994; Billett *et. al.*, 1997b) for such systems is simply (58) with the states replaced by the corresponding vector values and operators reinterpreted for nonlinear systems. Hence we have

$$\mathbf{U}^{n+1} = \mathbf{L}^W_{i,j,\Delta t}(\mathbf{U}^n, \mathbf{L}^G_{j,\frac{1}{2}\Delta t}(\mathbf{U}^n), \mathbf{L}^G_{i,\frac{1}{2}\Delta t}(\mathbf{U}^n)) \tag{61}$$

In (Billett, 1994; Billett *et. al.*, 1997b) an alternative extension of (55) was also suggested in which a Weighted Average *State* operator was employed instead of the WAF operator, and used before the Godunov operators. See (Billett, 1994) for another extension of flux (55) to nonlinear systems.

4. Unsplit Three-Dimensional WAF-Type Schemes

4.1. SCHEMES FOR THE LINEAR ADVECTION EQUATION

We now extend the two-dimensional unsplit WAF-type schemes of the last section to the linear advection equation in three space dimensions:

$$u_t + f_x + g_y + h_z = 0 \tag{62}$$

where $f = au, g = bu$ and $h = cu$ are flux functions. We assume here the wavespeed components a, b and c are constant and positive.

A general WAF intercell flux in three space dimensions can be defined in an analogous way to the two-dimensional flux (46).

Definition 6 *For a general grid (structured or unstructured), we can define a general* WAF *flux* $\mathbf{e}_A$ *across boundary* A *as:*

$$\mathbf{e}_A = \frac{1}{t_2 - t_1}\frac{1}{V(I)}\int_{t_1}^{t_2}\int_I \mathbf{e}\cdot\mathbf{n}\; dx\; dy\; dz\; dt \tag{63}$$

where I is the three dimensional space integration range, $V(I)$ is the volume of I, $\mathbf{e} = (au, bu, cu)$ is the general flux function and $\mathbf{n}$ is the outward facing unit vector normal to the boundary.

We consider here the case of Cartesian grids. Let Δx, Δy and Δz be the grid spacing, and Δt be the timestep. We use the conservative formula in the form:

$$\begin{aligned} u^{n+1}_{i,j,k} &= u^n_{i,j,k} - \frac{\Delta t}{\Delta x}\left[f_{i+\frac{1}{2},j,k} - f_{i-\frac{1}{2},j,k}\right] \\ &\quad -\frac{\Delta t}{\Delta y}\left[g_{i,j+\frac{1}{2},k} - g_{i,j-\frac{1}{2},k}\right] - \frac{\Delta t}{\Delta z}\left[h_{i,j,k+\frac{1}{2}} - h_{i,j,k-\frac{1}{2}}\right] \end{aligned} \tag{64}$$

where $f_{i+\frac{1}{2},j,k}$, $g_{i,j+\frac{1}{2},k}$ and $h_{i,j,k+\frac{1}{2}}$ are the intercell fluxes in the x, y and z directions respectively. From (63), $f_{i+\frac{1}{2},j,k}$ can be written as

$$f_{i+\frac{1}{2},j,k} = \frac{1}{t_2 - t_1}\frac{1}{V(I)}\int_{t_1}^{t_2}\int_I f(u^\star(x,y,x,t))\; dx\; dy\; dz\; dt \tag{65}$$

where $u^\star$ is the exact solution to equation (62) with the data u^n at time level n as initial conditions. The fluxes $g_{i,j+\frac{1}{2},k}$ and $h_{i,j,k+\frac{1}{2}}$ follow similarly. As the space integration range for $f_{i+\frac{1}{2},j,k}$, we consider the natural extension of the two-dimensional space integration range (49), which can be written as

$$I = \left[-\frac{1}{2}\Delta x, \frac{1}{2}\Delta x\right] \times [0, \Delta y] \times [0, \Delta z] \tag{66}$$

The integration ranges for $g_{i,j+\frac{1}{2},k}$ and $h_{i,j,k+\frac{1}{2}}$ follow by symmetry. Define also the directional Courant numbers as $\nu_x = a\frac{\Delta t}{\Delta x}$, $\nu_y = b\frac{\Delta t}{\Delta y}$ and $\nu_z = c\frac{\Delta t}{\Delta z}$. We assume all three to be less than unity.

Consider first an extension of the second order two dimensional flux (55). Working from the experience of the two dimensional case, we use the midpoint rule in time and exact integration in all three space directions in (65). The flux is then

$$f_{i+\frac{1}{2},j,k} = \frac{1}{\Delta x \Delta y \Delta z}\int_I f(u^\star(x,y,z,\frac{1}{2}\Delta t))\; dx\; dy\; dz \tag{67}$$

Performing the integral gives the flux in full as:

$$\begin{aligned} f_{i+\frac{1}{2},j,k} &= \frac{1}{8}(1+\nu_x)(2-\nu_y)(2-\nu_z)f_{i,j,k} + \frac{1}{8}(1-\nu_x)(2-\nu_y)(2-\nu_z)f_{i+1,j,k} \\ &\quad +\frac{1}{8}(1+\nu_x)\nu_y(2-\nu_z)f_{i,j-1,k} + \frac{1}{8}(1-\nu_x)\nu_y(2-\nu_z)f_{i+1,j-1,k} \end{aligned}$$

$$+\frac{1}{8}(1+\nu_x)(2-\nu_y)\nu_z f_{i,j,k-1}+\frac{1}{8}(1-\nu_x)(2-\nu_y)\nu_z f_{i+1,j,k-1}$$
$$+\frac{1}{8}(1+\nu_x)\nu_y\nu_z f_{i,j-1,k-1}+\frac{1}{8}(1-\nu_x)\nu_y\nu_z f_{i+1,j-1,k-1} \tag{68}$$

The fluxes $g_{i,j+\frac{1}{2},k}$ and $h_{i,j,k+\frac{1}{2}}$ follow by symmetry. Substituting the fluxes into the conservative formula (64) gives a scheme with a twenty point compact stencil, which is not reproduced here for reasons of space. The following theorem has been proved in (Billett *et. al.*, 1997a).

Theorem 4 *Let p be some positive integer. Any scheme written in the form*

$$u^{n+1}_{i,j,k}=\sum_{\alpha,\beta,\gamma}A_{\alpha,\beta,\gamma}u^n_{i+\alpha,j+\beta,k+\gamma}$$

for the solution of (62) is p^{th} order accurate in space and time if and only if

$$\sum_{\alpha,\beta,\gamma}A_{\alpha,\beta,\gamma}=(-\nu_x)^q(-\nu_y)^r(-\nu_y)^s \tag{69}$$

for all integer triples (q,r,s) such that $q\geq 0$, $r\geq 0$, $s\geq 0$ and $q+r+s\leq p$.

An interesting corollary follows from this Theorem.

Corollary 2 *The minimum stencil size for a pth order accurate scheme in three space dimensions is*

$$N=\frac{(p+3)!}{p!\ 3!}$$

Proof 4 *The proof of this requires two separate inductive proofs: the first to show from Theorem (4) that*

$$N=\sum_{j=1}^{p+1}\left(\sum_{i=1}^{j}i\right)$$

and the second to show that this is equivalent to the formula in the corollary.

Thus a first order scheme in three space dimensions requires a minimum stencil of 4 points, a second order scheme requires 10 points and a third order scheme requires 20 points. Corollaries 1 and 2 can be combined, together with the one dimensional equivalent, to give

Corollary 3 *The minimum stencil size for a pth order accurate scheme in d space dimensions is*

$$N = \frac{(p+d)!}{p!\ d!}$$

This formula is symmetrical in p and d, so the minimum stencil for a pth order accurate scheme in d space dimensions is the same as the minimum stencil for a dth order accurate scheme in p space dimensions.

Application of conditions (69) to the scheme resulting from (68) shows that it is second order accurate in space and time. Numerical experiments (Billett, 1994; Billett *et. al.*, 1996) suggest that this scheme is stable provided

$$\max\{\nu_x, \nu_y, \nu_z\} \leq 1 \tag{70}$$

and

$$\nu_x + \nu_y + \nu_z \leq 2 \tag{71}$$

To construct the final scheme, we need a monotone first order scheme. We now present a version of the first order flux (51) for the three dimensional linear equation (62). This can be done be using the midpoint rule in time, the midpoint rule in space perpendicular to the boundary, and exact integration in space in both directions parallel to the boundary. The intercell flux then becomes

$$f_{i+\frac{1}{2},j,k} = \frac{1}{\Delta y\ \Delta z} \int_{I \bigcap \{x=0\}} f(u(0, y, z, \frac{1}{2}\Delta t))\ dy\ dz \tag{72}$$

This can be written explicitly as:

$$\begin{aligned} f_{i+\frac{1}{2},j,k} &= \frac{1}{4}(2-\nu_y)(2-\nu_z) f_{i,j,k} + \frac{1}{4}\nu_y(2-\nu_z) f_{i,j-1,k} \\ &\quad + \frac{1}{4}(2-\nu_y)\nu_z f_{i,j,k-1} + \frac{1}{4}\nu_y \nu_z f_{i,j-1,k-1} \end{aligned} \tag{73}$$

The other fluxes, $g_{i,j+\frac{1}{2},k}$ and $h_{i,j,k+\frac{1}{2}}$, follow by symmetry. Substituting these into the conservative formula (64) gives a scheme with an eight point upwind stencil. Application of conditions (69) to the scheme resulting from (68) shows that it is first order accurate in space and time. The scheme is monotone under

$$\max\{\nu_x, \nu_y, \nu_z\} \leq \frac{2}{3} \tag{74}$$

since all the coefficients are positive under this condition, but this condition may not be necessary for monotonicity. Condition (74) may turn out to be the stability condition for the scheme, though numerical experiments indicate that this scheme has the same stability region as the second order scheme presented above. In practice, we use (74) to ensure stability for both first and second order schemes. Note that the scheme resulting from (73) is *not* the three dimensional CTU scheme as derived via the approach used by Colella (1990) in two dimensions, as might have been expected.

4.2. NONLINEAR SYSTEMS OF CONSERVATION LAWS

We now consider nonlinear systems of hyperbolic conservation laws, such as the three dimensional Euler equations (1)—(5). It is easy to see that (68) can be written as

$$f_{i+\frac{1}{2},j,k} = \frac{1}{2}(1+\nu_x) f\left(u^{(gh)}_{i,j,k}\right) + \frac{1}{2}(1-\nu_x) f\left(u^{(gh)}_{i+1,j,k}\right) \tag{75}$$

where

$$\begin{aligned} u^{(gh)}_{i,j,k} &= \frac{1}{4}(2-\nu_y)(2-\nu_z)u_{i,j,k} + \frac{1}{4}\nu_y(2-\nu_z)u_{i,j-1,k} \\ &\quad + \frac{1}{4}(2-\nu_y)\nu_z u_{i,j,k-1} + \frac{1}{4}\nu_y\nu_z u_{i,j-1,k-1} \end{aligned} \tag{76}$$

The flux (75) is just a Lax-Wendroff flux on the intermediate state $u^{(gh)}$. We replace this with a limited WAF flux and write, in operator notation,

$$u^{n+1} = \mathbf{L}^{W}_{i,j,k,\Delta t}(u^n, u^{(gh)}, u^{(fh)}, u^{(fg)}) \tag{77}$$

where $u^{(fh)}$ and $u^{(fg)}$ are intermediate states for the g and h fluxes respectively. The state $u^{(gh)}_{i,j,k}$ is the CTU scheme with timestep $\frac{1}{2}\Delta t$ in the $y-z$ plane. It can be interpreted in two ways. We can follow (Billett *et. al.*, 1996; Billett, 1994) and write

$$u^{(gh)}_{i,j,k} = \bar{u}_{i,j,k} - \frac{\frac{1}{2}\Delta t}{\Delta y}\left[g(\bar{u}_{i,j,k}) - g(\bar{u}_{i,j-1,k})\right] \tag{78}$$

where

$$\bar{u}_{i,j,k} = u^n_{i,j,k} - \frac{\frac{1}{2}\Delta t}{\Delta z}\left[h(u^n_{i,j,k}) - h(u^n_{i,j,k-1})\right] \tag{79}$$

Thus $u^{(gh)}_{i,j,k}$ can be interpreted as the result of operator splitting Godunov's method in the $y-z$ plane with timestep $\frac{1}{2}\Delta t$. In operator notation we can write

$$u^{(gh)}_{i,j,k} = \mathbf{L}^{G}_{j,\frac{1}{2}\Delta t}(\mathbf{L}^{G}_{k,\frac{1}{2}\Delta t}(u^n)) \tag{80}$$

Obviously, the order of the operators can be reversed. An alternative interpretation to $u_{i,j,k}^{(gh)}$ is to use the operator form of the two dimensional CTU scheme, i.e we write (80) in conservative form

$$u_{i,j,k}^{(gh)} = u_{i,j,k}^n - \frac{\frac{1}{2}\Delta t}{\Delta y}\left[g(\hat{u}_{i,j,k}) - g(\hat{u}_{i,j-1,k})\right] - \frac{\frac{1}{2}\Delta t}{\Delta z}\left[h(\tilde{u}_{i,j,k}) - h(\tilde{u}_{i,j,k-1})\right] \tag{81}$$

with

$$\hat{u}_{i,j,k} = u_{i,j,k}^n - \frac{\frac{1}{4}\Delta t}{\Delta z}\left[h(u_{i,j,k}^n) - h(u_{i,j,k-1}^n)\right] \tag{82}$$

$$\tilde{u}_{i,j,k} = u_{i,j,k}^n - \frac{\frac{1}{4}\Delta t}{\Delta y}\left[g(u_{i,j,k}^n) - g(u_{i,j-1,k}^n)\right] \tag{83}$$

Equation (81) is simply the conservative form in the $y - z$ plane with Godunov fluxes based on the states $\hat{u}$ and $\tilde{u}$ and a timestep $\frac{1}{2}\Delta t$. The states $\hat{u}$ and $\tilde{u}$ are found via one dimensional sweeps of the Godunov scheme in the z and y directions respectively using a timestep of $\frac{1}{4}\Delta t$. For general curvilinear grids we can thus write (80) in operator notation as

$$u^{(gh)} = \mathbf{L}_{j,k,\frac{1}{2}\Delta t}^{G}(u^n, \mathbf{L}_{k,\frac{1}{4}\Delta t}^{G}(u^n), \mathbf{L}_{j,\frac{1}{4}\Delta t}^{G}(u^n)) \tag{84}$$

Note that (84) is just (59) with timestep $\frac{1}{2}\Delta t$. States $u^{(fh)}$ and $u^{(fg)}$ follow by symmetry.

The two interpretations of $u_{i,j,k}^{(gh)}$ lead to two interpretations of the complete scheme, which can be written in full as

$$\begin{aligned} u^{n+1} \;=\; & \mathbf{L}_{i,j,k,\Delta t}^{W}(u^n, \mathbf{L}_{k,\frac{1}{2}\Delta t}^{G}(\mathbf{L}_{j,\frac{1}{2}\Delta t}^{G}(u^n)), \\ & \mathbf{L}_{k,\frac{1}{2}\Delta t}^{G}(\mathbf{L}_{i,\frac{1}{2}\Delta t}^{G}(u^n)), \mathbf{L}_{j,\frac{1}{2}\Delta t}^{G}(\mathbf{L}_{i,\frac{1}{2}\Delta t}^{G}(u^n))) \end{aligned} \tag{85}$$

and

$$\begin{aligned} u^{n+1} \;=\; & \mathbf{L}_{i,j,k,\Delta t}^{W}(u^n, \mathbf{L}_{j,k,\frac{1}{2}\Delta t}^{G}(u^n, \mathbf{L}_{k,\frac{1}{4}\Delta t}^{G}(u^n), \mathbf{L}_{j,\frac{1}{4}\Delta t}^{G}(u^n)), \\ & \mathbf{L}_{i,k,\frac{1}{2}\Delta t}^{G}(u^n, \mathbf{L}_{k,\frac{1}{4}\Delta t}^{G}(u^n), \mathbf{L}_{i,\frac{1}{4}\Delta t}^{G}(u^n)), \\ & \mathbf{L}_{i,j,\frac{1}{2}\Delta t}^{G}(u^n, \mathbf{L}_{j,\frac{1}{4}\Delta t}^{G}(u^n), \mathbf{L}_{i,\frac{1}{4}\Delta t}^{G}(u^n))) \end{aligned} \tag{86}$$

We refer to these schemes (and their nonlinear extensions) as WAF1 and WAF2 respectively. Numerical oscillations in regions of rapid gradient change are controlled via the limiter functions in the WAF operators. As in the two dimensional case, the use of limiter functions in the WAF fluxes does not make these schemes TVD, but numerical experiments have shown that the limiter functions do eliminate most of the spurious oscillations associated with second order schemes. By the results of Goodman and LeVeque (1985), a TVD version of the scheme would be at most first order accurate, and thus too diffusive for practical applications.

For nonlinear systems, such as the three dimensional Euler equations (1)—(5), we simply replace u in (85) and (86) by its vector equivalent, and interpret the operators as applying to the system of equations to get two schemes for systems:

$$\begin{aligned}\mathbf{U}^{n+1} = {} & \mathbf{L}^{W}_{i,j,k,\Delta t}(\mathbf{U}^{n}, \mathbf{L}^{G}_{k,\frac{1}{2}\Delta t}(\mathbf{L}^{G}_{j,\frac{1}{2}\Delta t}(\mathbf{U}^{n})), \\ & \mathbf{L}^{G}_{k,\frac{1}{2}\Delta t}(\mathbf{L}^{G}_{i,\frac{1}{2}\Delta t}(\mathbf{U}^{n})), \mathbf{L}^{G}_{j,\frac{1}{2}\Delta t}(\mathbf{L}^{G}_{i,\frac{1}{2}\Delta t}(\mathbf{U}^{n})))\end{aligned} \tag{87}$$

and

$$\begin{aligned}\mathbf{U}^{n+1} = {} & \mathbf{L}^{W}_{i,j,k,\Delta t}(\mathbf{U}^{n}, \mathbf{L}^{G}_{j,k,\frac{1}{2}\Delta t}(\mathbf{U}^{n}, \mathbf{L}^{G}_{k,\frac{1}{4}\Delta t}(\mathbf{U}^{n}), \mathbf{L}^{G}_{j,\frac{1}{4}\Delta t}(\mathbf{U}^{n})), \\ & \mathbf{L}^{G}_{i,k,\frac{1}{2}\Delta t}(\mathbf{U}^{n}, \mathbf{L}^{G}_{k,\frac{1}{4}\Delta t}(\mathbf{U}^{n}), \mathbf{L}^{G}_{i,\frac{1}{4}\Delta t}(\mathbf{U}^{n})), \\ & \mathbf{L}^{G}_{i,j,\frac{1}{2}\Delta t}(\mathbf{U}^{n}, \mathbf{L}^{G}_{j,\frac{1}{4}\Delta t}(\mathbf{U}^{n}), \mathbf{L}^{G}_{i,\frac{1}{4}\Delta t}(\mathbf{U}^{n})))\end{aligned} \tag{88}$$

Scheme (88) will be more expensive than (87) in terms of CPU time, since 12 Riemann problems per cell per timestep need to be solved for (88), whereas only 8 Riemann problems per cell per timestep need to be solved for (87). However, (88) is a more symmetrical scheme, and should thus give more symmetrical results.

Schemes (87) and (88) are illustrated geometrically in Figures 3 and 4. The aim is to transmit data from $\mathbf{U}^{n}$ at the bottom,left-hand, near corner of a cube to $\mathbf{U}^{n+1}$ at the top, right-hand, far corner of the cube. Each operator is equivalent to transmitting data along one edge of the cube. The symmetry of (88) is obvious: all edges of the cube in Fig. 4 are covered, with symmetrical distribution of weightings on Δt. Scheme (87) is obviously less symmetrical: four edges of the cube are unused. A more symmetrical scheme similar to (87) is illustrated in Fig 5. Part (a) is used on odd timesteps, part (b) is used on even timesteps. This can be written in operator notation as:

$$\begin{aligned}\mathbf{U}^{n+1} = {} & \mathbf{L}^{W}_{i,j,k,\Delta t}(\mathbf{U}^{n}, \mathbf{L}^{G}_{k,\frac{1}{2}\Delta t}(\mathbf{L}^{G}_{j,\frac{1}{2}\Delta t}(\mathbf{U}^{n})), \\ & \mathbf{L}^{G}_{i,\frac{1}{2}\Delta t}(\mathbf{L}^{G}_{k,\frac{1}{2}\Delta t}(\mathbf{U}^{n})), \mathbf{L}^{G}_{j,\frac{1}{2}\Delta t}(\mathbf{L}^{G}_{i,\frac{1}{2}\Delta t}(\mathbf{U}^{n})))\end{aligned} \tag{89}$$

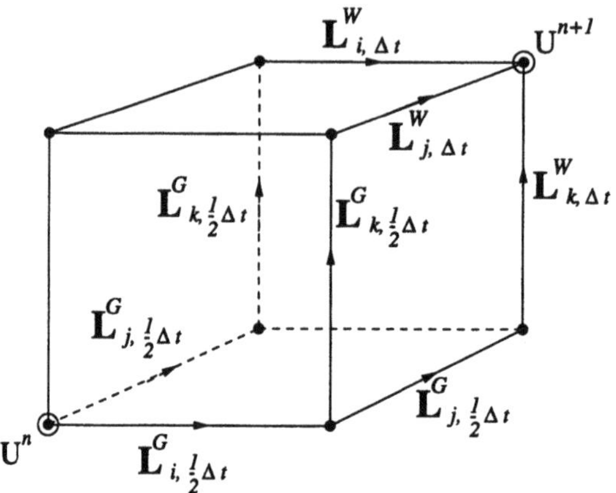

Figure 3. Cube representation of scheme (87).

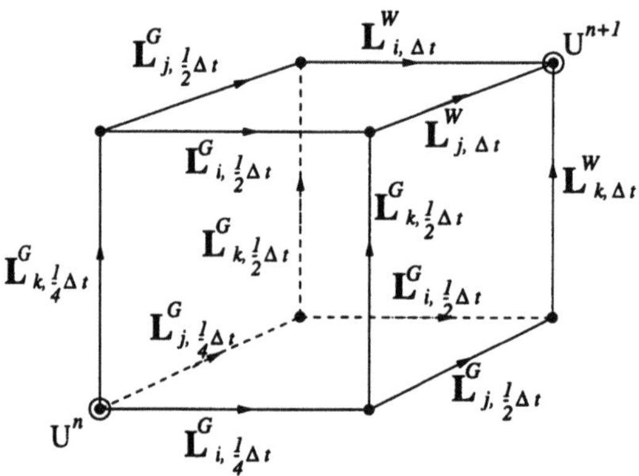

Figure 4. Cube representation of scheme (88).

for odd timesteps and

$$\begin{aligned} \mathbf{U}^{n+1} &= \mathbf{L}^W_{i,j,k,\Delta t}(\mathbf{U}^n, \mathbf{L}^G_{j,\frac{1}{2}\Delta t}(\mathbf{L}^G_{k,\frac{1}{2}\Delta t}(\mathbf{U}^n)), \\ &\qquad \mathbf{L}^G_{k,\frac{1}{2}\Delta t}(\mathbf{L}^G_{i,\frac{1}{2}\Delta t}(\mathbf{U}^n)), \mathbf{L}^G_{i,\frac{1}{2}\Delta t}(\mathbf{L}^G_{j,\frac{1}{2}\Delta t}(\mathbf{U}^n))) \end{aligned} \tag{90}$$

for even timesteps. This should produce more symmetrical results than (87), but requires the solution of 3 Riemann problems per cell per timestep.

4.3. BOUNDARY CONDITIONS

Boundary conditions are implemented as follows. For any dataset $\mathbf{Q}$ two *surfaces* of cells are attached to each boundary outside the computational

(a)

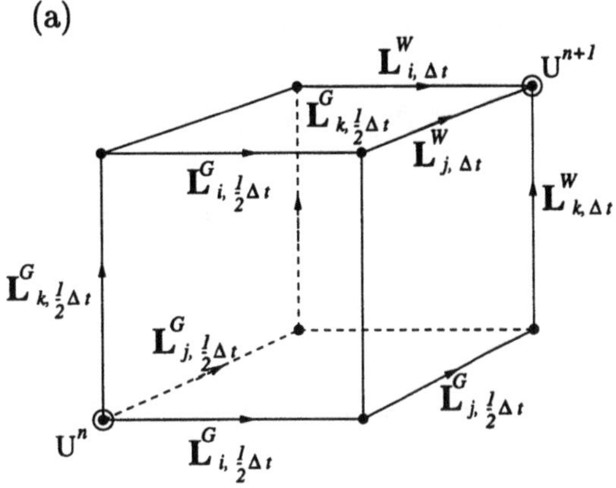

(b)

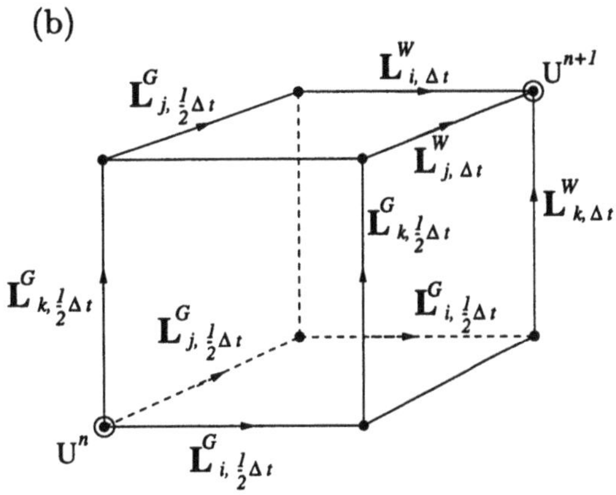

Figure 5. Cube representation of scheme (89),(90).

domain. Immediately before an operator $\mathbf{L}$ is applied to a dataset $\mathbf{Q}_0$, data is placed into each of the two planes of cells for each $\mathbf{Q}_j$ and for each boundary that affects the operation of $\mathbf{L}$. The data placed in the boundary cells depends on the type of boundary (transmissive or reflective) and the data just inside the boundary. For transmissive boundaries, the data for the boundary cell is simply copied from the matching cell across the boundary. For inviscid calculations, reflective boundaries are treated in exactly the same way *except* the component of velocity perpendicular to the boundary is reflected in the boundary. For viscous calculations, all components of velocity would be reflected at reflective boundaries. For example, if $\mathbf{L}^{G}_{x,\frac{1}{2}\Delta t}(\mathbf{U}^n, \bar{\mathbf{U}})$ is applied to a Cartesian grid, with boundary 0 reflective and boundary MX transmissive, we set

$$\bar{\rho}_{0,j,k} = \bar{\rho}_{1,j,k} \quad \bar{u}_{0,j,k} = -\bar{u}_{1,j,k}$$

$$\bar{p}_{0,j,k} = \bar{p}_{1,j,k} \quad \bar{v}_{0,j,k} = \bar{v}_{1,j,k}$$

$$\bar{w}_{0,j,k} = \bar{w}_{1,j,k}$$

and

$$\bar{\rho}_{MX+1,j,k} = \bar{\rho}_{MX,j,k} \quad \bar{u}_{MX+1,j,k} = \bar{u}_{MX,j,k}$$

$$\bar{p}_{MX+1,j,k} = \bar{p}_{MX,j,k} \quad \bar{u}_{MX+1,j,k} = \bar{u}_{MX,j,k}$$

$$\bar{w}_{MX+1,j,k} = \bar{w}_{MX,j,k}$$

for all j, k, where $\bar{\rho}$,$\bar{u}$,$\bar{v}$,$\bar{w}$ and $\bar{p}$ are the primitive variables associated with $\bar{\mathbf{U}}$. Note that the boundary conditions are applied just before $\mathbf{L}$ is used, even if $\mathbf{Q}_1, ..., \mathbf{Q}_p$ are intermediate datasets obtained from the use of other operators.

This boundary condition procedure is used for both the unsplit schemes discussed in the last section and the operator split schemes described in §2. Note that for the unsplit schemes, data never needs to be set in the 'corner' cells (e.g $i, j, k = 0, 0, 0$): applying boundary conditions between intermediate operators automatically takes care of corner data.

4.4. MEMORY REQUIREMENTS

The aim of this section is to discuss how to minimise the storage used when implementing the unsplit schemes scheme for nonlinear systems of equations. Minimisation of storage is important in three dimensional calculations because the number of cells required for even a relatively coarse grid will test the storage capabilities of many modern workstations. We consider a system of N equations here. The largest array required by the unsplit schemes is an array storing N conserved variables for each of $MX \times MY \times MZ$ cells, so it contains $N \times MX \times MY \times MZ$ real numbers in total. We denote this array by CV. This, along with any other arrays of the same size, will dominate the storage requirements of the algorithm. To minimise the number of these large arrays, we take advantage of property (31) of the operators, as follows.

Consider first scheme (85). The authors believe that the most memory efficient implementation of this algorithm requires two more arrays the size of CV, which we denote by **T1** and **T2** here. The steps performed in the algorithm are illustrated in Fig. 6. The operator notation (26) is used on the figure with data sets replaced with arrays; their meaning in this context is obvious. When a '$*$' appears after the superscript, e.g $\mathbf{L}^{q*}_{s_1,dt}(\mathbf{A}_0, \mathbf{A}_1)$, with

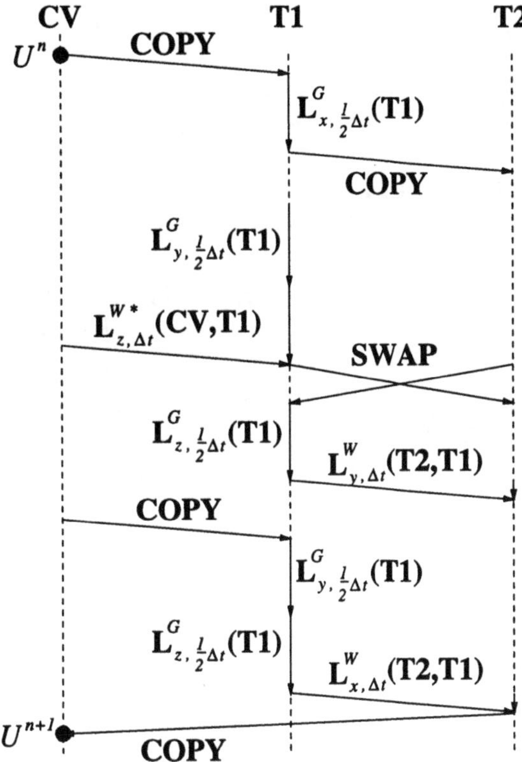

Figure 6. The use of large arrays by the WAF1 scheme.

$\mathbf{A}_0$ and $\mathbf{A}_1$ arrays, the operator is applied in the same way as before with scheme q, except the result is left in $\mathbf{A}_1$ rather than $\mathbf{A}_0$. The labels 'COPY' and 'SWAP' have obvious meaning.

The steps used in our implementation of scheme (86) are illustrated Fig. 7. Three extra arrays, called **T1**, **T2** and **T3** here, are needed for this scheme.

5. Numerical Experiments

The schemes discussed in previous sections have been implemented and tested for the three dimensional Euler equations of Gas Dynamics. In all results presented below, the ratio of specific heats, γ, is taken as a constant value of 1.4. Riemann solver adaption between three approximate Riemann solvers (a linearised Riemann solver, a two rarefaction Riemann solver and a two shock Riemann solver), described in (Toro, 1997), is used in all numerical experiments. The Courant number used for each scheme is always taken as 90% of the largest stable Courant number; i.e for the operator

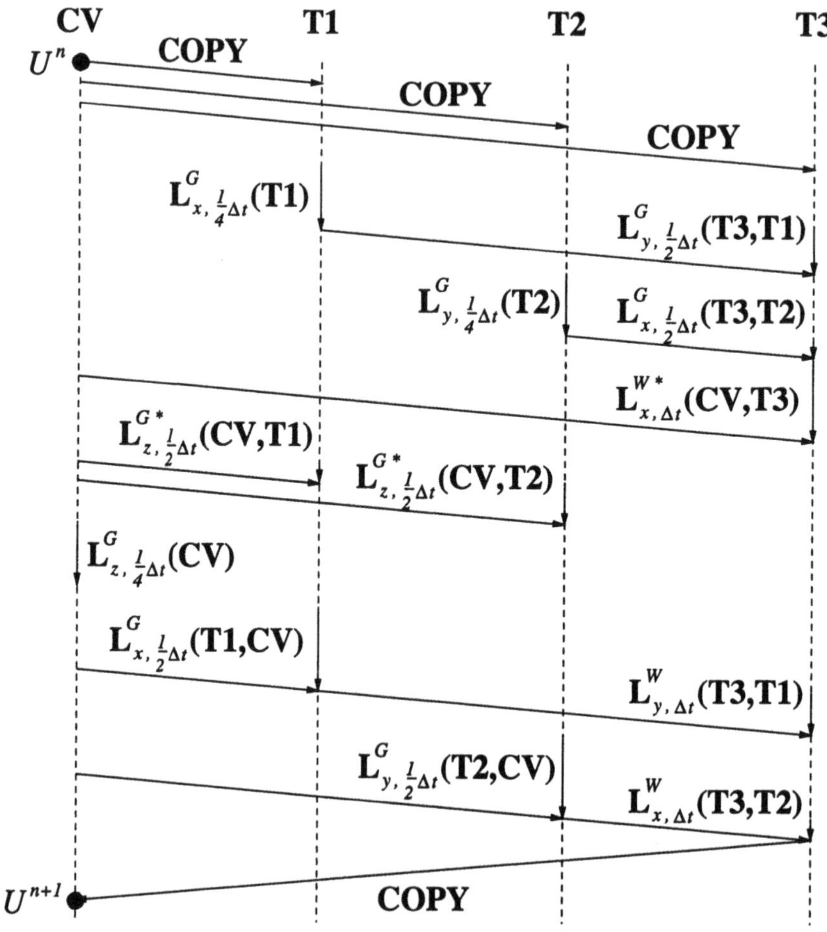

Figure 7. The use of large arrays by the WAF2 scheme.

split scheme we take 0.9, for the unsplit WAF schemes we take 0.6 and for the MUSCL-Hancock scheme we take 0.3.

5.1. EXPLOSION PROBLEM

The first numerical results presented here are for an explosion problem. The computational domain consists of a cube of sidelength 1.0 filled with a regular Cartesian grid. The initial conditions consists of two constant states separated by the surface of a sphere of radius 0.15 centred in the middle of the domain. The state inside the sphere is taken as

$$\rho_{IN} = 1.0 \qquad u_{IN} = 0.0 \qquad p_{IN} = 1.0$$

and the state outside the sphere is

$$\rho_{OUT} = 0.125 \qquad u_{OUT} = 0.0 \qquad p_{OUT} = 0.1$$

These are the left and right states used in the one dimensional Sod's problem (Sod, 1978). As time progresses, a spherical shock wave forms and moves away from the centre; this is followed by a spherical contact wave. At the same time, a spherical rarefaction wave moves *inwards* from the initial spherical surface and eventually reflects from itself at the centre. The results presented below are taken at time $t = 0.15$ just after the reflection of the rarefaction.

This problem was chosen for its spherical symmetry, which allows the numerical results to be compared with numerical results from the one dimensional Euler equations with a source terms to account for the spherical geometry:

$$\mathbf{U}_t + \mathbf{F}(\mathbf{U})_r = \mathbf{S}(\mathbf{U}) \tag{91}$$

where

$$\mathbf{U} = \begin{pmatrix} \rho \\ \rho u \\ E \end{pmatrix}, \qquad \mathbf{F} = \begin{pmatrix} \rho u \\ \rho u^2 + p \\ u(E+p) \end{pmatrix}, \qquad \mathbf{S} = -\frac{2}{r}\begin{pmatrix} \rho u \\ \rho u^2 \\ u(E+p) \end{pmatrix} \tag{92}$$

where t represents time, r represents the distance from the centre of the problem, ρ is the density, u is the velocity component away from the centre, p is the pressure and

$$E = \frac{1}{2}\rho u^2 + \rho e \tag{93}$$

is the total energy, and e is the internal energy. The system is closed by the ideal equation of state (7). This system is solved numerically by use of *time operator splitting*. The integration of the solution from time level n to timelevel $n+1$ consists of two stages. We first solve

$$\mathbf{U}_t + \mathbf{F}(\mathbf{U})_r = 0 \tag{94}$$

using one time step the WAF scheme with VAN LEER limiter using the solution $\mathbf{U}^n$ at timelevel n as initial data to give an intermediate solution $\bar{\mathbf{U}}$. The timestep is calculated with a CFL number of 0.9. We then solve the ODE

$$\mathbf{U}_t = \mathbf{S}(\mathbf{U}) \tag{95}$$

with $\bar{\mathbf{U}}$ as initial condition using the Euler method. The timestep is taken as the timestep used in the first stage. This gives the solution $\mathbf{U}^{n+1}$ at timelevel $n+1$. These equations are solved on a grid of 1000 cells so that the solutions can be assumed to be converged.

We compare the results from system (91),(92) with strips of data from the multidimensional results along the lines $y = z = 0.5$ and $x = y = z$, which we refer to as *x-strips* and *diagonal strips* respectively. In the following figures, the 1D solution is represented by a solid line and symbols represent the strips from the 3D solutions.

Figures 8 and 9 shows plots of density along the x and diagonal strips from the operator split WAF scheme, the WAF1 unsplit scheme, the WAF2 unsplit scheme and the MUSCL-Hancock unsplit scheme respectively. All the WAF fluxes computed for this problem were computed by first finding a weighted average state, and computing a flux from this state. The corresponding velocity, pressure and Mach number plots are shown in figures 10—15. It is clear from the figures that there is very little difference in the accuracy of the results from different schemes except at the centre of the domain, where there is a 'dip' caused by the reflection of the rarefaction. None of the schemes resolve this dip adequately. However, the operator split scheme does slightly better than all three unsplit schemes. Note that similar problems were found (Billett *et. al.*, 1997b) when the two dimensional schemes were used to compute solutions to an equivalent problem for the Shallow Water equations.

An encouraging feature of these results is the resolution of the shock wave along the diagonal strips by the unsplit WAF schemes. Even though only one dimensional limiters are used to damp spurious oscillations, there are no significant spurious oscillations behind the shock wave. The small overshoots that do exist, most visible in the pressure plots in figure 13, are no larger than on the operator split results. Overall the accuracy of the unsplit WAF schemes is comparable to the operator split scheme and the MUSCL-Hancock scheme.

We also present some two dimensional slices through the centre of the domain for each scheme. The density and pressure profiles on these slices are shown in figures 16 and 17.

The symmetry of the schemes can be seen by studying the contours between the shock wave (outer circle) and the contact wave (next circle in), or in the centre of the domain where the rarefaction is reflecting from itself. The lack of symmetry of the operator split scheme can be seen in both density and pressure plots, particularly at the centre of the domain. The unsplit WAF schemes do better, but are surprisingly similar; one would have expected much more symmetry from WAF2 than from WAF1. The MUSCL-Hancock scheme gives cross section comparable to the unsplit WAF schemes.

Some timings have been taken of the schemes solving this test problem. We compare the timings of the schemes by computing, for each scheme k, the parameter τ_k defined as

$$\tau_k = \left(\frac{t_k}{t_s}\right)\left(\frac{N_s}{N_k}\right)\left(\frac{\nu_s}{\nu_k}\right) S \tag{96}$$

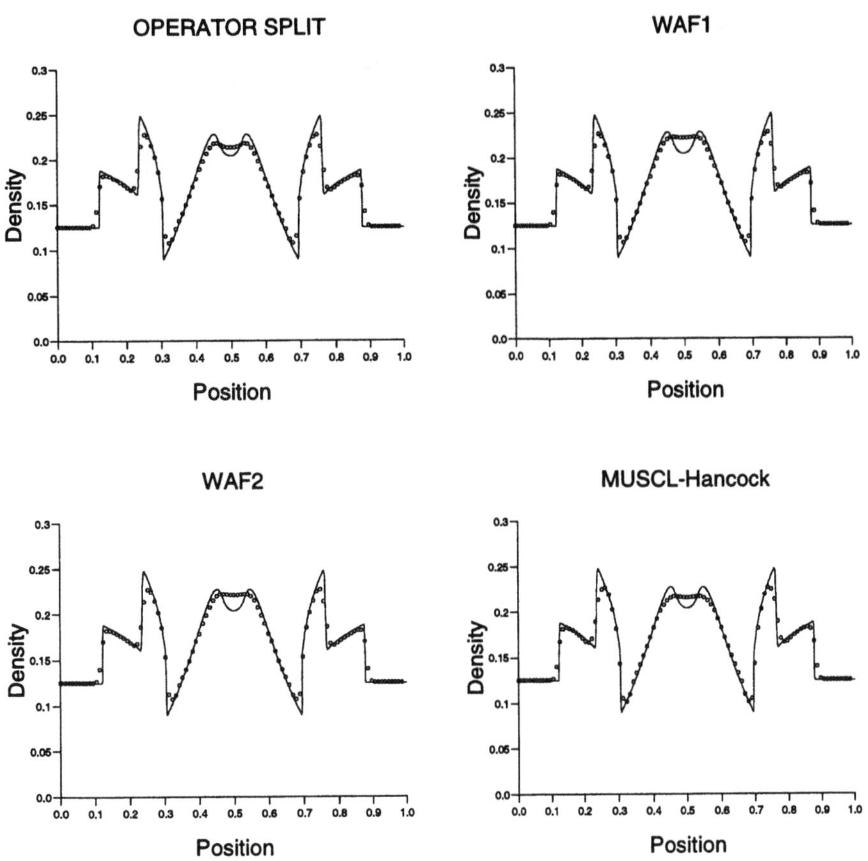

Figure 8. Results from the explosion test: Density profile through the center of the explosion along the line $y = z = 0.5$.

where the subscript s represents values for the operator split scheme. Here t_k is the CPU time taken by scheme k *minus* the time taken for input/output and setting the initial conditions. The other parameters are N_k, the number of timesteps taken for scheme k, N_s, the number of timesteps taken by the operator split scheme, ν_k, the maximum stable Courant number for scheme k, ν_s, the maximum stable Courant number for *each sweep* of the operator split scheme and S, the number of sweeps in the operator split scheme. The parameter τ_k is effectively an estimate of the number sweeps per timestep in an operator split scheme of the same cost as scheme k. Obviously, the lower τ_k is, the more efficient scheme k is.

The values of τ obtained for the schemes considered here are shown in

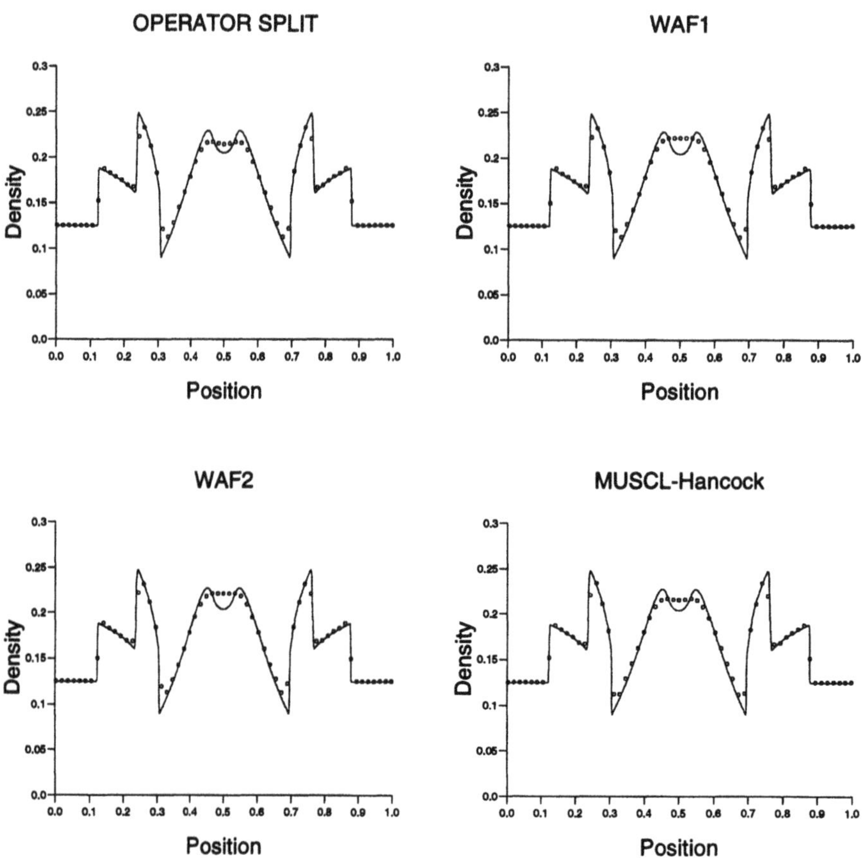

Figure 9. **Results from the explosion test: Density profile through the center of the explosion along the line $x = y = z$. The position axis is labeled with the center of the domain at 0.5.**

table (3). Clearly operator split schemes are much cheaper in terms of CPU time than any of the three unsplit schemes considered here. Split schemes become even more attractive when the splitting of Shang *et. al* (1995) is taken into account. This splitting takes six sweeps per timestep but has a maximum Courant number of 2ν, where ν is the maximum Courant number for the stability of each sweep. So this scheme would have $\tau = 3$. However, the unsplit WAF schemes are clearly competitive when compared with the MUSCL-Hancock scheme.

Another interesting fact that was found in the timings is related to the Riemann solver adaption used here. Approximately 95% of the Riemann

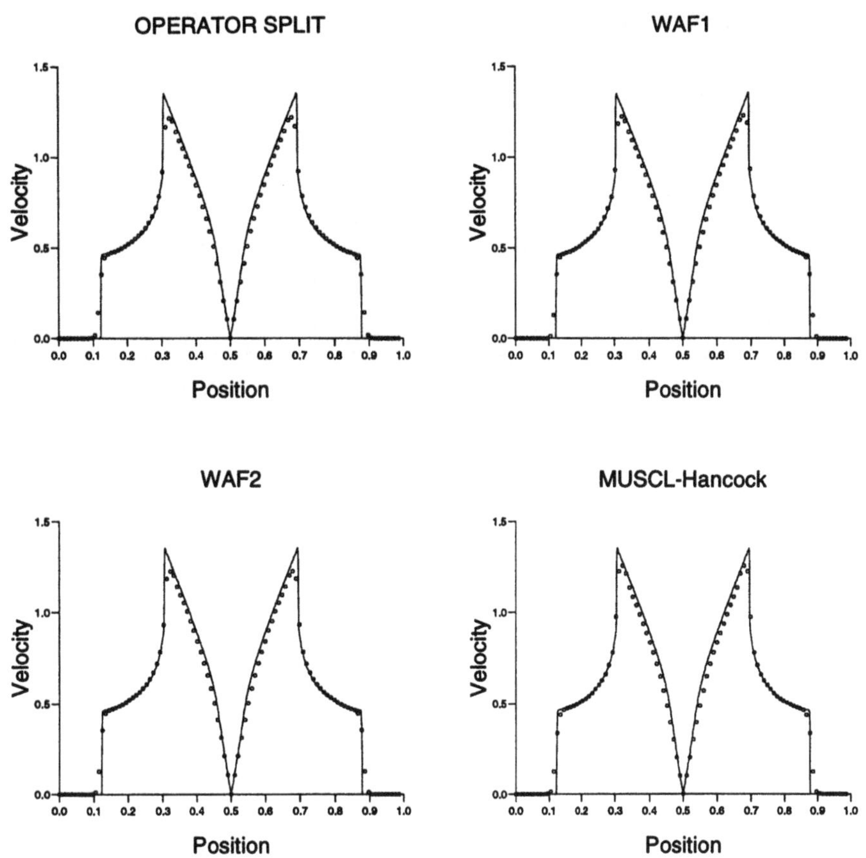

Figure 10. Results from the explosion test: Absolute velocity profile through the center of the explosion along the line $y = z = 0.5$.

problems were solver by the linearised Riemann solver, the cheapest of the three. The total cost of solving Riemann problems took only 15% of the CPU time used by the operator split WAF scheme (neglecting time for input/output and initialisation). By far the most expensive procedure in the operator split scheme was the routine that computes the states 'between' the contact and shear waves after limiting. Our timings indicate this took 45% of the CPU time. Any advances in the efficiency of operator split WAF must involve simplifying this procedure (or the coding of this procedure). For the unsplit WAF schemes, this procedure obviously takes proportionately less time, but this and the solution of Riemann problems are the two most time consuming parts of the schemes.

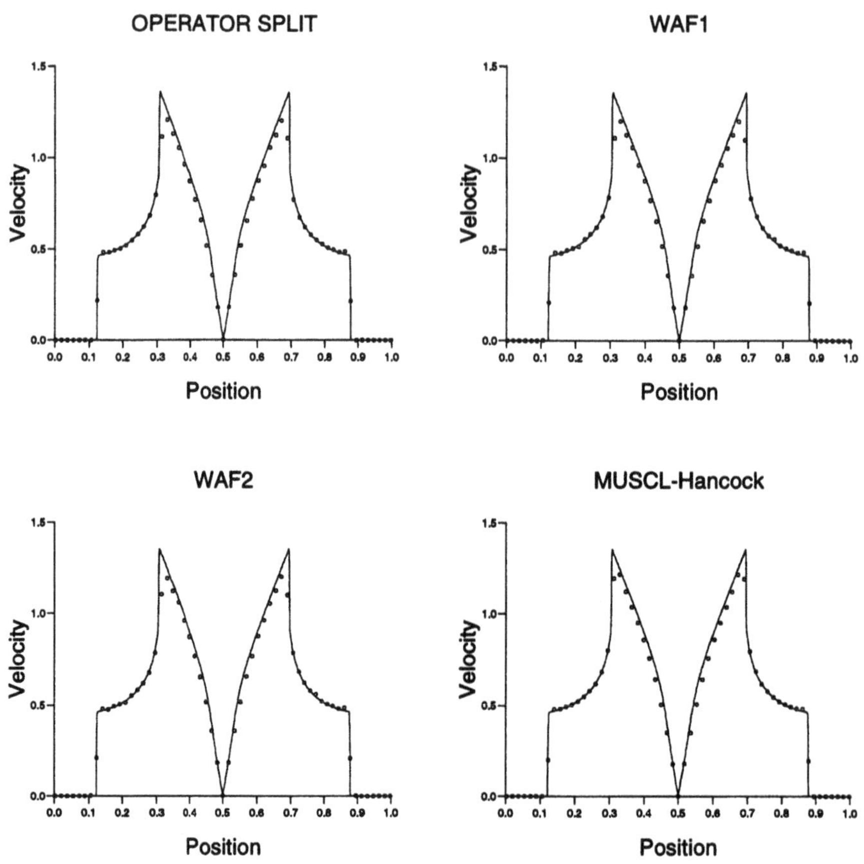

Figure 11. Results from the explosion test: Velocity profile through the center of the explosion along the line $x = y = z$. The position axis is labeled with the center of the domain at 0.5.

5.2. THE REFLECTION OF A STRONG SHOCK FROM A CONE

The WAF1 scheme has been implemented for the solution of the three dimensional Euler equations on curvilinear grids, using the ideas discussed in §2.6. The code has been tested on the reflection of strong shocks from cones whose axis is aligned with the direction of shock propagation. In order to avoid small cells, and the corresponding small timesteps, the geometry was modified from its ideal setup by adding a thin tube to the apex of the cone, aligned to the axis. This should not affect the overall flow features of the flow, but makes the calculations significantly cheaper. Both the unsplit code, and the operator split code used for comparison, were based on

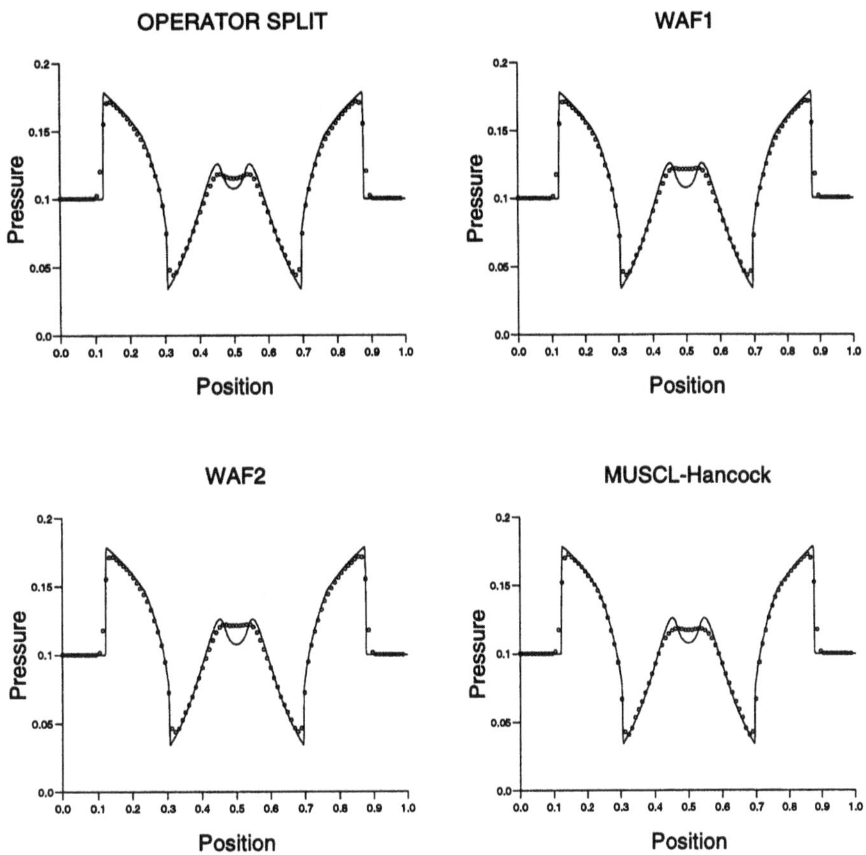

Figure 12. Results from the explosion test: Pressure profile through the center of the explosion along the line $y = z = 0.5$.

the true WAF approach, i.e no weighted average state was computed first. This is because if was found in the course of the code development that the weighted average state approach can become unstable in the region of strong shock waves, whereas the average flux approach will behave satisfactorily in the same situation. The authors have not been able to explain this phenomenon.

A discussion and experimental results concerning the reflection of shocks from cones, cylinders and spheres has been published by Bryson and Cross (1960). Further discussion of the physics of shock reflection can be found in the book by Ben-Dor (1992). Following the example of Bryson and Cross, we consider a shock of Mach number 3.68. Eight calculations were performed,

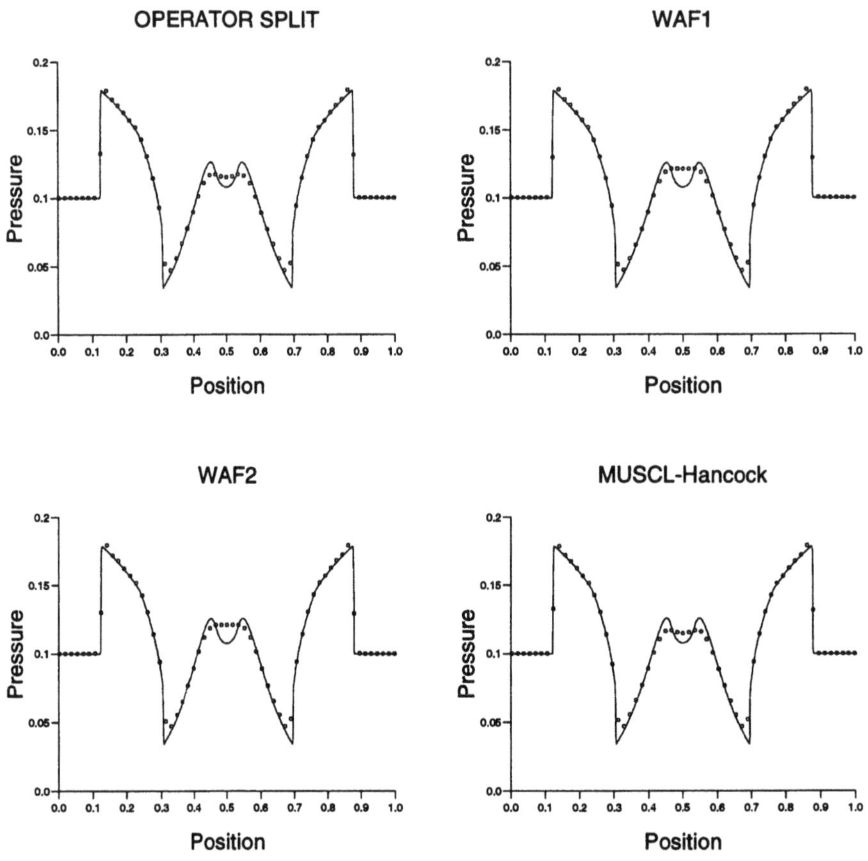

Figure 13. Results from the explosion test: Pressure profile through the center of the explosion along the line $x = y = z$. The position axis is labeled with the center of the domain at 0.5.

with the semi-apex angle of the cone varying from 15 degrees to 50 degrees in steps of 5 degrees. A grid of $100 \times 80 \times 36$ cells was used for these computations: 100 cells parallel to the cone axis, 80 cells perpendicular to the axis and 36 cells around the axis. Figure 18 shows the density and pressure profiles of half of a cross section of the flow computed by the WAF1 scheme when the semi-apex angle was 30 degrees. Figure 19 shows the equivalent plot from an operator split WAF code.

In the figure, the *incident shock* is vertical and moves from left to right. The features of the flow include a curved *reflected shock* which meets the incident shock at a point called the *Triple point.* A third shock extends

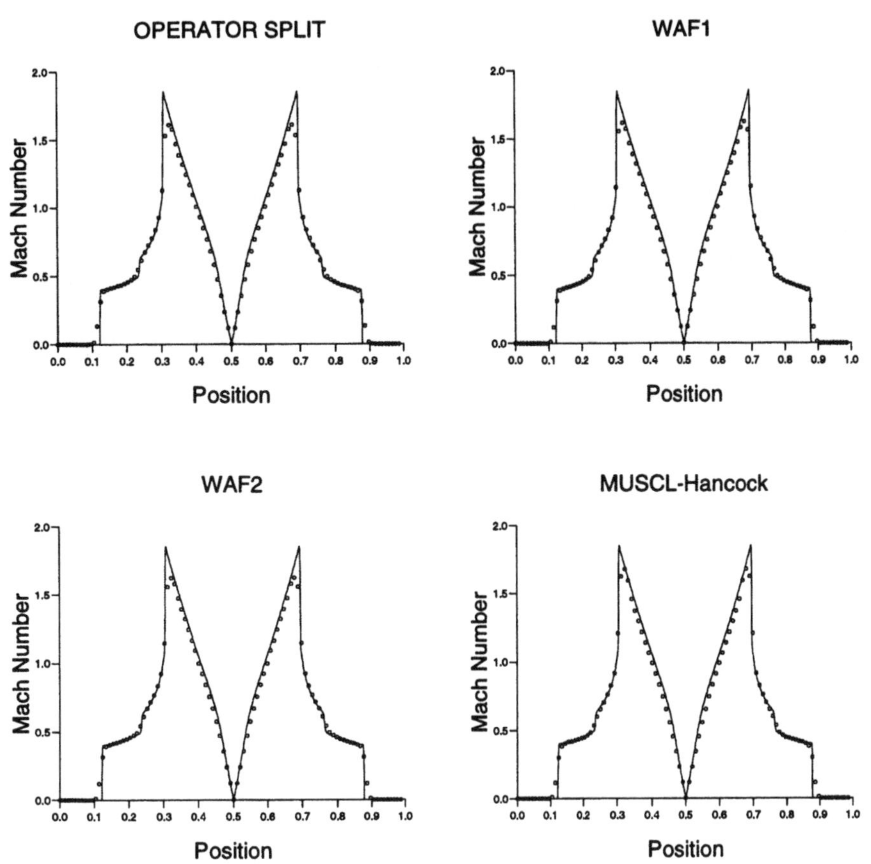

Figure 14. Results from the explosion test: Mach number profile through the center of the explosion along the line $y = z = 0.5$.

from the triple point to the surface of the cone, and is called the *Mach stem*. The existence of a Mach stem means that this is a *Mach Reflection*. A fourth wave, a combined contact and shear, extends from the triple point into the region between the reflected shock and the cone, as can be seen in the density plots. It is absent from the pressure plot, since pressure is constant across such waves. The locus of the triple point as time increases is called the *shock-shock*. The relection of a shock from a cone in this manner is self-similar, so the shock-shock is a straight line. The angle between this line and the axis of the cone is called the shock-shock angle and varies with shock Mach number and the semi-apex angle of the cone. The difference between the shock-shock angle and the semiapex angle for our eight runs

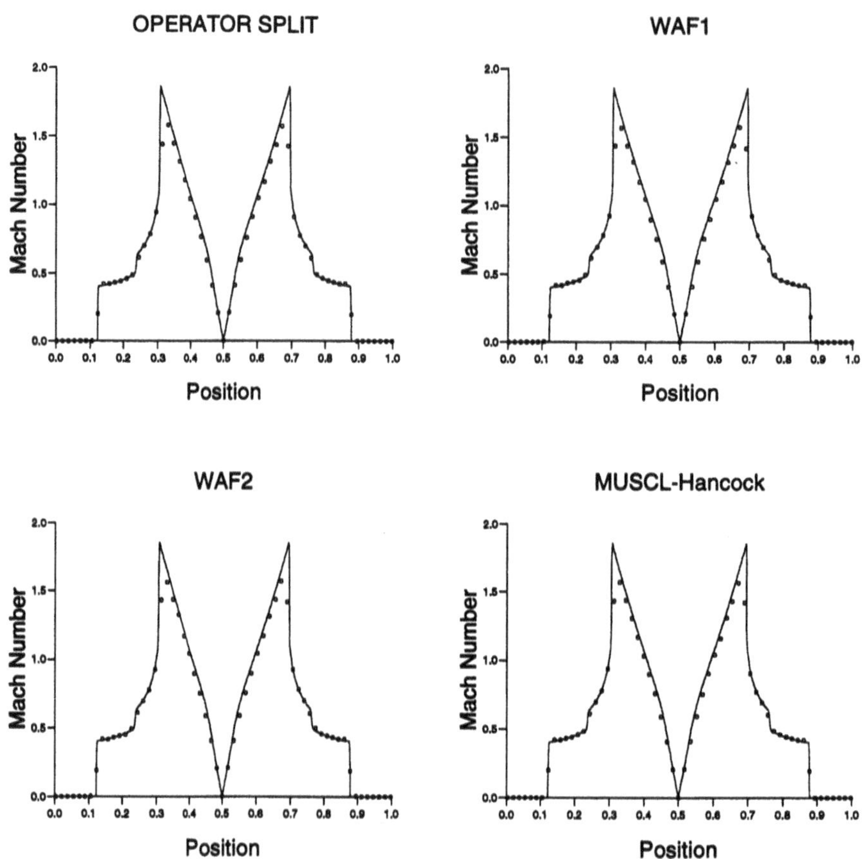

Figure 15. Results from the explosion test: Mach number profile through the center of the explosion along the line $x = y = z$. The position axis is labeled with the center of the domain at 0.5.

has been plotted against semi-apex angle in figure 20. These angles were measured by hand from contour plots. These measurements are slightly lower than the theoretical values shown in (Bryson *et. al.*, 1960) for all semi apex angles, with the difference being greater for smaller semi apex angles. Bryson and Cross point out that their experimental results also have these angles slightly lower than the theoretical results. They suggest that this is due to the fact that the theory used to predict these values does not take into account the possibility of regular Mach reflection. Further differences between our results and the theoretical results will also occur due to our modification of the geometry. The graph in figure 20 thus implies that the

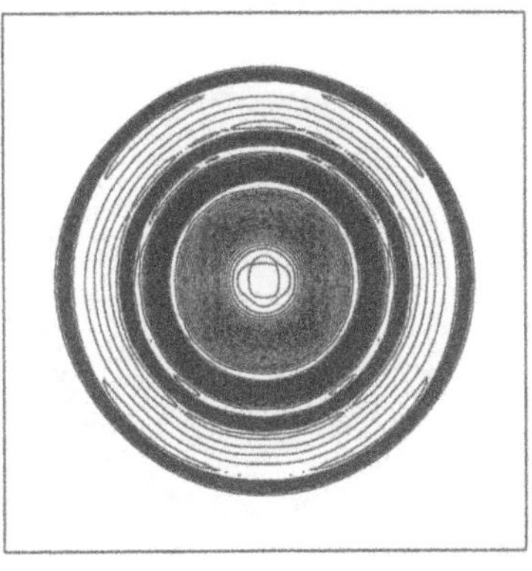

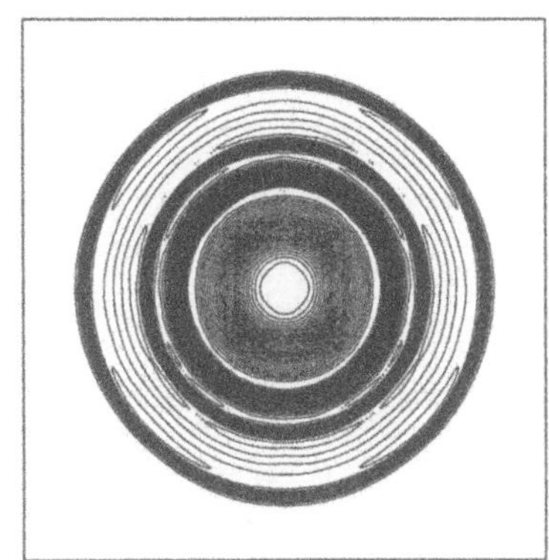

(a) Operator Split Scheme

(b) WAF1 scheme

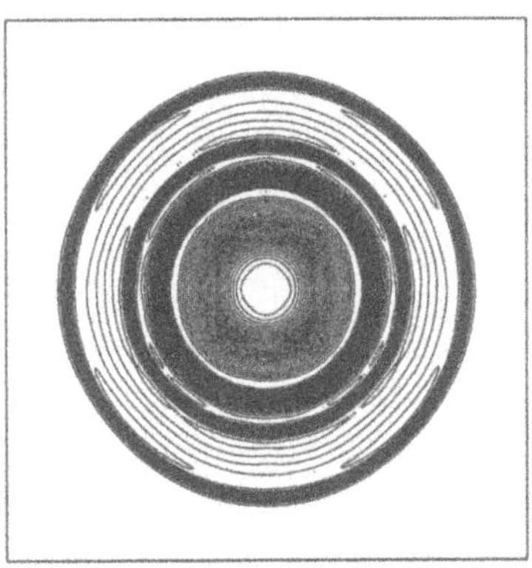

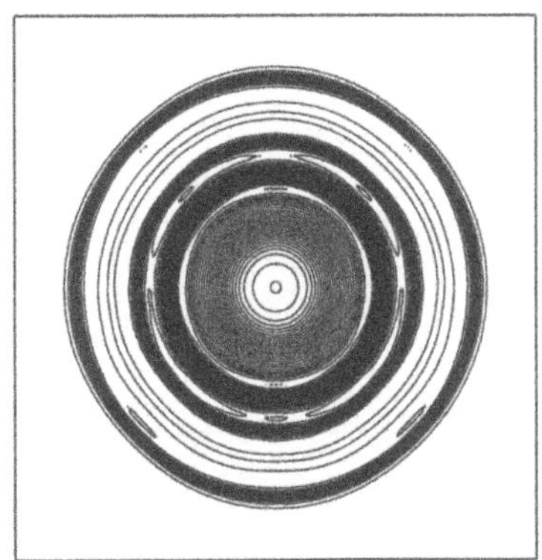

(c) WAF2 scheme (d) MUSCL-Hancock scheme

Figure 16. **Results from the explosion test: Density profiles of two dimensional sections through the center of the domain.**

WAF1 method is capturing the main features of the flow reasonably. For a semi-apex angle of 50 degrees, the contour plots on the results from the WAF1 scheme did not show a Mach stem, thus implying regular reflection.

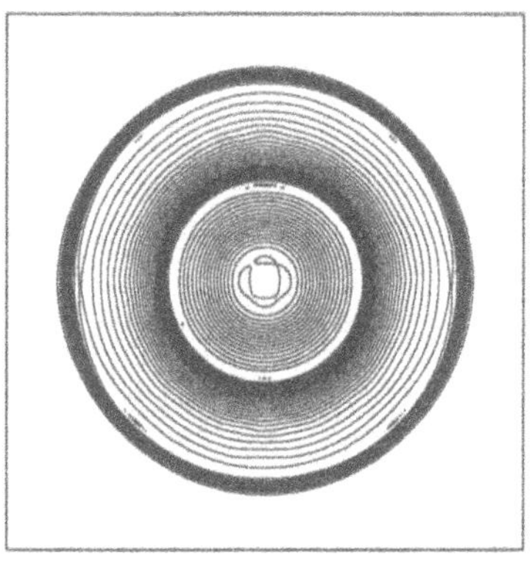

(a) Operator Split Scheme

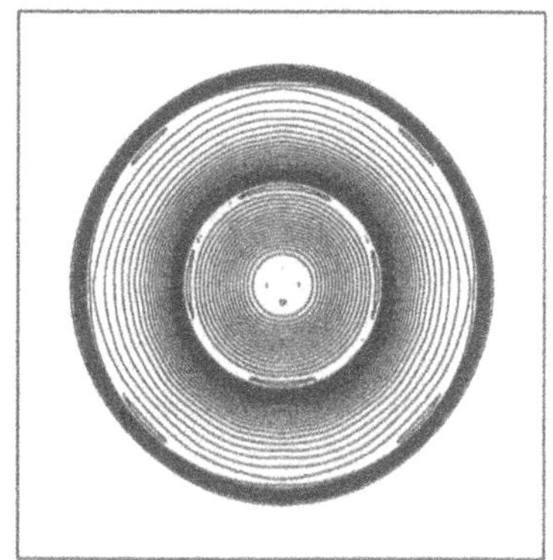

(b) WAF1 scheme

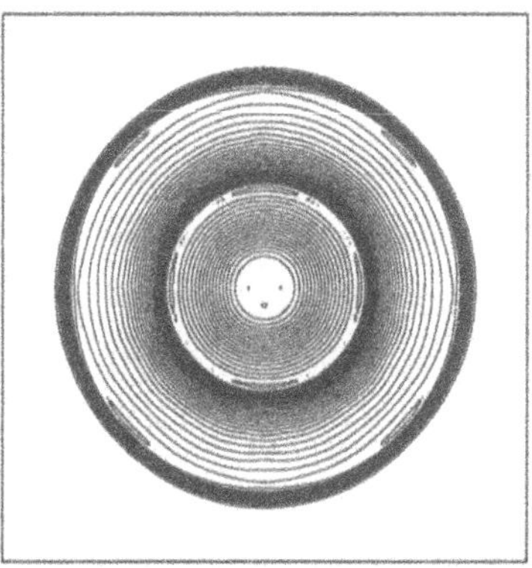

(c) WAF2 scheme

(d) MUSCL-Hancock scheme

Figure 17. Results from the explosion test: Pressure profiles of two dimensional sections through the center of the domain.

6. Conclusions

The aim of this chapter has been to discuss the use of unsplit WAF schemes for the solution of the three dimensional Euler equations of Gas Dynamics. The construction of these schemes in both two and three dimensions has

TABLE 3. A comparison of timings for the four schemes used on the explosion test. The parameter τ, defined in equation (96), is shown for each scheme.

Scheme	τ
Operator Split	5.0
WAF1	6.8
WAF2	9.8
MUSCL-Hancock	8.7

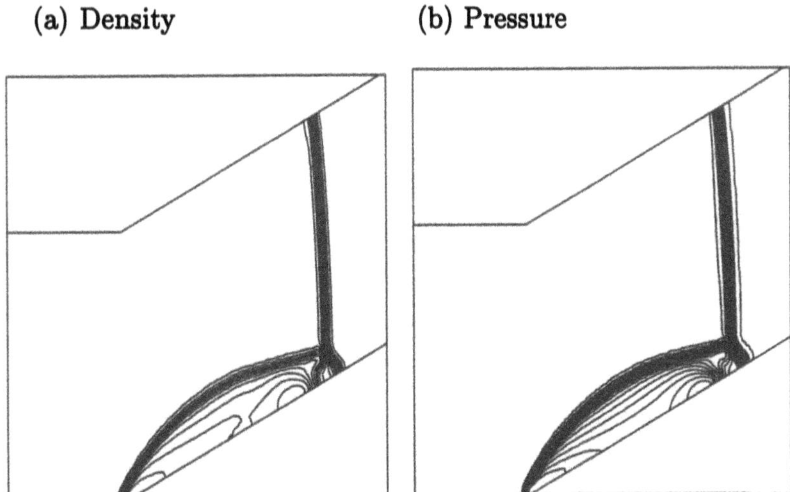

Figure 18. Two dimensional cross sections of the reflection of a Mach 3.68 shock from a cone with a semi-apex angle of 30 degrees. The WAF1 scheme was used for this calculation, on a curvilinear grid of $100 \times 80 \times 36$ cells.

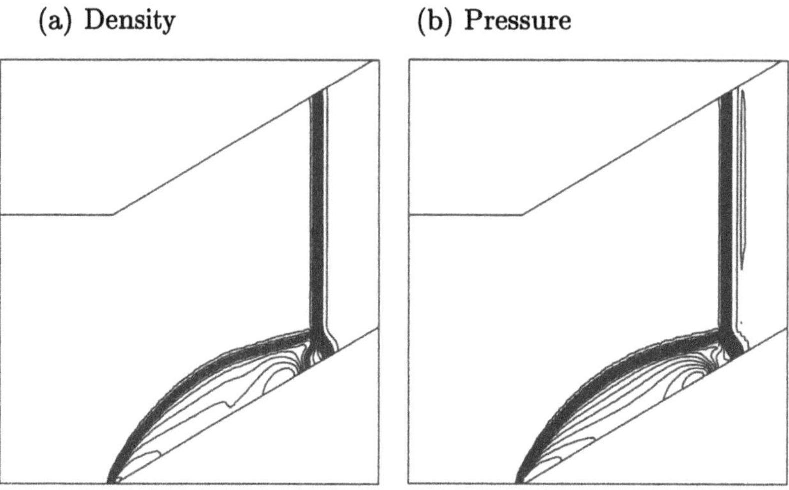

Figure 19. Two dimensional cross sections of the reflection of a Mach 3.68 shock from a cone with a semi-apex angle of 30 degrees. An operator split WAF scheme was used for this calculation, on a curvilinear grid of $100 \times 80 \times 36$ cells.

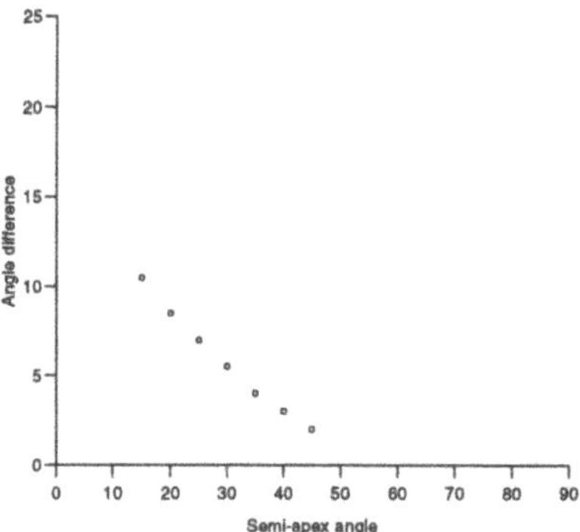

Figure 20. The difference between shock-shock angle and semi-apex angle, plotted against semi-apex angle for the reflection of a Mach 3.68 shock from a cone.

been reviewed. We presented numerical solutions to the three dimensional Euler equations on a Cartesian grid using an explosion problem as a test, comparing two unsplit WAF-schemes, an operator split WAF scheme and the three dimensional unsplit MUSCL-Hancock scheme. All four schemes are comparable in terms of accuracy. The unsplit WAF schemes are less efficient than the operator split WAF scheme, but comparable in cost to the MUSCL-Hancock scheme. The use of one of the unsplit WAF schemes on curvilinear grids has been tested on the reflection of strong shock waves from a cone. The result compares well with theory.

There is room for further development of the WAF approach. One important topic that has been brought to our attention during the course the work presented here is the inadequacy of the weighted average *state* approach to remain stable in the presence of strong shocks. The reasons for this need to be investigated.

The structure of the unsplit WAF schemes offers a framework for construction of other unsplit schemes. For example, more efficient schemes could be produced by replacing the Godunov scheme in the predictor stages with cheaper first order schemes, provided this does not compromise stability or accuracy.

References

Baines, M. J. (1997) Multidimensional Upwinding. In *Numerical Methods for Wave Propagation Problems; with The Harten Memorial Lecture by P. L. Roe.* (Editors: Toro, E. F. and Clarke, J. F.) Kluwer Academic Publishers, to appear 1998.

Ben-Artzi, M. and Falcovitz, J. (1984). A Second Order Godunov-Type Scheme for Compressible Fluid Dynamics. *J. Comput. Phys.*, 55:1–32.

Ben-Dor, G. *Shock Wave Reflection Phenomena.* Springer-Verlag, 1992.

Billett, S.J. (1994) A Class of Upwind Methods for Conservation Laws. Ph.D thesis, College of Aeronautics, Cranfield University, U.K.

Billett, S.J. and Toro, E.F. (1995a) Numerical Methods for Overlapping Grids and Moving Boundaries. In *Sixth International Symposium on Computational Fluid Dynamics*, volume I, pages 111–116, Lake Tahoe, Nevada, U.S.A.

Billett, S.J. and Toro, E.F. (1995b) TVB Limiter Functions for Multidimensional Schemes. In *Sixth International Symposium on Computational Fluid Dynamics*, volume I, pages 117–122, Lake Tahoe, Nevada, U.S.A.

Billett, S.J. and Toro, E.F. (1996) Implementing a Three-Dimensional Finite Volume WAF-Type Scheme for the Euler Equations. In *ECCOMAS proceedings*, Paris. John Wiley and Sons, 732-738.

Billett, S.J. and Toro, E.F. (1997a) On the Accuracy and Stability of Explicit Schemes for Multidimenional Linear Homogeneous Advection Equations. *J. Comput. Phys.*, 131:247–250.

Billett, S.J. and Toro, E.F. (1997b) On WAF-Type Schemes for Multidimensional Hyperbolic Conservation Laws. *J. Comput. Phys.*, 130:1–24.

Boden, E.P. and Toro, E.F. (1997) A Combined Chimera-AMR technique for Hyperbolic PDEs. In Djilali, editor, *Proc. 5th Annual Conference of the CFD Society of Canada*, Victoria University.

Bryson, A. E. and Gross, R. W. F. (1960) Diffraction of Strong Shocks by Cones, Cylinders and Spheres. *J. F. M.*, 10:1–16.

Colella, P. (1985) A Direct Eulerian MUSCL Scheme for Gas Dynamics. *SIAM J. Sci. Stat. Comput*, 6:104–117.

Colella, P. (1990) Multidimensional Upwind Methods for Hyperbolic Conservation Laws. *J. Comput. Phys*, 87:171–200.

Colella, P. and Woodward, P.R. (1984) The Piecewise Parabolic Method (PPM) method for Gas Dynamical Simulation. *J. Comput. Phys.*, 54:174–201.

Godunov, S.K. (1959) A Difference Method for the Numerical Calculation of Discontinuous Solutions of Hydrodynamic Equations. *Mat. Sb.*, 47:271–306.

Goodman, J.B. and LeVeque R.J. (1985) On the Accuracy of Stable Schemes for 2D Scalar Conservation Laws. *Math. Comp.*, 45(21):15–21.

Gottlieb, J.J. and Groth, C.P.T. (1988) Assessment of Riemann Solvers for Unsteady One-Dimensional Inviscid Flows of Perfect Gases. *J. Comput. Phys.*, 78:437–458.

Harten, A. (1983) High Resolution Schemes for Hyperbolic Conservation Laws. *J. Comput. Phys*, 49:357–393.

Harten, A., Lax, P.D. and van Leer, B. (1983) On Upstream Differencing and Godunov-Type Schemes for Hyperbolic Conservation Laws. *SIAM Review*, 25(1):35–61.

Karni, S. (1994) Multicomponent Flow Calculations Using A Consistent Primitive Algorithm. *J. Comput. Phys.*, 112(1):31–43.

Lax, P. and Wendroff, P. (1960) Systems of Conservation Laws. *Comm. Pure Appl. Math.*, XIII:217–237.

LeVeque, R. J. (1985) A Large Timestep Generalization of Godunov's Method for Systems of Conservation Laws. *SIAM J. Numer. Anal.*, 22(6):1051–1073.

LeVeque, R.J. (1988) High Resolution Finite Volume Methods on Arbitrary Grids via Wave Propagation. *J. Comput. Phys*, 78:36–63.

LeVeque, R.J. (1997) Wave Propagation Algorithms for Multidimensional Hyperbolic Systems. *J. Comput. Phys.*, 131:327–353.

Morel, A.T. (1997) *A Genuinely Multidimensional High-Relolution Scheme for the Shallow Water Equations*. PhD thesis, Seminar for Applied Mathematics, ETH Zurich.

Osher, S. and Solomon, F. (1982) Upwind Difference Schemes for Hyperbolic Conservation Laws *Math. Comp.* , 38,158:339-374.

Quirk, J.J. (1994) An Alternative to Unstructured Grids for Computing Gas Dynamic Flows Around Arbitrarily Complex Two Dimensional Bodies. *Computers and Fluids*, 23(1):125–142.

Roe, P.L. (1981) Approximate Riemann Solvers, Parameter Vectors, and Difference Schemes. *J. Comput. Phys*, 43:357–372.

Roe, P.L. (1986) Characteristic Based Schemes for the Euler Equations. *Ann. Rev. Fluid Mech.*, Annual Reviews:337–365.

Roe, P.L. (1991) *Progress in Multidimensional Upwinding*, pages 273–277. Springer.

Saltzman, J. (1994) An Unsplit 3D Upwind Method for Hyperbolic Conservation Laws. *J. Comput. Phys*, 115:153–168.

Shang, J.S. (1995) A Fractional Step Method for solving 3D, Time-Domain Maxwell Equations. *J. Comput. Phys.*, 118:109–119.

Sod, G.A. (1978) A Survey of Several Finite Difference Methods for Systems of Nonlinear Hyperbolic Conservation Laws. *J. Comput. Phys.*, 27:1–31.

Speares, W. (1991) A Finite Volume Approach to the Weighted Average Flux Method. MSc thesis, College of Aeronautics, Cranfield Institute of Technology, U.K..

Speares, W. and Toro, E.F. (1995) A High Resolution Algorithm for Time-Dependent Shock Dominated Problems with Adaptive Mesh Refinement. *Journal of Flight Sciences and Space Research*, 19:267–281.

Strang, G. (1968) On the Construction and Comparison of Difference Schemes. *SIAM J. Numer. Anal.*, 5(3):506–517.

Sweby, P.K. (1984) High Resolution Schemes Using Flux Limiters for Hyperbolic Conservation Laws. *SIAM J. Numer. Anal.*, 21:995–1011.

Toro, E.F. (1986) A New Numerical Technique for Quasi-Linear Hyperbolic Systems of Conservation Laws. Technical Report 8708, Cranfield CoA.

Toro, E.F. and Roe, P.L. (1987) A Hybridised High-Order Random Choice Method for Quasi-Linear Hyperbolic Systems. In Gronig, editor, *Proc. 16th Intern. Symp. on Shock Tubes and Waves*, pages 701–708, Aachen, Germany.

Toro, E.F. (1989a) A Fast Riemann Solver with Constant Covolume Applied to the Random Choice Method. *Int. J. Numer. Meth. Fluids*, 9:1145–1164.

Toro, E.F. (1989b) A Weighted Average Flux Method for Hyperbolic Conservation Laws. *Proc. Roy. Soc. Lond.*, A423:401–418.

Toro, E.F. (1991) A Linearized Riemann Solver for the Time-Dependent Euler Equations of Gas Dynamics. *Proc. Roy. Soc. Lond.*, A434:683–693.

Toro, E.F. (1992) The Weighted Average Flux Method Applied to the Time-Dependent Euler Equations. *Phil. Trans. Roy. Soc. Lond.*, A341:499–530.

Toro, E.F and Billett, S.J. (1993) A Unified, Riemann Problem Based Extension of the Warming-Beam and Lax-Wendroff Methods. In H. Daiguji, editor, *Proceedings of the Fifth International Symposium on Computational Fluid Dynamics*, volume III, pages 243–248, Tohoku University, Sendai, Japan.

Toro, E.F., Spruce, M. and Speares, W. (1994) Restoration of the Contact Surface in the HLL-Riemann solver. *Shock Waves*, 4:25–34.

Toro, E.F. (1995a) On Adaptive Primitive-Conservative Schemes for Conservation Laws. In M.M.Hafez, editor, *Sixth International Symposium on Computational Fluid Dynamics: A Collection of Technical Papers*, volume 3, pages 1288–1293, Lake Tahoe, Nevada, USA, September 4-8.

Toro, E.F. (1995b) Some IVPs for which Conservative Methods Fail Miserably. In M.M.Hafez, editor, *Sixth International Symposium on Computational Fluid Dynamics: A Collection of Technical Papers*, volume 3, pages 1294–1299, Lake Tahoe, Nevada, USA, September 4-8.

Toro, E.F and Billett, S.J. (1995) Multidimensional WAF-Type Schemes for Conservation Laws. In W. H. Hui, Y. Kwok, and J.R. Chasnov, editors, *The First Asian Computational Fluid Dynamics Conference*, volume 1, pages 347–352, Hong Kong, January 16-19.

Toro, E.F and Billett, S.J. (1997) A Unified, Riemann Problem Based Extension of the Warming-Beam and Lax-Wendroff Methods. *IMA J. Numer. Anal*, 17:161–102, 1997.

Toro, E. F. (1997) *Riemann Solvers and Numerical Methods for Fluid Dynamics.* Springer-Verlag.

van Leer, B. (1977) Towards the Ultimate Conservative Difference Scheme IV. A New Approach to Numerical Convection. *J. Comput. Phys*, 23:276–299.

van Leer, B. (1979) Towards the Ultimate Conservative Difference Scheme V. A Second Order Sequel to Godunov's Method. *J. Comput. Phys*, 32:101–136.

van Leer, B. (1985) On the Relation between the Upwind-Differencing Schemes of Godunov, Enguist-Osher and Roe. *SIAM J. Sci. Stat. Comput.*, 5(1):1–20.

van Leer, B. (1992) Progress in Multi-Dimensional Upwind Differencing. Technical Report CR-189708/ICASE 92-43, NASA.

Warming, R.F. and Beam, R.W. (1976) Upwind Second Order Difference Schemes with Applications in Aerodynamic Flows. *AIAA Journal*, 24:1241–1249.

Woodward, P. and Colella, P. (1984) The Numerical Simulation of Two-Dimensional Fluid Flow with Strong Shocks. *J. Comput. Phys*, 54:115–173.

Yanenko, N.N. (1971) *The Method of Fractional Steps.* Springer Verlag, New York.

Yang, G., Causon, D.M., Ingram, D.M., and Saunders, R. (1996) A Cartesian Cut Cell Method for Compressible 3-D Flows Involving both Static and Moving Bodies. In *ECCOMAS proceedings*, Paris. John Wiley and Sonsi, 485-491.

SEMI-IMPLICIT METHODS FOR FREE SURFACE ENVIRONMENTAL FLOWS

L. BONAVENTURA AND V. CASULLI
Laboratorio di Matematica Applicata
Dipartimento di Ingegneria Civile ed Ambientale
Università degli Studi di Trento
Mesiano di Povo 38050 (TN), Italy

Abstract. In this paper various semi-implicit discretization methods for the equations of large scale free surface flows are outlined. The Eulerian-Lagrangian discretization of advection and diffusion is introduced. Alternating direction, one time level semi-implicit and fully implicit splitting methods are discussed for 2D and 3D hydrostatic flows. An application to the hydrodynamics of the Lagoon of Venice is also presented.

1. Introduction

The hydrodynamical description of large scale geophysical flows is of great importance in environmental sciences, both for theoretical reasons and for practical applications such as the prediction of pollutant transport. Various approximations of the Navier-Stokes equations have been used to model the dynamics of large scale flows. On a geophysical scale the vertical acceleration and the vertical viscosity forces are often small, when compared to the gravity acceleration and to the pressure gradient in the vertical direction. Consequently, the hydrostatic approximation is well justified.

In the following, the computational aspects of some numerical methods for the Navier-Stokes equations with hydrostatic pressure will be reviewed and discussed. The key feature of these methods is the semi-implicit time discretization, which allows for a significant relaxation of the stability restrictions on the discretization timestep. The resulting algorithms consist of simple computational steps which can be efficiently implemented for flows in domains with complex geometry.

E.F. Toro and J.F. Clarke (eds.), Numerical Methods for Wave Propagation, 125–144.

For a deeper discussion of the modelling equations, the reader may refer to [12], [18]. A more detailed presentation of the numerical methods introduced here can be found in [4], [5], [7], [8], together with the results of their stability analysis. Other numerical methods for the equations presented here are discussed for instance in [3], [10], [11], [14], [16], [17], [18].

2. Semi-Implicit, Eulerian-Lagrangian Methods

The discretization of the viscous and nonlinear advective terms poses the first major difficulty in the numerical treatment the hydrodynamical equations. The well known upwind method can be applied, both with explicit and implicit time discretization (see e.g. [1], [4]). Upwind methods, however, are only first order accurate, and the truncation error has the form of a directionally dependent diffusion term. Hence, in advection dominated problems, drastically different numerical prediction may result from different spatial orientation of the computational grid. The Eulerian Lagrangian method allows for an improvement of the stability and the accuracy of the approximation. In order to introduce this scheme, we consider, first, a linear, constant coefficient advection-diffusion equation in the Lagrangian form

$$\frac{dc}{dt} = \nu\Big(\frac{\partial^2 c}{\partial x^2} + \frac{\partial^2 c}{\partial y^2} + \frac{\partial^2 c}{\partial z^2}\Big) - \gamma c, \tag{1}$$

where the substantial derivative d/dt indicates that the time rate of change is calculated along the streamline defined by

$$\frac{dx}{dt} = u, \qquad \frac{dy}{dt} = v, \qquad \frac{dz}{dt} = w. \tag{2}$$

If the computational domain is discretized into a set of N_x by N_y by N_z cells of sides Δx, Δy and Δz, respectively, a natural explicit discretization of equation (1) is taken to be

$$\begin{aligned}
&\frac{c^{n+1}_{i,j,k} - c^n_{i-a,j-b,k-d}}{\Delta t} \\
&= \nu\Big(\frac{c^n_{i-a+1,j-b,k-d} - 2c^n_{i-a,j-b,k-d} + c^n_{i-a-1,j-b,k-d}}{\Delta x^2} \\
&\quad + \frac{c^n_{i-a,j-b+1,k-d} - 2c^n_{i-a,j-b,k-d} + c^n_{i-a,j-b-1,k-d}}{\Delta y^2} \\
&\quad + \frac{c^n_{i-a,j-b,k-d+1} - 2c^n_{i-a,j-b,k-d} + c^n_{i-a,j-b,k-d-1}}{\Delta z^2}\Big) \\
&\quad -\gamma c^n_{i-a,j-b,k-d},
\end{aligned} \tag{3}$$

where the Courant numbers a, b and d are given by $a = u\frac{\Delta t}{\Delta x}$, $b = v\frac{\Delta t}{\Delta y}$ and $d = w\frac{\Delta t}{\Delta z}$. Equation (3) is not only a simple algorithm, but also accounts correctly for the advection and diffusion process, since the value of c at time t^{n+1} in the node (i, j, k) is related with the value of c at time t^n in $(i-a, j-b, k-d)$, which is advected and diffused to (i, j, k) in the time lapse Δt. In general,since a, b and d are not integers, $(i-a, j-b, k-d)$ is not a grid point and an interpolation formula must be used to define $c^n_{i-a,j-b,k-d}$ and its neighbors in equation (3). Thus, $c_{i-a,j-b,k-d}$ is approximated, for example, by a trilinear interpolation. If the time step Δt satisfies the inequality

$$\Delta t \leq \frac{1}{\frac{2\nu}{\Delta x^2} + \frac{2\nu}{\Delta y^2} + \frac{2\nu}{\Delta z^2} + \gamma}, \tag{4}$$

then $c^{n+1}_{i,j,k}$ can be regarded as a weighted average between zero and values of c at time step t^n. Thus, it is bounded above and below by the maximum and by the minimum, respectively, of zero and c^n at the surrounding grid points (discrete *max-min* property, see [13]). Scheme (3) is therefore stable when inequality (4) is satisfied. This method, though explicit, becomes unconditionally stable for pure convection equations ($\nu = \gamma = 0$). Moreover, the artificial diffusion, which can be regarded as the interpolation error, is reduced with respect to the one given by the explicit upwind method. Further elimination of the numerical diffusion can be achieved by using a higher order interpolation formula (see e.g. [16]).

In order to derive finite difference methods which are stable at a minimal computational cost, an implicitness factor θ for each of the terms in (1) can be introduced, so as to obtain the following semi-implicit Eulerian-Lagrangian approximation

$$\begin{aligned}
&\frac{c^{n+1}_{i,j,k} - c^n_{i-a,j-b,k-d}}{\Delta t} \\
&= \theta_1 \nu \Big(\frac{c^{n+1}_{i+1,j,k} - 2c^{n+1}_{i,j,k} + c^{n+1}_{i-1,j,k}}{\Delta x^2} + \frac{c^{n+1}_{i,j+1,k} - 2c^{n+1}_{i,j,k} + c^{n+1}_{i,j-1,k}}{\Delta y^2} \Big) \\
&\quad + \theta_2 \nu \frac{c^{n+1}_{i,j,k+1} - 2c^{n+1}_{i,j,k} + c^{n+1}_{i,j,k-1}}{\Delta z^2} - \theta_3 \gamma c^{n+1}_{i,j,k} \\
&\quad + (1-\theta_1)\nu \Big(\frac{c^n_{i-a+1,j-b,k-d} - 2c^n_{i-a,j-b,k-d} + c^n_{i-a-1,j-b,k-d}}{\Delta x^2} \\
&\qquad + \frac{c^n_{i-a,j-b+1,k-d} - 2c^n_{i-a,j-b,k-d} + c^n_{i-a,j-b-1,k-d}}{\Delta y^2} \Big) \\
&\quad + (1-\theta_2)\nu \frac{c^n_{i-a,j-b,k-d+1} - 2c^n_{i-a,j-b,k-d} + c^n_{i-a,j-b,k-d-1}}{\Delta z^2} \\
&\quad - (1-\theta_3)\gamma c^n_{i-a,j-b,k-d},
\end{aligned} \tag{5}$$

where θ_1 is the implicitness factor for the horizontal diffusion terms, θ_2 is the implicitness factor for the vertical diffusion term and θ_3 is the implicitness factor for the sink term. If $\theta_1 = \theta_2 = \theta_3 = 0$ is chosen, then the finite difference scheme (5) reduces to the simple explicit Eulerian-Lagrangian method (3). If $\theta_1 = \theta_2 = \theta_3 = 1$ is chosen, the finite difference scheme (5) becomes a fully implicit Eulerian-Lagrangian method, which possesses the discrete *max-min* property for any Δt, but requires the solution of a seven-diagonal system of $N_x N_y N_z$ equations at each time step. In general, some of the θ can be taken to be zero, and some can be taken to be one. The resulting finite difference formula is a semi-implicit scheme which possesses the discrete *max-min* property, provided the time step Δt satisfies the following inequality

$$\Delta t \leq \frac{1}{(1-\theta_1)\left(\frac{2\nu}{\Delta x^2} + \frac{2\nu}{\Delta y^2}\right) + (1-\theta_2)\frac{2\nu}{\Delta z^2} + (1-\theta_3)\gamma}. \tag{6}$$

A convenient choice for θ_1, θ_2 and θ_3, can be obtained by analyzing inequality (6). If, for example, γ is very large and is the dominating term at the denominator of (6), then a reasonable choice for θ_1, θ_2, and θ_3 is $\theta_1 = 0$, $\theta_2 = 0$ and $\theta_3 = 1$. In so doing, condition (6) becomes less restrictive at very little computational cost. Note, in fact, that no matrix inversion is required in this case. As a second example, if the vertical viscosity term imposes a more severe limitation on the time step (because Δz is much smaller than both Δx and Δy), then a convenient choice for θ_1, θ_2, and θ_3 is $\theta_1 = 0$, $\theta_2 = 1$ and $\theta_3 = 1$. In so doing, condition (6) becomes less restrictive and the overall computational effort is reduced to solving $N_x N_y$ linear tridiagonal systems with N_z equations each. In this case the tridiagonal systems, which are symmetric and positive definite, can be solved by a direct method. Finally, if more severe limitation in inequality (6) is imposed by the horizontal diffusion terms, a convenient choice for θ_1, θ_2, and θ_3 is $\theta_1 = 1$, $\theta_2 = 0$ and $\theta_3 = 1$. In this way, at every time step one must solve N_z linear five-diagonal systems with $N_x N_y$ equations. All these systems are symmetric and positive definite. Thus, they can be conveniently solved by a preconditioned conjugate gradient method.

Of course, if neither the diffusion terms, nor the sink term pose a severe restriction on the time step, use of $\theta_1 = \theta_2 = \theta_3 = \frac{1}{2}$ should be considered since the resulting Eulerian-Lagrangian method will have a higher order of accuracy. It should be emphasized, finally, that under condition (6) the resulting numerical solution is unique, possesses the discrete *max-min* property and therefore it is also stable. The stability of the method, however, can be established, usually, with less restrictive assumptions. Thus, for example, use of $\theta_1 = \theta_2 = \theta_3 = \frac{1}{2}$ leads to a method which is unconditionally stable.

3. The Two-Dimensional Shallow Water Equations

The circulation in well-mixed estuaries and coastal embayments can be satisfactorily represented by the vertically averaged shallow water equations. The shallow water equations can be derived from the Navier-Stokes equations by introducing the hydrostatic assumption and by vertically integrating these equations. It is assumed that the fluid density is constant and that the free surface elevation and the bottom profile can be represented by single valued functions $z = \eta(x, y, t)$ and $z = -h(x, y)$, respectively. The resulting two-dimensional equations are

$$\frac{dU}{dt} - fV = -g\eta_x + \frac{1}{H}\left[(\nu H U_x)_x + (\nu H U_y)_y\right] + \frac{\tau_x^s - \gamma U}{H} \tag{7}$$

$$\frac{dV}{dt} + fU = -g\eta_y + \frac{1}{H}\left[(\nu H V_x)_x + (\nu H V_y)_y\right] + \frac{\tau_y^s - \gamma V}{H} \tag{8}$$

$$\eta_t + (UH)_x + (VH)_y = 0, \tag{9}$$

where U, V denote the vertically averaged horizontal velocities. $H(x, y, t)$ denotes the total water depth, $H(x, y, t) = h(x, y) + \eta(x, y, t)$, (τ_x^s, τ_y^s) are the wind stresses, f is the Coriolis parameter, ν is a nonnegative eddy viscosity coefficient which is determined from a specific turbulence model and γ is a nonnegative bottom friction coefficient resulting from vertical averaging and of the form $\gamma = \frac{g\sqrt{U^2+V^2}}{C_z^2}$, where C_z is the Chezy friction coefficient. When $\nu = 0$, equations (7)-(9) form a quasilinear hyperbolic system of partial differential equations in three independent variables, x, y and t. When $\nu > 0$ this system is no longer hyperbolic but the stability of any explicit numerical scheme will strongly depend on its hyperbolic terms. A characteristic analysis of the hyperbolic part of the governing equations has been carried out in [4], showing the dependence of the characteristic cone upon the celerity $\sqrt{gH}$, which arises from the coefficient of η_x in the first momentum equation (7), the coefficient of η_y in the second momentum equation (8) and the coefficient of U_x and V_y in the continuity equation (9). These derivatives must be discretized implicitly in order to obtain methods whose stability is independent of the celerity. Various semi-implicit methods based on this principle will be discussed next.

3.1. ALTERNATING DIRECTION SEMI-IMPLICIT (ADI) METHODS

The basic alternating direction algorithm is a two levels scheme involving the solution of a set of simple tridiagonal systems. The discretization is carried out on a staggered C-type grid, consisting of $N_x \times N_y$ rectangular cells of length Δx and width Δy. In the first time level, assuming $U^{n-\frac{1}{2}}$, η^n and V^n to be known throughout the mesh, the finite difference analogue of the x-momentum and continuity equations can be formulated as follows

$$U^{n+\frac{1}{2}}_{i+\frac{1}{2},j} = FU^{n-\frac{1}{2}}_{i+\frac{1}{2},j} - g\frac{\Delta t}{\Delta x}\left(\eta^{n+\frac{1}{2}}_{i+1,j} - \eta^{n+\frac{1}{2}}_{i,j}\right) - \Delta t\frac{\gamma^n_{i+\frac{1}{2},j}}{H^n_{i+\frac{1}{2},j}}U^{n+\frac{1}{2}}_{i+\frac{1}{2},j} \quad (10)$$

$$\eta^{n+\frac{1}{2}}_{i,j} = \eta^n_{i,j} - \frac{\Delta t}{2\Delta x}\left(H^n_{i+\frac{1}{2},j}U^{n+\frac{1}{2}}_{i+\frac{1}{2},j} - H^n_{i-\frac{1}{2},j}U^{n+\frac{1}{2}}_{i-\frac{1}{2},j}\right)$$
$$-\frac{\Delta t}{2\Delta y}\left(H^n_{i,j+\frac{1}{2}}V^n_{i,j+\frac{1}{2}} - H^n_{i,j-\frac{1}{2}}V^n_{i,j-\frac{1}{2}}\right). \quad (11)$$

In equation (10) F is an explicit, nonlinear finite difference operator corresponding to the spatial discretization of advective, Coriolis and viscous terms. A particular form for F can be chosen in a variety of ways, such as for example by using an explicit Eulerian-Lagrangian scheme. Moreover, in order to avoid nonphysical negative values for the total depths H, one takes

$$H^n_{i\pm\frac{1}{2},j} = \max\left(0, h_{i\pm\frac{1}{2},j} + \frac{\eta^n_{i,j} + \eta^n_{i\pm1,j}}{2}\right), \quad (12)$$

$$H^n_{i,j\pm\frac{1}{2}} = \max\left(0, h_{i,j\pm\frac{1}{2}} + \frac{\eta^n_{i,j} + \eta^n_{i,j\pm1}}{2}\right). \quad (13)$$

For each j, equations (10) and (11) constitute a linear system of at most $2N_x$ equations. This system will be further decomposed into a set of smaller tridiagonal systems. Specifically, formal substitution of the expressions for $U^{n+\frac{1}{2}}_{i\pm\frac{1}{2},j}$ in (11) yields

$$\eta^{n+\frac{1}{2}}_{i,j} - g\frac{\Delta t^2}{2\Delta x^2}\left[\frac{(H^n_{i+\frac{1}{2},j})^2}{H^n_{i+\frac{1}{2},j} + \gamma^n_{i+\frac{1}{2},j}\Delta t}(\eta^{n+\frac{1}{2}}_{i+1,j} - \eta^{n+\frac{1}{2}}_{i,j}) - \frac{(H^n_{i-\frac{1}{2},j})^2}{H^n_{i-\frac{1}{2},j} + \gamma^n_{i-\frac{1}{2},j}\Delta t}(\eta^{n+\frac{1}{2}}_{i,j} - \eta^{n+\frac{1}{2}}_{i-1,j})\right]$$
$$= \eta^n_{i,j} - \frac{\Delta t}{2\Delta y}\left(H^n_{i,j+\frac{1}{2}}V^n_{i,j+\frac{1}{2}} - H^n_{i,j-\frac{1}{2}}V^n_{i,j-\frac{1}{2}}\right) \quad (14)$$
$$-\frac{\Delta t}{2\Delta x}\left(\frac{(H^n_{i+\frac{1}{2},j})^2}{H^n_{i+\frac{1}{2},j} + \gamma^n_{i+\frac{1}{2},j}\Delta t}FU^{n-\frac{1}{2}}_{i+\frac{1}{2},j} - \frac{(H^n_{i-\frac{1}{2},j})^2}{H^n_{i-\frac{1}{2},j} + \gamma^n_{i-\frac{1}{2},j}\Delta t}FU^{n-\frac{1}{2}}_{i-\frac{1}{2},j}\right).$$

For each j, equations (14) constitute a linear tridiagonal system of N_x equations with unknowns $\eta_{i,j}^{n+\frac{1}{2}}$. These systems are symmetric and positive definite. Therefore they all have a unique solution which can easily be determined by a direct method. Once the $\eta_{i,j}^{n+\frac{1}{2}}$ are computed, the new values for $U_{i+\frac{1}{2},j}^{n+\frac{1}{2}}$ are evaluated explicitly from equation (10). In summary, the first time level requires the solution of N_y linear tridiagonal system of N_x equations with unknowns $\eta_{i,j}^{n+\frac{1}{2}}$.

Next, one proceeds to the second level of calculation by finite differencing the y-momentum and continuity equations as follows

$$V_{i,j+\frac{1}{2}}^{n+1} = FV_{i,j+\frac{1}{2}}^{n} - g\frac{\Delta t}{\Delta y}\left(\eta_{i,j+1}^{n+1} - \eta_{i,j}^{n+1}\right) - \Delta t\frac{\gamma_{i,j+\frac{1}{2}}^{n+\frac{1}{2}}}{H_{i,j+\frac{1}{2}}^{n+\frac{1}{2}}}V_{i,j+\frac{1}{2}}^{n+1} \quad (15)$$

$$\eta_{i,j}^{n+1} = \eta_{i,j}^{n+\frac{1}{2}} - \frac{\Delta t}{2\Delta x}\left(H_{i+\frac{1}{2},j}^{n+\frac{1}{2}}U_{i+\frac{1}{2},j}^{n+\frac{1}{2}} - H_{i-\frac{1}{2},j}^{n+\frac{1}{2}}U_{i-\frac{1}{2},j}^{n+\frac{1}{2}}\right)$$
$$-\frac{\Delta t}{2\Delta y}\left(H_{i,j+\frac{1}{2}}^{n+\frac{1}{2}}V_{i,j+\frac{1}{2}}^{n+1} - H_{i,j-\frac{1}{2}}^{n+\frac{1}{2}}V_{i,j-\frac{1}{2}}^{n+1}\right). \quad (16)$$

For each i, equations (15) and (16) constitute a linear system of at most $2N_y$ equations. This system will be further decomposed into a set of smaller tridiagonal systems. Specifically, formal substitution of the expressions for $V_{i,j\pm\frac{1}{2}}^{n+1}$ in (16) yields, for each i, a linear tridiagonal system with unknowns $\eta_{i,j}^{n+1}$. These systems are symmetric and positive definite, so that their unique solution can easily be determined by a direct method. Finally, the new values for $V_{i,j+\frac{1}{2}}^{n+1}$ are evaluated explicitly from equation (15). In summary, the second time level requires the solution of N_x linear tridiagonal system of N_y equations with unknowns $\eta_{i,j}^{n+1}$.

An important feature of this method is its capability of treating the flooding and the uncovering of tidal flats at no additional computational effort. Indeed, the occurrence at some time of zero value for the total depth H on a side of a cell implies infinite bottom friction. Thus, zero velocity, and zero flow rate will result until, at a later time, a positive value for H is obtained. Of course, zero total depth on each side of a cell results in zero variation of the water surface elevation at the cell center. A cell with zero water depth on the four sides is considered to be a dry cell.

When an explicit Eulerian-Lagrangian discretization is used for the advective and horizontal viscosity terms, the stability condition is given by

$$\Delta t \leq \frac{1}{\frac{2\nu}{\Delta x^2} + \frac{2\nu}{\Delta y^2}}. \tag{17}$$

Several ADI methods have been developed and implemented, but they all share a major disadvantage: a source of inaccuracy, called ADI effect arises when these methods are used with large time steps in flow domain characterized by complex geometries. For this reason, at a slightly higher computational cost, a more robust, one time level semi-implicit method can be applied.

3.2. SEMI-IMPLICIT FINITE DIFFERENCE METHODS

A semi-implicit scheme for equations (7)-(9), like ADI methods, is derived by requiring that the stability of the method does not depend upon the wave celerity $\sqrt{gH}$. On a staggered grid this is accomplished by using at each time step only one level of calculations in which the friction terms, the gradient of surface elevation in the momentum equations and the velocity in the continuity equation are discretized implicitly. The remaining terms are discretized explicitly. The basic semi-implicit algorithm can be formulated as follows

$$U^{n+1}_{i+\frac{1}{2},j} = FU^n_{i+\frac{1}{2},j} - g\frac{\Delta t}{\Delta x}\left(\eta^{n+1}_{i+1,j} - \eta^{n+1}_{i,j}\right) - \Delta t\frac{\gamma^n_{i+\frac{1}{2},j}}{H^n_{i+\frac{1}{2},j}}U^{n+1}_{i+\frac{1}{2},j} \tag{18}$$

$$V^{n+1}_{i,j+\frac{1}{2}} = FV^n_{i,j+\frac{1}{2}} - g\frac{\Delta t}{\Delta y}\left(\eta^{n+1}_{i,j+1} - \eta^{n+1}_{i,j}\right) - \Delta t\frac{\gamma^n_{i,j+\frac{1}{2}}}{H^n_{i,j+\frac{1}{2}}}V^{n+1}_{i,j+\frac{1}{2}} \tag{19}$$

$$\begin{aligned}\eta^{n+1}_{i,j} = \eta^n_{i,j} &- \frac{\Delta t}{\Delta x}\left(H^n_{i+\frac{1}{2},j}U^{n+1}_{i+\frac{1}{2},j} - H^n_{i-\frac{1}{2},j}U^{n+1}_{i-\frac{1}{2},j}\right)\\ &-\frac{\Delta t}{\Delta y}\left(H^n_{i,j+\frac{1}{2}}V^{n+1}_{i,j+\frac{1}{2}} - H^n_{i,j-\frac{1}{2}}V^{n+1}_{i,j-\frac{1}{2}}\right),\end{aligned} \tag{20}$$

where F is an explicit, nonlinear finite difference operator corresponding to the spatial discretization of advective, Coriolis and viscous terms (using, for example, an explicit Eulerian-Lagrangian scheme). If the computational region is divided into $N_x \times N_y$ finite difference cells, equations (18)-(20) constitute a system of $3N_xN_y$ equations and $3N_xN_y$ unknowns $U^{n+1}_{i+\frac{1}{2},j}$, $V^{n+1}_{i,j+\frac{1}{2}}$ and $\eta^{n+1}_{i,j}$. This system has to be solved at each time step to determine, recursively, values of the field variables from given initial and boundary conditions. From a computational point of view, since most of the computational effort will be devoted to the solution of system (18)-(20), we will first reduce this system to a smaller one in which $\eta_{i,j}^{n+1}$ are the only unknowns. Specifically, elimination of $U_{i\pm\frac{1}{2},j}^{n+1}$ and $V_{i,j\pm\frac{1}{2}}^{n+1}$ from the continuity equation (20) yields

$$\begin{aligned}
\eta_{i,j}^{n+1} & -g\frac{\Delta t^2}{\Delta x^2}\left[\frac{(H_{i+\frac{1}{2},j}^n)^2}{H_{i+\frac{1}{2},j}^n+\gamma_{i+\frac{1}{2},j}^n\Delta t}(\eta_{i+1,j}^{n+1}-\eta_{i,j}^{n+1})\right. \\
&\qquad \left.-\frac{(H_{i-\frac{1}{2},j}^n)^2}{H_{i-\frac{1}{2},j}^n+\gamma_{i-\frac{1}{2},j}^n\Delta t}(\eta_{i,j}^{n+1}-\eta_{i-1,j}^{n+1})\right] \\
& -g\frac{\Delta t^2}{\Delta y^2}\left[\frac{(H_{i,j+\frac{1}{2}}^n)^2}{H_{i,j+\frac{1}{2}}^n+\gamma_{i,j+\frac{1}{2}}^n\Delta t}(\eta_{i,j+1}^{n+1}-\eta_{i,j}^{n+1})\right. \\
&\qquad \left.-\frac{(H_{i,j-\frac{1}{2}}^n)^2}{H_{i,j-\frac{1}{2}}^n+\gamma_{i,j-\frac{1}{2}}^n\Delta t}(\eta_{i,j}^{n+1}-\eta_{i,j-1}^{n+1})\right] \\
& =\eta_{i,j}^n-\frac{\Delta t}{\Delta x}\left[\frac{(H_{i+\frac{1}{2},j}^n)^2}{H_{i+\frac{1}{2},j}^n+\gamma_{i+\frac{1}{2},j}^n\Delta t}FU_{i+\frac{1}{2},j}^n\right. \\
&\qquad \left.-\frac{(H_{i-\frac{1}{2},j}^n)^2}{H_{i-\frac{1}{2},j}^n+\gamma_{i-\frac{1}{2},j}^n\Delta t}FU_{i-\frac{1}{2},j}^n\right] \\
& -\frac{\Delta t}{\Delta y}\left[\frac{(H_{i,j+\frac{1}{2}}^n)^2}{H_{i,j+\frac{1}{2}}^n+\gamma_{i,j+\frac{1}{2}}^n\Delta t}FV_{i,j+\frac{1}{2}}^n\right. \\
&\qquad \left.-\frac{(H_{i,j-\frac{1}{2}}^n)^2}{H_{i,j-\frac{1}{2}}^n+\gamma_{i,j-\frac{1}{2}}^n\Delta t}FV_{i,j-\frac{1}{2}}^n\right].
\end{aligned} \tag{21}$$

Equations (21) constitute a linear five-diagonal system of $N_x \times N_y$ equations for $\eta_{i,j}^{n+1}$. Since $H_{i\pm\frac{1}{2},j}^n \geq 0$ and $H_{i,j\pm\frac{1}{2}}^n \geq 0$, this system is symmetric and positive definite, thus it can be solved efficiently by a preconditioned conjugate gradient method which, in combination with a classical red/black technique, is suitable for vector computations. Also with this method the implementation of flooding and uncovering of tidal flats is quite straightforward (see [4]). When an explicit Eulerian-Lagrangian discretization is used for the advective and horizontal viscosity terms, the stability condition is given, as before, by equation (17).

3.3. FULLY IMPLICIT SPLITTING METHODS

In order to derive an efficient method which is unconditionally stable, a fully implicit, fractional step algorithm can be devised as follows. In a first fractional time step, only the advection-diffusion operator is considered, and is discretized implicitly by using, for example, an implicit Eulerian-Lagrangian scheme. Two linear, five-diagonal, symmetric and positive definite systems are obtained for the provisional values $\tilde{U}^{n+1}_{i+\frac{1}{2},j}$ and $\tilde{V}^{n+1}_{i,j+\frac{1}{2}}$, respectively. Once these two systems are solved, one proceeds to the second fractional step by including the propagation operator as follows

$$U^{n+1}_{i+\frac{1}{2},j} = \tilde{U}^{n+1}_{i+\frac{1}{2},j} - g\frac{\Delta t}{\Delta x}\left(\eta^{n+1}_{i+1,j} - \eta^{n+1}_{i,j}\right) - \Delta t \frac{\gamma^n_{i+\frac{1}{2},j}}{H^n_{i+\frac{1}{2},j}} U^{n+1}_{i+\frac{1}{2},j} \tag{22}$$

$$V^{n+1}_{i,j+\frac{1}{2}} = \tilde{V}^{n+1}_{i,j+\frac{1}{2}} - g\frac{\Delta t}{\Delta y}\left(\eta^{n+1}_{i,j+1} - \eta^{n+1}_{i,j}\right) - \Delta t \frac{\gamma^n_{i,j+\frac{1}{2}}}{H^n_{i,j+\frac{1}{2}}} V^{n+1}_{i,j+\frac{1}{2}} \tag{23}$$

$$\begin{aligned} \eta^{n+1}_{i,j} = \eta^n_{i,j} &- \frac{\Delta t}{\Delta x}\left(H^n_{i+\frac{1}{2},j} U^{n+1}_{i+\frac{1}{2},j} - H^n_{i-\frac{1}{2},j} U^{n+1}_{i-\frac{1}{2},j}\right) \\ &- \frac{\Delta t}{\Delta y}\left(H^n_{i,j+\frac{1}{2}} V^{n+1}_{i,j+\frac{1}{2}} - H^n_{i,j-\frac{1}{2}} V^{n+1}_{i,j-\frac{1}{2}}\right). \end{aligned} \tag{24}$$

Computationally, elimination of $U^{n+1}_{i+\frac{1}{2},j}$ and $V^{n+1}_{i,j+\frac{1}{2}}$ from the continuity equation (24) yields, again, a linear five-diagonal system of equations for the water surface elevation $\eta^{n+1}_{i,j}$. A typical equation of this system has the same structure of that in (21), with $\tilde{U}^{n+1}_{i+\frac{1}{2},j}$ and $\tilde{V}^{n+1}_{i,j+\frac{1}{2}}$ replacing $FU^n_{i+\frac{1}{2},j}$ and $FV^n_{i,j+\frac{1}{2}}$, respectively. Thus it is symmetric, positive definite, and can be solved efficiently by a preconditioned conjugate gradient method. The main difference between semi-implicit and fully implicit splitting methods is in the treatment of the viscous terms. Indeed, a semi-implicit method can be regarded as a splitting method where the first fractional step is fully explicit, while the second one is fully implicit. Of course, when an Eulerian-Lagrangian approach is used, and $\nu = 0$, the two methods are equivalent.

4. A Three-Dimensional Hydrostatic Model

The Navier-Stokes equations for an incompressible, hydrostatic flow with a free surface η can be written as

$$\frac{du}{dt} - fv = -g\eta_x + (\nu u_x)_x + (\nu u_y)_y + (\nu u_z)_z \tag{25}$$

$$\frac{dv}{dt} + fu = -g\eta_y + (\nu v_x)_x + (\nu v_y)_y + (\nu v_z)_z \tag{26}$$

$$u_x + v_y + w_z = 0 \tag{27}$$

$$\eta_t + \left[\int_{-h}^{\eta} udz\right]_x + \left[\int_{-h}^{\eta} vdz\right]_y = 0, \tag{28}$$

where $u(x, y, z, t,), v(x, y, z, t)$ and $w(x, y, z, t)$ are the velocity components in the horizontal x, y and vertical z-directions. One may refer to [5] for a complete derivation of equations (25)-(28). Under the assumption that the free surface is almost flat horizontal, the tangential stress boundary conditions are approximated as follows

$$\nu u_z = \tau_x^s \ , \qquad \nu v_z = \tau_y^s. \tag{29}$$

Similarly, the boundary conditions at the sediment-fluid interface are given by

$$\nu u_z = \gamma u \ , \qquad \nu v_z = \gamma v. \tag{30}$$

4.1. ALTERNATING DIRECTION SEMI-IMPLICIT METHOD

The discretization of equations (25)-(28) is carried out on a staggered C-type grid, consisting of rectangular boxes of length Δx, width Δy and variable height Δz_k. The alternating direction semi-implicit methods derived in section 3.1 can be extended to the $3D$ model equations. This two levels scheme will require only the solution of a large set of simple tridiagonal systems. In the first level, the finite difference analogue of the x-momentum and the free surface equation can be formulated in vector notation as follows

$$\mathbf{A}_{i+\frac{1}{2},j}^{n}\mathbf{U}_{i+\frac{1}{2},j}^{n+\frac{1}{2}} = \mathbf{G}_{i+\frac{1}{2},j}^{n-\frac{1}{2}} - g\frac{\Delta t}{\Delta x}(\eta_{i+1,j}^{n+\frac{1}{2}} - \eta_{i,j}^{n+\frac{1}{2}})\mathbf{\Delta Z}_{i+\frac{1}{2},j}^{n} \tag{31}$$

$$\eta_{i,j}^{n+\frac{1}{2}} = \eta_{i,j}^{n} - \frac{\Delta t}{2\Delta x}\left[\left(\mathbf{\Delta Z}_{i+\frac{1}{2},j}^{n}\right)^T \mathbf{U}_{i+\frac{1}{2},j}^{n+\frac{1}{2}} - \left(\mathbf{\Delta Z}_{i-\frac{1}{2},j}^{n}\right)^T \mathbf{U}_{i-\frac{1}{2},j}^{n+\frac{1}{2}}\right]$$
$$-\frac{\Delta t}{2\Delta y}\left[\left(\mathbf{\Delta Z}_{i,j+\frac{1}{2}}^{n}\right)^T \mathbf{V}_{i,j+\frac{1}{2}}^{n} - \left(\mathbf{\Delta Z}_{i,j-\frac{1}{2}}^{n}\right)^T \mathbf{V}_{i,j-\frac{1}{2}}^{n}\right], \tag{32}$$

where $\mathbf{U}$, $\mathbf{V}$, $\mathbf{\Delta Z}$, $\mathbf{G}$ and $\mathbf{A}$ are defined as:

$$\mathbf{U}^{n+\frac{1}{2}}_{i+\frac{1}{2},j} = \begin{pmatrix} u^{n+\frac{1}{2}}_{i+\frac{1}{2},j,M} \\ u^{n+\frac{1}{2}}_{i+\frac{1}{2},j,M-1} \\ \vdots \\ u^{n+\frac{1}{2}}_{i+\frac{1}{2},j,m} \end{pmatrix}, \quad \mathbf{G}^{n-\frac{1}{2}}_{i+\frac{1}{2},j} = \begin{pmatrix} \Delta z^n_M F u^{n-\frac{1}{2}}_{i+\frac{1}{2},j,M} + \Delta t \tau^s_x \\ \Delta z^n_{M-1} F u^{n-\frac{1}{2}}_{i+\frac{1}{2},j,M-1} \\ \vdots \\ \Delta z^n_m F u^{n-\frac{1}{2}}_{i+\frac{1}{2},j,m} \end{pmatrix},$$

$$\mathbf{V}^{n}_{i,j+\frac{1}{2}} = \begin{pmatrix} v^{n}_{i,j+\frac{1}{2},M} \\ v^{n}_{i,j+\frac{1}{2},M-1} \\ \vdots \\ v^{n}_{i,j+\frac{1}{2},m} \end{pmatrix}, \quad \mathbf{\Delta Z} = \begin{pmatrix} \Delta z_M \\ \Delta z_{M-1} \\ \vdots \\ \Delta z_m \end{pmatrix},$$

$$\mathbf{A} = \begin{pmatrix} \Delta z_M + a_{M-\frac{1}{2}} & -a_{M-\frac{1}{2}} & & 0 \\ -a_{M-\frac{1}{2}} & \Delta z_{M-1} + a_{M-\frac{1}{2}} + a_{M-\frac{3}{2}} & -a_{M-\frac{3}{2}} & \\ \cdot & \cdot & \cdot & \\ & \cdot & \cdot & \cdot \\ 0 & & -a_{m+\frac{1}{2}} & \Delta z_m + a_{m+\frac{1}{2}} + \gamma \Delta t \end{pmatrix}$$

with $a_k = \frac{\nu_k \Delta t}{\Delta z_k}$. Here, m and M denote the limit of k-index representing the bottom and the top finite-difference stencil, respectively. $\Delta z_{i+\frac{1}{2},j,k}$ and $\Delta z_{i,j+\frac{1}{2},k}$ are, in general, the thickness of the k-th water layer. If, however, a vertical face of the finite difference box is not fully filled (because either the bottom or the free surface crosses a vertical face of the box) then $\Delta z_{i+\frac{1}{2},j,k}$ and/or $\Delta z_{i,j+\frac{1}{2},k}$ are defined to be the wetted height of the corresponding face. Of course, some of the Δz can be allowed to be zero. The height of the surface layer depends on the position of the free surface, and since the free surface changes with time, Δz also depends on the time level n. Moreover, F is an explicit, nonlinear finite difference operator, which includes the contributions arising from the Eulerian-Lagrangian discretization of the advective, Coriolis and horizontal eddy viscosity terms. The values of u above free surface and below bottom have been eliminated by means of an appropriate discretization of the boundary conditions (29),(30). For each j, equations (31) and (32) constitute a linear system of at most $N_x(N_z+1)$ equations and $N_x(N_z+1)$ unknowns. This system will be further decomposed into a set of smaller tridiagonal systems. Specifically, formal substitution of the expressions for $\mathbf{U}^{n+\frac{1}{2}}_{i\pm\frac{1}{2},j}$ from (31) into (32) yields

$$\eta^{n+\frac{1}{2}}_{i,j} \quad - \quad g\frac{\Delta t^2}{2\Delta x^2}\Big[\big[(\mathbf{\Delta Z})^T \mathbf{A}^{-1} \mathbf{\Delta Z}\big]^n_{i+\frac{1}{2},j}(\eta^{n+\frac{1}{2}}_{i+1,j} - \eta^{n+\frac{1}{2}}_{i,j})$$

$$
\begin{aligned}
&\quad - \left[(\mathbf{\Delta Z})^T \mathbf{A}^{-1} \mathbf{\Delta Z}\right]^n_{i-\frac{1}{2},j} (\eta^{n+\frac{1}{2}}_{i,j} - \eta^{n+\frac{1}{2}}_{i-1,j})\Bigg] \\
&= \eta^n_{i,j} - \frac{\Delta t}{2\Delta x}\left[\left[(\mathbf{\Delta Z})^T \mathbf{A}^{-1}\mathbf{G}\right]^{n-\frac{1}{2}}_{i+\frac{1}{2},j} - \left[(\mathbf{\Delta Z})^T \mathbf{A}^{-1}\mathbf{G}\right]^{n-\frac{1}{2}}_{i-\frac{1}{2},j}\right] \\
&\quad - \frac{\Delta t}{2\Delta y}\left[\left(\mathbf{\Delta Z}^n_{i,j+\frac{1}{2}}\right)^T \mathbf{V}^n_{i,j+\frac{1}{2}} - \left(\mathbf{\Delta Z}^n_{i,j-\frac{1}{2}}\right)^T \mathbf{V}^n_{i,j-\frac{1}{2}}\right].
\end{aligned} \tag{33}
$$

Since $\mathbf{A}$ is positive definite, $\mathbf{A}^{-1}$ is also positive definite, consequently $(\mathbf{\Delta Z})^T\mathbf{A}^{-1}\mathbf{\Delta Z}$ is a non-negative number. Hence, for every j, equations (33) constitute a linear tridiagonal system of N_x equations for $\eta^{n+\frac{1}{2}}_{i,j}$. This system is symmetric and positive definite. Thus, it has a unique solution which can be determined by a direct method. Once the $\eta^{n+\frac{1}{2}}_{i,j}$ have been determined, equations (31) constitute a set of N_x tridiagonal systems with N_z equations each. All these systems are independent from each other, symmetric and positive definite. Thus, they can be conveniently solved by a direct method to determine $\mathbf{U}^{n+\frac{1}{2}}_{i+\frac{1}{2},j}$.

Finally, the vertical component of the velocity is determined from the discretized continuity equation and the boundary condition $w^{n+\frac{1}{2}}_{i,j,m-\frac{1}{2}} = 0$. In summary, the first time level requires the solution of N_y linear tridiagonal systems (33) of N_x equations with unknowns $\eta^{n+\frac{1}{2}}_{i,j}$, and N_xN_y linear tridiagonal systems (31) of N_z equations with unknowns $\mathbf{U}^{n+\frac{1}{2}}_{i+\frac{1}{2},j}$.

Next, one proceeds to the second level of calculation by finite differencing the y-momentum and the free surface equation as follows:

$$
\mathbf{A}^{n+\frac{1}{2}}_{i,j+\frac{1}{2}}\mathbf{V}^{n+1}_{i,j+\frac{1}{2}} = \mathbf{G}^n_{i,j+\frac{1}{2}} - g\frac{\Delta t}{\Delta y}(\eta^{n+1}_{i,j+1} - \eta^{n+1}_{i,j})\mathbf{\Delta Z}^n_{i,j+\frac{1}{2}} \tag{34}
$$

$$
\begin{aligned}
\eta^{n+1}_{i,j} = \eta^{n+\frac{1}{2}}_{i,j} &- \frac{\Delta t}{2\Delta x}\left[\left(\mathbf{\Delta Z}^{n+\frac{1}{2}}_{i+\frac{1}{2},j}\right)^T \mathbf{U}^{n+\frac{1}{2}}_{i+\frac{1}{2},j} - \left(\mathbf{\Delta Z}^{n+\frac{1}{2}}_{i-\frac{1}{2},j}\right)^T \mathbf{U}^{n+\frac{1}{2}}_{i-\frac{1}{2},j}\right] \\
&- \frac{\Delta t}{2\Delta y}\left[\left(\mathbf{\Delta Z}^{n+\frac{1}{2}}_{i,j+\frac{1}{2}}\right)^T \mathbf{V}^{n+1}_{i,j+\frac{1}{2}} - \left(\mathbf{\Delta Z}^{n+\frac{1}{2}}_{i,j-\frac{1}{2}}\right)^T \mathbf{V}^{n+1}_{i,j-\frac{1}{2}}\right].
\end{aligned} \tag{35}
$$

The values of v above free surface and below bottom in (34) have been eliminated by means of an appropriate discretization of the boundary conditions (29), (30). For each i, equations (34) and (35) constitute a linear system of at most $N_y(N_z + 1)$ equations. As before, this system is further decomposed into a set of smaller tridiagonal systems. Specifically, formal substitution of the expressions for $\mathbf{V}^{n+1}_{i,j\pm\frac{1}{2}}$ from (34) into (35), yields a linear tridiagonal system of N_y equations for $\eta^{n+1}_{i,j}$ for each i. This system is symmetric and positive definite. Once the $\eta^{n+1}_{i,j}$ have been determined, equations (34), which constitute a set of N_y tridiagonal systems with N_z equations each, are solved to determine $\mathbf{V}^{n+1}_{i,j+\frac{1}{2}}$.

Finally, the vertical component of the velocity is determined from the discretized continuity equation. In summary, the second time level requires the solution of N_x linear tridiagonal system of N_y equations with unknowns $\eta^{n+1}_{i,j}$, and N_xN_y linear tridiagonal systems of N_z equations with unknowns $\mathbf{V}^{n+1}_{i,j+\frac{1}{2}}$.

When an explicit Eulerian-Lagrangian discretization is used for the advective and horizontal viscosity terms, the stability condition is given, as before, by equation (17).

Note, finally, that when the vertical spacing Δz is taken to be large enough so that both the bottom and the free surface always fall within one vertical layer, this algorithm reduces to a two-dimensional ADI method, which is consistent with the two-dimensional, vertically integrated shallow water equations (7) - (9).

4.2. A SEMI-IMPLICIT FINITE DIFFERENCE METHOD

The alternating direction semi-implicit method derived in the previous section only requires the solution of three diagonal systems, but should be used with small time steps in order to reduce the ADI effect. For applications where large time steps become essential, a more robust, one time level semi-implicit scheme can be derived. In vector notation, the semi-implicit discretization for the momentum equations (25), (26), and for the free surface equation (28) takes the following form

$$\mathbf{A}^n_{i+\frac{1}{2},j}\mathbf{U}^{n+1}_{i+\frac{1}{2},j} = \mathbf{G}^n_{i+\frac{1}{2},j} - g\frac{\Delta t}{\Delta x}(\eta^{n+1}_{i+1,j} - \eta^{n+1}_{i,j})\mathbf{\Delta Z}^n_{i+\frac{1}{2},j} \tag{36}$$

$$\mathbf{A}^n_{i,j+\frac{1}{2}}\mathbf{V}^{n+1}_{i,j+\frac{1}{2}} = \mathbf{G}^n_{i,j+\frac{1}{2}} - g\frac{\Delta t}{\Delta y}(\eta^{n+1}_{i,j+1} - \eta^{n+1}_{i,j})\mathbf{\Delta Z}^n_{i,j+\frac{1}{2}} \tag{37}$$

$$\begin{aligned}\eta^{n+1}_{i,j} = \eta^n_{i,j} &- \frac{\Delta t}{\Delta x}\left[\left(\mathbf{\Delta Z}^n_{i+\frac{1}{2},j}\right)^T\mathbf{U}^{n+1}_{i+\frac{1}{2},j} - \left(\mathbf{\Delta Z}^n_{i-\frac{1}{2},j}\right)^T\mathbf{U}^{n+1}_{i-\frac{1}{2},j}\right] \\ &- \frac{\Delta t}{\Delta y}\left[\left(\mathbf{\Delta Z}^n_{i,j+\frac{1}{2}}\right)^T\mathbf{V}^{n+1}_{i,j+\frac{1}{2}} - \left(\mathbf{\Delta Z}^n_{i,j-\frac{1}{2}}\right)^T\mathbf{V}^{n+1}_{i,j-\frac{1}{2}}\right],\end{aligned} \tag{38}$$

where $\mathbf{U}$, $\mathbf{V}$, $\mathbf{\Delta Z}$, $\mathbf{G}$ and $\mathbf{A}$ are defined as:

$$\mathbf{U}^{n+1}_{i+\frac{1}{2},j} = \begin{pmatrix} u^{n+1}_{i+\frac{1}{2},j,M} \\ u^{n+1}_{i+\frac{1}{2},j,M-1} \\ \vdots \\ u^{n+1}_{i+\frac{1}{2},j,m} \end{pmatrix}, \quad \mathbf{V}^{n+1}_{i,j+\frac{1}{2}} = \begin{pmatrix} v^{n+1}_{i,j+\frac{1}{2},M} \\ v^{n+1}_{i,j+\frac{1}{2},M-1} \\ \vdots \\ v^{n+1}_{i,j+\frac{1}{2},m} \end{pmatrix}, \quad \mathbf{\Delta Z} = \begin{pmatrix} \Delta z_M \\ \Delta z_{M-1} \\ \vdots \\ \Delta z_m \end{pmatrix},$$

$$\mathbf{G}^{n}_{i+\frac{1}{2},j} = \begin{pmatrix} \Delta z^n_M F u^n_{i+\frac{1}{2},j,M} + \tau^s_x \Delta t \\ \Delta z^n_{M-1} F u^n_{i+\frac{1}{2},j,M-1} \\ \vdots \\ \Delta z^n_m F u^n_{i+\frac{1}{2},j,m} \end{pmatrix}, \quad \mathbf{G}^{n}_{i,j+\frac{1}{2}} = \begin{pmatrix} \Delta z^n_M F v^n_{i,j+\frac{1}{2},M} + \tau^s_y \Delta t \\ \Delta z^n_{M-1} F v^n_{i,j+\frac{1}{2},M-1} \\ \vdots \\ \Delta z^n_m F v^n_{i,j+\frac{1}{2},m} \end{pmatrix},$$

$$\mathbf{A} = \begin{pmatrix} \Delta z_M + a_{M-\frac{1}{2}} & -a_{M-\frac{1}{2}} & & 0 \\ -a_{M-\frac{1}{2}} & \Delta z_{M-1} + a_{M-\frac{1}{2}} + a_{M-\frac{3}{2}} & -a_{M-\frac{3}{2}} & \\ \cdot & \cdot & \cdot & \\ & \cdot & \cdot & \cdot \\ 0 & & -a_{m+\frac{1}{2}} & \Delta z_m + a_{m+\frac{1}{2}} + \gamma \Delta t \end{pmatrix}$$

with $a_k = \dfrac{\nu_k \Delta t}{\Delta z_k}$.

The values of u and v above free surface and below bottom have been eliminated by means of a proper discretization of the boundary conditions (29), (30). The linear system (36)-(38) consists of at most $N_x N_y (2N_z + 1)$ equations. Computationally, since a linear system of $N_x N_y (2N_z + 1)$ equations can be quite large even for modest values of N_x, N_y and N_z, the system (36)-(38) is decomposed into a set of $2N_x N_y$ independent tridiagonal systems of N_z equations, and one five diagonal system of $N_x N_y$ equations. Specifically, formal substitution of the expressions for $\mathbf{U}^{n+1}_{i\pm\frac{1}{2},j}$ and $\mathbf{V}^{n+1}_{i,j\pm\frac{1}{2}}$ from (36) and (37) into (38) yields

$$\begin{aligned} \eta^{n+1}_{i,j} &- g\frac{\Delta t^2}{\Delta x^2}\Bigg[\left[(\mathbf{\Delta Z})^T \mathbf{A}^{-1} \mathbf{\Delta Z}\right]^n_{i+\frac{1}{2},j} (\eta^{n+1}_{i+1,j} - \eta^{n+1}_{i,j}) \\ &\qquad - \left[(\mathbf{\Delta Z})^T \mathbf{A}^{-1} \mathbf{\Delta Z}\right]^n_{i-\frac{1}{2},j} (\eta^{n+1}_{i,j} - \eta^{n+1}_{i-1,j})\Bigg] \\ &- g\frac{\Delta t^2}{\Delta y^2}\Bigg[\left[(\mathbf{\Delta Z})^T \mathbf{A}^{-1} \mathbf{\Delta Z}\right]^n_{i,j+\frac{1}{2}} (\eta^{n+1}_{i,j+1} - \eta^{n+1}_{i,j}) \end{aligned}$$

$$-\left[(\mathbf{\Delta Z})^T\mathbf{A}^{-1}\mathbf{\Delta Z}\right]^n_{i,j-\frac{1}{2}}(\eta^{n+1}_{i,j}-\eta^{n+1}_{i,j-1})\Bigg]$$

$$= \eta^n_{i,j} - \frac{\Delta t}{\Delta x}\left[\left[(\mathbf{\Delta Z})^T\mathbf{A}^{-1}\mathbf{G}\right]^n_{i+\frac{1}{2},j} - \left[(\mathbf{\Delta Z})^T\mathbf{A}^{-1}\mathbf{G}\right]^n_{i-\frac{1}{2},j}\right]$$
$$- \frac{\Delta t}{\Delta y}\left[\left[(\mathbf{\Delta Z})^T\mathbf{A}^{-1}\mathbf{G}\right]^n_{i,j+\frac{1}{2}} - \left[(\mathbf{\Delta Z})^T\mathbf{A}^{-1}\mathbf{G}\right]^n_{i,j-\frac{1}{2}}\right]. \quad (39)$$

Since $\mathbf{A}$ is positive definite, $\mathbf{A}^{-1}$ is also positive definite, consequently $(\mathbf{\Delta Z})^T\mathbf{A}^{-1}\mathbf{\Delta Z}$ is a non-negative number. Hence equations (39) constitute a linear five-diagonal system of N_xN_y equations for $\eta^{n+1}_{i,j}$. This system is symmetric and positive definite. Thus, it has a unique solution which can be determined very efficiently by a preconditioned conjugate gradient method.

Once the new free surface location has been determined, equations (36) and (37) constitute a set of $2N_xN_y$ linear, tridiagonal systems with N_z equations each. All these systems are independent from each other, symmetric and positive definite. Thus, they can be conveniently solved by a direct method.

Finally, the vertical component of the velocity at the new time level is computed by discretizing the continuity equation (27) and by setting $w^{n+1}_{i,j,m-\frac{1}{2}} = 0$. The numerical algorithm presented above includes the simulation of the flooding and the drying of tidal flats. To this purpose at each time step the new water depths $H^{n+1}_{i+\frac{1}{2},j}$ and $H^{n+1}_{i,j+\frac{1}{2}}$ are defined by (12)-(13), with the understanding that an occurrence of the zero value for the total depth H simply means a dry point which may be flooded at a later time when H becomes positive. The vertical grid spacings $\mathbf{\Delta Z}^n_{i+\frac{1}{2},j}$ and $\mathbf{\Delta Z}^n_{i,j+\frac{1}{2}}$ are updated accordingly.

When an explicit Eulerian-Lagrangian discretization is used for the advective and horizontal viscosity terms, the stability condition for this method depends upon the explicit discretization of the viscous terms and is given by equation (17), as in the two dimensional case.

Note, finally, that when the vertical spacing Δz is taken to be large enough so that both the bottom and the free surface always fall within one vertical layer, this algorithm reduces to the two-dimensional semi-implicit numerical method (18)-(20), which is consistent with the two-dimensional, vertically integrated shallow water equations (7) - (9).

4.3. FULLY-IMPLICIT SPLITTING METHODS

In order to derive an efficient method which is unconditionally stable, a fully implicit splitting method can be derived for the 3D model. Specifically, the advective and the horizontal viscosity terms are discretized implicitly in a first fractional time step by using, for example, an implicit Eulerian-Lagrangian scheme. For each vertical level k two linear, five-diagonal, symmetric and positive definite systems are obtained for the provisional values $\tilde{u}^{n+1}_{i+\frac{1}{2},j,k}$ and $\tilde{v}^{n+1}_{i,j+\frac{1}{2},k}$, respectively. Once these systems are solved, one proceeds to the second fractional step by including the propagation operator as follows

$$\mathbf{A}^n_{i+\frac{1}{2},j}\mathbf{U}^{n+1}_{i+\frac{1}{2},j} = \tilde{\mathbf{U}}^n_{i+\frac{1}{2},j} - g\frac{\Delta t}{\Delta x}(\eta^{n+1}_{i+1,j} - \eta^{n+1}_{i,j})\mathbf{\Delta Z}^n_{i+\frac{1}{2},j} \tag{40}$$

$$\mathbf{A}^n_{i,j+\frac{1}{2}}\mathbf{V}^{n+1}_{i,j+\frac{1}{2}} = \tilde{\mathbf{V}}^n_{i,j+\frac{1}{2}} - g\frac{\Delta t}{\Delta y}(\eta^{n+1}_{i,j+1} - \eta^{n+1}_{i,j})\mathbf{\Delta Z}^n_{i,j+\frac{1}{2}} \tag{41}$$

$$\begin{aligned}\eta^{n+1}_{i,j} = \eta^n_{i,j} &- \frac{\Delta t}{\Delta x}\left[\left(\mathbf{\Delta Z}^n_{i+\frac{1}{2},j}\right)^T\mathbf{U}^{n+1}_{i+\frac{1}{2},j} - \left(\mathbf{\Delta Z}^n_{i-\frac{1}{2},j}\right)^T\mathbf{U}^{n+1}_{i-\frac{1}{2},j}\right] \\ &- \frac{\Delta t}{\Delta y}\left[\left(\mathbf{\Delta Z}^n_{i,j+\frac{1}{2}}\right)^T\mathbf{V}^{n+1}_{i,j+\frac{1}{2}} - \left(\mathbf{\Delta Z}^n_{i,j-\frac{1}{2}}\right)^T\mathbf{V}^{n+1}_{i,j-\frac{1}{2}}\right].\end{aligned} \tag{42}$$

Computationally, elimination of $\mathbf{U}^{n+1}_{i+\frac{1}{2},j}$ and $\mathbf{V}^{n+1}_{i,j+\frac{1}{2}}$ from the continuity equation (42) yields, again, a linear five-diagonal system of equations for the surface elevation $\eta^{n+1}_{i,j}$. This system has the generic equation of the form (39), thus it is symmetric and positive definite. The main difference between semi-implicit and fully implicit splitting methods is in the treatment of the horizontal viscous terms. Note, finally, that when the vertical spacing Δz is taken to be sufficiently large so that $m = M = 1$, this algorithm reduces to the fully implicit splitting method described in Section 3 for the two-dimensional vertically integrated shallow water equations.

5. Applications

The previous methods have been extensively applied to several sites (see, e.g., [9],[15]). We present here some applications to the Lagoon of Venice. The Lagoon of Venice, for example, is a very complex sea water basin whose area is about 50 km^2 and which consists of several inter-connected narrow channels with a maximum width of 1 km, and up 50 m deep encircling large and flat shallow areas. Additionally, several tidal marshes with a bathymetry of only $20 - 40$ cm above sea level require proper treatment of flooding and drying. The Lagoon is connected to the Adriatic Sea through three narrow inlets, namely Lido, Malamocco, and Chioggia (see Fig. 1). The city of Venice is situated upon the largest island near the Lido inlet. Tides propagate from the Adriatic Sea into the Lagoon through the three inlets. In the numerical model the Lagoon has been covered with a $N_x = 672$ by $N_y = 846$ finite-difference mesh of equal $\Delta x = \Delta y = 50$ m. At the three inlets, an M_2 tide of 0.5 m amplitude and 12 lunar hour period has been specified. The integration time step is chosen to be $\Delta t = 15\ min$ and the computations have been carried out by solving, at each time step, a corresponding linear, five-diagonal system of $N_x N_y = 568,512$ equations. When 200 vertical layers are used the computational box gets filled with over 100 million gridpoints, of which only 1.2% are active. The corresponding calculations run at a speed which is about 5 time faster than real time on a 4.2 Megaflops workstation.

6. Conclusions

Several numerical methods for free surface, hydrostatic, large scale flows have been presented and discussed. ADI methods require the solution of a set of linear, tridiagonal systems at each time step. They are not sufficiently accurate for the discretization of flows in regions with complex geometries. Semi-implicit methods require the solution of one large linear five diagonal system at each step. This system is symmetric and positive definite, hence use of a fast preconditioned conjugate gradient method is appropriate. A stability condition due to the explicit discretization of the viscous terms is required. An unconditionally stable method can be derived by using a fully implicit splitting approach. These algorithms apply to large, complex flow regions and to three dimensional flow problems. Based on the same principles, discretization methods for nonhydrostatic flows and for the primitive equations of atmospheric dynamics have also been developed in [2] and [6].

References

1. Bertolazzi, E., Casulli, V., "Semi-Implicit Numerical Methods for Convection-Diffusion Equations", in *Advanced Mathematical Tools in Metrology,* (Eds. Ciarlini P., Cox M.G., Monaco R., Pavese F.), 57-66, World Scientific, Singapore, 1994
2. Bonaventura, L., Casulli, V., "A Finite-Difference, Semi-Implicit Scheme for Primitive Equations of Atmospheric Dynamics", in *Numerical Methods in Laminar and Turbulent Flow,* (Eds. Taylor C., Durbetaki P.),Vol. 9, Part II, 1585-1595, Pineridge Press, Swansea, 1995
3. Blumberg, A.F. and Mellor, G.L., "A Description of a Three Dimensional Coastal Ocean Circulation Model", in *Three Dimensional Coastal Ocean Circulation Models, Coastal and Estuarine Sciences,* Vol. 4, (Ed. Heaps, N.S.), AGU, Washington, DC, 1-16 (1987).
4. Casulli, V., "Semi-implicit Finite Difference Methods for the Two-Dimensional Shallow-Water Equation", *Jour. of Computational Physics,* Vol.86, 56-74 (1990).
5. Casulli, V.,"Numerical Methods for Free Surface Hydrodynamics", *Lecture Notes,* unpublished, 1993
6. Casulli, V., "Recent Developments in Semi-Implicit Numerical Methods for Free Surface Hydrodynamics", in *Advances in Hydro-Science and Engineering* , Vol.2, 2174-2181, Tsinghua Univ. Press, Beijing, 1995
7. Casulli, V. and Cattani E., "Stability, Accuracy and Efficiency of a Semi-implicit Method for Three-Dimensional Shallow-Water Flow", *Computers Math. Applic.,*

Vol.27, 99-112 (1994).
8. Casulli, V. and Cheng, R.T., "Semi-implicit finite difference methods for three-dimensional shallow-water flow", *Int. Jour. for Numerical Methods in Fluids*, Vol.15, 629-648 (1992).
9. Cheng, R.T., Casulli, V. and Gartner, J.W. "Tidal, Residual, Inter-tidal Mud-flat (TRIM) Model with Applications to San Francisco Bay", *Estuarine, Coastal Shelf Science*, Vol. 36, 235-280 (1993).
10. Davies, A.M. and Aldridge, J.N., "A stable algorithm for bed friction in three-dimensional shallow sea modal models", *Int. Jour. for Numerical Methods in Fluids*, Vol.14, 477-493 (1992).
11. Duwe, K.C., Hewer, R.R. and Backhaus, J.O., "Results of a semi-implicit two-step method for the simulation of markedly nonlinear flow in coastal seas", *Continental Shelf Research*, Vol.2, No.4, 255-274 (1983).
12. Gill, A.E., "*Atmosphere - Ocean Dynamics* ", Academic Press, S.Diego, 1982
13. Greenspan, D. and Casulli, V., "*Numerical Analysis for Applied Mathematics, Science, and Engineering*", Addison Wesley, 1988.
14. Leendertse, J.J., "A new approach to three-dimensional free-surface flow modelling", *Rep. R-3712-NETH/RC*, Rand Corporation, Santa Monica, California (1989).
15. Signell, R.P. and Butman, B., "Modeling tidal exchange and dispersion in Boston Harbor", *Jour. of Geophysical Research*, Vol.97, No. C10, 15591-15606 (1992).
16. Staniforth A. and Coté J., "Semi-Lagrangian Integration Schemes for Atmospheric Models-A Review", *Monthly Weath. Rev.*, Vol. 119, 2206-2223,1991
17. Stelling, G.S., "On the Construction of Computational Methods for Shallow Water Flow Problems", *Raijkswaterstaat Communications*, No. 35, The Hague (1984).
18. Weinan, T., "*Shallow Water Hydrodynamics*" , Elsevier , Amsterdam, 1992

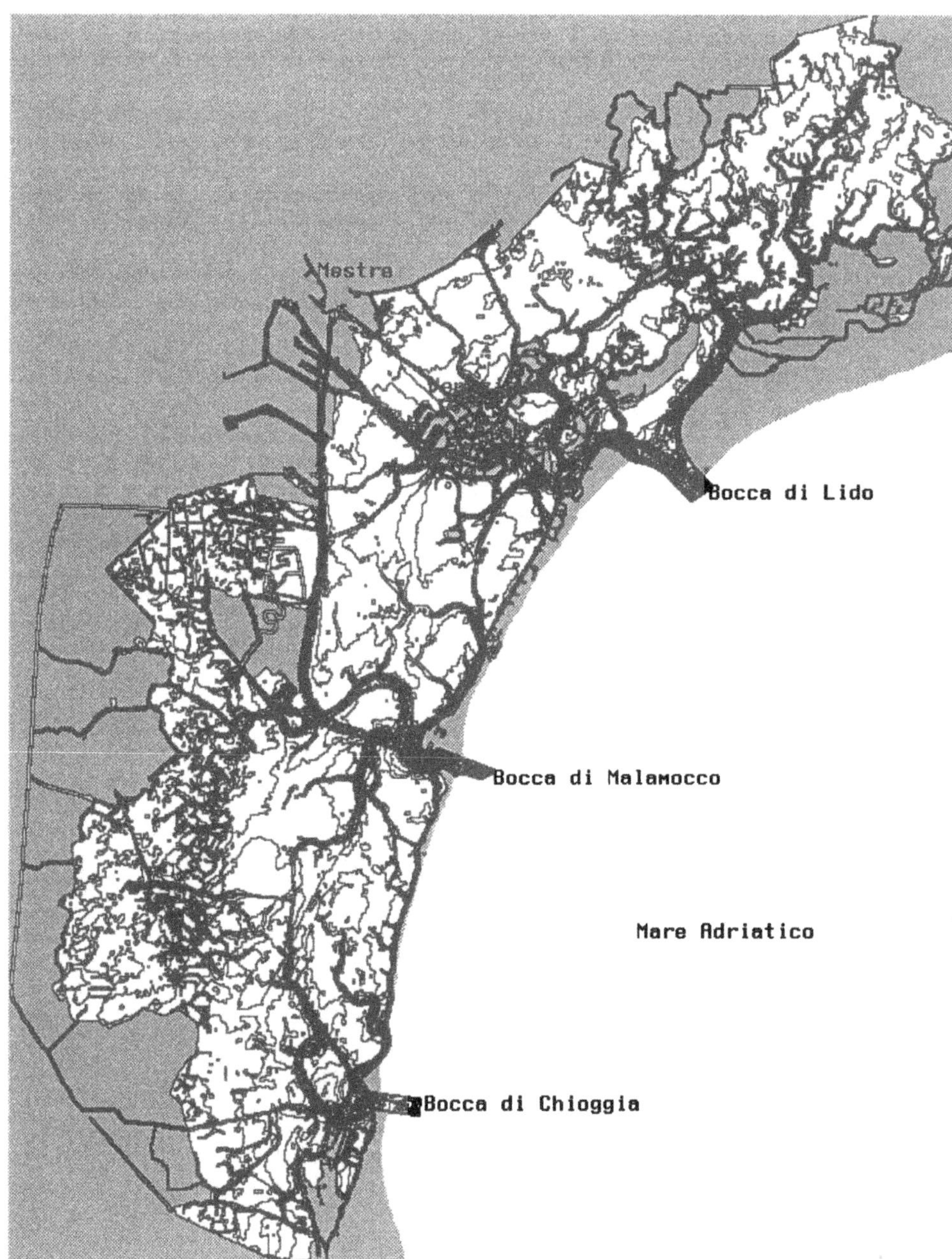

Figure 1. Computational grid for the Lagoon of Venice

ON APPLICATIONS OF HIGH RESOLUTION SHOCK CAPTURING METHODS TO UNSTEADY FLOWS

D.M. CAUSON, D.M. INGRAM AND G. YANG
Centre for Mathematical Modelling and Flow Analysis
the Manchester Metropolitan University
Chester Street
Manchester M1 5GD.

Abstract. Two applications of modern high resolution shock capturing methods to high speed compressible aeronautical flows are presented. The first application concerns the phenomenon of surge interaction in a twin side-by-side intake configuration. Unsteadiness was introduced by prescribing a pressure disturbance at the exit plane of one of the intakes. The case reported corresponds to static ground-running of the engines. The calculations were carried out on a body fitted structured grid using a finite volume method of the Godunov-type with MUSCL reconstruction and showed good agreement with available data. The second application demonstrates the utility of using a Cartesian cut cell method as an alternative to structured or unstructured methods for flows around complex geometries either stationary or in relative motion. The flow solver used in the first application was modified to facilitate both body motion and cell merging in order to maintain strict conservation and numerical stability.

1. Introduction

Numerical predictions of transient flows are important in many practical situations. Although calculation methods for steady flows employing high resolution numerical schemes are now firmly established, the simulation of unsteady flows involving moving shock waves or one or more bodies in relative motion is still a challenging problem in computational fluid dynamics. Typical examples can be found in many applications, including aircraft engine transients e.g. surge, store separation, flutter analysis, flow past a helicopter rotor and sabot/projectile separation. A comprehensive survey

E.F. Toro and J.F. Clarke (eds.), Numerical Methods for Wave Propagation, 145–171.

can be found in the review paper of Edwards and Thomas[1]. Since such unsteady flows also present difficulty for the experimental fluid dynamicist, it is reasonable to examine the possibility of modelling them numerically.

In the case of engine surge, the primary transient flow feature is a shock wave, or hammer shock, which moves upstream through the engine intake. The case of a multiple-engined installation in which unsteady flow caused by a surge in one engine may give rise to a complementary surge in the adjacent unit is addressed in the first part of this paper. The numerical modelling requires a suitable representation of a typical surge overpressure signature to be imposed at the exit plane of one of the intakes. A high resolution Godunov-type method with MUSCL reconstruction was employed to study the resulting core flow and induced unsteady flow behaviour in the adjacent intake. The cases considered include static ground running of the engines and operation at flight mach numbers of 0.6 and 1.2 (only results corresponding to the static case are presented here). The amplitude of the imposed pressure disturbance varied between 100–200% of the mean exit static pressure. A review of the physical nature of the engine surge phenomenon together with experimental data and measurement techniques has been presented by Lotter et al[2]. Other relevant studies include those of Marshall[3], Evans and Truax[4] and Kurkov et al[5].

At present, there are two main approaches for dealing with compressible flows involving arbitrarily moving boundaries or bodies, both of which fall into either structured or unstructured grid categories. In unstructured mesh approaches, the usual strategy is to use a global unstructured mesh and incorporate periodic local or global remeshing to account for moving boundaries or bodies. Both finite element[6, 7] and finite volume solver[8] solvers have been applied for the discretisation. Usually, unstructured mesh methods apply an adaptive mesh strategy to achieve higher order accuracy, and hence the time step will be restricted by the smallest cell for an unsteady flow calculation. Therefore, unstructured mesh methods need larger memory requirements and CPU time. Furthermore, due to repeated interpolation during remeshing, unstructured mesh methods may suffer from additional diffusive effects. For structured mesh methods, on the other hand, one promising approach is the *Chimera*[9] or FAME[10] philosophy of overlapping several meshes, each of which is specific to an appropriate section of the geometry. This allows an efficient representation of boundary layers via directional clustering on 'O' or 'C' meshes, which can then be overlapped on a background Cartesian mesh.

In the second part of this paper, we present an alternative approach for the prediction of unsteady compressible flowfields involving both static and moving bodies. The method is based on a Cartesian cut cell technique and does not at any stage involve the generation of a field mesh. Rather, certain cells are flagged as being completely or partially cut by the body and these cells are singled out for special treatment as the body moves. The remainder of the flow/uncut cells are treated in a straightforward manner which requires no detailed explanation.

2. The Governing Equations

The Euler equations for three-dimensional, compressible flows may be written in integral form in a general moving reference frame as

$$\frac{\partial}{\partial t}\int_{V_t} \mathbf{U}\, dV + \oint_{S_t} \mathbf{H}\cdot\mathbf{n}\, dS = 0 \tag{1}$$

where $\mathbf{U}$ is the vector of conserved variables and $\mathbf{H}$ is the flux vector. $\mathbf{n}$ is the outward unit vector normal to the boundary S_t, which encloses the time-dependent volume V_t. $\mathbf{U}$ and $\mathbf{H}$ are given by

$$\mathbf{U} = \begin{pmatrix} \rho \\ \rho u \\ \rho v \\ \rho w \\ e \end{pmatrix}, \quad \mathbf{H} = \begin{pmatrix} \rho(\mathbf{v}-\mathbf{v}_s) \\ \rho u(\mathbf{v}-\mathbf{v}_s)+p\mathbf{i} \\ \rho v(\mathbf{v}-\mathbf{v}_s)+p\mathbf{j} \\ \rho w(\mathbf{v}-\mathbf{v}_s)+p\mathbf{k} \\ (e+p)(\mathbf{v}-\mathbf{v}_s)+p\mathbf{v}_s \end{pmatrix}, \tag{2}$$

where ρ, u, v, p and e are density, x-, y- and z- components of fluid velocity $\mathbf{v}$, pressure and total energy per unit volume, $\mathbf{i}$, $\mathbf{j}$ and $\mathbf{k}$ are the Cartesian unit base vectors and $\mathbf{v}_s$ is the velocity of the boundary of the control volume V_t. In the case of moving body problems, we use a stationary background Cartesian mesh, therefore, $\mathbf{v}_s = 0$ on the flow interfaces of a cell, whereas on a moving solid face, $\mathbf{v}_s$ is the velocity of the moving boundary. Finally, the governing equations are closed by the ideal gas equation of state,

$$p = (\gamma - 1)\left[e - \frac{\rho}{2}(u^2 + v^2 + w^2)\right]. \tag{3}$$

3. Numerical Modelling of Aircraft Engine Transients

Engine surge is a violent transient phenomenon characterised by strong pressure pulses, or hammer shocks, propagating up the intake. This leads to a breakdown of steady flow conditions, mass flow reversal and physical pulsations of the engine. There is the possibility of engine flame out or even permanent engine damage. Common causes of engine surge include over-fuelling, nozzle (reheat) malfunction, power offtake irregularities, bird strike or other foreign object ingestion and disturbed intake air flow during manoeuvres. The peak pressure behind the propagating shock front can easily exceed the normal operating total pressure inside the intake so this is a critical design load for the intake structure.

At the onset of engine surge, the static pressure at the engine compressor face rises suddenly and describes either a single cycle ('pop' surge) or several cycles of a complete pressure signature ('cyclic' or 'lock-in' surge). Figure 1 illustrates the essential features of one cycle of a surge overpressure signature. The initial peak surge static pressure referenced to the pre-surge mean static pressure at the engine compressor face is known as the surge overpressure ratio (OPR).

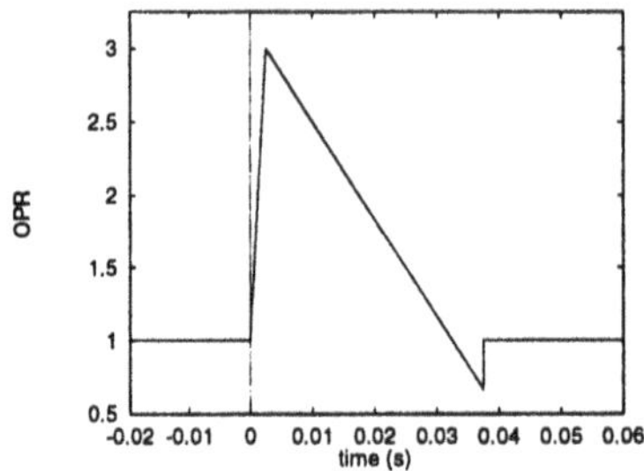

Figure 1. Simplified Surge Overpressure Signature

Engine surge is very complex to analyse, both experimentally and theoretically. Correlations have been derived experimentally for certain engine types which may be of value in preliminary design [3, 4, 5]. However, the peak surge pressure is dependent not only on a particular engine type but also on the precise initial location of the surge within the engine. Even if the maximum surge overpressure at the engine compressor face is known, estimating the transient pressure levels along the intake duct as the hammer shock travels upstream is a difficult task. Various workers have constructed novel shock wave generator (SWG) devices to simulate the hammer shock experimentally in scaled wind tunnel tests e.g.[2]. However, there appear to be few experimental or theoretical analyses relating specifically to the prediction of unsteady flow characteristics in an intake as a result of engine surge. The most closely related studies we have found concern unsteady flow in nozzles and ram-jet intake geometries caused by combustion instabilities [11, 12, 13, 14, 15, 16, 17]. For example, Hsieh and Wardlaw[12] investigated coupling between the intake flow field and combustion chamber pressure oscillations in a ram-jet. They prescribed the pressure fluctuations from the combustion chamber and examined the flow field in the intake alone. For validation purposes, we have carried out steady flow calculations using a nozzle configuration investigated experimentally by Sajben et al[16]. Our unsteady flow calculations were validated by comparing the results with those of Hsieh and Wardlaw who used this nozzle configuration for their unsteady flow calculations.

3.1. THE INTAKE GEOMETRY

The physical geometry consists of a pair of intakes in a side-by-side arrangement. Each intake duct has a rectangular section at its entrance plane which makes a smooth transition along an S-shaped centre-line to a circular section at each exit plane. A splitter plate positioned at the bifurcation point between the intakes provides aerodynamic shielding. Fixed compression ramps, which adjust the oblique shock wave onto the cowl lips at the design cruise Mach number, are located a short distance upstream of the intake entry plane.

A body-fitted structured grid with approximately 10^5 cells was generated using a transfinite interpolation technique[18]. A smooth transition from rectangular to circular section was achieved by mapping the four corners of the rectangular grid at each intake entry plane to corresponding points on the circumference of a circular section at each engine compressor face. Using appropriate non-linear stretching functions, the grid was extended a distance of four duct lengths upstream of the intake entry plane. The mesh spacings were graded to provide improved resolution around the bifurcation point. Figure 2 (a–d) illustrates the surface grid detail.

3.2. THE NUMERICAL SCHEME

The flow simulation requires a fully time-accurate calculation method which proceeds in two parts: a time-marched calculation to steady-state corresponding physically to steady flight or static ground-running conditions with both engines running normally, followed by a time-accurate phase commencing with the imposition of a suitable pressure signature boundary condition at the exit plane of one of the intakes, corresponding to the onset of surge in one engine. During the transient phase of the calculation, the boundary procedure imposed at the exit plane of the adjacent intake was that appropriate to an engine running normally. The intake in which a surge boundary procedure is imposed is called the *primary* intake; the adjacent intake is called the *secondary* intake. The purpose of this study was to assess the influence of an induced surge on the adjacent engine; a phenomenon known as surge interaction.

Approximation of the second term of (1) leads to the following semi-discrete finite volume discretisation for the cell centre mean value $\mathbf{U}_0$

$$\frac{\partial \mathbf{U}_0}{\partial t} = -\frac{1}{V_0}\sum_{i=1}^{6} \mathbf{H}_i.\mathbf{S}_i \qquad (4)$$

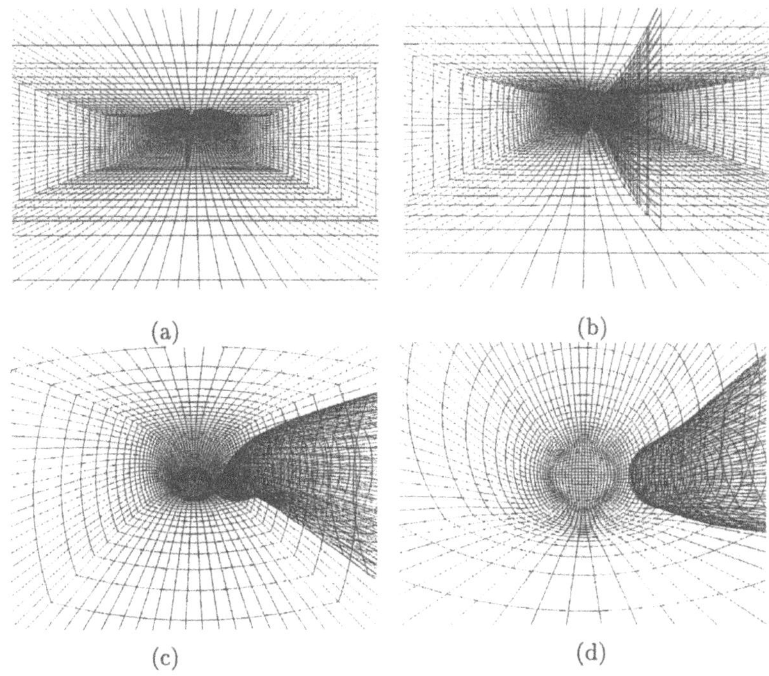

Figure 2. Discretised Intake Geometry

where $\mathbf{S}_i$ is the product of the ith cell face area and the unit outer normal vector (see Figure 3) and the fluxes on each cell face $H_i.\mathbf{S}_i$ have yet to be defined. We first consider the spatial discretisation i.e. the right hand side of (4). For clarity of exposition, we restrict attention to one space dimension. Later, we will extend the derived schemes to multi dimensions by operator-splitting.

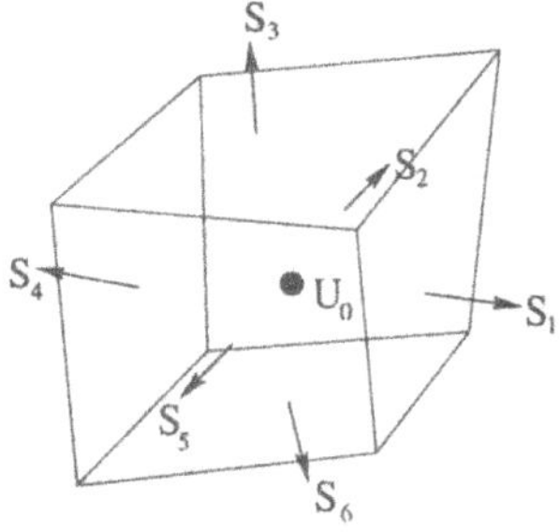

Figure 3. A typical finite volume cell

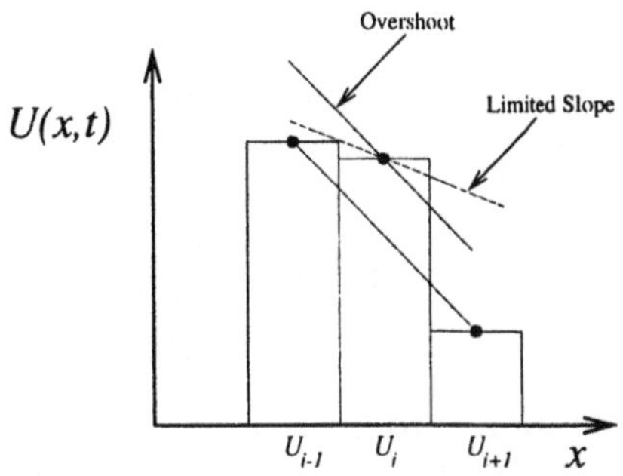

Figure 4. MUSCL reconstruction

3.2.1. *Spatial Discretisation*

The piecewise constant initial data is used to produce piecewise linear reconstructions within each cell. To maintain the positivity of physically non-negative variables like pressure and density, we limit non-physical interpolation where the solution is not smooth e.g. in the neighbourhood of shock waves. For example, in Figure 4, a gradient computed from a central difference approximation based on solution values to the left and right of cell U_i is limited so that it does not overshoot the original piecewise constant values on either side. This slope limiting approach is derived from the classical MUSCL (Monotonic Upstream Schemes for Conservation Laws) schemes of Van Leer[19]; variants of it form the basic reconstruction methodology for many Godunov-type schemes.

From the non-oscillatory reconstruction, the cell interface flux is determined by solving a Riemann problem defined by the interpolated values U_L and U_R on the left and right hand sides of each cell interface. This can be accomplished by using any available exact or approximate Riemann solver. We use an approximate Riemann solver based on the work of Harten, Lax and Van Leer[20], hereafter denoted by the acronym 'HLL'.

The steps involved in implementing a slope limited MUSCL reconstruction procedure and HLL Riemann solver for the spatial discretisation of (4) are:

1. *Gradient Vector*
   In each cell compute a gradient for each variable, e.g. for density

$$(\Delta\rho)_i = \frac{\rho_{i+1} - \rho_{i-1}}{2\Delta x},$$

   which provides a gradient vector $\mathbf{g}_i$.

2. *Slope Limiting*
   Limit every component $[j]$ of $\mathbf{g}_i$ such that

$$\begin{aligned} U[j]_{i+\frac{1}{2}} &\approx U[j]_i + \alpha_j g[j]_i . \frac{\Delta x}{2} \\ &< \max(U[j]_i, U[j]_{i+1}) \\ &> \min(U[j]_i, U[j]_{i+1}). \end{aligned}$$

   where α_j is a slope limiter to be defined. This is the interpolated value of $U[j]_i$ on the left side of the right hand cell interface $i + \frac{1}{2}$. The left hand interface is treated in a similar way.
   α_j is maximised subject to :
   (a) $0 \leq \alpha_j \leq 1$
   (b) $U[j]_i \pm \alpha_j g[j]_i . \Delta x/2$ does not cause under/over shoots at cell interface $i \pm \frac{1}{2}$

   For each cell interface at $i \pm \frac{1}{2}$, this procedure provides the left and right states U_L, U_R for the HLL Riemann solver.
3. *HLL Fluxes*
   We now find $U^*_{i\pm\frac{1}{2}} = U^*(U_L, U_R)$ where U^* is the approximate solution to the Riemann problem defined by $R(U_L, U_R)$ at cell interface $i\pm\frac{1}{2}$. We can then compute directly the required HLL fluxes $F^*_{i\pm 1/2} = F(U^*_{i\pm 1/2})$. Consider the following simplified Riemann fan bounded by the fastest and slowest moving waves S_L, S_R

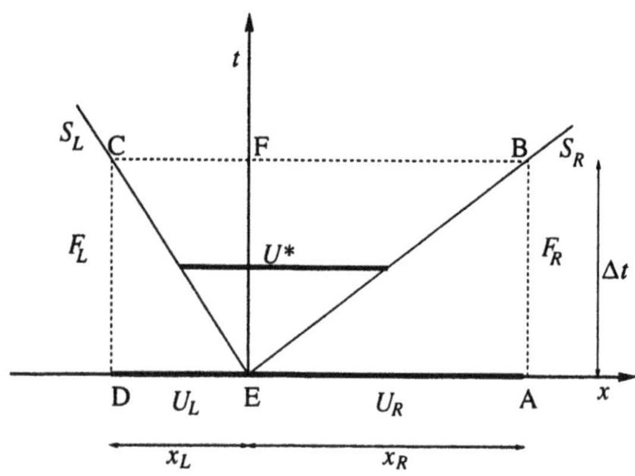

Figure 5. Wave diagram

Given á priori estimates of the wave speeds S_L and S_R, U^* can be determined from the conservation law in integral form

$$\oint_c F\,\mathrm{d}t - U\,\mathrm{d}x = 0.$$

Integrating round the contour $ABCD$ we have

$$F_R\Delta t - U^*(\Delta t S_R - \Delta t S_L) - F_L\Delta t + U_L\Delta t S_L + U_R\Delta t S_R = 0$$

where the L, R subscripts indicate that the quantity is to be evaluated on the basis of the left, right cell interface data, gives

$$U^* = \frac{F_R - F_L - S_L U_L + S_R U_R}{S_R - S_L}.$$

Correspondingly,

$$F(U^*) = \begin{cases} F_L & \text{if } s_L > 0 \\ F_R & \text{if } s_R < 0 \\ \frac{S_R F_L - S_L F_R + S_L S_R (U_R - U_L)}{S_R - S_L} & \text{otherwise} \end{cases} \quad (5)$$

This may be generalised for an arbitrary hexahedral cell of the type shown in Figure 3 by projection of the flux vector $\mathbf{H}$ into the direction of the outward pointing unit normal cell side vector $\hat{\mathbf{S}}_{i\pm\frac{1}{2}}$ of the $i\pm\frac{1}{2}$th cell interface giving e.g.

$$\mathbf{H}^*_{i+\frac{1}{2}}.\hat{\mathbf{S}}_{i+\frac{1}{2}} = \frac{s_R \mathbf{H}_L.\hat{\mathbf{S}}_{i+\frac{1}{2}} - s_L \mathbf{H}_R.\hat{\mathbf{S}}_{i+\frac{1}{2}} + s_L s_R(\mathbf{U}_R - \mathbf{U}_L)}{s_R - s_L} \quad (6)$$

4. Wave Speed Estimates

All that remains is to provide appropriate estimates for the wave speeds S_L, S_R. Various estimates are available[21]; the wave speed estimates we use are as follows

$$S_L = \min(u_L, u_R) - c_{max}$$

$$S_R = \max(u_L, u_R) + c_{max}$$

where $c_{max} = \max(c_L, \bar{c}, c_R)$ and

$$c_L = \sqrt{\frac{\gamma(p_L)}{\rho_L}},\ c_R = \sqrt{\frac{\gamma(p_R)}{\rho_R}}$$

$$\bar{c} = \sqrt{\frac{\gamma\bar{p}}{\bar{\rho}}},\ \bar{\rho} = \frac{1}{2}(\rho_L + \rho_R),\ \bar{p} = \bar{e}(\gamma - 1),\ \bar{e} = \frac{1}{2}(e_L + e_R).$$

3.3. TIME INTEGRATION

The solution can be advanced through one time step using any appropriate time-stepping scheme. We use the following two stage second order accurate Runge-Kutta variant attributed by Van Leer to Hancock[22].

$$U_o^{n+\frac{1}{2}} = U_o^n - \frac{\Delta t}{2V_o} \sum_{k=1}^{6} H\left(U_o^n + \mathbf{r}_{ok}.\mathbf{g}_o^n\right).\mathbf{S}_k \tag{7}$$

$$U_o^{n+1} = U_o^n - \frac{\Delta t}{V_o} \sum_{k=1}^{6} H\left(U_o^{n+\frac{1}{2}} + \mathbf{r}_{ok}.\mathbf{g}_o^n, U_p^{n+\frac{1}{2}} + \mathbf{r}_{pk}.\mathbf{g}_p^n\right).\mathbf{S}_k \tag{8}$$

where $\mathbf{g}_o^n$ is the limited slope, o denotes a generic cell in the mesh, p its neighbouring cell adjacent to cell side k and $\mathbf{r}_{ok}$ is the position vector from cell centre o to cell interface midpoint k. The notation $(.,.)$ in (8) indicates the solution of a Riemann problem $U^* = R(U_L, U_R)$ at the cell interface. No Riemann solutions are required in the first stage (7).

For multi-dimensional calculations, we use the operator split form of equations (7,8) with an operator sequence of the form

$$U_0^{n+1} = L_1(\frac{\Delta t}{2})L_2(\frac{\Delta t}{2})L_3(\Delta t)L_2(\frac{\Delta t}{2})L_1(\frac{\Delta t}{2}).U_0^n \tag{9}$$

where the split operators $L_1(\Delta t)$, $L_2(\Delta t)$ and $L_3(\Delta t)$ are defined similarly to that shown in (7,8) except that the summation is reduced to two terms corresponding to flux balances across opposite faces of the cell. The time step Δt is obtained from the usual CFL condition defined here in i, j, k notation for a structured grid

$$\Delta t = \nu \min(\Delta t_x, \Delta t_y, \Delta t_z) \tag{10}$$

where

$$\Delta t_x = \min_i \frac{V_{ijk}}{\left|\mathbf{v}_{ijk}.\mathbf{S}_{i+\frac{1}{2}jk}\right| + c_{ijk}\left|\mathbf{S}_{i+\frac{1}{2}jk}\right|} \tag{11}$$

$\mathbf{v}_{ijk}$ is the flow velocity, c_{ijk} is the local speed of sound and the Courant number ν was taken in our calculations to be unity. Δt_y and Δt_z are similarly defined.

3.4. VALIDATION

Steady and unsteady flow computations were carried out for the nozzle geometry investigated experimentally by Sajben et al.[16]. A detailed description of the geometry is given in reference[16]. The exit plane to reservoir pressure is denoted by p_r and the reservoir pressure by p_t. The calculated cases are as follows (1) steady flow for p_r values of $0.72, 0.82$ and

0.862 (only results for 0.82 are shown); (2) Unsteady flow in response to a single exit plane pressure pulse of amplitude $\Delta p/p_t = 0.1$ at $p_r = 0.82$ for comparison with the computed solutions of Hsieh and Wardlaw[12]. The reservoir condition was set at a total temperature $T_t = 292\,\text{K}$ and total pressure $p_t = 135\,\text{kPa}$. This corresponds to a Reynolds number of $\text{Re} = u_0 Y_0 \rho_0/\mu_0 = 822,400$ where u_0 is the velocity, Y_0 is the nozzle height, μ_0 is the dynamic viscosity and ρ_0 is the density taken at the inflow boundary at $x/H = -4.0$. Here, x is the axial distance along the nozzle and H is the throat height.

All calculations were carried out on a mesh having 80 points in the axial direction and 50 points in the vertical direction. The mesh was exponentially stretched in the vertical direction near the upper and lower walls in order to ensure the presence of at least two points in the laminar sub-layer. The axial mesh points were clustered near each end of the nozzle and in the throat region to improve the resolution of the terminal shock. The initial flow field was prescribed using a one dimensional steady duct solution. Near the upper and lower walls, the flow properties were modified using a one-seventh power law profile. The boundary conditions were implemented in a manner similar to that described by Hsieh and Wardlaw[12]. The numerical method used for solving the thin layer Navier-Stokes equations with a $k - \epsilon$ turbulence model was a convection diffusion split (CDS) version of the present algorithm described by Batten et al.[23].

3.5. STEADY FLOW

The Mach contours and wall pressure results shown in Figure 6 are in reasonable agreement with the experimental data of Sajben et al.[16]. The numerical solution is slightly oscillatory downstream of the terminal shock. However, the experimental data also showed a high sensitivity to small perturbations in the exit pressure near the terminal shock. As expected, the corresponding inviscid solutions predict a stronger terminal shock located further downstream. Comparisons at other p_r values were similar, indicating that the numerical method can adequately predict the nozzle flow field.

3.6. UNSTEADY FLOW

A single pressure pulse with a magnitude of 14% of the mean exit static pressure was applied to the nozzle exit at $p_r = 0.82$[12]. The pulse consisted of an amplitude $\Delta p/p_t = 0.1$ wave with a rise time of 0.2 ms, a duration of 0.4 ms and a recession time of 0.2 ms (see Figure 7). Calculations started with the imposition of the pressure pulse and terminated when an approximate steady state was recovered. Mach number contour plots are shown in Figure 8 corresponding to $p_r = 0.82$. Following the application of the

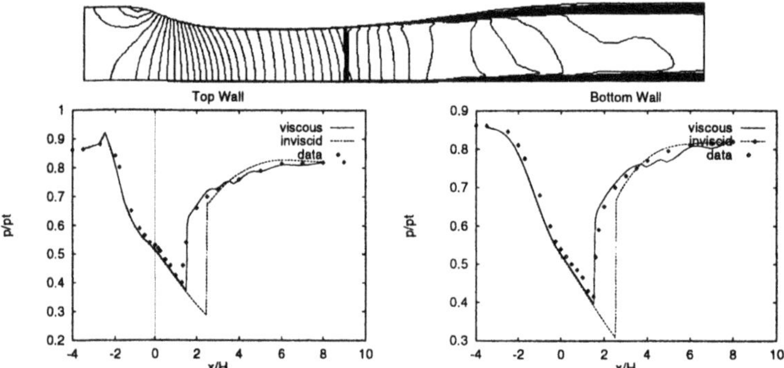

Figure 6. Nozzle flow, Iso-Mach lines (top), Pressure distribution on the top wall (left) and bottom wall (right), $p_r = 0.82$

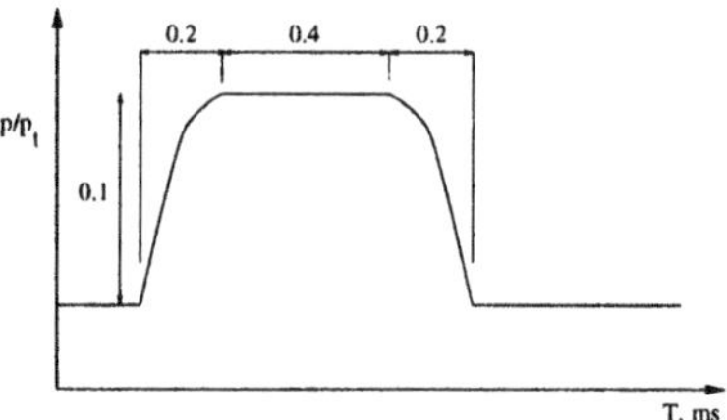

Figure 7. Prescribed Exit Pressure Disturbance

unsteady pressure pulse at the nozzle exit, a wave moves up the nozzle and interacts with the terminal shock. The top wall separation bubble lengthens and then shrinks in response to the passing wave. The terminal shock responds to the pressure wave by moving upstream and becoming weaker. Similar behaviour, though less marked, is seen in the bottom wall separation zone where the separation bubble is generally thinner. The results are in qualitative agreement with those of Hsieh and Wardlaw[12]. Unfortunately, no experimental data is available for the verification of either set of results. However, these calculations were useful for establishing the validity of the approach to be used for modelling the hammer shock in the intake calculations.

3.7. EULER CALCULATIONS

Although a viscous calculation for the unsteady intake flow would have been desirable, it was felt that an inviscid solution would provide a description of the essential features of the core flow. In each case a steady flow solution was obtained first by marching forward in time until a converged solution was obtained. Two cases were considered: static ground running of the

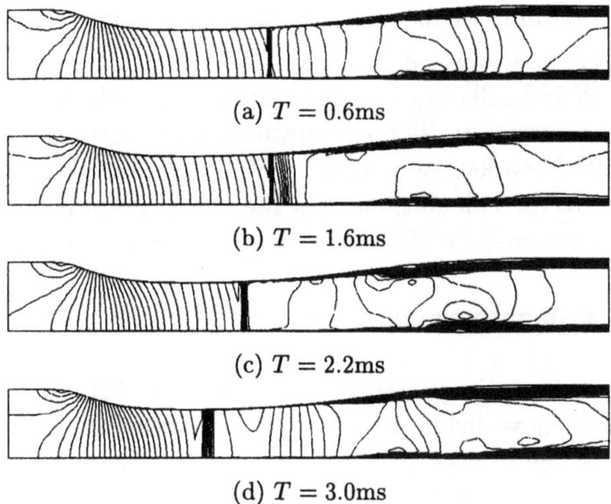

(a) $T = 0.6$ms

(b) $T = 1.6$ms

(c) $T = 2.2$ms

(d) $T = 3.0$ms

Figure 8. Flow Separation Characteristics of Nozzle Flow in Response to Exit Pressure Pulse

engines and Mach 0.6, to facilitate comparisons with available test data. Only results for the static case are presented here.

The boundary conditions imposed in the steady flow calculations in each case were as follows: a constant static pressure boundary condition, with extrapolation of density and momentum, was applied at free-air boundaries and a surface tangency condition $\mathbf{v}.\mathbf{n} = 0$, applied at solid boundaries. The free-air boundaries correspond to the top and sides of the computational domain upstream of the intake entry plane. The solid boundaries correspond to the surfaces of the intakes, and the lower plane upstream of the intake entry plane which represents the surface of the airframe. At each intake exit plane, a constant static pressure boundary condition was applied, corresponding to an assumed pressure recovery factor.

The transient phase of the calculation commenced when a single pop surge pressure pulse was imposed at the exit plane of the primary intake. The pulse corresponded to an OPR of 3.0 with a rise time of 2.5 ms and a recession phase of 30 ms (Figure 1). The solution process then became fully time-accurate with a new ordinate value being taken from the pressure signature at each successive time step. A surge in any one engine type will generally fall within a fairly narrow band of OPR's, so representing the surge by a single surge signature and OPR should be an acceptable assumption. The time step used in the transient phase of the calculations was calculated from the Courant condition (10). The time step was in the order of 10^{-5} s which is approximately two orders of magnitude smaller than the specified rise time, so it is unlikely that any significant smoothing of maxima and minima in the pressure signature occurred.

All of the calculations were carried out on a mesh having a total of 92,004 cells with 102 cells in the x direction, 22 cells in the y direction and 41 cells in the z direction. The calculations were carried out on a Silicon Graphics INDIGO workstation with 16 M bytes of dynamic memory. The transient phase of the calculations was the most compute-intensive in each case, taking approximately 25 hours of CPU time per run.

3.8. RESULTS FOR THE STATIC CASE AT OPR=3.0

The results for this static ground-running case are displayed in Figures 9 – 13. These figures show distributions of density, pressure and Mach number along the centre-line of each duct at successive times after the surge pressure signature has been applied. The origin of coordinates corresponds to the trailing edge of the compression ramp. In the density and Mach number graphs, the curves correspond to the primary and secondary intakes as indicated. The Mach number is based on the stream wise velocity component only: so the Mach number is negative at points at which u is negative. The four lines shown on each pressure plot correspond to static/total pressure in the primary/secondary intakes. In each case, the higher set of values are total pressure. The elapsed time is shown in each figure. Time zero corresponds to the commencement of the transient phase of the calculation.

At approximately 7.3 milliseconds, Figure 9 shows a shock wave propagating upstream from the exit plane of the primary intake (moving from right to left). The hammer shock is about one-third of the way up the intake from its exit plane. It is quite well resolved numerically and can be seen as a sharp rise in static pressure and density. Behind the shock wave, the gas is moving upstream with a velocity of approximately one-third of the local sound speed. Similar plots after approximately 14 and 21 milliseconds, shown in Figures 10, 11, correspond to points just before and just after the shock wave has reached the bifurcation point, respectively. The density at the exit plane of the primary intake can be seen to be falling in response to the fluctuations in the applied pressure signature. After 21 milliseconds, the gas is rushing out of the primary intake at around 300 m/s. It is interesting to observe that the shock wave emerging from the primary intake weakens in strength and a fairly weak rarefaction wave begins to move down the adjacent secondary intake. The progression of the rarefaction wave can be followed in Figures 12, 13 corresponding to times of approximately 28 and 35 milliseconds respectively. At the latter time frame, the rarefaction wave has reached the exit plane of the secondary intake. The static pressure attenuation which occurs after the hammer shock pulse emerges from the primary intake appears to confirm the shielding effect of the inter-intake splitter plate. Overall, there is no evidence in the CFD predictions to suggest that a complementary surge is likely to be induced in the other engine. This has been confirmed by comparisons with available test and flight data[24].

4. Cartesian Cut Cell Methods for Moving Body Problems

In order to deal with arbitrarily complex geometries, which in general may be stationary or moving relative to one and other, we use a Cartesian cut cell approach. Solid bodies are simply cut out of a background Cartesian mesh. Figures 14 and 15 show all possibilities for the four basic sub-types and special sub-types of cut cell. Essentially, a Cartesian cut cell mesh in two-dimensions can be generated as follows

1. *Construction of a background Cartesian mesh.*
2. *Finding the intersection points between the Cartesian mesh lines and the boundaries of solid bodies.* The cells partially cut by the boundaries are registered as cut cells and the sub-type, volume and other geometric information are determined.
3. *Locating solid cells.* Sweeps across the background mesh are then performed to identify which cells or rows of cells are bounded by solid or partially cut cells; these are registered as solid cells.

In practice a cut cell can become arbitrarily small and the time step may become excessively small so to avoid this a cell merging technique[25, 26] is implemented. A minimum acceptable cell volume V_{min} is specified and if the volume of a cut cell is smaller than V_{min}, a suitable neighbouring cell is found to merge with the cut cell.

The choice for V_{min} is based on a trade-off between the time step and resolution accuracy. In our calculations, V_{min} is taken to be one half the flow cell size.

4.1. NUMERICAL DISCRETISATION

The MUSCL-Hancock finite volume scheme described earlier is implemented here with apropriate modifications. The predictor step (8) for a moving grid/body with arbitrary cut cells is

$$(V\mathbf{U})_{ij}^{n+\frac{1}{2}} = (V\mathbf{U})_{ij}^{n} - \frac{\Delta t}{2} \sum_{k=1}^{m} \mathbf{H}(\mathbf{U}_k) \cdot \mathbf{S}_k^n \quad (12)$$

where V is cell volume, $\mathbf{S}$ is cell face area vector and m is the maximum number of cell faces. For a flow (or uncut) cell, $m = 4$; for a cut cell, $m = 3$ to 7.

The flux vector $\mathbf{H}(\mathbf{U}_k)$ is evaluated at the midpoints of cell faces following a linear reconstruction of the flow solution within each cell, via,

$$\mathbf{U}_k = \mathbf{U}_{ij}^n + \mathbf{r}_{ok} \cdot \mathbf{g}_{ij}^n \tag{13}$$

where $\mathbf{r}_{ok}$ is the normal distance vector from the cell centroid to face k and $\mathbf{g}_{ij}^n$ is a gradient vector in space.

The corrector step of the scheme is

$$(V\mathbf{U})_{ij}^{n+1} = (V\mathbf{U})_{ij}^n - \Delta t \sum_{k=1}^{m} \mathbf{H}\left(\mathbf{U}_k^L, \mathbf{U}_k^R\right) \cdot \mathbf{S}_k^{n+\frac{1}{2}} \tag{14}$$

where the upwind flux $\mathbf{H}\left(\mathbf{U}_k^L, \mathbf{U}_k^R\right)$ is obtained by solving a local Riemann problem normal to the cell interface using an improved version of the HLL Riemann solver proposed by Toro[27]. For a static or moving solid boundary (or face) of a cut cell, a different approach is used, based on an exact Riemann solution for a moving piston.

4.1.1. *Calculation of Gradients and Reconstruction Technique*
For cells near solid boundaries, a modified gradient calculation is needed. Reflection boundary conditions are used to define the variables in a fictional cell R as follows:

$$\begin{cases} \rho_R = \rho_{ij} \\ \mathbf{v}_R = \mathbf{v}_{ij} - 2\left(\mathbf{v}_{ij} \cdot \mathbf{n}\right)\mathbf{n} + 2\left(\mathbf{v}_s \cdot \mathbf{n}\right)\mathbf{n} \\ p_R = p_{ij} \\ e_R = p_R/(\gamma - 1) + \frac{1}{2}\rho_R |\mathbf{v}_R|^2 \end{cases} \tag{15}$$

We calculate the "fluid" gradients and "solid" gradients separately, i.e.,

$$\begin{aligned} \mathbf{U}_x^f &= G\left(\frac{\mathbf{U}_{i+1,j} - \mathbf{U}_{i,j}}{DX_{i+\frac{1}{2},j}}, \frac{\mathbf{U}_{i,j} - \mathbf{U}_{i-1,j}}{DX_{i-\frac{1}{2},j}}\right), \\ \mathbf{U}_y^f &= G\left(\frac{\mathbf{U}_{i,j+1} - \mathbf{U}_{i,j}}{DY_{i,j+\frac{1}{2}}}, \frac{\mathbf{U}_{i,j} - \mathbf{U}_{i,j-1}}{DY_{i,j-\frac{1}{2}}}\right) \end{aligned} \tag{16}$$

and

$$\begin{aligned} \mathbf{U}_x^s &= G\left(\frac{\mathbf{U}_R - \mathbf{U}_{i,j}}{DX_{i,R}}, \frac{\mathbf{U}_{i,j} - \mathbf{U}_{i-1,j}}{DX_{i-\frac{1}{2},j}}\right), \\ \mathbf{U}_y^s &= G\left(\frac{\mathbf{U}_{i,j+1} - \mathbf{U}_{i,j}}{DY_{i,j+\frac{1}{2}}}, \frac{\mathbf{U}_{i,j} - \mathbf{U}_R}{DY_{j,R}}\right) \end{aligned} \tag{17}$$

where G is the van Leer limiter:

$$G(a,b) = \frac{|b| + |a|b/a}{|a| + |b|} \tag{18}$$

and e.g. $DX_{i+\frac{1}{2},j} = X_{i+1,j} - X_{i,j}$.

A weighted average technique based on cell side lengths is then used to obtain unique gradients in the cut cell,

$$\mathbf{U}_x = \frac{\Delta Y_s \mathbf{U}_x^s + \Delta Y_f \mathbf{U}_x^f}{\Delta Y}, \ \mathbf{U}_y = \frac{\Delta X_s \mathbf{U}_y^s + \Delta X_f \mathbf{U}_y^f}{\Delta X} \tag{19}$$

where $\Delta X_f = |AB|, \Delta X_s = |BC|, \Delta Y_s = |CD|$ and $\Delta Y_f = |DE|$. ΔX and ΔY are the flow/uncut cell side lengths in the x and y directions, respectively.

Since $\Delta X_f + \Delta X_s = \Delta X$ and $\Delta Y_f + \Delta Y_s = \Delta Y$, we note that if $\Delta Y_f = \Delta Y$, $\Delta Y_s = 0$, so $\mathbf{U}_x = \mathbf{U}_x^f$; otherwise, if $\Delta Y_s = \Delta Y$, $\Delta Y_f = 0$, so $\mathbf{U}_x = \mathbf{U}_x^s$. $\mathbf{U}_x$ and $\mathbf{U}_y$ are components of a gradient vector in the cut cell, that is

$$\mathbf{g}_{ij} = \begin{pmatrix} \mathbf{U}_x \\ \mathbf{U}_y \end{pmatrix}. \tag{20}$$

Given the gradient vector $\mathbf{g}_{ij}$, a reconstructed solution vector $\mathbf{U}(x,y)$ can be found anywhere within the cut cell from

$$\mathbf{U}(x,y) = \mathbf{U}_{ij} + \mathbf{r} \cdot \mathbf{g}_{ij} \tag{21}$$

where $\mathbf{r}$ is the normal distance vector from the centroid to any specific interface or solid boundary.

Once the body moves, the cut cell information changes. In essence, these changes can be divided into four categories:

1. *Cut cell becomes solid cell.*
2. *Cut cell becomes an uncut flow cell.*
3. *Cut cell remains unchanged.*
4. *Uncut flow cell becomes a cut cell.*

Categories three and four do not cause any problems. However, where a cut cell becomes solid (category 1), the volume of the cell at the end of time step is zero; obviously this will lead to problems within the flow solver. For category two, failure to consider the new-born cell may result in strict conservation being lost. In general, these problems can be solved by using a cell merging technique [26, 28]. The basic idea is to merge a small cut cell with one or several neighbouring cells so that any interface between the merged cells is ignored and the waves are allowed to travel in the larger merged cell without reducing the global value of Δt.

For example, a time step, Δt, based on flow cell B will be too large for cut cell A which will also become solid after Δt (see Figure 16). To merge the two cells, we first compute the updates at cells A and B as usual,

$$\Delta(V\mathbf{U})_A = -\Delta t \sum_{k=1}^{m_A} \mathbf{H}_k \cdot \mathbf{S}_k,\ \Delta(V\mathbf{U})_B = -\Delta t \sum_{k=1}^{m_B} \mathbf{H}_k \cdot \mathbf{S}_k \tag{22}$$

Then, we ignore the interface between cell A and B, and update the merged cell C simply by combining the volume updates of cells A and B,

$$\Delta(V\mathbf{U})_C = \Delta(V\mathbf{U})_A + \Delta(V\mathbf{U})_B \tag{23}$$

The fluxes on interface $|cd|$ between cell A and B cancel out automatically since the flux calculation is conservative. The conserved variable $\mathbf{U}$ for cell C at time t^{n+1} is

$$V_C^{n+1}\mathbf{U}_C^{n+1} = V_A^n\mathbf{U}_A^n + V_B^n\mathbf{U}_B^n - \Delta t\left(\sum_{k=1}^{m_A} \mathbf{H}_k \cdot \mathbf{S}_k + \sum_{k=1}^{m_B} \mathbf{H}_k \cdot \mathbf{S}_k\right) \tag{24}$$

Although cut cell A vanishes, the mass, momentum and energy balances, will be transferred into neighbouring cells so that strict conservation is maintained automatically.

In order to prevent a single cut cell becoming solid without merging with a neighbouring cell, a suitable estimate for the time step is used

$$\Delta t_x = \frac{\sqrt{V_{min}}}{\max(|u_s|, |u|) + a},\ \Delta t_y = \frac{\sqrt{V_{min}}}{\max(|v_s|, |v|) + a} \tag{25}$$

$$\Delta t = \nu \min(\Delta t_x, \Delta t_y)\ (\nu \leq 1.0) \tag{26}$$

where a is the local sound speed, ν is the Courant number and u_s and v_s are the x- and y-components of the moving boundary velocity $\mathbf{v}_s$.

4.2. NUMERICAL RESULTS

4.2.1. *Static Body Problems*

The first example is double Mach reflection on an inclined ramp. The initial conditions are: an incident shock with $M_s = 8.7$ and $\gamma = 1.4$, moves right to left and interacts with a 27° inclined ramp. In this computation, a uniform Cartesian mesh of 220 × 140 cells was used on a computational domain of 11.0×7 units. Figure 17 shows the Cartesian mesh and density contours which compare favourably with relevant experimental results given by Deschambault and Glass [29]. All the flow features are well resolved.

The second example is a planar shock wave interaction with a NACA 0018 aerofoil at an angle of attack of 30° which has been studied experimentally by Mandella & Bershader[30] and numerically by Yee[31]. A plane shock at Mach 2 travels toward a NACA 0018 aerofoil located downstream at an angle of attack of 30°. A Cartesian mesh of 360×400 cells was used on a computational domain of 1.8×2.0 units. For comparison purposes, a sequence of frames depicting the diffraction process were obtained at approximately the same time frames as the ones in [31]. Figures 18 shows four sequential interferogrammes compared with computed density plots. The incident and reflected shocks, Mach stems and contact surfaces on both the lower and upper surfaces are extremely well resolved. The vortices at the leading edge and trailing edge of the aerofoil are also well-captured.

4.2.2. *Moving Body Problems*

A Mach 1.76 shock wave emerging from the open ended of a shock tube, has been studied extensively, both experimentally and numerically[32, 33, 26]. In the numerical studies, a Mach 1.76 shock wave was specified at the exit of the shock tube. In our simulation a moving piston was used to create a shock of the same strength moving down the tube. The piston was placed initially a distance $5.5D$ (where D is the diameter of the shock tube) upstream of the tube exit. The gas in front of the piston was assumed to be quiescent, and a post-shock state equivalent to Mach 1.76 was set behind the piston. The piston was set in motion impulsively at the speed of the post-shock gas. A Cartesian mesh of 450×150 cells was used on a computational domain of $9D \times 3D$. A sequence of shock diffraction plots were obtained at approximately the same times as the corresponding experimental ones in [33]. Figure 19 shows computed density flowfields at two time stages which can be compared with the experimental interogrammes (not shown). Figure 20 shows a comparison of calculated and measured overpressure variation with time on the centreline, 1.5 diameters downstream of the exit.

The second example is a 15° wedge flow at Mach 2. The geometry is a channel 1 unit high and 7.5 units long with a 15° wedge on its lower wall. A fixed wedge placed in a Mach 2 flow was calculated initially. The wedge was located at $x = 7.25$ units (origin at bottom left corner, see Figure 21) with height 0.2588 units. The attached shock is reflected at the upper wall of the channel and weakened by the expansion fan originating from the expansion corner. Downstream, multiple shock reflections occur at the upper and lower walls of the channel. A Cartesian mesh having 300×40 cells is used. Density contours are shown in Figure 21, where the results at three different times $t = 1, 2$ and 3 units are presented.

We now assume that the wedge suddenly moves at the same speed into a quiescent gas, and hence expect an identical flow field. The 15° wedge is initially located at $x = 1.25$ units. An attached oblique shock is gradually produced and reflected at the upper and lower walls. At time $t = 3$, the wedge is in exactly the same position as the fixed wedge shown in Figure 21. Figure 22 shows density contours at time frames $t = 1, 2$ and 3 units respectively. Comparing the two cases, we can see there is little difference between the wedge moving at $M_w = 2$ into quiescent gas and the fixed wedge placed in a free stream at $M_\infty = 2$.

Finally, a store separation problem is considered. A store is released from a cavity into a freestream flow at Mach 1.5. Inside the cavity, the flow is initially assumed stationary with the same pressure as the freestream. The trajectory of the projectile is prescribed. A uniform Cartesian mesh with 140×60 cells was used on a domain of 2.8×1.2 units. The translational and rotational velocities of the projectile centre of mass were $\mathbf{v}_c = [0.0, \ -0.03]^T$ units and $\omega_c = 0.001$ units respectively. Figure 23 shows the mesh and computed density contours at two different times.

5. Conclusions

A time-accurate Euler CFD code has been applied to study the unsteady flow arising in a twin side-by-side intake system as a result of a surge in one engine. The flow disturbance was modelled by applying a pressure pulse, representative of a pop surge, to the exit plane of one of the intakes. The calculations have shown that the level of dynamic flow distortion at the exit plane of the adjacent intake was not increased significantly and that a complementary surge in the adjacent engine would be unlikely to occur. The predictions compare favourably with available test data[24] and demonstrate the validity and value of modern high resolution shock capturing methods for modelling complex engine surge phenomena.

A Cartesian cut cell method for the computation of unsteady compressible flows involving both static and arbitrarily moving bodies has also been presented. A stationary background Cartesian mesh was used and bodies were allowed to move arbitrarily across the mesh-lines. For static body cases, the present method can be viewed as a viable alternative to unstructured mesh methods for simulating complex compressible flows around arbitrarily complicated, multi-element geometries. For moving body cases, the approach described deals with moving boundary problems without the requirment for a moving mesh, hence problems such as mesh distortion and body motion restrictions are avoided completely.

Since a Cartesian cut cell approach does not involve the generation of a field mesh in the usual sense, extension of the two dimensional Cartesian cut cell method to three dimensions can be accomplished in a straightforward manner. An implementation of the method in 3D is currently underway.

The extension of the inviscid method to low Reynolds number viscous flows is straightforward if mesh refinement techniques are implemented near solid boundaries. However, for high Reynolds number flows, further work is needed due to the lack of a preferred direction at solid boundaries.

References

1. JW Edwards and JL Thomas. Computational Methods for Unsteady Transonic Flows. Paper 86-0107, AIAA, 1986.
2. K Lotter, PA Mackrodt, and R Scherbaum. Engine surge simulation in a wind-tunnel model inlet ducts. In *16th Congress of the International Council of the Aeronautical Sciences (ICAS)*, 1988. ICAS Paper No 88-4.11.4.
3. FI Marshall. Prediction of inlet overpressure resulting from engine surge. *Journal of Aircraft*, 10(5):274–278, May 1973.
4. PJ Evans and PO Truax. YF-16 air induction system design loads associated with engine surge. *Journal of Aircraft*, 12(4):205–209, April 1975.
5. AP Kurkov, RH Soeder, and JE Moss. Investigation of the stall hammer shock at the engine inlet. *Journal of Aircraft*, 12(4):198–204, April 1975.
6. R Lohner. Adaptive Remeshing for Transient Problems. *Comp. Methods in Appl. Mech. and Eng.*, 75:195–214, 1989.
7. EJ Prodert, O Hassan, K Morgan, and J Peraire. An Adaptive Finite Element Method for Transient Compressible Flows with Moving Boundaries. *International Journal Numerical Methods Eng.*, 32(4):751–765, 1991.
8. JY Trepanier, M Reggio, M Paraschivoiu, and R Camarero. Unsteady Euler Solutions for Arbitrarily Moving Bodies and Boundaries. *AIAA J.*, 31(10):1869–1876, 1993.
9. JL Steger, FC Dougherty, and JA Benek. A Chimera Grid Scheme. In *Advances in Grid Generation, ASME FED-5*, pages 59–69, 1983.
10. TA Blaylock and SH Onslow. Application of the FAME Method to Store Release Prediction. In *Computational Fluid Dynamics '94, Stuttgart, Germany*, 1994.
11. T Hsieh, AB Wardlaw, P Collins, and T Coakley. Numerical investigation of unsteady inlet flowfields. *JAIAA*, 25(1):75–81, 1987.
12. T Hsieh and AB Wardlaw. Numerical simulation of unsteady flow in a ramjet inlet. In *14th Congress of the International Council of the Aeronautical Sciences (ICAS)*, pages 1005–1014. ICAS, 1984. ICAS Paper No 84-1.9.2.
13. TJ Bogar, M Sajben, and JC Kroutil. Characteristic frequencies of transonic diffuser flow. *JAIAA*, 21(9):1232–1240, 1983.
14. MS Liou and TJ Coakley. Numerical simulation of unsteady transonic flow in diffusers. Paper 82-1000, AIAA, 1982.
15. MS Liou, TJ Coakley, and MY Bergmann. Numerical simulation of transonic flow in diffusers. Paper 81-1240, AIAA, 1981.
16. M Sajben, TJ Bogar, and JC Kroutil. Forced oscillation experiments in supercritical diffuser flows with applications to ramjet instabilities. Paper 81-1487, AIAA, 1981.
17. JT Salman, TJ Bogar, and M Sajben. Laser velocimeter measurements in unsteady separated transonic diffuser flows. Paper 81-1197, AIAA, 1981.
18. JF Thompson, ZAU Warsi, and C Wayne-Mastin. *Numerical Grid Generation: Foundation and Applications.* Elsevier Science Publishing, 1985.
19. B van Leer. Towards the ultimate conservative difference scheme ii: Monotonicity and conservation combined in a second order scheme. *Journal of Computational*

Physics, 14:361–370, 1974.
20. A Harten, PD Lax, and B van Leer. On Upstream Differencing and Godunov-Type Schemes for Hyperbolic Conservation Laws. *SIAM Review*, 25(1):35–61, 1983.
21. B Einfeldt. On Godunov-Type Methods for the Euler Equations with General Equations of State. In *2nd International Conf. on Hyperbolic Problems, Aachen*, 1988.
22. B van Leer. On the Relation Between the Upwind-Differencing Schemes of Godunov, Engquist-Osher and Roe. *SIAM Journal on Scientific and Statistical Cmputing*, 5, 1984.
23. P Batten, DM Ingram, R Saunders, and DM Causon. An implicit viscous solver for the compressible navier-stokes equations. *Submitted to Computers & Fluids*, February 1993.
24. British Aerospace plc. Private communication.
25. DK Clarke, MD Salas, and HA Hassan. Euler Calculations for Multielement Airfoils Using Cartesian Grids. *AIAA Journal Vol.24, No.3*, 1986.
26. Y Chiang, B van Leer, and KG Powell. Simulation of Unsteady Inviscid Flow on an Adaptively Refined Cartesian Grid. In *30th Aerospace Sciences Meeting and Exhibit, Reno, Nevada*, 1992.
27. EF Toro, M Spruce, and W Speares. Restoration of the Contact Surface in the HLL Riemann Solver. *Shock Waves, Vol 4*, pages 25–34, 1994.
28. JJ Quirk. An Alternative to Unstructured Grids for Computing Gas Dynamic Flows Around Arbitrarily Complex Two-Dimensional Bodies. *Computers & Fluids, Vol 23, No. 1*, 23(1):125–142, 1994.
29. R. Deschambault and I. Glass. An Update on Non-Stationary Oblique Shock Wave Reflection: Actual Isopycnics and Numerical Experiments. *Journal of Fluid Mechanics*, 131:27-57, 1983.
30. M. Mandella and D. Bershader. Quantitative Study of Compressible Vortices: Generation, Structure and Interaction with Airfoil. *AIAA Paper no.87-0328*, 1987.
31. H. C. Yee. A Class of High- Resolution Explicit and Implicit Shock-Capturing Methods. *NASA TM 101088*, February, 1989.
32. EM Schmidt and S Duffy. Noise from shock tube facilities. *AIAA paper 85-0049*, 1985.
33. JCT Wang and Widhopf. Numerical simulation of blast flow fields using a high resolution tvd finite volume scheme. *Computers and Fluids*, 18(1), 1990.

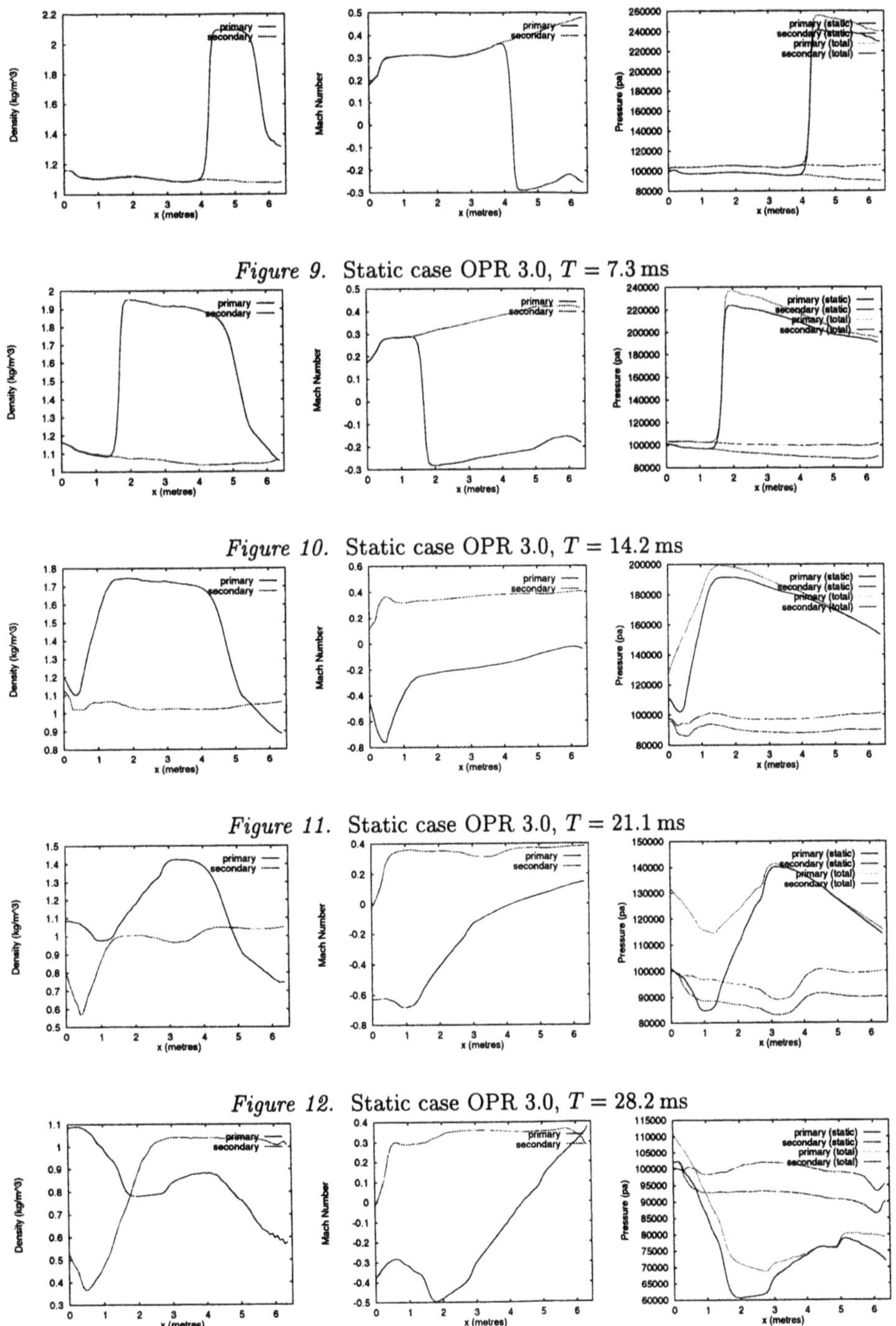

Figure 9. Static case OPR 3.0, $T = 7.3$ ms

Figure 10. Static case OPR 3.0, $T = 14.2$ ms

Figure 11. Static case OPR 3.0, $T = 21.1$ ms

Figure 12. Static case OPR 3.0, $T = 28.2$ ms

Figure 13. Static case OPR 3.0, $T = 35.8$ ms

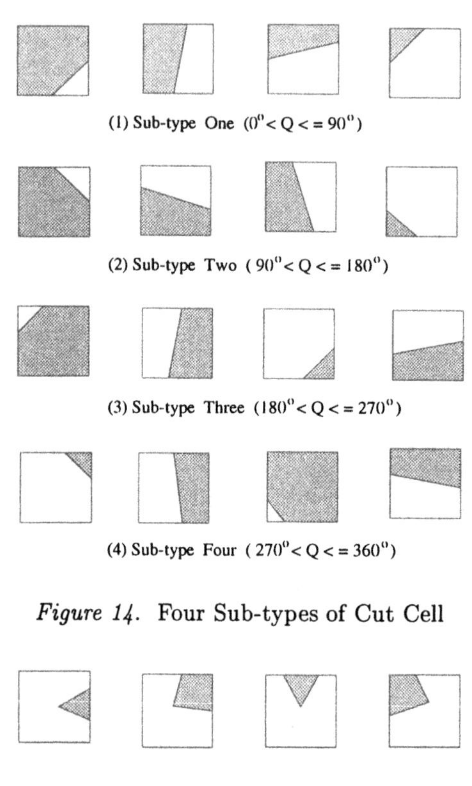

Figure 14. Four Sub-types of Cut Cell

Figure 15. Special Sub-type of Cut Cell

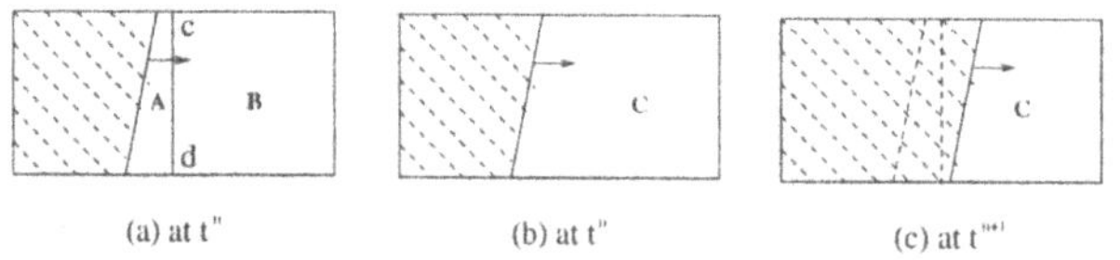

Figure 16. Cell Merging technique for a Moving Boundary

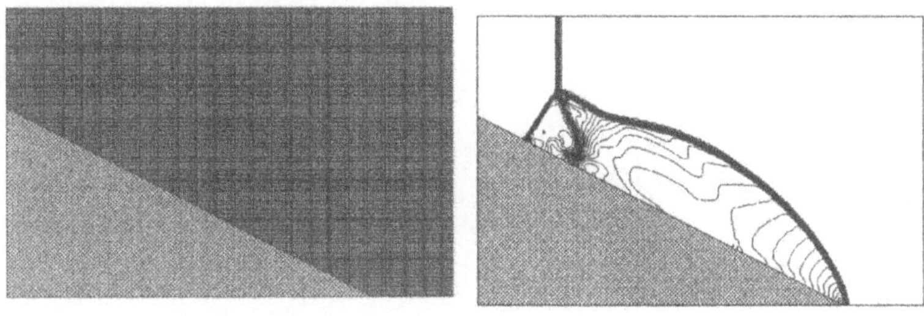

Figure 17. Cartesian Mesh and Density Contours for Double Reflection

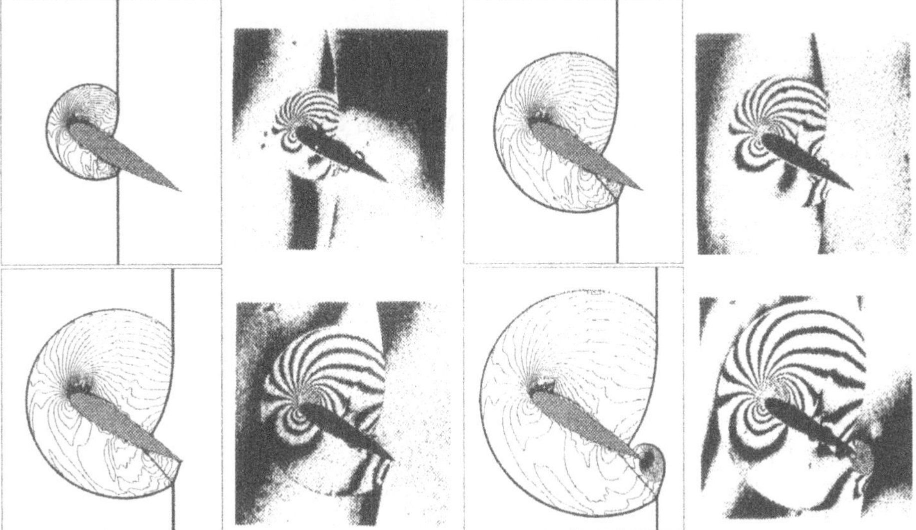

Figure 18. Shock Interaction with an Aerofoil, Computed Density Contours Compared with interferograms

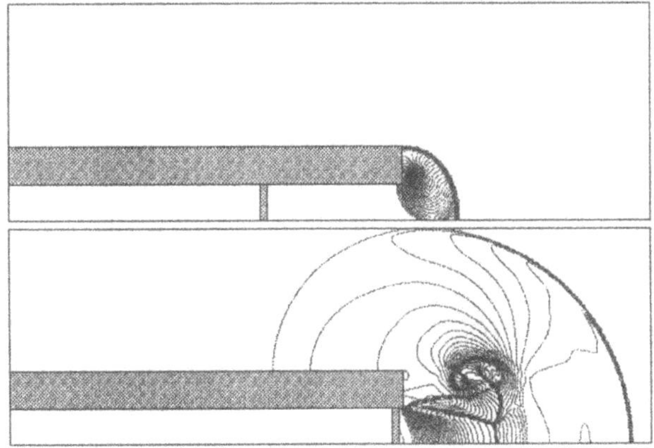

Figure 19. Computed Shock Wave Diffractions at $t = 0.2$ and $1.0ms$ after t_0

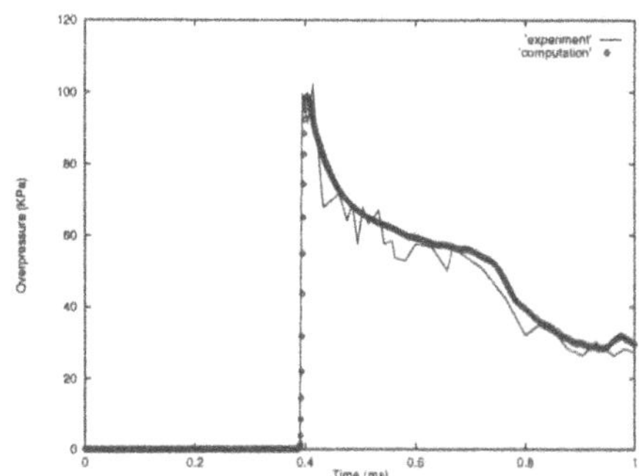

Figure 20. Comparison of Computed and Measured Overpressure on the centreline 1.5 diameters downstream of exit after t_0

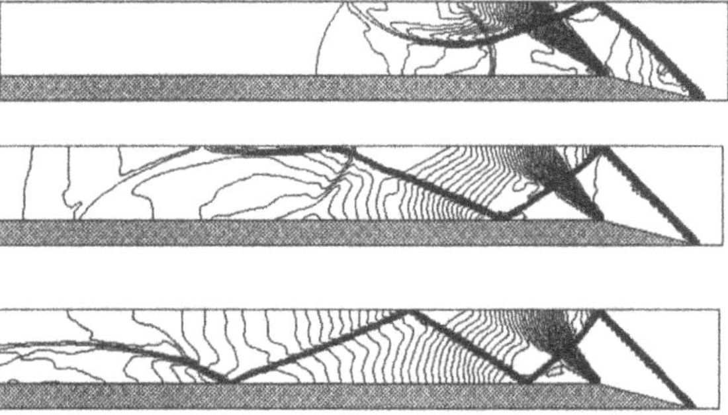

Figure 21. Density Contours at $t = 1$, 2 and 3 for Fixed Wedge

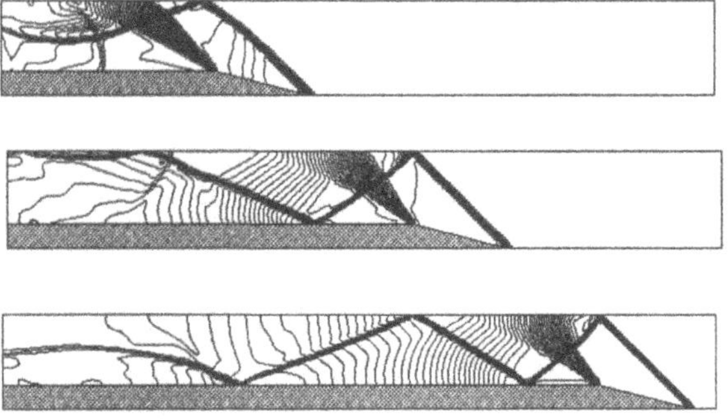

Figure 22. Density Contours at $t = 1$, 2 and 3 for Moving Wedge

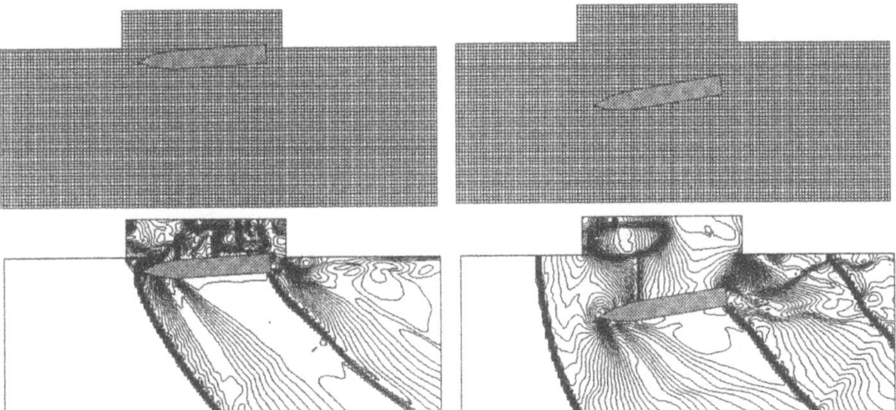

Figure 23. Mesh and Density Contours for Store Separation at two different times

WAVE PROPAGATION PHENOMENA IN THE THEORY OF SEDIMENTATION

Mathematical theory of gravitational solid-liquid separation processes

F. CONCHA
University of Concepción, Dept. of Metallurgical Engineering
Casilla 53-C, Correo 3, Concepción, Chile

AND

R. BÜRGER
University of Stuttgart, Institute of Mathematics A
Pfaffenwaldring 57, 70569 Stuttgart, Germany

1. Introduction

Sedimentation processes are employed in a variety of industrial applications in which a suspension, a mixture of a fluid and fine solid particles, is separated into its solid and liquid components under the influence of gravity. In the mining industry, large vessels, so-called *thickeners* of up to 100 m in diameter and 6 m in depth with a slightly conical bottom, are used for the settling of the suspension and for the consolidation of the sediment. In a *continuous* thickener, the control operations (feeding of fresh suspension, discharge of the concentrated sediment at the bottom and overflow of the supernatant clear liquid) take place continuously. Fig. 1 shows a schematic cross-section of a continuous thickener [14].

Although the research on *thickening* (by which the combined effects of sedimentation and compression are denoted) dates back to the early works by Mishler in 1912 and 1918, [49, 50] and by Coe and Clevenger in 1916, [23], who were the first to observe the existence of the three distinct zones indicated in Fig. 1, the operation, design and control of thickeners are still based on empirical evidence. Hence, appropriate mathematical models to change this state of matters are of high theoretical, practical and, eventually, economical importance, as they can help to increase thickener efficiency and thus to cut production costs. The formulation of such models and the determination of appropriate constitutive equations is a topic of current research (see, e.g., [11, 25, 29, 34, 37, 45, 58, 59]). In particular, the authors'

E.F. Toro and J.F. Clarke (eds.), Numerical Methods for Wave Propagation, 173–196.

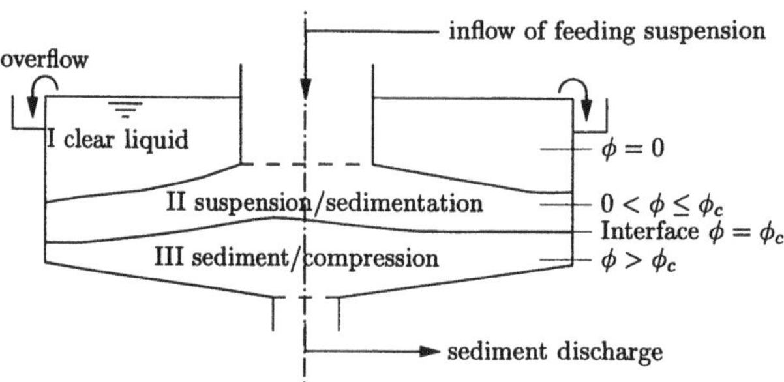

Figure 1. Schematic cross section of a continuous thickener. The suspension is assumed to be entirely *flocculated.* Then, the solid flocs perform *hindered settling* in regions where the solid volumetric concentration ϕ is less or equal than a critical value ϕ_c at which the particles begin to touch each other. For $\phi > \phi_c$, they form a *compressible sediment layer.*

interest in this research area arises from the application of sedimentation in Chilean copper mines, which are mostly located in arid desert areas where water is a scarce resource.

In this contribution, mathematical models for sedimentation are studied which yield the volumetric solid concentration ϕ as a function of time and space under a given initial state and feeding and discharge, i.e., initial and boundary conditions. These models emerge from a phenomenological theory of sedimentation [4, 26, 29]. The modelling equations of this theory are sketched in Section 2 in general form. The restriction of this theory to one space dimension corresponds to the concept of an *Ideal Thickener* (IT). Assuming further an *ideal* suspension, i.e., a non-flocculated suspension, which satisfies certain assumptions (see [16]), the scalar hyperbolic conservation law of Kynch's sedimentation theory [44] is recovered. The resulting model permits the prediction of continuous sedimentation processes by determining the solution of a Riemann problem with entropy boundary conditions. Section 3 reviews the results of the analysis of this model and its predictions of sedimentation behaviour. For a flocculated suspension, an additional diffusion term for the compression zone appears in the balance equation. The corresponding results of the mathematical analysis and some numerical examples are presented in Section 4. Even in one space dimension, the study of sedimentation gives rise to a variety of open problems in mathematical and numerical analysis. These questions and the extension of the existing results to several space dimensions are addressed in Section 5.

2. Phenomenological Model of Sedimentation

On the scale of industrial sedimentation vessels, a flocculated suspension can be described as a superposition of two continuous media with a solid and a liquid component. The modelling equations are obtained from the macroscopic mass balances for each component and the macroscopic linear momentum equation of the solid and of the mixture.

2.1. MODELLING EQUATIONS

Let ϕ denote the volumetric solid concentration and $\mathbf{v}_s$ and $\mathbf{v}_f$ the solid and fluid velocity, respectively. Then, in regions where the variables are continuous, the local mass balances for both components are

$$\frac{\partial \phi}{\partial t} + \nabla \cdot (\phi \mathbf{v}_s) = 0, \tag{1}$$

$$\frac{\partial (1-\phi)}{\partial t} + \nabla \cdot ((1-\phi)\mathbf{v}_f) = 0. \tag{2}$$

Introducing the volume-average velocity $\mathbf{q} := \phi \mathbf{v}_s + (1-\phi)\mathbf{v}_f$, equation (2) can be replaced by

$$\nabla \cdot \mathbf{q} = 0, \tag{3}$$

which is the continuity equation of the mixture. The linear momentum balance of the suspension is obtained by summing the corresponding equations of each component as given by Ungarish [59]:

$$\underbrace{\varrho(\phi)\left[\frac{\partial \mathbf{v}}{\partial t} + \mathbf{v} \cdot \nabla \mathbf{v}\right]}_{\{1\}} = -\nabla p - \varrho(\phi) g \mathbf{k} + \nabla \cdot \mathbf{T}^E \underbrace{-\nabla \cdot \left(\phi(1-\phi)\frac{\varrho_f \varrho_s}{\varrho(\phi)} \mathbf{v}_r \mathbf{v}_r\right)}_{\{2\}}. \tag{4}$$

Here, $\varrho(\phi) := \phi \varrho_s + (1-\phi)\varrho_f$ is the mass density of the mixture, where ϱ_s and ϱ_f are the solid and fluid mass densities, $\mathbf{v}$ is the mass-average velocity $\mathbf{v} := (\phi \varrho_s \mathbf{v}_s + (1-\phi)\varrho_f \mathbf{v}_f)/\varrho(\phi)$, p and $\mathbf{T}^E$ are the pressure and the extra stress tensor of the mixture, g is the acceleration of gravity, $\mathbf{k}$ is the upwards pointing unit vector and $\mathbf{v}_r := \mathbf{v}_s - \mathbf{v}_f$ is the relative solid-fluid velocity. By a dimensional analysis [14], terms $\{1\}$ and $\{2\}$ can be dropped, if we assume that the settling velocity of a single floc in an unbounded medium, the particle diameter and the kinematic viscosity of the mixture take the order-of-magnitude values of 10^{-4} [m/s], 10^{-5} [m] and 10^{-6} [m^2/s], respectively. Assuming that the suspension behaves as a generalized linear viscous fluid, the most general constitutive equation for the extra stress tensor in terms of the stretching is

$$\mathbf{T}^E = \left(k_{b\text{eff}} - \frac{2}{3}\mu_{\text{eff}}\right)(\nabla \cdot \mathbf{v})\mathbf{I} + \mu_{\text{eff}}\left[\nabla \mathbf{v} + (\nabla \mathbf{v})^T\right],$$

where $k_{b\text{eff}}$ is the effective bulk viscosity and $\mu_{\text{eff}} = \mu_{\text{eff}}(\phi)$ is the effective shear viscosity of the mixture. If the effective bulk viscosity is neglected, equation (4) can be rewritten in the simple form

$$0 = -\nabla p - \varrho(\phi) g\mathbf{k} + \nabla \cdot \left(\mu_{\text{eff}}(\phi)\widetilde{\mathbf{T}}\right) \tag{5}$$

where $\widetilde{\mathbf{T}} = \mathcal{S}(\mathbf{v}) := \nabla\mathbf{v} + (\nabla\mathbf{v})^T - \frac{2}{3}(\nabla \cdot \mathbf{v})\mathbf{I}$. Usually it is assumed that $\mu_{\text{eff}}(\phi) = \mu(\phi)\mu_0$, where μ_0 is the viscosity of the pure fluid and $\mu(\phi)$ a semi-empirical correlation, e.g. of the common type $\mu(\phi) = (1 - \phi/\phi_\infty)^{-2.5\phi_\infty}$, where ϕ_∞ is the maximum solid concentration. To complete the set of field equations, the linear momentum equation for the solid component should be given. Instead, and following the procedure in the Kynch theory of sedimentation [15, 16], the constitutive equation for the solid flux density vector $\mathbf{f} := \phi\mathbf{v}_s$ will be postulated, based on the results of the theory of one-dimensional dynamical sedimentation processes [29]:

$$\mathbf{f} = \phi\mathbf{q} + f_{bk}(\phi)\left(\mathbf{k} + \frac{\sigma_e'(\phi)}{\Delta\varrho\phi g}\nabla\phi\right). \tag{6}$$

Here, f_{bk} is Kynch's solid flux density function, σ_e is the solid effective stress and $\Delta\varrho := \varrho_s - \varrho_f$. By the definition of $\mathbf{q}$, $\mathbf{v}_r$, $\mathbf{v}$ and $\mathbf{f}$ are related by

$$\mathbf{v}_r = \frac{1}{\phi(1-\phi)}(\mathbf{f} - \phi\mathbf{q}) = \frac{f_{bk}(\phi)}{\phi(1-\phi)}\left(\mathbf{k} + \frac{\sigma_e'(\phi)}{\Delta\varrho\phi g}\nabla\phi\right).$$

The functions f_{bk} and σ_e are obtained from experiments. Since only touching flocs permit the transmission of solid stress, $\sigma_e(\phi)$ is assumed to satisfy

$$\sigma_e'(\phi) = 0 \ \text{ for } 0 \le \phi \le \phi_c, \quad \sigma_e'(\phi) > 0 \ \text{ for } \phi_c < \phi \le 1. \tag{7}$$

For f_{bk}, the function by Richardson and Zaki [53] is widely used:

$$f_{bk}(\phi) = u_\infty \phi(1-\phi)^{C+1}, \quad u_\infty < 0, \quad C > 0. \tag{8}$$

Similar formulae are presented in [32]. In general, we assume that

$$f_{bk}(0) = f_{bk}(\phi_\infty) = 0, \; f_{bk}(\phi) < 0 \text{ for } \phi \in (0, \phi_\infty), \; f_{bk}'(0) < 0, \; f_{bk}'(\phi_\infty) \ge 0 \tag{9}$$

is satisfied [19], and consider only physically relevant concentration values $\phi \in \mathcal{C} := [0, \phi_\infty]$. Now $\mathbf{v}$ can be rewritten in terms of $\mathbf{q}$, ϕ and $\nabla\phi$ as

$$\mathbf{v} = \mathbf{q} + \Delta\varrho\frac{\phi(1-\phi)}{\varrho(\phi)}\mathbf{v}_r = \mathbf{q} + \Delta\varrho\frac{f_{bk}(\phi)}{\varrho(\phi)}\left(\mathbf{k} + \frac{\sigma_e'(\phi)}{\Delta\varrho\phi g}\nabla\phi\right).$$

Since $\nabla \cdot \mathbf{q} = 0$, the tensor $\widetilde{\mathbf{T}}$ can then be rewritten in the form [14]

$$\widetilde{\mathbf{T}} = \mathcal{S}(\mathbf{q}) + \mathcal{S}(\mathbf{v} - \mathbf{q}) = \nabla\mathbf{q} + (\nabla\mathbf{q})^T + \mathbf{\Psi}\left(D_x^\alpha \phi\right), \quad |\alpha| \le 2,$$

where $\mathbf{\Psi}$ is a known function satisfying $\mathbf{\Psi}(D_x^\alpha 0) = 0$. We then obtain

$$\nabla \cdot \mathbf{T}^E = \mu_{\text{eff}}(\phi)\nabla^2 \mathbf{q} + \mu'_{\text{eff}}(\phi)\nabla\phi \cdot \left[\nabla\mathbf{q} + (\nabla\mathbf{q})^T\right] + \nabla \cdot \left[\mu_{\text{eff}}(\phi)\mathbf{\Psi}(D_x^\alpha \phi)\right].$$

Denoting by $\mathbf{f}_k(\phi;\mathbf{q}) := \phi\mathbf{q} + f_{bk}(\phi)\mathbf{k}$ the extended Kynch flux density function [16, 17] and by $a(\phi) := -f_{bk}(\phi)\sigma'_e(\phi)/(\Delta\varrho g\phi)$ a diffusion coefficient, the field equations (1), (3) and (5), together with the constitutive equation (6), yield the following system of equations for the unknown concentration distribution ϕ, the volume-average velocity field $\mathbf{q}$ and the pressure p:

$$\begin{aligned}
\frac{\partial\phi}{\partial t} + \nabla \cdot \mathbf{f}_k(\phi;\mathbf{q}) &= \nabla \cdot (a(\phi)\nabla\phi), && (10)\\
\nabla \cdot \mathbf{q} &= 0, && (11)\\
\mu_{\text{eff}}(\phi)\nabla^2\mathbf{q} - \nabla p &= \varrho(\phi)g\mathbf{k} - \mu'_{\text{eff}}(\phi)\nabla\phi \cdot \left[\nabla\mathbf{q} + (\nabla\mathbf{q})^T\right] \\
&\quad -\nabla \cdot \left[\mu_{\text{eff}}(\phi)\mathbf{\Psi}(D_x^\alpha\phi)\right] && (12)
\end{aligned}$$

Note that, for a pure fluid ($\phi \equiv 0$), equations (11) and (12) recover Stokes' system for incompressible flow. At discontinuities, equations (10)–(12) are replaced by the Rankine-Hugoniot conditions

$$\left[(\mathbf{f}_k - a(\phi)\nabla\phi) \cdot \mathbf{n}_I\right] = \sigma[\phi], \quad [\mathbf{q} \cdot \mathbf{n}_I] = 0, \quad \left[-p\mathbf{n}_I + \mu_{\text{eff}}(\phi)\widetilde{\mathbf{T}} \cdot \mathbf{n}_I\right] = 0,$$

where $[\cdot]$ denotes a jump with normal $\mathbf{n}_I$ and propagation velocity σ. Note that (10) is a degenerate parabolic equation since

$$a(\phi)\begin{cases} = 0 & \text{for} \quad \phi \le \phi_c: \quad \text{eqn (10) is } \textit{hyperbolic} \\ > 0 & \text{for} \quad \phi_c < \phi < 1: \quad \text{eqn (10) is } \textit{parabolic} \\ = 0 & \text{for} \quad \phi \ge 1: \quad \text{eqn (10) is } \textit{hyperbolic.} \end{cases}$$

In view of the industrial application, we are interested in solutions of (10)–(12) which imply hyperbolicity of (10) at least in a part of the computational domain. Hence the concentration distribution ϕ will be discontinuous in general and is expected to show wave propagation phenomena. These solutions were also the subject of study of the theory of *kinematic waves*, applied by Kluwick [38] and Schneider [56, 57] to sedimentation processes.

In Sections 3 and 4, we consider exact and numerical solutions of equations (10)–(12) under the assumption of inviscidity ($\mu_{\text{eff}} \equiv 0$), which corresponds to modelling the solid and liquid component as elastic fluids [29].

2.2. SEDIMENTATION IN AN IDEAL THICKENER

Consider a sedimentation process in an *Ideal Thickener* IT, see Fig. 2 [17]. An IT is a cylindrical vessel without wall effects, therefore the field variables

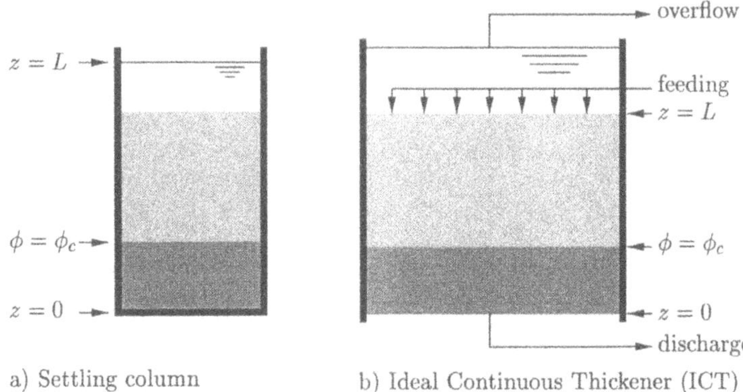

Figure 2. Schematic representation of an Ideal Thickener.

are horizontally constant. An IT can be operated continuously (ICT) with suspension fed in at $z = L$ and sediment discharged at $z = 0$, as in Fig. 2 b), or in batch mode as a settling column, Fig. 2 a). We consider now the space interval $\Omega := (0, L)$, the time interval $J := (0, T)$ and look for solutions on $\overline{Q_T}$ with the cylinder $Q_T := \Omega \times J$. From (11) and the Rankine-Hugoniot conditions we obtain $q(z, t) = q(t)$, hence the volume average velocity q can be prescribed externally. Assuming $\mu_{\text{eff}} \equiv 0$ and since $f_k(\phi, t) = q(t)\phi + f_{bk}(\phi)$, equations (10) and (12) reduce to a convective diffusion equation for ϕ and an equation for the total pressure p:

$$\begin{aligned} \frac{\partial \phi}{\partial t} + \frac{\partial}{\partial z} f_k(\phi, t) &= \frac{\partial}{\partial z}\left(a(\phi)\frac{\partial \phi}{\partial z}\right), \quad (z, t) \in \overline{Q_T}, \qquad (13) \\ \frac{\partial p}{\partial z} &= -\varrho(\phi) g. \end{aligned}$$

Decomposing p into an *excess pore pressure* p_e, an *effective solid stress* σ_e, for which a constitutive equation is assumed to be given, and a hydrostatic pressure, $p(z, t) = p_e(z, t) + \sigma_e(\phi(z, t)) + \varrho_f g(H - z)$, the pressure equation can be rewritten in these experimental rather than theoretical variables as

$$\frac{\partial p_e}{\partial z} = -\Delta\varrho\phi g - \sigma_e'(\phi)\frac{\partial \phi}{\partial z}.$$

Since $p(z, t)$ and $p_e(z, t)$ can be calculated *a posteriori* on $\overline{Q_T}$, once ϕ is computed, the pressure equation will not be included in the mathematical models for an IT. We are left with the parabolic degenerate quasilinear equation (13), together with a prescribed initial concentration distribution and a concentration prescribed at $z = L$:

$$\phi(z, 0) = \phi_0(z), \quad z \in \overline{\Omega}, \quad \phi_0(z) \in \mathcal{C}, \qquad (14)$$

$$\phi(L,t) \;=\; \phi_1(t),\;\; t\in \bar{J},\;\; \phi_1(t)\in \mathcal{C}. \tag{15}$$

The boundary condition at $z=0$ will be formulated separately for ideal and flocculated suspensions.

3. Sedimentation of Ideal Suspensions

3.1. INITIAL-BOUNDARY VALUE PROBLEM

Equation (13) can be further simplified by assuming that the mixture is an *ideal*, i.e. *non-flocculated* suspension [16]. In this case, the solid effective stress σ_e is assumed to be constant, $\sigma_e' \equiv 0$, and therefore $a \equiv 0$. Equation (13) reduces to the nonlinear hyperbolic conservation law

$$\frac{\partial \phi}{\partial t} + \frac{\partial}{\partial z}\left(q(t)\phi + f_{bk}(\phi)\right) = 0,\quad (z,t)\in \overline{Q_T} \tag{16}$$

of a Kynch sedimentation process [16, 28], which is solved with conditions (14) and (15) and with the following boundary condition at $z=0$, which prescribes the maximum concentration ϕ_∞:

$$\phi(0,t) = \phi_\infty,\;\; t\in \bar{J} \tag{17}$$

Here, we assume that ϕ_∞ is constant in time. Equations (14)–(17) form the initial-boundary value problem (IBVP) of sedimentation of ideal suspensions, what we call a *Kynch Sedimentation Process.* In the following two subsections, its solvability is dicussed in general, while in 3.4 and 3.6, generalized solutions of standardized IBVPs are determined, which permit the classification of batch and continuous sedimentation of ideal suspensions.

3.2. SOLVABILITY OF THE INITIAL-BOUNDARY VALUE PROBLEM

Solutions of the nonlinear equation (16) are discontinuous in general even if ϕ_0 and ϕ_1 are smooth, as is well known from Burgers' equation [47]. *Generalized solutions* of the IBVP are defined as weak solutions which satisfy an *entropy condition*, and are defined by a Kružkov entropy inequality [40]. Bustos *et al.* obtain the following existence and uniqueness result [19, 52]:

Theorem 1 *The IBVP* (14)–(17) *has exactly one generalized solution* $\phi \in L^\infty(Q_T)\cap BV(Q_T)$ *satisfying* $\phi(z,0)=\phi_0(z)$ *for almost all* $z\in\Omega$ *and the following inequality for all* $k\in\mathbb{R}$ *and for all* $\psi\in C_0^2(\overline{\Omega}\times J)$, $\psi\geq 0$:

$$\begin{aligned}
&\int_0^T\int_0^L \left\{|\phi-k|\frac{\partial\psi}{\partial t} + \operatorname{sgn}(\phi-k)\left[f_k(\phi,t)-f_k(k,t)\right]\frac{\partial\psi}{\partial z}\right\} dzdt \\
&\quad\geq \int_0^T \{\operatorname{sgn}(\phi_1(t)-k)\left[f_k((\gamma\phi)(L,t))-f_k(k,t)\right]\psi(L,t) \\
&\qquad -\operatorname{sgn}(\phi_\infty-k)\left[f_k((\gamma\phi)(0,t))-f_k(k,t)\right]\psi(0,t)\}\,dt
\end{aligned} \tag{18}$$

We set $I(a,b) := [\min\{a,b\}, \max\{a,b\}]$ and, for notational convenience, $\sigma(\phi_1,\phi_2;t) := (f_k(\phi_1,t) - f_k(\phi_2,t))/(\phi_1 - \phi_2)$. For constant q, the parameter t in the definition of σ will be omitted. As is well known, the local propagation speed $s(z_0,t_0)$ of a jump between two one-sided limits $\phi^-(z_0,t_0)$ and $\phi^+(z_0,t_0)$ at $(z_0,t_0) \in Q_T$ of the generalized solution is given by the Rankine-Hugoniot condition, and from inequality (18) it follows that Oleĭnik's jump entropy condition [47] is satisfied:

$$s(z_0,t_0) := \sigma\left(\phi^-(z_0,t_0), \phi^+(z_0,t_0); t_0\right), \tag{19}$$

$$\forall k \in I(\phi^-,\phi^+): \quad \sigma\left(\phi^-, k; t_0\right) \leq s \leq \sigma\left(\phi^+, k; t_0\right) \tag{20}$$

The *existence* of generalized solutions follows from the existence proof by Bardos *et al.* in [3] using the vanishing viscosity method, for which $q, f_{bk} \in C^1(\mathbb{R})$, $\phi_0 \in C^2(\overline{\Omega})$ and $\phi_1 \in C^2(\bar{J})$ are assumed to be satisfied [19]. However, in view of the proof in [48] applying a mollifier to incorporate smoothing the initial and boundary data into the vanishing viscosity regularization, these assumptions can be relaxed to $\phi_0 \in L^\infty(\overline{\Omega})$, $\phi_1 \in L^\infty(\bar{J})$.

3.3. ENTROPY BOUNDARY CONDITIONS

To prove the *uniqueness* of generalized solutions, we have to consider the way in which the boundary conditions (15) and (17) are satisfied. To illustrate this problem, note that the entropy solution of the following Riemann problem consists of a 'fan' of straight characteristic and discontinuity lines, which in general intersect the lines $z = 0$ and $z = L$ after finite time.

$$\frac{\partial \phi}{\partial t} + \frac{\partial}{\partial z}\left(q\phi + f_{bk}(\phi)\right) = 0, \quad q = const.; \quad (z,t) \in Q_T, \tag{21}$$

$$\phi_0(z) = \phi^+ \text{ for } z > C, \qquad \phi_0(z) = \phi^- \text{ for } z \leq C, \quad C \in \Omega \tag{22}$$

Hence, the IBVP of (21), (22) and corresponding boundary conditions

$$\phi(L,t) = \phi^+, \quad \phi(0,t) = \phi^- \quad \text{for } t \in J \tag{23}$$

extending the initial data is not well-posed for all times. To obtain well-posedness, conditions (23) are replaced by *set-valued* or so-called *entropy boundary* conditions, which consider that the values ϕ^+ and ϕ^- are not always assumed in a pointwise sense by the traces $(\gamma\phi)(0,t)$ and $(\gamma\phi)(L,t)$ of the generalized solution of (21)–(23). For the general case of time-dependent functions q, ϕ_1 and ϕ_∞, Bustos *et al.* [19] prove the following theorem using the results by Bardos *et al.* [3] and by Dubois and Le Floch [33].

Theorem 2 *A function $\phi \in L^\infty(Q_T) \cap BV(Q_T)$ is a generalized solution of the IBVP* (14)–(17) *if and only if $\phi(0,z) = \phi_0(z)$ f.a.a. $z \in \Omega$ for all*

$k \in \mathbb{R}$, $\psi \in C_0^2(Q_T)$, $\psi \geq 0$, the integral inequality

$$\int_0^T \int_0^L \left\{ |\phi - k| \frac{\partial \psi}{\partial t} + \operatorname{sgn}(\phi - k)\left[f_k(\phi, t) - f_k(k, t)\right] \frac{\partial \psi}{\partial z} \right\} dz dt \geq 0$$

is satisfied, and if the traces $(\gamma\phi)(L,t)$ and $(\gamma\phi)(0,t)$ belong to the sets of admissible states *at $z = L$, $\mathcal{E}_L(\phi_1(t))$, and at $z = 0$, $\mathcal{E}_0(\phi_\infty(t))$, respectively:*

$$(\gamma\phi)(L,t) \in \mathcal{E}_L(\phi_1(t)) := \{\phi \in \mathcal{C} | \sigma(\phi, k; t) \geq 0 \; \forall k \in I(\phi_1(t), \phi)\} \quad (24)$$

$$(\gamma\phi)(0,t) \in \mathcal{E}_0(\phi_\infty(t)) := \{\phi \in \mathcal{C} | \sigma(\phi, k; t) \leq 0 \; \forall k \in I(\phi_\infty(t), \phi)\} \quad (25)$$

The geometrical interpretation of the sets $\mathcal{E}_L(\phi_1(t))$ and $\mathcal{E}_0(\phi_\infty(t))$ is given in [13]. Clearly, we always have $\phi_1(t) \in \mathcal{E}_L(\phi_1(t))$ and $\phi_\infty(t) \in \mathcal{E}_0(\phi_\infty(t))$.

3.4. CLASSIFICATION OF BATCH SEDIMENTATION PROCESSES

Bustos and Concha [15, 27] classify all possible Kynch Batch Sedimentation Processes of homogeneous suspensions by determining generalized solutions of equation (21) with $q \equiv 0$, $\phi_0 = const.$, $\phi_1 \equiv 0$, $\phi_\infty = const$, and a flux density function f_{bk} with exactly one or exactly two inflection points. Except for the trivial case $\phi_0 = \phi_\infty$, here the boundary conditions are always assumed exactly, hence the IBVP

$$\frac{\partial \phi}{\partial t} + \frac{\partial f_{bk}(\phi)}{\partial z} = 0, \; \phi(z,0) = \phi_0 \text{ for } z \in \overline{\Omega}, \; \begin{cases} \phi(L,t) = 0 \text{ for } t \in \bar{J} \\ \phi(0,t) = \phi_\infty \text{ for } t \in \bar{J} \end{cases} \quad (26)$$

is well posed. In addition to (9), we assume that f_{bk} satisfies Kynch's [44] assumption $f'_{bk}(\phi_\infty) > 0$, which ensures that sedimentation terminates after finite time. Applying the solution technique developed by Ballou [2] and Cheng [20]–[22] which is a method of characteristics, Bustos and Concha [15] show that the generalized solution of the IBVP (26) belongs to one of the following five *modes of sedimentation* (MS), depending on ϕ_0 and ϕ_∞:

MS-1 The initial concentration ϕ_0 changes abruptly to ϕ_∞.

MS-2 The initial concentration ϕ_0 changes suddenly and then increases continuously to ϕ_∞.

MS-3 The concentration increases continuously from ϕ_0 to ϕ_∞.

MS-4 The initial concentration ϕ_0 changes suddenly and then increases continuously, followed by an abrupt increase to ϕ_∞.

MS-5 The initial concentration ϕ_0 increases continuously and then jumps to ϕ_∞.

The modes **MS-4** and **MS-5** occur only for exactly two inflection points. The generalized solution will be constructed for an **MS-2**. For the complete solution we refer to the works of Bustos and Concha [15, 16, 27].

Assume that a flux density function f_{bk} satisfying (9) and $f'_{bk}(\phi_\infty) > 0$ with exactly one inflection point at $\phi = \phi_a$ with $f''_{bk}(\phi) > 0$ for $\phi < \phi_a$ and $f''_{bk}(\phi) < 0$ for $\phi > \phi_a$ is given. For $\alpha < \phi_a$ we define $\alpha^* > \phi_a$ to be the the unique number satisfying $f'_{bk}(\alpha^*) = \sigma(\alpha, \alpha^*)$, and for $\beta > \phi_a$, β^{**} is defined as the unique number satisfying $\beta = (\beta^{**})^*$.

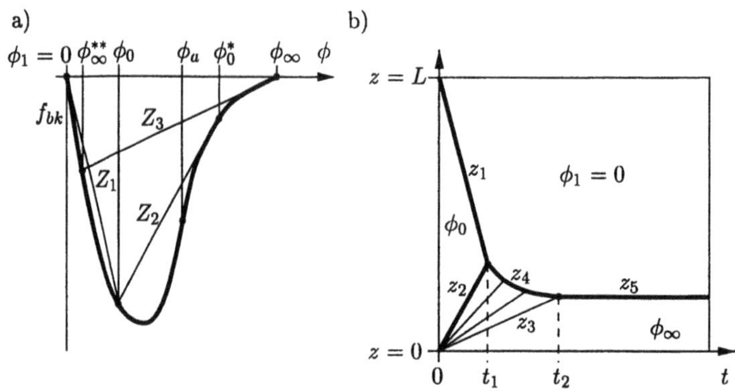

Figure 3. Construction of the generalized solution for batch sedimentation of an ideal suspension. a) Flux density function f_{bk} with auxiliary values. b) Settling plot. The fat lines correspond to discontinuities.

Now let $\phi_\infty^{**} < \phi_0 < \phi_a$ be satisfied, see Fig. 3 a). The location of the chord Z_1 in Fig. 3 a) indicates that the sedimentation process starts with a shock between ϕ_1 and ϕ_0 propagating into the settling column, while the tangential chord Z_2 implies that ϕ_0 and ϕ_∞ are linked by a shock between ϕ_0 and ϕ_0^* emerging at $z = 0$, followed by a rarefaction wave between ϕ_0^* and ϕ_∞. Defining the shock line $z_1(t) := L + \sigma(\phi_0, \phi_1)t$, the contact discontinuity $z_2(t) := f'_{bk}(\phi_0^*)t$, the continuity line $z_3(t) := f'_{bk}(\phi_\infty)t$ and $g := (f'_{bk})^{-1}|_{\phi > \phi_a}$, the generalized solution of the IBVP (26) is

$$\phi|_{[0,t_1]} = \begin{cases} \phi_1 & \text{for } z \geq z_1(t), \qquad g(\frac{z}{t}) \quad \text{for } z_3(t) \leq z < z_2(t), \\ \phi_0 & \text{for } z_2(t) \leq z < z_1(t), \quad \phi_\infty \quad \text{for } z < z_3(t), \end{cases}$$

where $t_1 := L/(f'_{bk}(\phi_0^*) - \sigma(\phi_1, \phi_0))$ is the time when $z_1(t)$ and $z_2(t)$ meet. The shock curve $z_4(t)$ in Fig. 3 b) satisfies for $t \geq t_1$ the ODE $z_4'(t) = \sigma\left(g\left(z_4(t)/t\right), \phi_1\right)$. It meets $z_3(t)$ at $t_2 := \phi_0 L/(\phi_\infty f'_{bk}(\phi_\infty))$, when $z_3(t)$ intersects with the stationary shock $z_5 := L\phi_0/\phi_\infty$ [15]. Hence, we have

$$\phi|_{(t_1,t_2]} = \begin{cases} \phi_1 & \text{for } z \geq z_4(t) \\ g(\frac{z}{t}) & \text{for } z_3(t) \leq z < z_4(t) \\ \phi_\infty & \text{for } z < z_3(t) \end{cases}, \quad \phi|_{(t_2,\infty)} = \begin{cases} \phi_1 & \text{for } z \geq z_5 \\ \phi_\infty & \text{for } z < z_5. \end{cases}$$

3.5. THE POLYGONAL APPROXIMATION METHOD

In the preceding example, most of the computational effort consists in the calculation of the function g. This is an easy task if f_{bk} is replaced by a polygon Π. Dafermos showed in [31] that the Cauchy problem

$$\frac{\partial u}{\partial t} + \frac{\partial}{\partial z}\Pi(u) = 0, \quad z \in \mathbb{R},\ t > 0; \quad u(z,0) = u_0(z), \quad z \in \mathbb{R}$$

has a solution for every real-valued polygon Π and every step function u_0. The solution is piecewise constant with straight shock lines satisfying the jump conditions (19) and (20).

Kunik *et al.* [41]–[43] formulate an efficient algorithm to calculate batch and, under certain conditions, continuous sedimentation processes with piecewise constant initial and boundary data, by approximating f_{bk} with a polygon Π. Instead of the method of characteristics of the preceding subsection, which can not be applied because Π is not differentiable, the representation formula by Cheng [21] is used. Cheng showed that the generalized solution minimizes or maximizes a certain known functional at each point in Q_T. Replacing f_{bk} by Π yields a finite test set for this functional, which is further reduced by the application of Dafermos' regularity result. This method permits an a-priori L^1-error estimate of the approximate solution in terms of the approximation error committed by replacing f_{bk} by Π and of the number of steps of the initial data, see [16].

3.6. CLASSIFICATION OF CONTINUOUS SEDIMENTATION PROCESSES

The classification of sedimentation processes was extended to continuous sedimentation [16, 17, 28]. We consider equation (21) with $q = const.$, a Riemann initial state and set-valued boundary conditions with $\phi_1 < \phi_\infty$:

$$\phi_0(z) := \begin{cases} \phi_1 \text{ for } z > C \\ \phi_\infty \text{ for } z < C \end{cases}; \quad \begin{cases} (\gamma\phi)(L,t) \in \mathcal{E}_L(\phi_1) \text{ for } t \in \bar{J} \\ (\gamma\phi)(0,t) \in \mathcal{E}_0(\phi_\infty) \text{ for } t \in \bar{J} \end{cases} \tag{27}$$

Here, $C \in \Omega$ is the initial sediment height. An ICT is said to *empty* if $\phi_\infty \neq (\gamma\phi)(0,t) \leq (\gamma\phi)(L,t)$, and to *overflow* if $(\gamma\phi)(L,t) \neq \phi_1$. Assuming a flux density function $f_k(\phi) := q\phi + f_{bk}(\phi)$ with $q < 0$, exactly one inflection point ϕ_a, $f_k''(\phi) > 0$ for $\phi < \phi_a$ and $f_k''(\phi) < 0$ for $\phi > \phi_a$, Bustos *et al.* show in [17] that the generalized solution of the IBVP (21), (27) is one of the following *modes of continuous sedimentation* (MCS):

MCS-1 The two constant states are separated by a shock.

MCS-2 The two constant states are separated by a contact discontinuity and a rarefaction wave.

MCS-3 The generalized solution is continuous and consists of two constant states separated by a rarefaction wave.

TABLE 1. Modes of continuous sedimentation of ideal suspensions for $f_k'(\phi_\infty) \leq 0$

	Location of ϕ_1	Admissible states at $z = L$	MCS	ICT behaviour
a)	$\phi_1 \leq \phi_\infty^{**}$	$\mathcal{E}_L(\phi_1) = \{\phi_1\}$	MCS-1	empties
b)	$\phi_\infty^{**} < \phi_1 < \phi_M^{**}$	$\mathcal{E}_L(\phi_1) = \{\phi_1\}$	MCS-2	empties
c)	$\phi_1 = \phi_M^{**}$	$\mathcal{E}_L(\phi_1) = \{\phi_1, \phi_M\}$	MCS-2	$\rightarrow$ steady state: $\phi(0,t) \overset{t\to\infty}{\longrightarrow} \phi_M$
d)	$\phi_M^{**} < \phi_1 < \phi_m$	$\mathcal{E}_L(\phi_1) = \{\phi_1\} \cup [\phi_1^0, \phi_M]$	MCS-2	overflows
e)	$\phi_m \leq \phi_1 < \phi_a$	$\mathcal{E}_L(\phi_1) = [\phi_m, \phi_M]$	MCS-2	overflows
f)	$\phi_a \leq \phi_1 < \phi_M$	$\mathcal{E}_L(\phi_1) = [\phi_m, \phi_M]$	MCS-3	overflows
g)	$\phi_1 = \phi_M$	$\mathcal{E}_L(\phi_1) = [\phi_m, \phi_M]$	MCS-3	empties
h)	$\phi_1 > \phi_M$	$\mathcal{E}_L(\phi_1) = [\phi_m, \phi_1^\sharp] \cup \{\phi_1\}$	MCS-3	empties

TABLE 2. Modes of continuous sedimentation of ideal suspensions for $f_k'(\phi_\infty) > 0$

	Location of ϕ_1	Admissible states at $z = L$	MCS	ICT behaviour
a)	$\phi_1 < \phi_s$	$\mathcal{E}_L(\phi_1) = \{\phi_1\}$	MCS-1	empties
b)	$\phi_1 = \phi_s$	$\mathcal{E}_L(\phi_1) = \{\phi_1, \phi_\infty\}$	MCS-1	steady state, $\phi(0,t) = \phi_\infty \ \forall t \geq 0$
c)	$\phi_s < \phi_1 \leq \phi_\infty^{**}$	$\mathcal{E}_L(\phi_1) = \{\phi_1\} \cup [\phi_1^0, \phi_\infty]$	MCS-1	overflows
d)	$\phi_\infty^{**} < \phi_1 \leq \phi_m$	$\mathcal{E}_L(\phi_1) = \{\phi_1\} \cup [\phi_1^0, \phi_\infty]$	MCS-2	overflows
e)	$\phi_m < \phi_1 < \phi_a$	$\mathcal{E}_L(\phi_1) = [\phi_m, \phi_\infty]$	MCS-2	overflows
f)	$\phi_a \leq \phi_1 < \phi_\infty$	$\mathcal{E}_L(\phi_1) = [\phi_m, \phi_\infty]$	MCS-3	overflows

The IBVP (21), (27) was solved in a similar way as the problem of batch sedimentation [17]. For $f_k'(\phi_\infty) \leq 0$, the result is summarized in Table 1 and given graphically in Fig. 4; for $f_k'(\phi_\infty) > 0$, see Table 2 and Fig. 5.

Here, $\phi_M < \phi_\infty$ denotes the local maximum of f_k, $\phi_m < \phi_M$ is its local minimum, $\phi_1^0 > \phi_1$ is defined as the smallest value $\phi \in (\phi_1, \phi_\infty]$ satisfying $f_k(\phi_1^0) = f_k(\phi_1)$, and $\phi_1^\sharp$ is the greatest value $\phi \in [0, \phi_1)$ satisfying $f_k(\phi_1^\sharp) = f_k(\phi_1)$. Finally, for the case $f_k'(\phi_\infty) > 0$, let $\phi_s := \phi_\infty^\sharp$.

Tables 1 and 2 also contain the sets $\mathcal{E}_L(\phi_1)$ for the different choices of ϕ_1. For the case $f_k'(\phi_\infty) \leq 0$, we have [19] $\mathcal{E}_0(\phi_\infty) = [0, \phi_M^{**}] \cup [\phi_M, \phi_\infty]$ for all choices of ϕ_1; for $f_k'(\phi_\infty) > 0$, we obtain $\mathcal{E}_0(\phi_\infty) = [0, \phi_s] \cup \{\phi_\infty\}$. Note that these sets are determined *a priori*, independently of the initial condition or the actual generalized solution the IBVP under consideration. For details on the construction, see [13, 14, 19, 33].

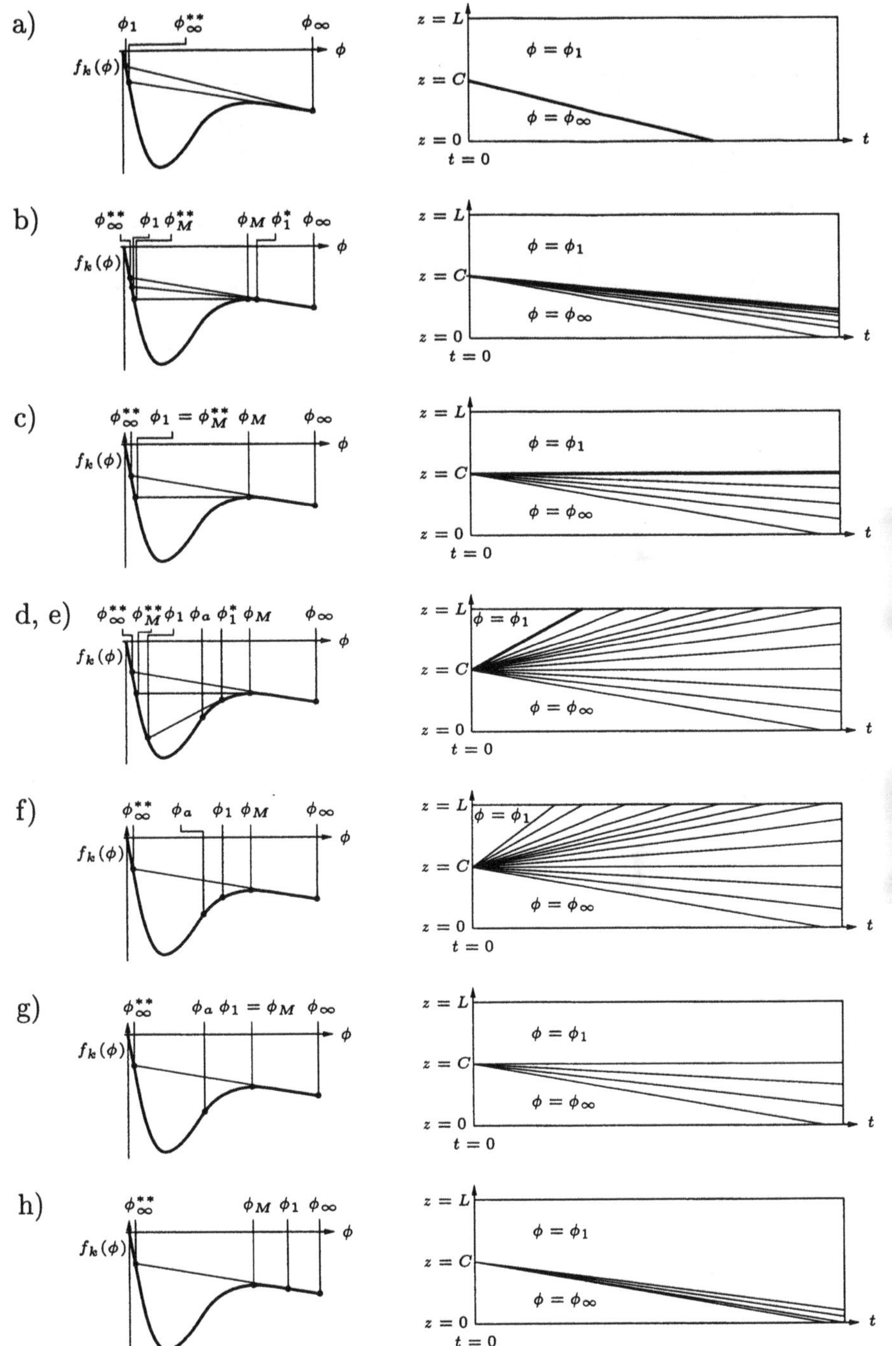

Figure 4. Modes of continuous sedimentation for $f_k'(\phi_\infty) \leq 0$. The fat lines in the settling plots correspond to discontinuities.

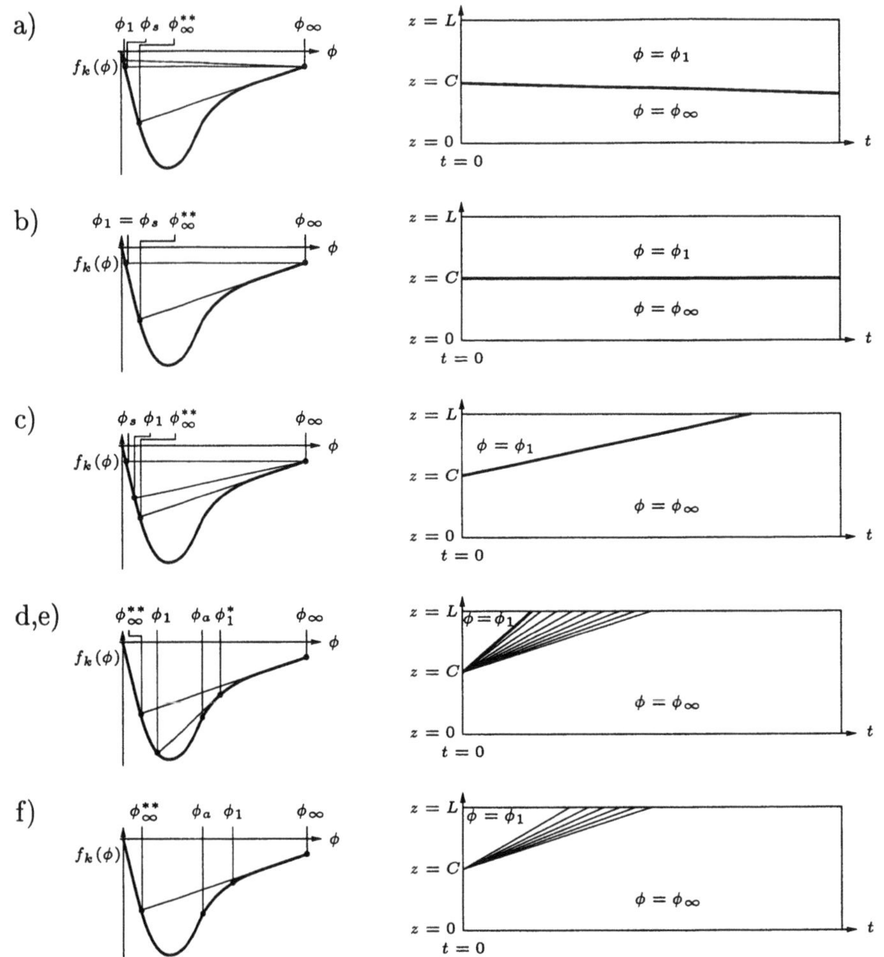

Figure 5. Modes of continuous sedimentation for $f_k'(\phi_\infty) > 0$. The fat lines in the settling plots correspond to discontinuities.

3.7. CONTROL OF CONTINUOUS SEDIMENTATION

We have shown now how the generalized solution of the IBVP of sedimentation of ideal suspensions can be determined for piecewise constant initial and boundary data. In this case the exact location and local propagation speed of the sediment level is always known, which was used in the control model for continuous sedimentation by Bustos *et al.* in [18]. They prove that the steady state of Table 2 and Fig. 5 b) can always be recovered after a perturbation of the feeding flux density by solving two IBVPs at known times and with parameters q and ϕ_1 that can be calculated a priori.

4. Sedimentation of Flocculated Suspensions

4.1. SOLVABILITY OF THE INITIAL-BOUNDARY VALUE PROBLEM

Sedimentation of a flocculated suspension involves the formation of a compressible sediment layer at the bottom of the IT, therefore the term *sedimentation with compression* is also widely used for this phenomenon. We consider the IBVP of equation (13), the initial condition (14), the boundary condition (15), and the condition

$$f_{bk}(\phi) + a(\phi)\frac{\partial \phi}{\partial z}\bigg|_{z=0} = 0, \quad 0 < t \leq T, \tag{28}$$

which is equivalent to reducing the solid volume flux at $z = 0$ to the convective outflow part: $\phi v_s = q(t)\phi(0,t)$. This IBVP was analyzed by Bürger and Wendland [7, 8], [12]–[14]. With the *functional superposition* defined for $u \in BV(Q_T) \cap L^\infty(Q_T)$ by $\widehat{g(u)}(z,t) := \int_0^1 (\tau u^+(z,t) + (1-\tau)u^-(z,t))\, d\tau$ where u^+, u^- denote the approximate limits of u with respect to the normal to a discontinuity at (z,t) and otherwise equal $u(z,t)$, their existence result [12] obtained by vanishing viscosity following Wu and Wang [61]–[63] is:

Theorem 3 *Assume that* $\phi_0 \in C^{2+\beta}(\overline{\Omega})$ *and* $\phi_1 \in C^{1+\beta}(\bar{J})$ *satisfy conditions* (14) *and* (15) *and that initial and boundary conditions are first order compatible. If* $q \in C^1(\bar{J})$, $q(t) \leq 0$ *for* $t \in \bar{J}$, $f_{bk} \in C^2(C)$, $f_{bk}(\phi) \leq 0$ *for* $\phi \in C$, $\sigma_e \in C^2$ *and conditions* (7) *are satisfied, the IBVP* (13)–(15), (28) *has a generalized solution* $\phi \in BV(Q_T) \cap L^\infty(Q_T)$ *satisfying:*

1.) $\widehat{\sqrt{a(\phi)}}\frac{\partial \phi}{\partial z} \in L^2(Q_T)$, *which ensures that* $\widehat{a(\phi)}\frac{\partial \phi}{\partial z}$ *is an absolutely continuous measure.*

2.) For $A(v) := \int_0^v a(\tau)d\tau$, $\forall\, k \in \mathbb{R}$, $\forall \psi \in C_0^\infty((0,L] \times (0,T])$, $\psi \geq 0$*:*

$$\int_0^T \int_0^L \left\{ |\phi - k| \frac{\partial \psi}{\partial t} + \mathrm{sgn}(\phi - k) \left[f_k(\phi,t) - f_k(k,t) - \widehat{a(\phi)}\frac{\partial \phi}{\partial z} \right] \frac{\partial \psi}{\partial z} \right\} dzdt$$
$$\geq -\int_0^T \Bigg\{ -\mathrm{sgn}(\phi_1(t) - k) \left[f_k(\gamma\phi, t) - f_k(k,t) + \gamma\left(\widehat{a(\phi)}\frac{\partial \phi}{\partial z}\right) \right] \psi(L,t)$$
$$+ (\mathrm{sgn}(\gamma\phi - k) - \mathrm{sgn}(\phi_1(t) - k))(A(\gamma\phi) - A(k))\frac{\partial \psi}{\partial z}(L,t) \Bigg\}\, dt.$$

3.) For almost all $z \in \overline{\Omega}$, $(\gamma\phi)(z,0) = \phi_0(z)$.

4.)

$$\gamma\left(f_{bk}(\phi) - \widehat{a(\phi)}\frac{\partial \phi}{\partial z} \right)\bigg|_{z=0} = 0 \text{ for almost all } t \in \bar{J}. \tag{29}$$

The stability result from [12] implies uniqueness of generalized solutions:

Theorem 4 *If* $\phi, \eta \in BV(Q_T) \cap L^\infty(Q_T)$ *are generalized solutions of the IBVP* (13), (15), (28) *and initial conditions* $\phi(z,0) = \phi_0(z)$ *and* $\eta(z,0) = \eta_0(z)$, *then we have* $\|\phi(\cdot,t) - \eta(\cdot,t)\|_1 \leq \|\phi_0 - \eta_0\|_1$ *for* $t \in \bar{J}$.

4.2. JUMP AND ENTROPY BOUNDARY CONDITIONS

The proof of Theorem 4 follows the classical proofs by Kružkov [40] and Vol'pert and Hudjaev [60] for Cauchy problems and uses the Rankine-Hugoniot type jump condition and a specific entropy boundary condition valid for equations of the type of (13) as well as the nonpositivity of q.

Vol'pert and Hudjaev [60] proposed a jump condition for equation (13) which coincides with the conditions (19) and (20) for a hyperbolic equation. Wu and Yin [64] show that this condition is wrong and present the correct jump condition given in the following theorem [13]. This issue is also discussed in the recent work by Evje and Hvistendahl Karlsen [35].

Theorem 5 *If ϕ^- and ϕ^+ are the approximate limits of a jump of a generalized solution of equation* (13) *with respect to its normal at* $(z_0, t_0) \in Q_T$, *then for all ϕ between ϕ^- and ϕ^+, we have $a(\phi) = 0$. The local jump propagation velocity is given by* (19) *for $0 \leq \phi^-, \phi^+ < \phi_c$ and by*

$$s(z_0, t_0) = \frac{1}{\phi^\pm - \phi_c} \left[f_k(\phi^\pm, t_0) - f_k(\phi_c, t_0) + \lim_{z \to z_0^\mp} \widehat{a(u)} \frac{\partial \phi}{\partial z} \right] \tag{30}$$

for $0 \leq \phi^\pm < \phi^\mp = \phi_c$. For $0 < \phi^-, \phi^+ < \phi_c$, the corresponding jump entropy condition is given by (20), *and for $0 < \phi^\pm < \phi^\mp = \phi_c$, by*

$$\sigma(\phi^\pm, k; t_0) \lesseqgtr s \lesseqgtr \sigma(\phi_c, k; t_0) + \frac{1}{k - \phi_c} \lim_{z \to z_0^\mp} \widehat{a(u)} \frac{\partial \phi}{\partial z} \quad \forall k \in [\phi^\pm, \phi_c].$$

Hence, jumps are excluded in the interior of the compression zone where $a(\phi) > 0$ is valid. As the propagation speed of the sediment level at $\phi = \phi_c$ depends also on the derivative of the generalized solution and can not be determined a priori for transient states, a control model in the sense of [18] has not yet been formulated for sedimentation with compression.

While condition (29) implies that the boundary condition at $z = 0$, (28), is satisfied almost everywhere, condition (15) must be reformulated in a set-valued manner. The analogue of Theorem 2 for sedimentation with compression yields that, to obtain well-posedness of the IBVP (13)–(15), (28) condition (15) has to be replaced by the entropy boundary condition

$$(\gamma\phi)(L, t) \begin{cases} = \phi_1(t) & \text{if} \quad \phi_1(t) > \phi_c \\ \in \mathcal{E}_L(\phi_1(t)) & \text{if} \quad \phi_1(t) \leq \phi_c, \quad \mathcal{E}_L(\phi_1(t)) \text{ from (24)}, \end{cases} \tag{31}$$

hence only boundary values $\phi_1(t)$ for which (13) is parabolic are always assumed exactly by the generalized solution. See [13, 63] for details.

4.3. NUMERICAL ALGORITHM

In [9]–[11], we present the following *operator splitting* algorithm to simulate batch and continuous sedimentation of flocculated suspensions by calculating numerical solutions of the IBVP (13), (14), (28), (31).

We use a rectangular grid with parameters $\Delta z := L/H$, $\Delta t := T/N$, $z_j := j\Delta z$ for $j = -2, -1, 0, \ldots, H+2$, $t_n := n\Delta t$ for $n = 0, 1, 2, \ldots, N$ and set $\lambda := \Delta t/\Delta z$. If at $t = t_n$, an approximate solution $v(z, t_n)$ of the IBVP is given, calculate $v(z, t_{n+1})$ in the following way [9]:

1. Calculate $\hat{v}(\cdot, t_{n+1})$ from

$$\frac{\partial w}{\partial \tau} = \frac{\partial}{\partial z}\left(a(w)\frac{\partial w}{\partial z}\right), \quad w(z,0) = \begin{cases} \phi_0(z) \text{ for } n = 0 \\ v(z, t_n) \text{ for } n > 0 \end{cases}, \quad z \in \overline{\Omega},$$

$$w(0,\tau) = \begin{cases} \phi_0(0) \text{ for } n = 0 \\ v(0, t_n) \text{ for } n > 0 \end{cases}, \quad w(L,\tau) = \begin{cases} \phi_0(L) \text{ for } n = 0 \\ v(L, t_n) \text{ for } n > 0, \end{cases}$$

   $\tau \in [0, \Delta t]$, evaluated at $\tau = \Delta t$, i.e. $\hat{v}(\cdot, t_n) \equiv w(\cdot, \Delta t)$.

2. If $q(t_n) < 0$, then calculate $\tilde{v}(\cdot, t_n)$ from

$$\frac{\partial w}{\partial \tau} + q(t_n)\frac{\partial w}{\partial z} = 0, \quad w(z,0) = \begin{cases} \hat{v}(z, t_n) \text{ for } z < L \\ \phi_1(t_n) \text{ for } z \geq L \end{cases}$$

   evaluated at $\tau = \Delta t$, otherwise $\tilde{v}(\cdot, t_n) \equiv \hat{v}(\cdot, t_n)$.

3. Calculate $v(z, t_{n+1}) \approx \phi(z, t_{n+1})$ from

$$\frac{\partial w}{\partial \tau} + \frac{\partial}{\partial z} f_{bk} = 0, \quad w(z,0) = \tilde{v}(z, t_n), \quad z \in \overline{\Omega},$$

   $f_{bk}(w)|_{z=0} = 0$, $\tau \in [0, \Delta t]$, $w(L, \tau) \in \widetilde{\mathcal{E}}_L(\phi_1(\tau + t_n))$, where $\widetilde{\mathcal{E}}_L(\phi_1(\tau + t_n))$ is defined by (24) but with respect to f_{bk} instead of f_k.

The numerical treatment of these steps is as follows:

1. The diffusion step is solved by an implicit finite difference discretization for $j = 1, \ldots, K_n = \max\{1 \leq j \leq H : v_j^n \geq \phi_c\}$:

$$a(\phi)\frac{\partial \phi}{\partial z} \approx \theta a_j^{n+1}\left(\frac{\partial \phi}{\partial z}\right)^{n+1} + (1-\theta)a_j^n\left(\frac{\partial \phi}{\partial z}\right)^n, \quad 0 < \theta \leq 1.$$

2. The convective equation is solved by the second order upwind method:

$$\tilde{v}_j^{n+1} = \hat{v}_j^n - \frac{1}{2}\lambda \cdot q(t_n) \cdot \left[-3\hat{v}_j^n + 4\hat{v}_{j+1}^n - \hat{v}_{j+2}^n\right], \quad j = -2, \ldots, H.$$

3. The *non-oscillatory central difference method* by Nessyahu and Tadmor [51] is used for the remaining nonlinear hyperbolic equation:

$$\begin{aligned} v_j^{n+1/2} &:= \tilde{v}_j^n - \lambda(f')_j^n/2, \quad j = -1, \ldots, H \\ v_j^{n+1} &= \frac{1}{2}\left[\tilde{v}_{j-1}^n + \tilde{v}_{j+1}^n\right] + \frac{1}{4}\left[(v')_{j-1}^n - (v')_{j+1}^n\right] \\ &\quad -\frac{\lambda}{2}\left[f_{bk}\left(v_{j+1}^{n+1/2}\right) - f_{bk}\left(v_{j-1}^{n+1/2}\right)\right], \quad j = 0, \ldots, H-1. \end{aligned}$$

The *numerical derivatives* are calculated by a *minmod limiter*:

$$\begin{aligned} \mathrm{MM}(x,y,z) &:= \begin{cases} \operatorname{sgn} x \min\{|x|,|y|,|z|\} & \text{if } \operatorname{sgn} x = \operatorname{sgn} y = \operatorname{sgn} z \\ 0 & \text{otherwise,} \end{cases} \\ (v')_j^n &:= \mathrm{MM}\left(\kappa(\tilde{v}_{j+1}^n - \tilde{v}_j^n), (\tilde{v}_{j+1}^n - \tilde{v}_{j-1}^n)/2, \kappa(\tilde{v}_j^n - \tilde{v}_{j-1}^n)\right) \\ (f')_j^n &:= f'_{bk}(v_j^n)\cdot(v')_j^n; \quad \kappa \in [1,4). \end{aligned}$$

This method is second order accurate and TVD for Cauchy problems if $\lambda \max_u |f'_{bk}(u)| \leq (\sqrt{1+2\kappa-\kappa^2}-1)/\kappa$. In the following examples, $\phi_1(t)$ can always be prescribed, i.e. $v_j^{n+1} := \phi_1(t_{n+1})$ for $j \geq H$.

4.4. NUMERICAL EXAMPLES

This method was used to simulate numerous batch and continuous sedimentation processes [9]–[11], [14]. Here, we give two examples of batch settling and of filling up, steady state and emptying of an ICT. A *steady state* Φ is a stationary solution of equation (13) with $q = const.$, that is, of the ODE

$$\frac{\mathrm{d}}{\mathrm{d}z} f_k(\Phi) = -\frac{\mathrm{d}}{\mathrm{d}z}\left(\frac{f_{bk}(\Phi)\sigma'_e(\Phi)}{\Delta\varrho\Phi g}\frac{\mathrm{d}\Phi}{\mathrm{d}z}\right) \Rightarrow f_k(\Phi) + \frac{f_{bk}(\Phi)\sigma'_e(\Phi)}{\Delta\varrho\Phi g}\frac{\mathrm{d}\Phi}{\mathrm{d}z} = C \quad (32)$$

with a prescribed discharge concentration $\Phi(0) = \Phi_D > \phi_c$, hence the integration constant is $C = q\Phi_D$. The remaining first-order ODE is integrated from $z = 0$ to the height λ of the sediment, where $\Phi(\lambda) = \phi_c$ is valid. Above that, $\Phi(z) = \Phi_1$; the value Φ_1 is obtained from setting $s(\lambda) = 0$ and solving (30) for $\phi^+ = \Phi_1$, whose right-hand side is obtained from (32) (see [11, 29]).

We select the flux density function f_{bk} given in (8) and an effective solid stress function of the type $\sigma_e(\phi) = a_1 \exp(a_2\phi)$ for $\phi > \phi_c$ and use the parameters determined by Becker [5] for ore tailings from copper mining:

$$\left.\begin{array}{lllllllll} \varrho_s &=& 2500\,[\mathrm{kg/m^3}] & u_\infty &=& 6.05\times10^{-4}\,[\mathrm{m/s}] & a_2 &=& 17.9 \\ \varrho_f &=& 1000\,[\mathrm{kg/m^3}] & C &=& 11.59 & \phi_c &=& 0.23 \\ g &=& 9.8\,[\mathrm{m/s^2}] & a_1 &=& 5.35\,[\mathrm{N/m^2}] & & & \end{array}\right\} \quad (33)$$

Fig. 6 shows the resulting model functions. Furthermore, we used the parameters $\theta = 0.5$, $\kappa = 1.3$, $\Delta z = 0.015\,[\mathrm{m}]$, $L = 6\,[\mathrm{m}]$ and $\lambda = 333.\bar{3}\,[\mathrm{s/m}]$.

The first example (Fig. 7) shows the batch settling of a uniform suspension of concentration $\phi_0 = 0.13$ during $T = 72$ hours.

In the second example, we consider continuous sedimentation with $q = -1.5\times10^{-5}\,[\mathrm{m/s}]$ of an initially empty ($\phi_0 \equiv 0$) ICT. We intend to fill it up, to operate it for a time at the steady state given by q and $\Phi_D = 0.39$, from which we calculate $\Phi_1 = 0.01077641$ and $\lambda \approx 2.54\,[\mathrm{m}]$, and to empty

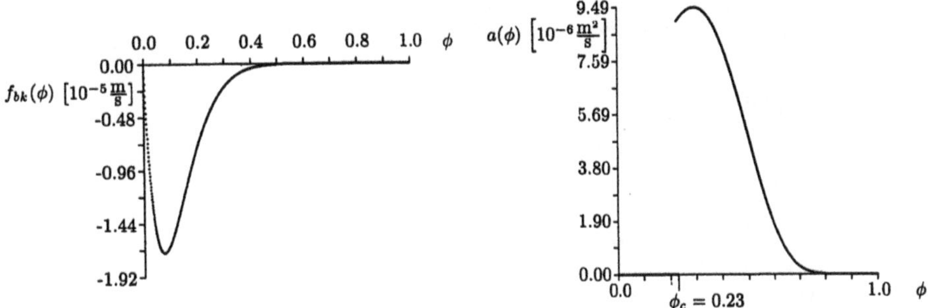

Figure 6. Kynch batch flux density function f_{bk} and diffusion coefficient $a(\phi)$ constructed with experimental parameters given in (33).

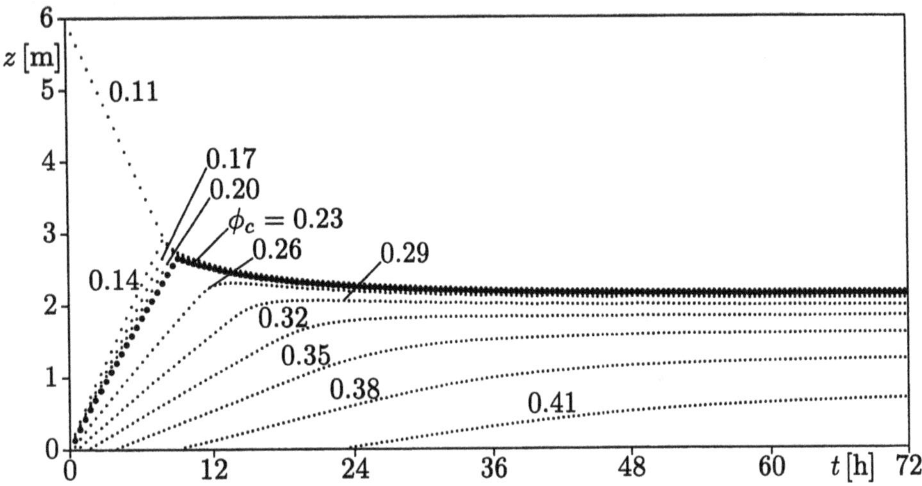

Figure 7. Settling plot for batch settling of a homogeneous suspension. The dotted lines correspond to the annotated concentrations, the fat dots to the critical concentration.

it. At $t = 0$, we start filling it up by setting $\phi_1(t) = 0.06$, which causes a rarefaction wave between the characteristic lines $z_1(t) = L + f_k'(0)t$ and $z_2(t) = L + f_k'(0.06)t$ to propagate into the thickener, see Fig. 8. After 19 hours, the sediment has surpassed the height λ of the steady state which is necessary to obtain a discharge concentration of $\phi_D = 0.39$, therefore ϕ_1 is changed to the value Φ_1. The sediment level decreases reaching the steady state approximately after 105 hours. After operating at approximate steady state for 24 hours, we decide to empty it by setting $\phi_1 = 0$ at $t = 129$ hours. The ICT has emptied by $T = 192$ hours, when the simulation ends. Note that the changes of $\phi_1(t)$ propagate into the ICT along the shock lines

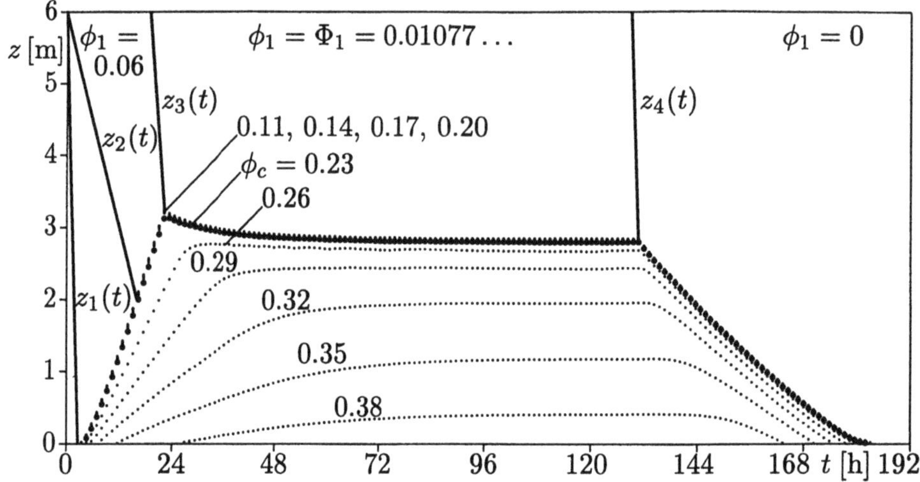

Figure 8. Settling plot for filling up, operation at steady state and emptying of an ICT. The dotted lines correspond to the annotated concentrations, the fat dots to the critical concentration.

$z_3(t) = L + (t - 19\,\mathrm{h})\sigma(0.06, \Phi_1)$ and $z_4(t) = L + (t - 129\,\mathrm{h})\sigma(\Phi_1, 0)$, as indicated in Fig. 8.

5. Discussion

In this contribution, we present a mathematical framework for the study of solid-liquid separation processes. We have shown that this research area is important to engineers who would greatly benefit from these results. On the other hand, sedimentation provides an interesting application for the analysis and numerics of hyperbolic conservation laws and, more general, for degenerate parabolic equations. The latter are also of interest in oil reservoir simulation models [35, 36].

Kynch Sedimentation Processes, also frequently referred to as *kinematical sedimentation theory*, are now well understood in one space dimension. Their main advantage is that they predict correctly most, though not all, phenomena observed in the settling of real suspensions by merely selecting an appropriate flux density function, and that exact solutions can be determined with reasonable effort. Kynch theory has also been applied by Kluwick [39] to study the propagation of finite waves of small amplitudes in ideal suspensions, by Schneider and Schaflinger [54]–[56] to the sedimentation beneath inclined walls, and by Anestis and Schneider [1, 57] to the centrifugation of ideal suspensions. Concha *et al.* [30] studied the sedimentation of a suspension with six different particle sizes. The extension of

Kynch theory to this case is a system of conservation laws which is genuinely nonlinear in the sense of Lax [46, 47]. Although the mathematical tools are available, exact solutions of this problem in the sense of [15], as presented here, have not yet been determined [16].

Nevertheless, the main shortcoming of Kynch theory of sedimentation remains that it applies only to ideal suspensions where concentration values can only propagate along straight lines. Curved sediment concentration lines as predicted in Fig. 7 and as observed in experiments [6, 58] for flocculated suspensions are excluded. This disadvantage is amended by the dynamic sedimentation model with compression.

However, even in one space dimension, a variety of open questions persist concerning the corresponding mathematical model and the suggested numerical algorithm. The convergence of the algorithm, designed on the basis of the total variation boundedness of the exact solution [7, 12] remains to be proved, possibly following the analysis by Evje and Hvistendahl Karlsen [35]. The algorithm yields reasonable results even in the case of a discontinuous coefficient $a(\phi)$ as used in subsection 4.4, therefore we suppose that the regularity assumptions made in Theorem 3 can be relaxed.

Currently, the authors' interest focusses on the formulation and analysis of a two-dimensional axisymmetric or a three-dimensional sedimentation model within the framework given by equations (10)–(12), which should consider the shape of the sedimentation vessel and the location of inlets and outlets, as sketched in Fig. 1. As pointed out by Schneider [56], the assumption of inviscidity implies that the concentration depends only on height, which can be seen easily by setting $\mu_{\text{eff}} \equiv 0$ and taking the curl of equation (12). The flow field itself, however, is two- or three-dimensional, and presents the problem that, for inclined walls, the conditions $\mathbf{f}\cdot\mathbf{n} = \mathbf{q}\cdot\mathbf{n} = 0$ can not be satisfied simultaneously. Schneider [56] develops a boundary layer theory to resolve the problem. Since measurements [5] show that the concentration in an industrial thickener is also dependent on radius in the axisymmetric case, the next step will be a study of equations (10)–(12) for non-vanishing viscosity.

Acknowledgement

The authors thank Prof. M.C. Bustos from the Department of Mathematical Engineering of the University of Concepción for reviewing the manuscript. The preparation of this chapter was possible through support from Fundación Andes, Chile (project C-13131 at the University of Concepción) and from the SFB 404 (Collaborative Research Programme 404) of the German Research Foundation (DFG) at the University of Stuttgart.

References

1. Anestis, G. and Schneider, W. (1983) Application of the Theory of Kinematic Waves to Centrifugation, *Ingenieur-Archiv* **53**, pp. 399–407
2. Ballou, D.P. (1970) Solutions to Non-linear Hyperbolic Cauchy Problems without Convexity Conditions, *Trans. Amer. Math. Soc.* **152**, pp. 441–460
3. Bardos, C., Le Roux, A.Y. and Nedelec, J.C. (1979) First Order Quasilinear Equations with Boundary Conditions, *Comm. Partial Diff. Eqns.* **4** (9), pp. 1017–1034
4. Bascur, O. (1976) *Modelo fenomenológico de suspensiones en sedimentación*, Engineering Thesis, University of Concepción.
5. Becker, R. (1982) *Espesamiento continuo, diseño y simulación de espesadores*, Engineering Thesis, University of Concepción.
6. Been, K. and Sills, G.C. (1981) Self-Weight Consolidation of Soft Soils: An Experimental and Theoretical Study, *Geotechnique* **31**, 519–535 (1981).
7. Bürger, R. (1996) *Ein Anfangs-Randwertproblem einer quasilinearen entarteten parabolischen Gleichung in der Theorie der Sedimentation mit Kompression*, Doctoral Thesis, University of Stuttgart.
8. Bürger, R. (1997) A Mathematical Model for Sedimentation with Compression, Proceedings of the *Conference on Analysis, Numerics and Applications of Differential and Integral Equations*, Stuttgart, October 9–11, 1996. Addison-Wesley-Longman; to appear.
9. Bürger, R. and Concha, F. (1997) Numerical Simulation of Continuous Sedimentation of Flocculated Suspensions, Proceedings of *Numerical Modelling in Continuum Mechanics: The Third Summer Conference*, Prague, Czech Republic, September 8–11, 1997; to appear.
10. Bürger, R. and Concha, F. (1997) Simulation of the Transient Behaviour of Flocculated Suspensions in a Continuous Thickener, Proceedings of the *XX International Mineral Processing Congress*, Aachen, Germany, September 21–26, 1997, GDMB, Clausthal-Zellerfeld, Vol. **4**, pp. 91–101.
11. Bürger, R. and Concha, F. (1997) Mathematical Model and Numerical Simulation of the Settling of Flocculated Suspensions, Preprint 97/28, Sonderforschungsbereich 404, University of Stuttgart.
12. Bürger, R. and Wendland, W.L. (1997) Existence, Uniqueness and Stability of Generalized Solutions of an Initial-Boundary Value Problem for a Degenerating Quasilinear Parabolic Equation, Preprint 97/39, Sonderforschungsbereich 404, University of Stuttgart; to appear in *J. Math. Anal. Appl.*
13. Bürger, R. and Wendland, W.L. (1997) Entropy Boundary and Jump Conditions in the Theory of Sedimentation with Compression, Preprint 97/40, Sonderforschungsbereich 404, University of Stuttgart; to appear in *Math. Meth. Appl. Sci.*
14. Bürger, R. and Wendland, W.L. (1997) Mathematical Problems in Sedimentation, Proceedings of *Numerical Modelling in Continuum Mechanics: The Third Summer Conference*, Prague, Czech Republic, September 8–11, 1997; to appear.
15. Bustos, M.C. and Concha, F. (1988) On the Construction of Global Weak Solutions in the Kynch Theory of Sedimentation, *Math. Meth. Appl. Sci.* **10**, pp. 245–264
16. Bustos, M.C. and Concha, F. (1996) Kynch Theory of Sedimentation, Chapter 2 (pp. 7–49) in Tory, E. (ed.), *Sedimentation of Small Particles in a Viscous Fluid*, Computational Mechanics Publications, Southampton, UK.
17. Bustos, M.C., Concha, F. and Wendland, W.L. (1990) Global Weak Solutions to the Problem of Continuous Sedimentation of an Ideal Suspension, *Math. Meth. Appl. Sci.* **13**, pp. 1–22
18. Bustos, M.C., Paiva, F. and Wendland, W.L. (1990) Control of Continuous Sedimentation of Ideal Suspensions as an Initial and Boundary Value Problem, *Math. Meth. Appl. Sci.* **12**, pp. 533–548
19. Bustos, M.C., Paiva, F. and Wendland, W.L. (1996) Entropy Boundary Conditions in the Theory of Sedimentation of Ideal Suspensions, *Math. Meth. Appl. Sci.* **19**,

pp. 679–697
20. Cheng, K.S. (1981) Asymptotic Behaviour of Solutions of a Conservation Law without Convexity Conditions, *J. Diff. Eqns.* **40**, pp. 343–376
21. Cheng, K.S. (1983) Constructing Solutions of a Single Conservation Law, *J. Diff. Eqns.* **49**, pp. 344–358
22. Cheng, K.S. (1986) A Regularity Theorem for a Nonconvex Scalar Conservation Law, *J. Diff. Eqns.* **61**, pp. 79–127
23. Coe, H.S. and Clevenger, G.H. (1916) Methods for Determining the Capacity of Slimesettling Tanks, *Trans. AIME* **55**, pp. 356–385
24. Concha, F. and Barrientos, A. (1993) A Critical Review of Thickener Design Methods, *KONA* No. 11, pp. 79–104
25. Concha, F., Barrientos, A. and Bustos, M.C. (1995) Phenomenological Model of High Capacity Thickening, Proceedings of the *19th International Mineral Processing Congress (XIX IMPC)*, San Francisco, USA, November 1995, Ch.14, pp. 75–79
26. Concha, F. and Bascur, O. (1977) Phenomenological Model of Sedimentation. Proceedings of the *12th International Mineral Processing Congress (XII IMPC)*, São Paulo, Brazil, **4**, pp. 29–46
27. Concha, F. and Bustos, M.C. (1991) Settling Velocities of Particulate Systems: 6. Kynch Sedimentation Processes: Batch Settling, *Int. J. Min. Proc.* **32**, pp. 193–212
28. Concha, F. and Bustos, M.C. (1992) Settling Velocities of Particulate Systems: 7. Kynch Sedimentation Processes: Continuous Thickening, *Int. J. Min. Proc.* **34**, pp. 33–51
29. Concha, F., Bustos, M.C. and Barrientos, A. (1996) Phenomenological Theory of Sedimentation, Chapter 3 (pp. 51–96) in Tory, E. (ed.), *Sedimentation of Small Particles in a Viscous Fluid*, Computational Mechanics Publications, Southampton, UK.
30. Concha, F., Lee, C.H. and Austin, L.G. (1992) Settling Velocities of Particulate Systems: 8. Batch Sedimentation of Polydispersed Suspensions of Spheres, *Int. J. Min. Proc.* **35**, pp. 159–175
31. Dafermos, C.M. (1972) Polygonal Approximations of Solutions of the Initial Value Problem for a Conservation Law, *J. Math. Anal. Appl.* **38**, pp. 33–41
32. Davis, R.H. and Acrivos, A. (1985) Sedimentation of Noncolloidal Particles at Low Reynolds Numbers, *Ann. Rev. Fluid Mech.* **17**, pp. 91–118
33. Dubois, F. and Le Floch, P. (1988) Boundary Conditions for Nonlinear Hyperbolic Systems of Conservation Laws, *Journal of Diff. Eqns* **71**, pp. 93–122
34. Eckert, W.F., Masliyah, J.H., Gray, M.R. and Fedorak, P.M. (1996) Prediction of Sedimentation and Consolidation, *AIChE J.* **42**, No. 4, pp. 960–972
35. Evje, S. and Hvistendahl Karlsen, K. (1997) Viscous Splitting Approximation of Mixed Hyperbolic-Parabolic Convection-Diffusion Equations, Report, Institut Mittag-Leffler, Stockholm, Sweden.
36. Evje, S., Hvistendahl Karlsen, K., Lie, K.-A. and Risebro, N.H. (1997) Front Tracking and Operator Splitting for Nonlinear Degenerate Convection-Diffusion Equations, Report, Institut Mittag-Leffler, Stockholm, Sweden.
37. Green, M.D., Eberl, M. and Landman, K.A. (1996) Compressive Yield Stress of Flocculated Suspensions: Determination via Experiment, *AIChE J.* **42**, No. 8, pp. 2308–2318
38. Kluwick, A. (1977) Kinematische Wellen, *Acta Mechanica* **26**, pp. 15–46
39. Kluwick, A. (1983) Small Amplitude Finite-Rate Waves in Suspensions of Particles in Fluids, *ZAMM* **63**, pp. 161–171
40. Kružkov, S.N. (1970) First Order Quasilinear Equations in Several Independent Variables, *Math. USSR Sbornik*, **10**, No. 2, pp. 217–243
41. Kunik, M. (1989) *Über die schwachen Lösungen skalarer hyperbolischer Erhaltungsgleichungen in der Theorie der Sedimentation*, Doctoral Thesis, University of Stuttgart.

42. Kunik, M. (1993) A Solution Formula for a Non-Convex Scalar Hyperbolic Conservation Law with Monotone Initial Data, *Math. Meth. Appl. Sci.* **16**, pp. 895–902
43. Kunik, M., Wendland, W.L., Paiva, F. and Bustos, M.C. (1993) The Polygonal Approximation in the Kynch Theory of Sedimentation, Preprint 93-07, Department of Mathematical Engineering, University of Concepción.
44. Kynch, G.J. (1952) A Theory of Sedimentation, *Trans. Farad. Soc.* **48**, pp. 166–176
45. Landman, K.A. and White, L.R. (1992) Determination of the Hindered Settling Factor for Flocculated Suspensions, *AIChE J.* **38**, No.2, pp. 184–192
46. Lax, P.D. (1957) Hyperbolic Systems of Conservation Laws II, *Comm. Pure Appl. Math.* **10**, pp. 537–566
47. Le Veque, R.J. (1992) *Numerical Methods for Conservation Laws*, Second Ed., Birkhäuser Verlag, Basel.
48. Málek, J., Nečas, J., Rokyta, M. and Růžička, M. (1996) *Weak and Measure-valued Solutions to Evolutionary PDEs*, Chapman & Hall, London.
49. Mishler, R.T. (1912) Settling Slimes at the Tigre Mill, *Eng. Mining J.* **94** (14), pp. 643–646
50. Mishler, R.T. (1918) Methods for Determining the Capacity of Slime-Settling Tanks, *Trans. AIME* **58**, pp.102 ff
51. Nessyahu, H. and Tadmor, E. (1990) Non-Oscillatory Central Differencing for Hyperbolic Conservation Laws, *J. Comp. Phys.* **87**, pp. 408–463
52. Paiva, F. and Bustos, M.C. (1994) Existence of the Entropy Solution for the Simulation of the Sedimentation in a Continuous Thickener, *Revista Notas*, Sociedad Matemática de Chile, **XII, XIII** (1), pp. 47–56
53. Richardson, J.F. and Zaki, W.N. (1954) The Sedimentation of Uniform Spheres under Conditions of Viscous Flow, *Chem. Engrg. Science* **3**, pp. 65–73
54. Schaflinger, U. (1985) Experiments on Sedimentation Beneath Downward-Facing Inclined Walls, *Int. J. Multiphase Flow* **11**, pp. 189-199
55. Schaflinger, U. (1985) Influence of Non-Uniform Particle Size on Settling Beneath Downward-Facing Inclined Walls, *Int. J. Multiphase Flow* **11**, pp. 783–796
56. Schneider, W. (1982) Kinematic-Wave Theory of Sedimentation Beneath Inclined Walls, *J. Fluid Mech.* **120**, pp. 323–346
57. Schneider, W. (1985) Kinematic Wave Description of Sedimentation and Centrifugation Processes, pp. 326–337 in: Meier, G.E.A. and Obermaier, F. (eds.), *Flow of Real Fluids*, Lecture Notes in Physics, Springer Verlag, Berlin.
58. Tiller, F.M., Hsyung, N.B. and Cong, D.Z. (1995) Role of Porosity in Filtration: XII. Filtration with Sedimentation, *AIChE J.* **41**, No. 5, pp. 1153–1164
59. Ungarish, M. (1993) *Hydrodynamics of Suspensions*, Springer-Verlag.
60. Vol'pert, A.I. and Hudjaev, S.I. (1969) Cauchy's Problem for Degenerate Second Order Quasilinear Parabolic Equations, *Math. USSR Sbornik* **7**, No. 3, pp. 365–387
61. Wu, Z. (1982) A Note on the First Boundary Value Problem for Quasilinear Degenerate Parabolic Equations, *Acta Math. Sc.* **4**, No.2, pp. 361–373
62. Wu, Z. (1983) A Boundary Value Problem for Quasilinear Degenerate Parabolic Equations, MRC Technical Summary Report #2484, University of Wisconsin, USA.
63. Wu, Z. and Wang, J. (1982) Some Results on Quasilinear Degenerate Parabolic Equations Of Second Order, Proceedings of the *1980 Beijing Symposium on Differential Geometry and Differential Equations*, Vol. 3, Science Press, Beijing, Gordon & Breach, Science Publishers Inc., pp. 1593–1609
64. Wu, Z. and Yin, J. (1989) Some Properties of Functions in BV_x and Their Applications to the Uniqueness of Solutions for Degenerate Quasilinear Parabolic Equations, *Northeastern Math. J.* **5** (4), pp. 395–422

DIFFERENCE APPROXIMATIONS OF ACOUSTIC AND ELASTIC WAVE EQUATIONS

DUGALD B. DUNCAN
Department of Mathematics
Heriot-Watt University
Edinburgh UK, EH14 4AS

Abstract. Computer simulation of linear wave propagation has applications in areas including ultrasonic testing, seismology and electromagnetic scattering. Realistic problems, often involving complicated multi-dimensional geometries, require large amounts of computing resources, and so it is essential to use the most efficient schemes available.

We describe the construction of accurate and efficient finite difference approximations of linear wave equations and discuss properties of the schemes, including mesh dependent anisotropy, and their suitability for use on parallel computers. We illustrate how these schemes can be coupled with overlapping meshes to solve acoustic and elastic (forward) scattering problems involving multiple cracks and inclusions with arbitrary locations and orientations.

1. Introduction

Two main applications of computer simulation of linear wave propagation are

- *Forward Scattering:* simulate the solution of PDEs or integral equations which describe scattering of linear waves from an object of known size, shape and location.
- *Inverse Problem:* reconstruct the shape of an object by recording and processing waves scattered by it.

We examine the forward scattering problem in this paper and note that is important as a predictor of wave interactions with structures (e.g. electromagnetic interactions with antennas, seismic interactions with rock structures) and also as a step in understanding and simplifying the inverse prob-

E.F. Toro and J.F. Clarke (eds.), Numerical Methods for Wave Propagation, 197–210.

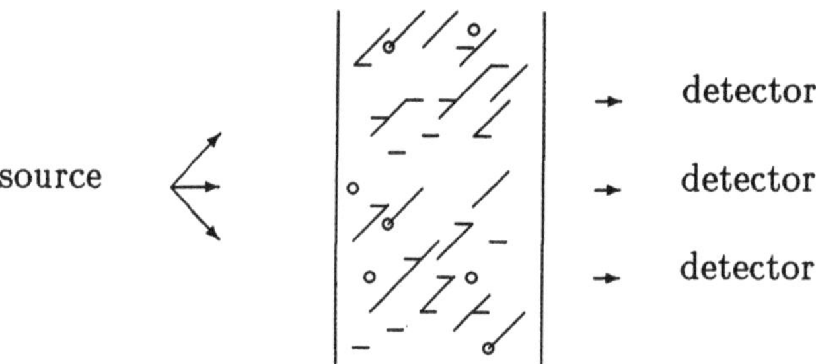

Figure 1. Typical heterogeneous slab composed of cracks and inclusions embedded in an otherwise homogeneous medium.

lem. Solving inverse problems is often the ultimate goal of work in this area since it has many practical applications, including radar, sonar, seismic surveying, ultrasonic testing etc.

An example of the application of forward scattering simulation comes in joint work with C. MacBeth of the British Geological Survey. The immediate aim is to simulate elastic (seismic) waves propagating in heterogeneous media containing distributions of various types of cracks and inclusions. The next step is to use the information obtained from these simulations to derive a homogeneous equivalent medium which approximates the macroscopic behaviour of the waves through the cracked media. Finally the homogeneous description can be used to simplify the inverse problems solved in processing seismic surveying data. The main practical difficulty in solving this forward scattering problem by difference methods is in incorporating a large number of cracks of arbitrary length and orientation at arbitrary positions on a finite difference/element grid and it is this that we examine in the rest of the paper.

2. Background

Linear wave equations are often formulated naturally with second degree time and space derivatives. The acoustic or scalar wave equation (in 2 dimensional Cartesian space),

$$u_{tt} = u_{xx} + u_{yy} ,$$

is probably the simplest example. We also consider the elastic wave equation (EWE) which for homogeneous, isotropic media can be written in "mass $\times$ acceleration = force" form as

$$\begin{aligned} \rho u_{tt} &= \sigma^{xx}_x + \sigma^{xy}_y , \\ \rho v_{tt} &= \sigma^{xy}_x + \sigma^{yy}_y \end{aligned}$$

where u, v are the displacements in the x, y directions, the stresses are defined by

$$\sigma^{xx} = (\lambda + 2\mu)u_x + \lambda v_y \ , \sigma^{yy} = (\lambda + 2\mu)v_y + \lambda u_x \ , \sigma^{xy} = \mu(u_y + v_x)$$

and the material properties are the density ρ and the Lamé elasticities λ, μ. This can be collapsed into the more compact form

$$\begin{aligned} u_{tt} &= P^2 u_{xx} + (P^2 - S^2)v_{xy} + S^2 u_{yy} \ , \\ v_{tt} &= S^2 v_{xx} + (P^2 - S^2)u_{xy} + P^2 v_{yy} \end{aligned} \tag{1}$$

where the compressional and shear wave speeds

$$P = \sqrt{(\lambda + 2\mu)/\rho} \quad \text{and} \quad S = \sqrt{\mu/\rho}$$

appear explicitly in the equations.

It is also possible and common to reduce the second order equations to a first order system of hyperbolic equations with canonical form

$$\underline{w}_t = A\underline{w}_x + B\underline{w}_y \ . \tag{2}$$

For the EWE (1), (2) is composed of $\underline{w} = (\rho u_t, \rho v_t, \sigma^{xx}, \sigma^{xy}, \sigma^{yy})^T$ with

$$A = \left(\begin{array}{cc|ccc} 0 & 0 & 1 & 0 & 0 \\ 0 & 0 & 0 & 0 & 1 \\ \hline P^2 & 0 & 0 & 0 & 0 \\ D^2 & 0 & 0 & 0 & 0 \\ 0 & S^2 & 0 & 0 & 0 \end{array} \right) \quad \text{and} \quad B = \left(\begin{array}{cc|ccc} 0 & 0 & 0 & 0 & 1 \\ 0 & 0 & 0 & 1 & 0 \\ \hline 0 & D^2 & 0 & 0 & 0 \\ 0 & P^2 & 0 & 0 & 0 \\ S^2 & 0 & 0 & 0 & 0 \end{array} \right) \tag{3}$$

where $D^2 = P^2 - 2S^2 > 0$.

There are various ways to approximate the solution of the equations above, and a comprehensive collection of papers on many different methods for seismic wave propagation can be found in [10]. Another collection with more general applications is [4]. We only consider finite difference schemes here, but note that there is a growing body of work on the use of pseudospectral methods in wave propagation, and these schemes appear to be well suited to some common geometrical configurations [8, 16]. Unfortunately, they do not yet appear to be suitable for the seismic problem outlined in the Introduction.

2.1. EXPLICIT AND IMPLICIT DIFFERENCE METHODS

It is tempting to argue that one should reduce higher order hyperbolic problems to the canonical first order form (2), and then approximate their solution by a generic numerical solver, in the same way as higher order ODEs are reduced to $\underline{y}' = \underline{f}(t, \underline{y})$ to allow standard library software to be used. Generic methods for (2) like second central differencing, Lax-Wendroff [13, p.181] and 2-4 Kreiss-Oliger [11] schemes do indeed exist, but one can obtain much more efficient methods by tailoring the approximation to the particular properties of the wave equation being studied by balancing and cancelling space and time errors in the components of the solution.

This is usually done by constructing schemes from wider and wider space difference stencils to obtain higher order accuracy [6], which in the extreme gives a pseudospectral approximation [8]. Some examples of high order accurate finite difference approximations tailored for the EWE can be found in [12, 23] (these are second order accurate in time and fourth order in space using 25 point stencils in 2D) and in [5] (fourth order in both time and space). fourth and sixth order accurate (both time and space) schemes for the acoustic wave equation using products of nearest neighbour difference operators are given in [7].

The high order accurate schemes mentioned above are explicit, in that the solution at each mesh point is updated by combining information from its near neighbours at previous time levels without the need for the solution of a (linear) system of algebraic equations. Standard Fourier analysis for uniform grid explicit schemes shows that the maximum stable time step size is directly proportional to the space mesh size. This is not as bad a restriction as it might first appear, since the spatial and temporal variation of the solutions of linear wave problems are directly proportional, and optimal results are obtained when the space and time resolution are comparable. This, and the prohibitive expense of solving a large system of linear equations at each step to advance in time, effectively rules out implicit difference methods as unnecessary and inefficient for linear wave problems. Roughly speaking, for schemes in two space dimensions with mesh size proportional to $1/N$, an explicit method requires $O(N^2)$ operations to advance one time step, while an implicit scheme requires $O(N^3)$.

The advantage of explicit methods extends to parallel implementation of the algorithms. Implicit schemes require expensive global data communications, but explicit methods only use local information and so require only local communications. This minimises the communications overhead, and allows more efficient use of parallel computers. Some timing results for acoustic wave schemes on a parallel computer composed of T800 Transputers are given in [7] and an overview of parallel methods for FDTD (finite difference time domain) electromagnetic wave propagation can be found in [15].

Other considerations in the design of difference schemes include the problems of grid anisotropy and the related problem of frequency dependent propagation speeds. Grid anisotropy (in isotropic wave equations) causes waves to travel at different speeds depending on their orientation on the grid, and the extreme speeds tend to occur along grid axes and symmetry lines. See [21] for a general discussion and [7] for work on the acoustic wave equation. It is common to "cure" these problems by using finer grids, higher order accuracy and schemes whose leading error terms are isotropic, and [9] suggests the use of schemes with guaranteed bounds on the group velocity error over a range of frequencies.

2.2. CRACKS ON A GRID

We now return to the multiple crack example from seismology given in the Introduction. There are two main ways to incorporate cracks and other objects into a finite difference/element grid.

The first is to use a global (irregular) mesh that fits around all the cracks and other boundaries, and this is attractive because it lends itself to standard finite element coding. Much work on grid generation has appeared [20]. However when combined with explicit schemes for wave equations this approach suffers from linear stability problems, since the maximum stable time step size τ is restricted by

$$\tau \searrow 0 \text{ as Minimum Radius Element} \searrow 0 \ .$$

In other words the smallest and narrowest elements set the time step size, which is inefficiently small over the rest of the domain. (Recall that we have already discounted the use of implicit schemes as inefficient.) There are also difficulties in deriving high order accurate schemes for irregular grids, (2nd order accuracy is reported in the literature [14] for Maxwell's equations) and problems with spurious reflections at drastic changes in mesh element shape and size [1].

The second method is to use small sections of uniform grid round each of the cracks, and an underlying uniform grid for the bulk of the region being modelled [3]. Information is passed between these overlapping grids using interpolation. A great deal of work has been done with overlapping grids for various problems ranging from shock capture to steady state heat transfer (see for example [3, 22]), although their use in multi-dimensions appears to be running ahead of what can proved about their properties.

Stability results for difference approximations of hyperbolic equations with multiple interfaces and boundaries rely on the GKS stability theory and various authors [2, 17, 19] have used it to examine stability of hyperbolic problems with mesh refinement and overlap. This work is complicated enough for 1D problems, and does not appear to have been extended to practical cases in 2D and 3D. Convergence analysis of wave schemes on irregular grids in more than one dimension is also complicated, and results for electromagnetics have only appeared recently in [14].

3. Approximation of 1D Wave Equation

To illustrate some of the main points above and to avoid unnecessary detail, let us now turn to the 1D scalar or acoustic wave equation

$$u_{tt} = u_{xx}, \quad t > 0, \ x \in (0,1) \ . \tag{4}$$

The solution $u(x,t)$ can be found given initial data $u(x,0)$, $u_t(x,0)$ and boundary conditions $u(0,t) = 0 = u(1,t)$. Using the notation

$$\begin{aligned} u_j^m & \quad \approx u(jh, m\tau) \\ \tau & \quad \text{time step size} \\ h = 1/N & \quad \text{space element size} \\ r = \tau/h & \quad \text{mesh ratio,} \end{aligned}$$

the standard finite difference approximation using second central differences is

$$u_j^{m+1} - 2u_j^m + u_j^{m-1} = r^2(u_{j+1}^m - 2u_j^m + u_{j-1}^m) \tag{5}$$

for $j = 1, \ldots, N-1$; $m \geq 1$; given u_j^{-1}, u_j^0 from initial data and $u_0^m \equiv 0 \equiv u_N^m$ from boundary conditions. This approximation (5) can be written in matrix/vector form as

$$\underline{u}^{m+1} - 2\underline{u}^m + \underline{u}^{m-1} = r^2 A \ \underline{u}^m \tag{6}$$

where $\underline{u} = (u_1, \ldots, u_{n-1})$ and $A = \text{tridiag}(1,\text{-}2,1)$.

Difference schemes must be stable if they are to be useful, and for scheme (6) stability can be determined by examining the one step version

$$\begin{pmatrix} \underline{u}^{m+1} \\ \underline{u}^m \end{pmatrix} = \left(\begin{array}{c|c} 2I + r^2 A & -I \\ \hline I & 0 \end{array}\right) \begin{pmatrix} \underline{u}^m \\ \underline{u}^{m-1} \end{pmatrix} .$$

The eigenvalues μ of the composite matrix above satisfy:

$$\mu^2 - 2\mu + 1 = \mu r^2 \lambda(A)$$

and the scheme is exponentially unstable unless $|\mu| = 1$, which is equivalent to requiring that the eigenvalues $\lambda(A)$ are real and satisfy $-4 \leq r^2\lambda(A) \leq 0$. This applies to the tridiag(1,-2,1) form arising from (5), as well as to more general choices of A.

3.1. IRREGULAR GRID

As a very simplistic model of an irregular grid, let us consider what happens if the mesh consists of locally uniform regions joined by a localised irregular

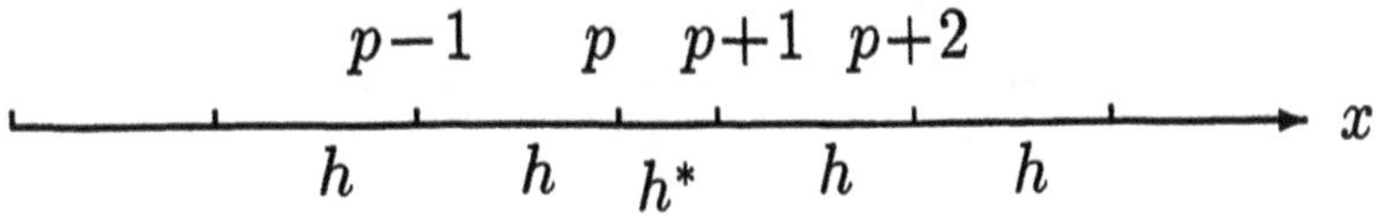

$p-1 \quad p \quad p+1 \quad p+2$

$h \quad h \quad h^* \quad h \quad h \qquad x$

Figure 2. One bad interval of size $h^* = \epsilon h$ in an otherwise uniform grid of size h.

section. In particular we look at the effect of having one interval of size h^* in an otherwise uniform mesh of size h as shown in Figure 2. A reasonable irregular mesh scheme which is derived from a mass lumped piecewise linear finite element method gives

$$\begin{aligned}
D_t^2 u_j^m &= r^2(u_{j+1}^m - 2u_j^m + u_{j-1}^m)\,, \quad j \neq p, p+1\,, \\
D_t^2 u_p^m &= \frac{2\tau^2}{h+h^*}\left(\frac{u_{p+1}^m - u_p^m}{h^*} - \frac{u_p^m - u_{p-1}^m}{h}\right), \\
D_t^2 u_{p+1}^m &= \frac{2\tau^2}{h^*+h}\left(\frac{u_{p+2}^m - u_{p+1}^m}{h} - \frac{u_{p+1}^m - u_p^m}{h^*}\right).
\end{aligned}$$

where the second time difference operator is defined by $D_t^2 u_j^m = u_j^{m+1} - 2u_j^m + u_j^{m-1}$.

Setting $h^* = \epsilon h, \quad 0 < \epsilon \leq 1$, we obtain the matrix/vector form of this scheme

$$D_t^2 \underline{u}^m = r^2 \begin{pmatrix}
\vdots\cdots & \cdots & \cdots & \cdots & \cdots & \cdots\vdots \\
\vdots\ 1 & -2 & 1 & & & \vdots \\
\vdots & c_1 & -c_2 & c_3 & & \vdots \\
\vdots & & c_3 & -c_2 & c_1 & \vdots \\
\vdots & & & 1 & -2 & 1\ \vdots \\
\vdots\cdots & \cdots & \cdots & \cdots & \cdots & \cdots\vdots
\end{pmatrix} \underline{u}^m \equiv A\underline{u}^m$$

where $\underline{u} = (u_1, u_2, \ldots, u_{n-1})$ and

$$c_1 = \frac{2}{1+\epsilon}\,, c_2 = \frac{2}{\epsilon}\,, c_3 = \frac{2}{(1+\epsilon)\epsilon}\,.$$

The eigenvalues of matrix A above are real when $0 < \epsilon \leq 1$ since there exists a real matrix B such that BAB^{-1} is symmetric. (In this case B is diagonal with elements 1, except for the elements corresponding to the two irregular rows which are both $1/\sqrt{c_1}$.) Using properties of eigenvalues of

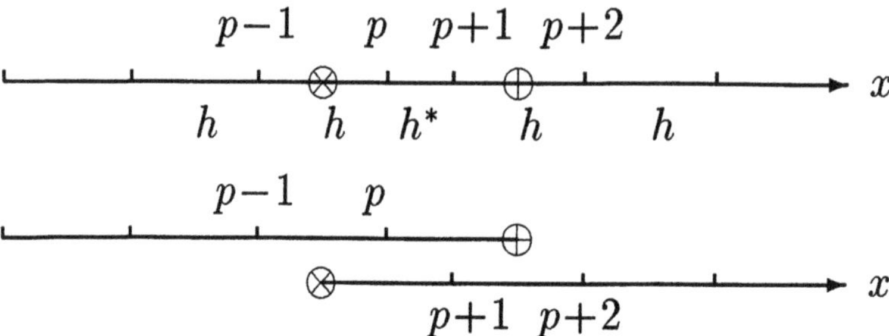

Figure 3. A grid with one bad interval of size $h^* = \epsilon h$ split into two uniform overlapping grids.

symmetric matrices, it is relatively easy to show that for the scheme to be stable, the mesh ratio r is limited by

$$r^2 \leq Ch^*/h = C\epsilon$$

as the irregular element gets smaller relative to the regular ones. Thus the time step must be reduced when the relative size of the irregular interval is reduced as noted in Section 2.2.

3.2. OVERLAPPING GRID

Now consider the case where an overlapping grid is used is join two regions of uniform grid which are offset from one another. As shown in Figure 3, we have the right grid shifted by $h - h^*$ relative to the left grid. Once the grids are separated as shown we use the standard finite difference scheme away from the join to obtain

$$D_t^2 u_j^m = r^2(u_{j+1}^m - 2u_j^m + u_{j-1}^m) \quad j \neq p, p+1$$

and on the left and right uniform grids we can use the uniform grid scheme for $j = p$ and $j = p+1$:

$$\begin{aligned} D_t^2 u_p^m &= r^2(u_{p-1}^m - 2u_p^m + u_\oplus^m) \ , \\ D_t^2 u_{p+1}^m &= r^2 \left(u_\otimes^m - 2u_{p+1}^m + u_{p+2}^m \right) \ . \end{aligned}$$

We need the values of u at $\oplus$ and $\otimes$ (see Figure 3) and polynomial interpolation is the most obvious way to do this. For example, using the simplest linear interpolation gives

$$u_\oplus = \epsilon u_{p+1} + (1-\epsilon)u_{p+2} \quad \text{and} \quad u_\otimes = \epsilon u_p + (1-\epsilon)u_{p-1} \ ,$$

and we will see later later that higher order interpolation is desirable.

Let us first investigate the stabilty of the overlapping grid scheme with linear interpolation as outlined above. The wave equation in matrix form is

$$D_t^2 \underline{u}^m = r^2 \left(\begin{array}{cccc|cccc} \ddots & \ddots & & & & & & \\ \ddots & \ddots & 1 & & & & & \\ & 1 & -2 & & \epsilon & 1-\epsilon & & \\ \hline & 1-\epsilon & \epsilon & & -2 & 1 & & \\ & & & & 1 & \ddots & \ddots & \\ & & & & & \ddots & & \end{array} \right) \underline{u}^m \equiv A\underline{u}^m$$

where $\underline{u} = (u_1, u_2, \ldots, u_{n-1})$ and the eigenvalues of matrix above are real when $0 < \epsilon \leq 1$ since there is a reasonably simple similarity transform such that BAB^{-1} is symmetric. Using the Gerschgorin circle theorem, it is then easy to show that the stability limit is independent of $\epsilon \in (0,1]$. Similar results are obtained when one sided quadratic interpolation is used.

We have only discussed stability so far and now report the results of a representative numerical experiment to investigate the convergence of overlapping grid approximations. We solve (4) on a uniform mesh to act as a reference, and then on a mesh made up of three uniform sections overlapping in two places. The results are shown in Figure 4, and we see that the uniform mesh second order convergence is not obtained when linear interpolation is used, but it is recovered when one sided quadratic interpolation is used. This issue is discussed in [3], and the obvious point is that the interpolation must be high enough order accurate to maintain the overall accuracy of the approximations.

3.3. FIRST ORDER FORM AND STAGGERED GRIDS

We have so far considered the second order form of the acoustic wave equation, and now we recall that it is also common to use wave equations in the first order form and illustrate some of the problems that can arise combining overlapping grids with the first order form of the acoustic wave equation.

We recast the acoustic wave equation (4)in first order form as:

$$\frac{\partial}{\partial t}\begin{pmatrix} f \\ g \end{pmatrix} = \begin{pmatrix} 0 & 1 \\ 1 & 0 \end{pmatrix} \frac{\partial}{\partial x}\begin{pmatrix} f \\ g \end{pmatrix} \tag{7}$$

where

$$f(x,t) \equiv u_t \ , \quad g(x,t) \equiv u_x \ .$$

Using first central differences for the derivatives we get:

$$f_j^{m+1} - f_j^{m-1} = r\left(g_{j+1}^m - g_{j-1}^m\right) ,$$

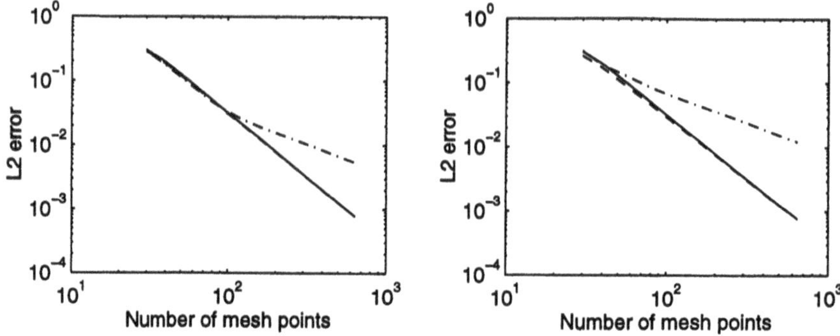

Figure 4. Error against number of mesh points for the problem described at the end of section 3.2, using a uniform mesh (solid line), an overlapping mesh with linear interpolation (dash-dot line), and with quadratic interpolation (dash line). The left plot has an overlap of $0.3h$ and the right one $0.5h$.

$$g_j^{m+1} - g_j^{m-1} = r\,(f_{j+1}^m - f_{j-1}^m)$$

for $j = 1, \ldots, N-1$ where the mesh ratio $r = \tau/h$ and we note that the numerical solution splits into 4 disjoint sets of solutions (ev = even, od = odd)

$$\{f_{\rm od}^{\rm od}, g_{\rm ev}^{\rm ev}\}\ ,\{f_{\rm ev}^{\rm od}, g_{\rm od}^{\rm ev}\}\ ,\{f_{\rm od}^{\rm ev}, g_{\rm ev}^{\rm od}\}\ ,\{f_{\rm ev}^{\rm ev}, g_{\rm od}^{\rm od}\}\ ,$$

and it is only necessary to solve for one of these combinations, since no useful information is gained by doing more. The staggered grid scheme for the $\{f_{\rm od}^{\rm ev}, g_{\rm ev}^{\rm od}\}$ component is often rewritten on a half size mesh as

$$\begin{aligned} f_j^{m+1/2} - f_j^{m-1/2} &= r\,(g_{j+1/2}^m - g_{j-1/2}^m)\ , \\ g_{j+1/2}^{m+1} - g_{j+1/2}^m &= r\,(f_{j+1}^{m+1/2} - f_j^{m+1/2}) \end{aligned} \tag{8}$$

for $j = 1, \ldots, N-1$.

Note that the skew diagonal form of the matrix in (7) also appears (in block form) in the matrices (3) arising in the elastic wave equation and in those for Maxwell's equations (see [13, p.181]). The skew diagonal form is important in the construction of efficient difference schemes, since as noted above we obtain disjoint sets of solutions and need only solve for one on them. This leads naturally to the use of staggered grids where the different components are approximated on grids translated relative to one another. The elastic wave schemes [12, 23] use staggered grids, and examples from electromagnetism can be found in [14, 15]. Staggered grid EWE schemes are much more complicated than our illustration since in 2D there are 5 solution components on 4 staggered grids, and 9 solution components on 7 grids in 3D.

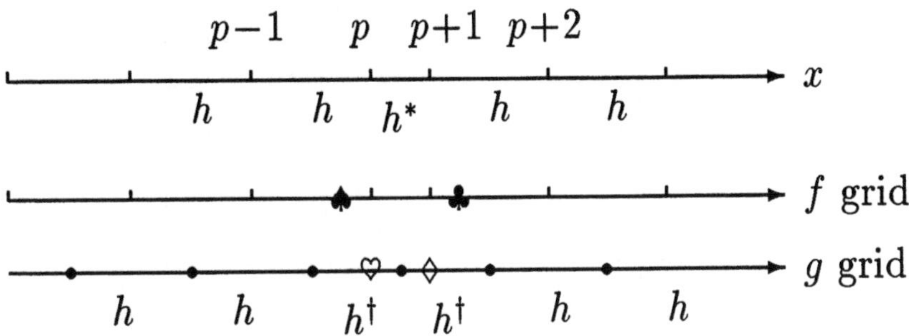

Figure 5. One bad interval of size $h^* = \epsilon h$ on the upper reference grid leads to one bad interval of size h^* in the f grid, and two of size $h^\dagger = (1+\epsilon)h/2$ in the staggered g grid. The g grid points are half way between those on the f grid.

The combination of first order form staggered grid schemes with overlapping (or just irregular) grids gets very messy to set up, but is still an efficient way to deal with the multiple crack problem. The effect of introducing a single irregular grid interval to a staggered grid scheme is illustrated in Figure 5. The scheme (8) above holds everywhere except around the irregular interval where we extend the overlapping uniform grid idea to get:

$$\begin{aligned} f_p^{m+1/2} - f_p^{m-1/2} &= r\left(g_\diamondsuit^m - g_{p-1/2}^m\right), \\ f_{p+1}^{m+1/2} - f_{p+1}^{m-1/2} &= r\left(g_{p+3/2}^m - g_\heartsuit^m\right), \\ g_{p+1/2}^{m+1} - g_{p+1/2}^m &= r\left(f_\clubsuit^{m+1/2} - f_\spadesuit^{m+1/2}\right), \end{aligned}$$

using interpolation to get f at $\clubsuit$ & $\spadesuit$ and g at $\heartsuit$ & $\diamondsuit$.

4. Example

Here we illustrate the use of overlapping grids in 2D linear elastodynamics. We model a homogeneous, isotropic material containing randomly sized and oriented planar cracks with displacement free surfaces. The computational region has an underlying uniform square grid of 400×200 points, with absorbing boundary conditions at the top and bottom, and periodic conditions on the left and right boundaries.

Each crack in the simulation lies at the centre of a local uniform grid with the same mesh size as the underlying grid. A typical mesh around a crack is shown in Figure 6. In the simulation we use a simple second order approximation of (1) composed of second central difference operators in

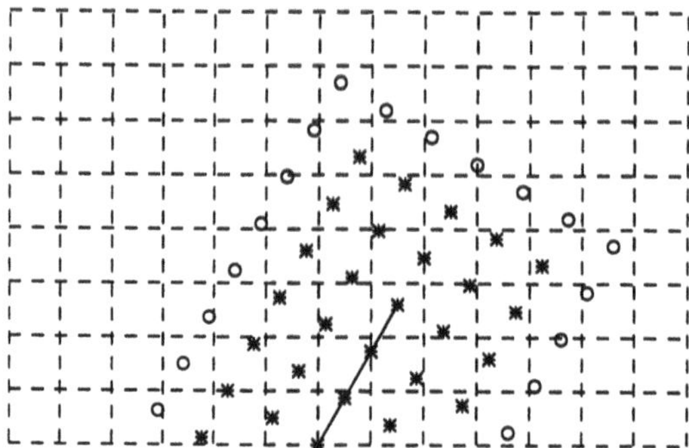

Figure 6. Section of underlying grid (dashed) and the overlapping local grid (∗ and ∘) around a crack (solid).

space and time. The scheme is applied over most of the underlying grid, and part of the local grids around each crack (points marked ∗ in Figure 6) with boundary conditions applied on the crack surface. Linear interpolation is used to transfer information from the underlying grid to the crack grid (points marked ∘ in Figure 6) and interpolation from the crack grid is used for underlying grid points falling in mesh elements adjacent to the crack.

Figure 7 shows eight snapshots of the displacement wave field obtained by the method outline above. A narrow, plane compressional wave is fired from the top of a region to interact with the group of cracks. Diffraction, reflection and mode conversion all occur repeatedly and there is no sign of instability. However, if the displacement free cracks are replaced by stress free cracks there are some stability problems, but it is not clear if they are caused by the overlapping grid or stress free boundary approximation (which is notoriously difficult to stabilise [18]).

5. Conclusions

Many aspects of the use of overlapping grids remain to be analysed and resolved, not least their stability in multi-dimensional situations. However, overlapping grids look like a promising way to use high order finite difference schemes designed for uniform grids in the solution of linear hyperbolic systems in complicated domains.

References

1. Z. Băzant (1978), Spurious reflection of elastic waves in nonuniform finite element grids, *Comp. Meth. in Appl. Mech. and Eng.*, **Vol. no. 16**, pp. 91-100
2. M. J. Berger (1985), Stability of interfaces with mesh refinement, *Math. Comp.*, **Vol. no. 45**, pp. 301-318

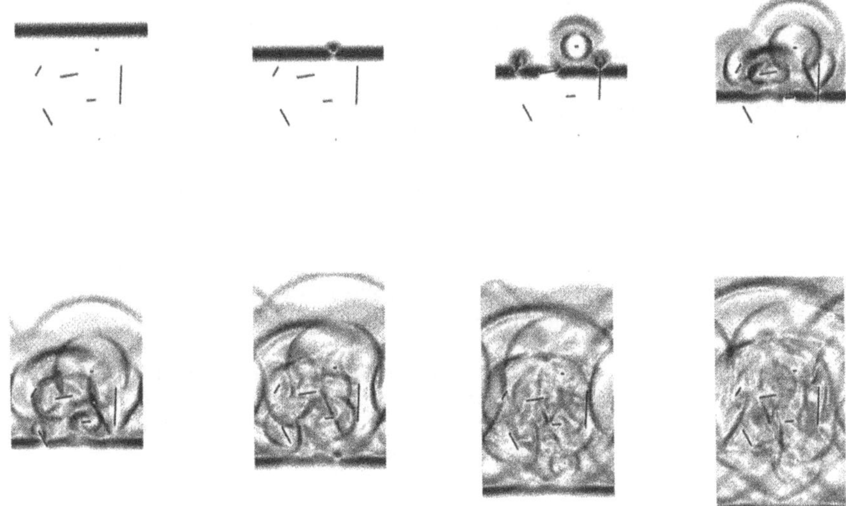

Figure 7. Snapshots of the wave field of an incident compressional wave interacting with a collection of cracks.

3. G. Cheshire and W. D. Henshaw (1990), Composite overlapping meshes for the solution of partial differential equations, *J. Comp. Phys.*, **Vol. no. 90**, pp. 1-64
4. G. Cohen and L. Halpern and P. Joly (1991), Mathematical and numerical aspects of wave propagation phenomena, Society for Industrial and Applied Mathematics.
5. G. Cohen (1986), Schemes, fourth order in time and space, for 2-D elastodynamics equations, Expanded abstracts, 58th Annual Meeting, Society of Exploration Geophysicists, pp. 1288-1291
6. M. A. Dablain (1986), The application of high-order differencing to the scalar wave equation, *Geophysics*, **Vol. no. 51**, pp. 54-66
7. D. B. Duncan and M. A. M. Lynch (1991), Jacobi iteration in implicit difference schemes for the wave equation, *SIAM J. Numer. Anal.*, **Vol. no. 28**, pp. 1661-1679
8. B. Fornberg (1987), The pseudospectral method: comparisons with finite differences for the elastic wave equation, *Geophysics*, **Vol. no. 52**, pp. 483-501
9. O. Holberg (1987), Computational aspects of the choice of operator and sampling interval for numerical differentiation in large-scale simulation of wave phenomena, *Geophysical Prospecting*, **Vol. no. 35**, pp. 629-655
10. K.R. Kelly and K.J. Marfurt (1990), Numerical modelling of seismic wave propagation, *Geophysics reprint series number 13*, Society of Exploration Geophysicists.
11. H. Kreiss and J. Oliger (1973), Methods for the approximate solution of time dependent problems, *GARP Publications series number 10*, Global Atmospheric Research Programme.
12. A.R. Levander (1988), Fourth-order finite-difference P-SV seismograms, *Geophysics*, **Vol. no. 53**, pp. 1425-1436
13. A. R. Mitchell and D. F. Griffiths (1980), The finite difference method in partial differential equations, John Wiley & Sons.
14. P. Monk and E. Süli (1994), A convergence analysis of Yee's scheme on nonuniform grids, *SIAM J. Numer. Anal.*, **Vol. no. 31**, pp. 393-412

15. S. Nguyen and B. Zook and Xiaodong Zhang (1995), Parallelising FDTD methods for solving electromagnetic scattering problems, *SIAM News*, **Vol. no. 28**, pp. 10-11
16. P. Nielsen and F. If and P. Berg and O. Skovgaard (1995), Using the pseudospectral method on curved grids for 2D elastic forward modelling, *Geophysical Prospecting*, **Vol. no. 43**, pp. 369-395
17. K. Otto and M. Thuné (1989), Stability of a Runge-Kutta method for the Euler equations on a substructured domain, *SIAM J. Sci. Comput.*, **Vol. no. 10**, pp. 154-174
18. R. Stacey (1994), New finite-difference methods for free surfaces with a stability analysis, *Bulletin of the Seismological Society of America*, **Vol. no. 84**, pp. 171-184
19. G. Starius (1980), On composite mesh difference methods for hyperbolic differential equations, *Numer. Math.*, **Vol. no. 35**, pp. 241-255
20. J.F. Thomson and Z. U. A. Warsi and C. W. Mastin (1985), Numerical grid generation - foundations and applications, North Holland Publishing Co.
21. L. N. Trefethen (1982), Group velocity in finite difference schemes, *SIAM Rev.*, **Vol. no. 24**, pp. 113-136
22. J.Y. Tu and L. Fuchs (1995), Calculations of flows using three-dimensional overlapping grids and multigrid methods, *Int. J. Num. Meth. Eng.*, **Vol. no. 38**, pp. 259-282
23. J. Virieux (1986), P-SV wave propagation in heterogeneous media: velocity-stress finite-difference method, *Geophysics*, **Vol. no. 51**, pp. 889-901

APPROXIMATE RIEMANN SOLVERS FOR FLUID FLOW WITH MATERIAL INTERFACES

MANFRED F. GÖZ

Department of Chemical Engineering †
Princeton University
Princeton
NJ 08544
USA

AND

CLAUS-DIETER MUNZ

Institute for Aerodynamics and Gasdynamics
Stuttgart University
Pfaffenwaldring 21
70550 Stuttgart
Germany

Abstract

The construction of Godunov-type schemes at material interfaces is considered. The main building block of the flux calculation in a Godunov-type scheme is the exact or the approximate solution of Riemann problems. At a material interface the flux function is discontinuous due to the change of the equation of state. A regularization of this problem is introduced and a Roe linearization of the regularized Riemann problem is given. From its solution appropriate numerical fluxes at material interfaces are deduced. These can be applied directly in the framework of Godunov-type schemes for the Euler equations in the Lagrangian frame of reference or on a moving grid. In Eulerian coordinates fixed in space it must be combined with an appropriate tracking of the material interfaces.

†New address: Department of Mechanical Engineering, MVT/UST, Martin-Luther-University of Halle-Wittenberg, D-06099 Halle, Germany

E.F. Toro and J.F. Clarke (eds.), Numerical Methods for Wave Propagation, 211–235.

1. Introduction

A difficult problem in the numerical simulation of fluid flow consists in the sharp resolution of contact discontinuities representing material or phase interfaces. Modern high resolution upwind schemes are able to capture discontinuities within a narrow transition zone consisting of one or two interior grid cells for shock waves, while a contact discontinuity is usually smeared out a bit more. The conservation property of the numerical schemes leads necessarily to a smearing of all nonstationary interfaces as long as a grid fixed in space is used. Such a numerical damping is associated with severe difficulties in the simulation of the flow of different materials, because it may produce unphysical mixed states across the interface. In this case an equation of state has to be specified within the transition zone which is a numerical artifact without any physical meaning. Other problems where such a damping is intolerable occur in the simulation of combustion processes, in which the chemistry has a decisive influence on the hydrodynamics and a numerical smearing of the physical variables may bear wrong results.

Two approaches exist to tackle this problem. One is to use the equivalent formulation of the fluid equations in Lagrangian coordinates or in a moving frame of reference in which material or other interfaces remain stationary. In this situation the conservation property as well as the sharp resolution of the contacts can be achieved simultaneously. Godunov-type schemes for the Euler equations in Lagrangian coordinates meeting these requirements have been considered e.g. in [1], [2]. The multi-dimensional Lagrangian case, however, is a bit awkward. The grid may become distorted so strongly that a rezoning becomes necessary in order to avoid the matching of mesh points and the breakdown of the calculation [3]. The other approach is an explicit tracking of contacts and a local sub-cell resolution of the interface within a fixed Eulerian grid. The information about the location for the interface is then used to correct the numerical fluxes through the boundaries of that grid zone in which the interface has been localized. LeVeque and Shyne [4] explicitly introduce two different states in this grid zone. In order to avoid a strong time-step restriction according to the CFL condition, they use a large time step wave propagation method within this small subcell. Davis [5] gives up the conservation form locally, namely in the grid zone containing the interface which is tracked. Several other approaches of tracking interfaces are surveyed by Hyman [6].

In both approaches an appropriate calculation of the fluxes across the interfaces has to be determined. This is the topic of this paper. In the framework of Godunov-type schemes [7], [8] we propose numerical fluxes which are based on the approximate solution of Riemann problems. The presence of two different materials introduces an alteration of the equation of state at the interface between the different materials and, thus, in general a discontinuity in the flux function. We introduce in this paper an auxiliary variable and a corresponding conservation equation which may be thought of as providing for a smooth switching between the two different flux functions. The fluxes may be assumed to be smoothly continued into the whole space. Then, the derivative of the flux is well-defined at the previous discontinuity, such that the system can be linearized and all the formal manipulations leading to the approximation can be applied. We do not propose a new tracking algorithm in this paper but focus our consideration to obtain good flux calculations at material interfaces that can be used for all tracking methods.

The general features of the method presented are described in section 2 for the Euler equations in Lagrangian coordinates. In section 3 the exact solution of the Riemann problem for two different materials is described briefly. Godunov-type schemes are considered from section 4 on. At first, an appropriate Roe linearization and Roe mean values are derived; based on these, the simplest Godunov-type scheme, which is usually called the HLL scheme after Harten, Lax and van Leer [9] and the modification due to Einfeldt [10] are constructed in section 5. In section 6 the numerical results are discussed for some test problems. In the concluding section 7 we transform the conservation equation for the switching variable to Eulerian coordinates and suggest the use of the present approach for a frame of reference fixed in space in combination with a tracking method.

2. The Enlarged System of Equations

First we will consider the Euler equations in Lagrangian coordinates which will be abbreviated by EEL. In this case of a frame of reference that moves with the flow, any material interface becomes stationary in time. The usual conservation law form of these equations is

$$u_t + f(u)_m = 0, \tag{2.1}$$

where u is the vector of the conserved quantities and $f(u)$ is the flux:

$$u = \begin{pmatrix} V \\ v \\ E \end{pmatrix}, \quad f(u) = \begin{pmatrix} -v \\ p \\ vp \end{pmatrix}. \tag{2.2}$$

The Lagrangian coordinate used is the mass m and t is the time. The dependent variables V, v, E and p denote the specific volume $1/\rho$ with density ρ, velocity, specific total energy and pressure, respectively. The pressure p is related to the conserved quantities by an equation of state

$$p = p(V, \epsilon), \tag{2.3}$$

where ϵ is the specific internal energy, i.e.,

$$E = \epsilon + \frac{1}{2}v^2. \tag{2.4}$$

This is the typical situation for the flow of a single fluid. If two non-mixing materials occur, the situation changes and the system of conservation laws now reads as

$$u_t + f(u, m)_m = 0. \tag{2.5}$$

We assume that only one material interface occurs in the computational domain and hence the flux function is assumed to be smooth except at a point m_0 where it has a discontinuity. More precisely, we take

$$f(u, m) = \begin{cases} f_l(u), & m < m_0 \\ f_r(u), & m > m_0 \end{cases} \tag{2.6}$$

with $f_{l,r} \in C^2(\Re^3, \Re^3)$. The discontinuity in the flux function is due to the change of the equation of state by passing from one material to the other.

This discontinuity of the flux function at $m = m_0$ may be regularized by introducing an additional conserved variable q that controls the transition from one state to the other. This idea was used in [11] for a scalar equation. We define a new flux function

$$g(q, u) := (1 - q)f_l(u) + qf_r(u), \tag{2.7}$$

which coincides with the fluxes in the different materials for $q = 1$ and $q = 0$, respectively. The time evolution of q is determined by the conservation equation

$$q_t + k(q)_m = 0, \tag{2.8}$$

where the smooth flux function k is chosen in such a way that $k(0) = k(1) > 0$ and k is concave in $(0, 1)$. In this case the solution of the Riemann problem for (2.8) with the initial data

$$q(m, 0) = \begin{cases} 0, & m < m_0, \\ 1, & m > m_0. \end{cases} \tag{2.9}$$

is given by the stationary shock wave

$$q(m, t) = q(m, 0) \quad \text{for all } t > 0. \tag{2.10}$$

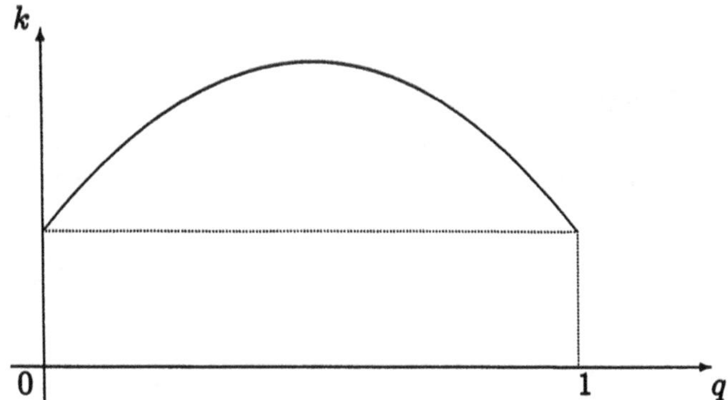

Figure 1. Graph of an appropriate flux function $k(q)$.

It is convenient to choose $k(q)$ such that it is symmetric with respect to the line $q = 1/2$; in particular, we want to have $\infty > k'(0) = -k'(1) > 0$. Fig. 1 shows the graph of such a flux function k.

Combining (2.1) with (2.7) and (2.8) gives an enlarged system of conservation equations

$$U_t + F(U)_m = 0, \qquad \text{with} \qquad U = \begin{pmatrix} q \\ u \end{pmatrix}, \;\; F(U) = \begin{pmatrix} k(q) \\ g(U) \end{pmatrix}. \tag{2.11}$$

The quasilinear form of this system is given by

$$U_t + A(U)U_m = 0, \tag{2.12}$$

where the Jacobian matrix A of the flux $F(U)$ has the special form

$$A(U) = \begin{pmatrix} k'(q) & 0 \\ \partial_q g & D_u g \end{pmatrix}. \tag{2.13}$$

Hence $\partial_q g = f_r(u) - f_l(u)$, while $D_u g$ is the original Jacobian of the flux for the EEL (2.1), but the pressure p is replaced by

$$P(u,q) := (1-q)p_l(u) + qp_r(u), \tag{2.14}$$

where $p_l(u)$ and $p_r(u)$ denote the equation of state of material 1 and 2, respectively.

The eigenvalues of the Jacobian matrix $A(U)$ are given by

$$a_0 = k'(q); \;\; a_1 = -C, \quad a_2 = 0, \;\; a_3 = C, \tag{2.15}$$

where C denotes the Lagrangian sound velocity with

$$C^2 = P\partial_\epsilon P - \partial_V P \tag{2.16}$$

and $\partial_\epsilon P$ and $\partial_V P$ denote the thermodynamic derivatives. For $q = 0$ and $q = 1$ the wave speeds a_1, a_2 and a_3 are those of material 1 and material 2 respectively. Hence the variable q may be considered as a switching function for the equation of state.

The directions of the waves in phase space are given by the eigenvectors

$$R^0 = \begin{pmatrix} 1 \\ -P_q/\alpha \\ P_q a_0/\alpha \\ (P + va_0)P_q/\alpha \end{pmatrix}, \qquad \text{where} \quad \alpha = a_0^2 - C^2, \tag{2.17}$$

$$R^1 = \begin{pmatrix} 0 \\ 1 \\ C \\ vC - P \end{pmatrix}, R^2 = \begin{pmatrix} 0 \\ 1 \\ 0 \\ -P_V/P_\epsilon \end{pmatrix}, R^3 = \begin{pmatrix} 0 \\ 1 \\ -C \\ -vC - P \end{pmatrix} \tag{2.18}$$

Here and in the following sections it is tacitly assumed that k is defined such that $\alpha \neq 0$ in order to avoid a degeneracy and to preserve the strict hyperbolicity.

The main building block of a Godunov-type scheme is the exact or approximate solution of the Riemann problem. This solution is used to incorporate the nonlinear wave propagation into the numerical scheme. For the enlarged EEL system (2.11) we consider the Riemann problem with initial data

$$U_0(m) = U(m, 0) = \begin{cases} (0, U_l)^T, & m < m_0 \\ (1, U_r)^T, & m > m_0 \end{cases} \tag{2.19}$$

With these inital data the solution of q is given by a stationary shock wave as described in (2.10). This shock wave with the jump of q from 0 to 1 coincides with the material interface and thus the solution of the Riemann problem for the enlarged system with the initial data (2.19) coincides with the solution of (2.5) for the initial data u_l and u_r. A (m, t)-diagram of this solution is given in Fig. 2. How to get the exact solution is shortly reviewed in the next section.

The advantage of the enlarged system is that approximations of the Riemann solution, such as the Roe linearization, can be developed without any lack of rigor. For the construction of the Roe linearization the equation of state has to be continuously differentiable with respect to its arguments ϵ

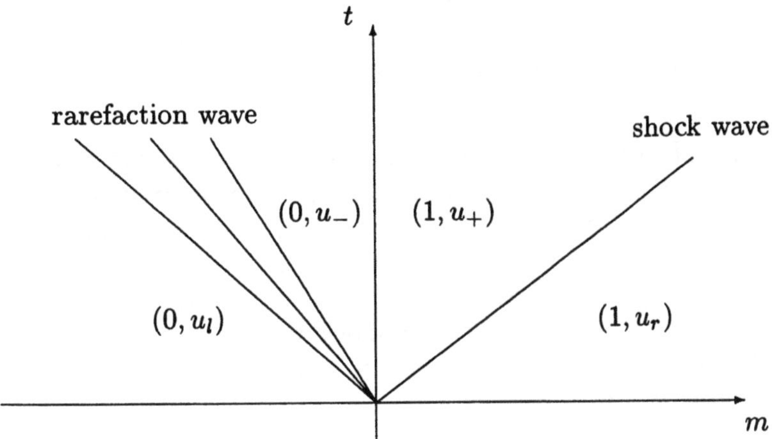

Figure 2. The (m,t)-diagram for a solution of the Riemann problem (2.11), (2.19).

and V. But at a material interface the equation of state changes its functional form with respect to the space location. Introducing the function q as a new dependent variable the flux becomes continuous with respect to all of its arguments and the switching of the equation of state at the material interface is done by this variable, i.e. by the solution itself. Due to the construction of the evolution equation for the switching function q it jumps at the interface to give the exact transition from one equation of state to the other.

3. Exact Solution of the Riemann Problem for Two Non-Mixed Materials

The exact solution of the Riemann problem for different materials is an extension of the one material solution as given in [12], [13], [14], and [15]. In the following we shortly review this procedure and mention the differences.

If the equation of state satisfies the convexity conditions, then the characteristic fields of the Euler equations are either genuinely nonlinear or linearly degenerate in the sense of Lax [16]. From this it follows that the solution of the Riemann problem consists of four constant states separated by three elementary waves. The intermediate states are connected by a contact discontinuity which is stationary in a Lagrangian frame of reference. Across this contact discontinuity the pressure as well as the velocity are continuous; let us denote their values by p_* and v_*, respectively. The values of the variables which undergo a jump are given an index '1' or '2' for the left and right intermediate state, respectively. To solve the Riemann problem one has to look in phase space how to connect the right and left state u_r and u_l. Because velocity and pressure are constant within the intermediate state, it is favorable to consider the projection of the phase space onto the (v, p) plane.

In the case of two different materials the structure of the solution is essentially the same. The different materials are separated by the contact discontinuity. The right as well as the left going waves propagate entirely within one material. If we assume, e.g., the left wave to be a shock wave, the Rankine-Hugoniot relations have to be valid across the discontinuity and establish a connection between shock velocity and the values of the physical quantities. If the shock velocity is eliminated in these expressions, the remaining two relations are

$$v_* - v_l = \sqrt{(p_* - p_l)(V_1 - V_l)}\,, \tag{3.1a}$$

$$\epsilon_1 - \epsilon_l = \frac{1}{2}(p_* + p_l)(V_1 - V_l)\,. \tag{3.1b}$$

Together with the equation of state for the left material these equations form a nonlinear system of three equations with four unknowns. If the EOS of the left material is given in the inverse form $\epsilon_1 = \epsilon_l(V_1, p_*)$, the internal energy ϵ_1 may be replaced in (3.1b). Subsequent elimination of V_1 from (3.1a) and (3.1b) results in a relation of the form

$$v_* = F_l(p_*; V_l, v_l, p_l) \tag{3.2}$$

which is valid for values of pressure $p_* > p_l$ only. But a relation of this type is also obtained in the case $p_* \leq p_l$ where a rarefaction wave occurs. This follows from the constancy of two Riemann invariants within the rarefaction fan and the EOS of the left material.

An analogous consideration for the right waves leads to a relationship

$$v_* = F_r(p_*; V_r, v_r, p_r)\,. \tag{3.3}$$

The relations (3.2), (3.3) gives a fixed-point problem

$$F_l(p_*; V_l, v_l, p_l) = F_r(p_*; V_r, v_r, p_r) \tag{3.4}$$

for the intermediate pressure, where within the function F_l the EOS of the left material and within F_r that of the right material is used only.

It is well known that for a single perfect gas the fixed-point problem can be written in a relatively simple form and may be solved numerically by iteration, see e.g. [17], [18]; for results on the existence and uniqueness of a solution see [12], [14]. For a single fluid with a real EOS, Colella and Glaz [19] propose to use the same procedure with a locally defined adiabatic exponent which is adjusted within an outer iteration.

These algorithms may be extended to the treatment of different materials using the structure outlined above. In the following we briefly describe the modifications of the fixed-point problem in the case of the two materials being different perfect gases. In this case the equations of state are given by

$$p_\mu(V, \epsilon) = (\gamma_\mu - 1)\epsilon_\mu / V_\mu, \qquad \mu = l,\, r, \tag{3.5}$$

where γ_l and γ_r denote the adiabatic exponent of left and right equation of states, respectively.

The formulas for the sound velocities simplify to

$$C_\mu^2 = \gamma_\mu p_\mu / V_\mu, \qquad \mu = l, r. \tag{3.6}$$

Because the contact discontinuity coincides with the material interface, the solution procedure follows that for the case of constant γ. The remaining two fields represent shock and/or rarefaction waves running completely in one of the two gases (cf. Fig. 2). The i-shock wave, $i = 1$ or 3 corresponding to a left or right running shock, respectively, has to obey the Rankine-Hugoniot relations as well as the Lax entropy conditions

$$i = 1: \quad -C_- < s < -C_l, \qquad i = 3: \quad C_r < s < C_+, \tag{3.7}$$

where

$$C_\pm^2 = \gamma_\pm p_* / V_\pm \tag{3.8a}$$

with

$$p_* = p_+ = p_- \,; \quad \gamma_+ = \gamma_r\,, \;\; \gamma_- = \gamma_l\,; \quad V_+ = V_2\,, \;\; V_- = V_1\,. \tag{3.8b}$$

In addition, $v_* = v_- = v_+$ $(v_- = v_1,\; v_+ = v_2)$. Rarefaction wave solutions are obtained by a similarity ansatz which gives the usual relations but now with the respective adiabate exponent.

Using the knowledge about the values at the contact, the two states u_l and u_r have to be connected properly. This leads to the definition of the variables (see, e.g. [18])

$$M_\mu = \sqrt{\frac{p_\mu}{V_\mu}} \cdot \Phi\!\left(\frac{p_*}{p_\mu}, \gamma_\mu\right), \qquad \mu = l,\, r\,, \tag{3.9}$$

where

$$M_l = -\frac{p_l - p_*}{v_l - v_*}, \qquad M_r = \frac{p_r - p_*}{v_r - v_*}, \tag{3.10}$$

$$\Phi(w,\gamma) = \begin{cases} \sqrt{\frac{\gamma-1}{2}(1+\beta w)}, & w \geq 1 \ \text{(shocks)} \\ \frac{\gamma-1}{2\sqrt{\gamma}} \cdot \frac{1-w}{1-w^\tau}, & 1 > w \geq 0 \ \text{(rarefactionwaves)} \end{cases} \tag{3.11}$$

and

$$\beta(\gamma) = \frac{\gamma+1}{\gamma-1}, \qquad \tau(\gamma) = \frac{\gamma-1}{2\gamma}. \tag{3.12}$$

Elimination of v_* by

$$v_* = (M_l v_l + M_r v_r + p_l + p_r)/(M_l + M_r), \tag{3.13}$$

gives the fixed-point equation

$$p_* = G(p_*) := [M_l M_r (v_l - v_r) + M_r p_l + M_l p_r]/(M_l + M_r). \tag{3.14}$$

Formally this is the same fixed-point problem as in the one-phase case, but now each of the M_μ depends on the corresponding γ_μ. It is uniquely solvable for $p_* \in [0, \infty)$, if $G(0) \geq 0$, i.e.,

$$v_l - v_r + \frac{2}{(\gamma_l - 1)(\gamma_r - 1)}[(\gamma_r - 1)C_l V_l + (\gamma_l - 1)C_r V_r] \geq 0. \tag{3.15}$$

Our procedure for solving this fixed-point problem numerically is based on Halter's [18] fast Riemann solver for $\gamma = const.$ and uses the Brent algorithm to solve the fixed-point problem. This algorithm requires reasonably good starting values, i.e., an estimate $p_* \in [p_{min}, p_{max}]$. As in the single-fluid case p_{min} is set to zero. For $v_l - v_r \leq 0$ it is $p_{max} = \max\{p_l, p_r\}$. If $v_l - v_r > 0$ and $p_{max} > \max\{p_l, p_r\}$, then an upper bound is easily found by replacing γ in an appropriate way by $\gamma_{max} = \max\{\gamma_l, \gamma_r\}$ or $\gamma_{min} = \min\{\gamma_l, \gamma_r\}$ in the single-material formula for p_{max}. This leads to

$$p_{max} = \frac{2\gamma_{max}^2 (v_l - v_r)^2 \rho_l \rho_r}{(1+\gamma_{min})(\sqrt{\rho_l} + \sqrt{\rho_r})^2} + 4\frac{p_l\sqrt{\rho_r} + p_r\sqrt{\rho_l}}{\sqrt{\rho_l} + \sqrt{\rho_r}} \tag{3.16}$$

as a starting value for the pressure iteration.

4. The Approximate Riemann Solver of Roe

The Roe method [20] consists of replacing the exact Riemann solution in the calculation of the numerical fluxes by the exact solution of a linearized problem. We consider this linearization for the enlarged system (2.7), (2.8):

$$U_t + A_{lr} U_m = 0, \quad U(m,0) = \begin{cases} U_l, & m < m_0 \\ U_r, & m > m_0 \,. \end{cases} \tag{4.1}$$

The Roe matrix $A_{lr} = A_{lr}(U_l, U_r)$ is required to be consistent with the Jacobian (2.13) in the sense that $A_{lr}(U,U) = A(U)$, to have real eigenvalues with a complete set of linearly independent eigenvectors and to satisfy the mean value property

$$F(U_l) - F(U_r) = A_{lr}(U_l - U_r). \tag{4.2}$$

This condition guaranties that the Roe method satisfies the internal conservation property and may be written in conservation form; for more details see [20] and [9].

The Riemann problem (4.1) is an initial value problem for a linear hyperbolic system with constant coefficients and may be solved by the characteristic theory. Its solution consists of four constant states separated by lines of discontinuity. The different states $U_0 := U_l$, U_1, U_2, and $U_3 := U_r$ are given by the formula

$$U_k = U_l + H(k-n)\beta_0 R^0_{Roe} + \sum_{j=1}^{k} \beta_j R^j_{Roe}, \tag{4.3}$$

where H denotes the step function $H(x) = 0$ for $x \le 0$, $H(x) = 1$ for $x > 0$, and n is the number of negative eigenvalues of A_{lr}. As above, R^j_{Roe}, $j = 0,\dots 3$, are the right eigenvectors of the Roe matrix and β_j the coefficients in the resolution

$$\Delta U \equiv U_r - U_l = \sum_{j=0}^{3} \beta_j R^j_{Roe}. \tag{4.4}$$

The approximate Riemann solution is then used to calculate an approximation of the flux between the grid zones. This numerical flux can be written as

$$g_{Roe}(U_l, U_r) = \frac{1}{2}(F(U_l) + F(U_r)) - \frac{1}{2}\sum_{j=0}^{3} |a_j^{Roe}| \beta_j R^j_{Roe}, \tag{4.5}$$

where a_j^{Roe} is the j-th eigenvalue of the Roe matrix. For further details on this method we refer to [20], [9] and for EEL to [1].

Here, we are interested in the derivation of the so-called Roe mean values upon which the Roe method is based. For the enlarged system (2.11) the mean value property (4.2) yields

$$\begin{aligned} \Delta k &= \overline{k_q}\Delta q, \\ -\Delta v &= -\Delta v, \\ \Delta P &= \overline{P_V}\Delta V + \overline{P_v}\Delta v + \overline{P_E}\Delta E + \overline{P_q}\Delta q, \\ \Delta(vP) &= \overline{vP_V}\Delta V + (\overline{P} + \overline{vP_v})\Delta v + \overline{vP_E}\Delta E + \overline{vP_q}\Delta q. \end{aligned} \tag{4.6}$$

Here, $\Delta v = v_r - v_l$, etc. and the barred quantities denote the elements of the matrix A_{lr}. Of course, (4.6b) is trivial. The first relation leads to $\overline{k_q} = 0$ at $m = m_0$ according to (2.9), (2.10) and the properties of k. To proceed, we assume the following factorization properties to hold: $\overline{vP_V} = \bar{v}\overline{P_V}$, $\bar{P} = P(\bar{V}, \bar{\epsilon}, \bar{q})$, $\overline{P_V} = P_V(\bar{V}, \bar{\epsilon}, \bar{q})$, etc. Then (4.6c) may be written in the form of

$$\begin{aligned} \Delta P &= \overline{P_V}\Delta V - \overline{vP_\epsilon}\Delta v + \overline{P_\epsilon}[\Delta\epsilon + \Delta(v^2)/2] + \overline{P_q}\Delta q \\ &= \overline{P_V}\Delta V + \overline{P_\epsilon}\Delta\epsilon + \overline{P_q}\Delta q + \overline{P_\epsilon}\left(\frac{v_l + v_r}{2} - \bar{v}\right)\Delta v. \end{aligned}$$

This suggests to choose

$$\bar{v} = \frac{1}{2}(v_l + v_r), \tag{4.7}$$

so that the velocity terms vanish in this pressure equation and we are left with

$$\Delta P = \overline{P_V}\Delta V + \overline{P_\epsilon}\Delta\epsilon + \overline{P_q}\Delta q. \tag{4.8}$$

Similarly, using (4.8), (4.6d) may be rewritten as

$$\bar{v}\Delta P + \bar{P}\Delta v = \Delta(vP) = \frac{1}{2}(v_l + v_r)\Delta P + \frac{1}{2}(P_l + P_r)\Delta v,$$

which again gives (4.7) and in addition

$$\bar{P} = \frac{1}{2}(P_l + P_r). \tag{4.9}$$

In order to satisfy (4.8), we first consider the EOS of perfect gases. It will be convenient to define an effective adiabate index, Γ, such that

$$P = (\Gamma - 1)\frac{\epsilon}{V}, \quad \Gamma(q) := (1-q)(\gamma_l - 1) + q(\gamma_r - 1) + 1 = \gamma_l + q(\gamma_r - \gamma_l). \tag{4.10}$$

Then, the following relations hold

$$\begin{array}{rcl}\Gamma_q &=& \gamma_r - \gamma_l, \;\; P_q = p_r - p_l = (\gamma_r - \gamma_l)\frac{\epsilon}{V} = (\gamma_r - \gamma_l)\frac{P}{\Gamma - 1}, \\ P_V &=& -\frac{P}{V}, \;\; P_\epsilon = \frac{P}{\epsilon} = \frac{\Gamma - 1}{V}, \;\; C^2 = \frac{\Gamma P}{V} \; [= \; (1-q)C_l^2 + qC_r^2],\end{array} \tag{4.11}$$

and (4.8) becomes

$$\Delta P = \overline{\left(\frac{P}{\Gamma - 1}\right)}\Delta\Gamma - \overline{\left(\frac{P}{V}\right)}\Delta V + \overline{\left(\frac{\Gamma - 1}{V}\right)}\Delta\epsilon. \tag{4.12}$$

Replacing $\Delta\epsilon$ by

$$\begin{aligned}\Delta\epsilon &= \frac{P_r V_r}{\Gamma_r - 1} - \frac{P_l V_l}{\Gamma_l - 1} \\ &= \frac{1}{2}\left[\left(\frac{P_l}{\Gamma_l - 1} + \frac{P_r}{\Gamma_r - 1}\right)\Delta V - \frac{P_r V_r + P_l V_l}{(\Gamma_r - 1)(\Gamma_l - 1)}\Delta\Gamma \right. \\ &\quad \left. + \left(\frac{V_r}{\Gamma_l - 1} + \frac{V_l}{\Gamma_r - 1}\right)\Delta P\right]\end{aligned}$$

leads to the condition

$$\begin{array}{rcl} 0 &=& \Delta P \cdot \left[\overline{\left(\frac{\Gamma-1}{V}\right)} \cdot \frac{1}{2}\left(\frac{V_r}{\Gamma_l - 1} + \frac{V_l}{\Gamma_r - 1}\right) - 1\right] \\ && + \Delta V \cdot \left[\overline{\left(\frac{\Gamma-1}{V}\right)} \cdot \frac{1}{2}\left(\frac{P_l}{\Gamma_l - 1} + \frac{P_r}{\Gamma_r - 1}\right) - \overline{\left(\frac{P}{V}\right)}\right] \\ && + \Delta\Gamma \cdot \left[\overline{\left(\frac{P}{\Gamma-1}\right)} - \overline{\left(\frac{\Gamma-1}{V}\right)} \cdot \frac{1}{2}\frac{P_r V_r + P_l V_l}{(\Gamma_r - 1)(\Gamma_l - 1)}\right].\end{array} \tag{4.13}$$

Now we claim that the individual terms vanish, such that we arrive at the relations

$$\begin{array}{rcl}\overline{\left(\frac{\Gamma-1}{V}\right)} &=& \frac{2(\Gamma_r-1)(\Gamma_l-1)}{V_r(\Gamma_r-1)+V_l(\Gamma_l-1)} = \left[\frac{1}{2}\left(\frac{V_r}{\Gamma_l-1} + \frac{V_l}{\Gamma_r-1}\right)\right]^{-1}, \\ \overline{\left(\frac{P}{V}\right)} &=& \overline{\left(\frac{\Gamma-1}{V}\right)}\frac{1}{2}\left(\frac{P_l}{\Gamma_l-1} + \frac{P_r}{\Gamma_r-1}\right) = \frac{P_l(\Gamma_r-1)+P_r(\Gamma_l-1)}{V_r(\Gamma_r-1)+V_l(\Gamma_l-1)}, \\ \overline{\left(\frac{P}{\Gamma-1}\right)} &=& \frac{P_r V_r + P_l V_l}{V_r(\Gamma_r-1)+V_l(\Gamma_l-1)} = \frac{\epsilon_r(\Gamma_r-1)+\epsilon_l(\Gamma_l-1)}{V_r(\Gamma_r-1)+V_l(\Gamma_l-1)} = \overline{\left(\frac{\epsilon}{V}\right)}.\end{array} \tag{4.14}$$

Finally, assuming the factorization property in the case of (4.14a) and (4.14b) the following mean values are obtained:

$$\begin{array}{rcl}\bar{V} &=& \frac{1}{2}(P_l + P_r) \cdot \frac{V_r(\Gamma_r-1)+V_l(\Gamma_l-1)}{P_l(\Gamma_r-1)+P_r(\Gamma_l-1)}, \\ \bar{\Gamma} - 1 &=& \frac{(\Gamma_l-1)(\Gamma_r-1)(P_l+P_r)}{P_l(\Gamma_r-1)+P_r(\Gamma_l-1)}.\end{array} \tag{4.15}$$

Notice that the mean value of the specific volume is reduced to $\bar{V} = \frac{1}{2}(V_r + V_l)$ in the case of constant Γ (cf. [1], [2]). Moreover, it can be checked easily that

$$\overline{\left(\frac{P}{\epsilon}\right)} = \overline{\left(\frac{\Gamma-1}{V}\right)} = \frac{\bar{P}}{\bar{\epsilon}} = \left[\frac{1}{2}\left(\frac{\epsilon_r}{P_r} + \frac{\epsilon_l}{P_l}\right)\right]^{-1}.$$

On the other hand, the factorization property does not hold for (4.14c):

$$\frac{\bar{P}}{\bar{\Gamma}-1} = \frac{1}{2}\left(\frac{P_l}{\Gamma_l - 1} + \frac{P_r}{\Gamma_r - 1}\right) = \frac{1}{2}\left(\frac{\epsilon_l}{V_l} + \frac{\epsilon_r}{V_r}\right) \neq \overline{\left(\frac{P}{\Gamma - 1}\right)}.$$

This means that in this situation the Roe matrix does not satisfy $A_{lr} = A(\bar{U})$, $\bar{U}$ being the Roe mean values. Such a behaviour is typical for general equations of state but is not important here, since the term $\overline{P_q}$ does occur in the first row of the Jacobian only which is decoupled from the other parts of the matrix, see (2.13). Therefore, the Roe conditions are still satisfied.

We also do not want to declare $\overline{(\epsilon/V)} = \bar{\epsilon}/\bar{V}$, because this would lead to $\bar{\epsilon}|_{\Gamma_l=\Gamma_r} = (\epsilon_l + \epsilon_r)/2$, which is not in accordance with the previously derived mean value for constant Γ. Instead, we define

$$\bar{\epsilon} = \frac{\bar{P}\bar{V}}{\bar{\Gamma}-1}. \tag{4.16}$$

Similarly for the mean sound velocity

$$\bar{C}^2 = \frac{\bar{\Gamma}\bar{P}}{\bar{V}} = \frac{(\Gamma_r - 1)V_l C_l^2 + (\Gamma_l - 1)V_r C_r^2}{V_r(\Gamma_r - 1) + V_l(\Gamma_l - 1)}, \tag{4.17}$$

which is also reduced to the usual one in case of constant Γ. From (4.15b) a mean value of q can be defined, if $\gamma_l \neq \gamma_r$:

$$(\gamma_r - \gamma_l) \cdot \bar{q} = (\gamma_r - \gamma_l) \cdot \left(\frac{P_l}{\Gamma_l - 1} q_l + \frac{P_r}{\Gamma_r - 1} q_r\right) / \left(\frac{P_l}{\Gamma_l - 1} + \frac{P_r}{\Gamma_r - 1}\right),$$

from which

$$\bar{q} = \frac{1}{2}(q_l + q_r) + \frac{\gamma_r - \gamma_l}{\gamma_r + \gamma_l - 2} \cdot \frac{q_l - q_r}{2} \tag{4.18}$$

is obtained.

We summarize the previous results in the

THEOREM 4.1. *In the case of two different non-mixed perfect gases, a Roe matrix $A_{lr} = A(U_l, U_r)$ can be constructed for the Euler equations in Lagrangian coordinates, that is determined by the mean values (4.7),*

(4.9), (4.15) for the primitive variables, by (4.14) for the thermodynamic derivatives and by (4.16)–(4.18) for the other dependent variables.

To complete the description of the Roe method, we note that the eigenvalues of the Roe matrix are given by

$$a_0^{Roe} = \overline{k_q} = 0; \quad a_1^{Roe} = -\bar{C}, \quad a_2^{Roe} = 0, \quad a_3^{Roe} = \bar{C}, \tag{4.19}$$

and that the corresponding right eigenvectors are given by (3.6) with the barred quantities being substituted. The system of the linear equations (4.4) is solved by the coefficients

$$\begin{aligned} \beta_0 &= q_r - q_l = \Delta q, \qquad \beta_1 = \tfrac{\Delta v}{\bar{C}} + \beta_3, \\ \beta_2 &= \Delta V - \tfrac{\Delta v}{\bar{C}} - 2\beta_3 - \tfrac{\overline{P_q}}{\bar{C}^2}\Delta q, \\ \beta_3 &= -\tfrac{\overline{P_\epsilon}}{2\bar{C}^2}\Delta E + \tfrac{\bar{v}\overline{P_\epsilon} - \bar{C}}{2\bar{C}^2}\Delta v - \tfrac{\overline{P_V}}{2\bar{C}^2}\Delta V - \tfrac{\overline{P_q}}{2\bar{C}^2}\Delta q. \end{aligned} \tag{4.20}$$

At a pure contact discontinuity $\Delta v = \Delta P = 0$ and (4.20) simplifies to

$$\begin{aligned} \beta_1 &= \beta_3 = -\tfrac{1}{2\bar{C}^2}\left(\overline{P_\epsilon}\Delta\epsilon + \overline{P_V}\Delta V + \overline{P_q}\Delta q\right) = -\tfrac{\Delta P}{2\bar{C}^2} = 0, \\ \beta_2 &= \Delta V - \tfrac{\Delta\gamma}{\bar{C}^2}\overline{\left(\tfrac{P}{\Gamma - 1}\right)}. \end{aligned} \tag{4.21}$$

We remark that the first component of the numerical flux (4.5) is constant $(= k(0) = k(1))$ due to the special construction of the additional conservation law which is used only to derive Roe mean values and to obtain an expression for the numerical flux of the Roe method approximating the physical flux.

For general EOS', modifications of the above mean values for the thermodynamic derivatives, $\overline{P_V}$ and $\overline{P_\epsilon}$, can be used as discussed for EEL in [1]. Another way is the extension to general equation of states as proposed for a single fluid by Collela and Glaz [19]. There an effective adiabatic exponent is defined and adjusted in an outer iteration. This can be done in the same way for different materials using the formulas given above.

5. The Method of Harten, Lax, and van Leer

Harten, Lax, and van Leer [9] described a simple Godunov-type scheme where the approximate Riemann solution consists of three constant states only, the undisturbed left and right states and one intermediate state:

$$W((m - m_0)/t; U_l, U_r) = \begin{cases} U_l, & (m - m_0)/t < a_l \\ U_{lr}, & a_l < (m - m_0)/t < a_r \\ U_r, & a_r < (m - m_0)/t \end{cases} \tag{5.1}$$

For the enlarged system (2.1) the intermediate state in the approximate Riemann solution (5.1) is given by

$$U_{lr} = \frac{a_r U_r - a_l U_l}{a_r - a_l} - \frac{F(U_r) - F(U_l)}{a_r - a_l}, \tag{5.2}$$

where the first component reads as

$$q_{lr} = \begin{cases} 0, & m < m_0 \\ a_r/(a_r - a_l), & m = m_0 \\ 1, & m > m_0 . \end{cases} \tag{5.3}$$

This is obtained in the usual way from the consistency condition with the integral conservation.

The numerical flux of the HLL scheme is given by

$$g_{HLL}(U_l, U_r) = \frac{a_r^+ F(U_l) - a_l^- F(U_r)}{a_r^+ - a_l^-} + \frac{a_r^+ a_l^-}{a_r^+ - a_l^-}(U_r - U_l), \tag{5.4a}$$

with

$$a_r^+ = \max(0, a_r), \qquad a_l^- = \min(0, a_l), \tag{5.4b}$$

and the first component

$$g_{HLL}^q = k(0) + \frac{a_r^+ a_l^-}{a_r^+ - a_l^-} \cdot \Delta q. \tag{5.5}$$

An essential part of the HLL Riemann solution is to find appropriate a priori estimates for the signal velocities a_l and a_r. For EEE, this topic has been considered by Einfeldt [10] and for EEL by Munz [1]. Using the signal velocities based on the Roe mean values, negative values for the specific volume may occur in the EEL case near strong compressions. This is avoided by introducing some corrections which may also be applied to the extended system (2.11)([1]).

A positively conservative scheme is obtained, if the signal velocities are chosen to be

$$a_l = -\sqrt{\left(a_1^{Roe}\right)^2 + \eta^2}, \qquad a_r = \sqrt{\left(a_3^{Roe}\right)^2 + \eta^2}, \tag{5.6a}$$

where

$$\eta = \eta_0 \frac{\max(0, v_l - v_r)}{V_r + V_l}, \qquad \eta_0 = \text{const.} \geq 1. \tag{5.6b}$$

The appropriate resolution of material boundaries cannot be achieved by the HLL Riemann solution, because there the two intermediate constant states of the exact Riemann solution, separated by the material interface, are replaced by one average state. Einfeldt [10] proposed a modification by adding to the HLL-intermediate state a correction term that is linear in m/t. This scheme can be regarded as a generalization of the Roe method. For the Euler equations in Eulerian as well as in Lagrangian coordinates the modified HLL scheme turned out to be identical to the Roe method, if the signal velocities a_l and a_r are defined to be the smallest and largest Roe mean value ([1], [10]). For the enlarged EEL system this modification becomes

$$W((m-m_0)/t; U_l, U_r) = \begin{cases} U_l, & (m-m_0)/t < a_l \\ U_{lr} + b\left(\frac{m-m_0}{t} - d\right)\cdot\theta, & a_l < (m-m_0)/t < a_r \\ U_r, & a_r < (m-m_0)/t \end{cases} \tag{5.7a}$$

where

$$\theta = \beta_0(\bar{U})R^0(\bar{U}) + \beta_2(\bar{U})R^2(\bar{U}) \tag{5.7b}$$

and R^0, R^2 are the zeroth and second right eigenvectors of the Jacobian, $\bar{U}$ is any mean value, β_0 and β_2 are the coefficients of the projection of $U_r - U_l$ onto $R^0(\bar{U})$ and $R^2(\bar{U})$, respectively, as given by (4.20). For $\bar{U}$ we will choose the Roe mean values derived above.

The consistency of the approximate Riemann solution with the integral conservation property leads to

$$d = \frac{a_r + a_l}{2}, \tag{5.8a}$$

while the condition that at a single contact discontinuity the numerical flux should equal the exact flux gives

$$b = \frac{2}{a_r - a_l}. \tag{5.8b}$$

Then, the numerical flux has the form

$$g_{HLLEM} = g_{HLL} - \frac{a_r a_l}{a_r - a_l}\cdot\theta. \tag{5.9}$$

Because

$$R^0(\bar{U}) = \left(1, \frac{\overline{P_q}}{\bar{C}^2}, 0, -\frac{\bar{P}\overline{P_q}}{\bar{C}^2}\right)^T, \tag{5.10}$$

the terms, which occur additionally compared with the case of constant γ, are proportional to

$$\overline{P_q}|_{m_0} = (\gamma_r - \gamma_l)\cdot\frac{P_r V_r + P_l V_l}{(\gamma_r - 1)V_r + (\gamma_l - 1)V_l}. \tag{5.11}$$

The necessity of these $\Delta\gamma$ terms will be seen clearly from the test problems in the next section.

Summarizing, we can say that *the HLLEM scheme for different materials contains the usual Einfeldt modification for a single material plus an additional term stemming from the regularization by the EOS-switching variable q; this additional modification is effective at the material interface only.*

6. Numerical Results

The well-known Sod problem [21] is a one-fluid shock tube problem with the initial data

$$(\rho, v, p) = \begin{cases} (1.0, 0.0, 1.0), & m < m_0 \\ (0.125, 0.0, 0.1), & m > m_0 . \end{cases} \tag{6.1}$$

and $\gamma = 1.4$ overall. In addition we introduce a jump of γ at m_0 as

$$\gamma = \begin{cases} 1.4 & for\ m < m_0 \\ 5/3 & for\ m > m_0 \end{cases} \tag{6.2}$$

and vice versa

$$\gamma = \begin{cases} 5/3 & for\ m < m_0 \\ 1.4 & for\ m > m_0 \end{cases} \tag{6.3}$$

which corresponds to the fluid flow of the two different gases, air and hydrogen. The exact solution of the one-fluid Sod problem (6.1) consists of four constant states separated by three elementary waves : A rarefaction wave moves to the left, a stationary contact discontinuity in the Lagrangian frame of reference is located at $m = m_0$, and a shock wave of moderate strength propagates to the right. Qualitatively this solution does not change in the two-fluid case with the values of γ as given by (6.2),(6.3). The material interface coincides with the contact discontinuity.

Using Lagrangian coordinates, the material interface becomes stationary. In this case the conservation property and the exact resolution of the interface can be combined in the sense that no intermediate numerical state is introduced. For the HLL method this is not valid because in the approximate Riemann solution the two intermediate states are replaced by one average state. Hence a mixing of the two phases is incorporated inherently. The numerical approximation introduces two intermediate states, one right and one left of the interface. Therefore the HLL method is not favourable to calculate material interfaces. In the following we show results using the HLLEM scheme. At the interface the approximate Riemann solver proposed in this paper is applied. These results are compared with those obtained by the HLLEM scheme for a general but continuously differentiable equation of state.

For the calculation the mass interval [0.0, 0.9] was partitioned into 100 equidistant grid zones, and the initial jump was positioned at $m_0 = 0.7$. The figures show the values of density, velocity, pressure, and internal energy at $t = 0.5$. The exact solution is plotted by a solid line (—), while the numerical results are indicated by open circles (∘ ∘ ∘). The time step sizes were chosen to satisfy the CFL condition

$$\Delta t = \sigma \Delta m / (\max_i C_i) \tag{6.4}$$

with $\sigma = 0.8$ and $\Delta m = 0.009$. Here $max\ C_i$ is estimated using the approximate values at the previous time level. The exact solution is available in terms of the fixed-point problem (3.14). Its iterative solution can be obtained using the Brent algorithm [18]. The starting value has been calculated as indicated in section 3.

We consider Godunov-type schemes which are able to resolve a one-fluid contact discontinuity in the Lagrangian frame of reference exactly in the sense that no intermediate state is generated. As shown in [1], this is the Godunov scheme using the exact solution of the Riemann problem, the Roe method using the exact solution of the linearized Riemann problem and the HLLEM scheme. In the following discussion of test results we focus on the HLLEM scheme, which may be considered as a generalization of the Roe method and becomes identical to it when the signal velocities for the HLLEM method are chosen to be the smallest and largest Roe eigenvalues. Fig. 3 shows the numerical results of the HLLEM scheme for the initial values (6.1) and (6.3). Inside each material the usual flux calculation is applied, while for the approximate Riemann problem at the interface the calculation is performed using the signal velocities as given by (4.19) and (5.6). The pressure is constant across the material interface as stated by the exact solution. The density plot clearly indicates that no intermediate state is introduced. There is some small undershoot right at the interface. This is the same undershoot as observed in one-fluid calculations [1]; it is generated from the discontinuous initial values by approximation errors in resolving the break-up of the discontinuity in the different elementary waves. In [22], Menikoff studied such errors when shock waves interact or when a shock wave is incident on a material interface. This pressure drop is stationary and does not evolve in time. These results clearly indicate that the flux calculation at the interface works very well.

The next figure 4 shows results where the original HLLEM method is applied even at the material interface. Here the one-fluid calculation for a general equation of state is applied as given in [1] to this discontinuous flux situation. The main difference in the Roe mean values is that in this case the Lagrangian sound velocities C_r and C_l are calculated with respect to the equation of state in the corresponding material and the Roe mean value is given by

$$\overline{C}^2 = \frac{V_l C_l^2 + V_r C_r^2}{V_r + V_l}$$

instead of (4.17). This method is not able to capture the constant pressure across the interface. At the interface an unphysical pressure wiggle is generated. The density plot clearly shows the resolution of the density jump without any intermediate state, but a strong under- and overshooting occurs at the left and right of the interface, respectively. The results for the Roe method are quite similar.

All results up to now have been produced with the first order accurate approach. The method developed above for the desired resolution of material interfaces applies also to higher-order Godunov-type schemes as obtained via the MUSCL approach of van Leer [15]. We show some results for the HLLEM scheme in Fig. 5 for the two-phase Sod problem (6.1), (6.2). The values of γ have been interchanged. The slope calculation is performed by applying Sweby's slope calculation (see, e.g. [1], [22]) to the primitive variables with slope parameter $s_i = 1.4$. In the second-order case the resolution of both the shock and rarefaction wave is greatly improved, while the good resolution of the interface is maintained. For comparison Fig. 6 shows the result of a calculation where the modification of the Riemann solver at material interface is not applied. Here the typical non-physical jump of the pressure at the material interface is clearly to be seen.

7. Conclusions

The construction of Godunov-type schemes for the Euler equations in Lagrangian coordinates has been extended to the treatment of material boundaries. We have constructed Roe mean values and, based on these, have developed a method for the accurate resolution of an interface without introducing an unphysical equation of state. The various approximation schemes have been tested extensively by a two-phase variant of the Sod problem for two perfect gases.

A key ingredient of the methods described is the introduction of a new dependent variable which serves as a switching variable. To extent this approach to Eulerian coordinates we have to transform the equation (2.5) for the switching variable to Eulerian coordinates and make the replacement

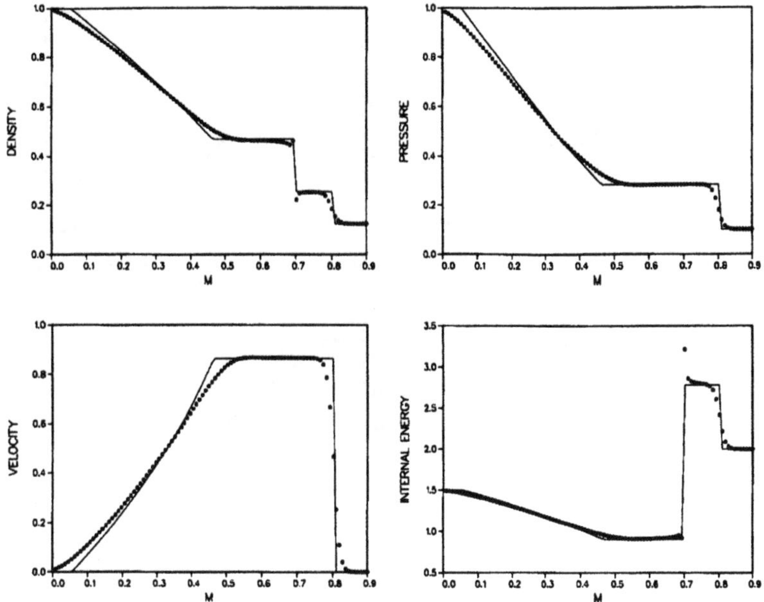

Figure 3. First-order HLLEM scheme with regularization.

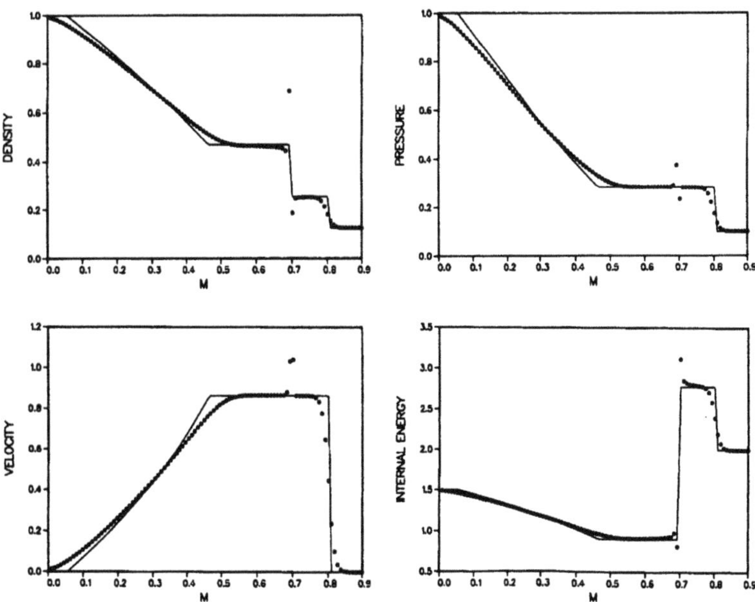

Figure 4. First-order HLLEM scheme without regularization.

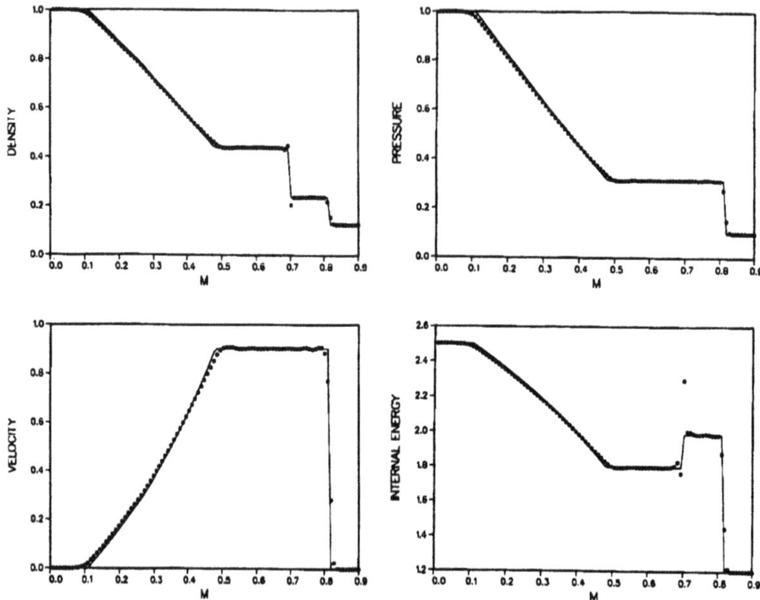

Figure 5. **Second-order HLLEM scheme with regularization.**

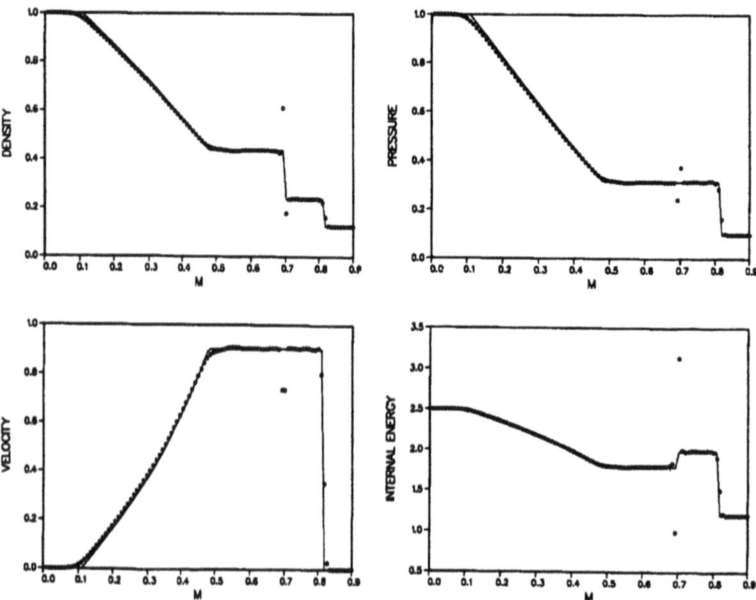

Figure 6. **Second-order HLLEM scheme without regularization.**

(2.3) for the equation of state. Since the transformation is given by the relations

$$q(m,t) = \bar{q}(x(m,t),t), \quad m = \int_{x(0,t)}^{x(m,t)} \bar{\rho}(s,t)\,\mathrm{ds} \;\; \forall\, t, \quad m_x = \bar{\rho}, \tag{7.1}$$

$$x(m,t) = x(m,0) + \int_0^t \bar{v}(m,s)\,\mathrm{ds}; \quad \partial_t \leftarrow \partial_t + \bar{v}\partial_x, \tag{7.2}$$

where the barred quantities denote the Eulerian variables, we obtain

$$\bar{q}_t + \bar{v}\bar{q}_x + m_x^{-1} k(\bar{q})_x = 0. \tag{7.3}$$

Inserting the last relation of (7.1) and using the conservation equation for the density,

$$0 = \bar{\rho}(\bar{q}_t + \bar{v}\bar{q}_x) + k(\bar{q})_x = -\bar{q}[\bar{\rho}_t + (\bar{\rho}\bar{v})_x], \tag{7.4}$$

leads to the corresponding conservation equation in Eulerian coordinates

$$\partial_t(\rho q) + \partial_x[\rho v q + k(q)] = 0, \tag{7.5}$$

where we now have omitted the bars. It is natural to introduce another conserved quantity such that we finally obtain

$$\partial_t r + \partial_x[v r + k(r/\rho)] = 0 \qquad \text{with} \quad r = \rho q. \tag{7.6}$$

Thus, the situation for the Euler equations in Eulerian coordinates looks similar to that for the Lagrangian formulation: The solution structure is the same, only the eigenvalues get a shift by the velocity v, and the contact is not stationary any longer.

A scheme in conservation form cannot resolve the time development of this contact on a grid fixed in space without introducing a numerical mixing. To treat a material interface accurately in this situation, a tracking method has to be introduced which keeps the interface sharp. Tracking methods have been reviewed in [23] and recently proposed in [4] and [5]. The evaluation of the numerical flux through the material interface should then be given by an approximate Riemann solver for the Euler equations that is constructed in a similar way based on (2.5), (2.7) and (7.6).

REFERENCES

[1] C.-D. MUNZ, *On Godunov-type schemes for Lagrangian gas dynamics*, SIAM J. Numer. Anal., 31 (1994), pp. 17-24.
[2] M. F. GÖZ AND C.-D. MUNZ, *Simple Godunov-type schemes for Lagrangean gas dynamics*, Notes Numer. Fluid Mech. 43, A. Donato and F. Oliveri, eds., Vieweg-Verlag, Braunschweig, Germany, 1993, pp. 298-305.
[3] J. K. DUKOWICZ AND J. W. KODIS, *Accurate conservative remapping (rezoning) for arbitrary Lagrangian-Eulerian computations*, SIAM J. Sci. Stat. Comput., 8 (1987), pp. 305-321.
[4] R. J. LEVEQUE AND K. M. SHYNE, *Shock tracking based on high resolution wave propagation methods*, Technical Report No. 92-01, ETH Zürich, Switzerland.
[5] S. F. DAVIS, *An interface tracking method for hyperbolic systems of conservation laws*, Appl. Numer. Math., 10 (1992), pp. 447-472.
[6] J. M. HYMAN, *Numerical methods for tracking interfaces*, Physica D, 12 (1984), pp. 396-407.
[7] S. K. GODUNOV, *Finite-diference method for numerical computation of discontinuous solutions of the equations of fluid dynamics*, Mat. Sbornik, 47 (1959), pp. 279-308.
[8] R. J. LEVEQUE, *Numerical Methods for Conservation Laws*, Birkhäuser Verlag, Basel, Boston, Berlin (1992)
[9] A. HARTEN, P. D. LAX, AND B. VAN LEER, *On upstream differencing and Godunov-type schemes for hyperbolic conservation laws*, SIAM Rev., 25 (1983), pp. 35-62.
[10] B. EINFELDT, *On Godunov-type methods for gas dynamics*, SIAM J. Numer. Anal., 25 (1988), pp. 294-318.
[11] T. GIMSE AND N. H. RISEBRO, *Riemann problems with a discontinuous flux function*, Notes Numer. Fluid Mech., B. Enquist and B. Gustafsson, eds., Vieweg-Verlag, Braunschweig/ Wiesbaden, Germany, 1992, pp. 488-502.
[12] J. SMOLLER, Shock Waves and Reaction-Diffusion Equations, Springer, New York/ Heidelberg/ Berlin, 1983.
[13] R. MENIKOFF AND B. J. PLOHR, *The Riemann problem for fluid flow of real materials*, Rev. Mod. Phys., 61 (1989), pp. 75-130.
[14] R. SMITH, *The Riemann problem in gas dynamics*, Trans. Am. Math. Soc., 249 (1979), pp. 1-50.
[15] B. VAN LEER, *Towards the ultimate conservative difference scheme V. A second order sequel to Godunov's method*, J. Comput. Phys., 32 (1979), pp. 101-136.

[16] P. D. LAX, *Hyperbolic systems of conservation laws II*, Comm. Pure Appl. Math., 10 (1957), pp. 537-566.
[17] A. J. CHORIN AND J. E. MARSDEN, A Mathematical Introduction to Fluid Mechanics, Springer, New York/ Heidelberg/ Berlin, 1979.
[18] E. HALTER, *A fast solver for Riemann problems*, Math. Meth. Appl. Sci., 7 (1985), pp. 101-107.
[19] P. COLLELA AND H. M. GLAZ, *Efficient solution algorithms for the Riemann problem for real gases*, J. Comput. Phys., 59 (1985), pp. 264-289.
[20] P. L. ROE, *Approximative Riemann solvers, parameter vectors, and difference schemes*, J. Comput. Phys., 43 (1981), pp. 357-372.
[21] G. A. SOD, *A survey of several finite difference methods for systems of nonlinear hyperbolic conservation laws*, J. Comput. Phys., 27 (1978), pp. 1-31.
[22] R. MENIKOFF, *Errors when shock waves interact due to numerical shock width*, SIAM J. Sci. Comput., 15(5) (1994), pp. 1227-1242.
[23] C.-D. MUNZ, *On the numerical dissipation of high resolution schemes*, J. Comput. Phys., 77 (1988), pp. 18-39.

FORMULATION OF THE ECMWF FORECAST MODEL

MARIANO HORTAL
European Centre for Medium Range Weather Forecasts
Reading, UK

1. Presentation

The European Centre for Medium-Range Weather Forecasts (ECMWF) is an international organization supported by eighteen European States.

The Centre's principal objectives are:

- the development of numerical methods for medium-range weather forecasting;
- the preparation, on a regular basis, of medium-range weather forecasts for distribution to the meteorological services of the Member States;
- scientific and technical research directed to the improvement of these forecasts;
- collection and storage of appropriate meteorological data.

In addition, the Centre makes available a proportion of its computing facilities to its Member States for their research. It assists in implementing the programmes of the World Meteorological Organization. It provides advanced training to the scientific staff of the Member States in the field of numerical weather prediction.

Originally a COST (European Cooperation in Science and Technology) project, the Centre was established in 1973 by a Convention.

The first operational medium-range weather forecasts were issued in September 1979.

ECMWF predicts the behaviour of the atmosphere in the medium-range up to ten days ahead. In this time the future state of the atmosphere at any point can be influenced by phenomena at very distant geographical locations. Many applications of medium-range forecasting, for example ship routing, or pollution dispersion, are not confined to limited areas of the globe. Therefore the whole atmosphere must be included in the model; a model for medium-range forecasting must be global and must describe the

E.F. Toro and J.F. Clarke (eds.), Numerical Methods for Wave Propagation, 237–251.

atmosphere from the earth's surface to a height of some 30 Km or more. The discretization we can afford depends on the power of the computer we have available and how efficiently we use this power.

Some important factors influencing the evolution of the atmosphere occur on a very small scale. These include the heating of the soil by the sun, the turbulence of the air near the ground and at high levels in the atmosphere, for example when air flows over mountains, and cumulus cloud systems. These cannot be represented properly by the discretization we can afford in even the most powerful computers available. We must represent their effects by taking into account their influence on the behaviour of the parameters of the large scales. This "parameterization" is one of the areas where much effort has to be put in order to improve our forecasts.

Since September 1991, thanks to the availability of a powerful computer, together with the introduction of the semi-Lagrangian semi-implicit scheme to solve the evolution equations, we have been able to use a spectral forecast model with triangular truncation at 213 wavenumbers and 31 levels in the vertical. The corresponding representation in physical space contains 138346 points per level, a total of 4,288,726 points in the atmosphere plus 553,384 points in the soil. The number of computations required for a ten-day forecast is of the order of $2*10^{13}$.

From its beginning ECMWF has been heavily involved with large supercomputer systems. In 1977, the Centre's scientists had access to machine number one in the CRAY1 series. In October 1978, at the time of the opening of the permanent headquarters at Shinfield Park in Reading, ECMWF took delivery of its own supercomputer. This (a CRAY1-A, serial number 9) was the first CRAY supercomputer system in Europe. ECMWF delivered the first operational medium- range weather forecast to its Member States on 1 July 1979.

Since that time ECMWF has continued to be in the forefront of the practical use of large computer systems. The CRAY 1-A was a single processor computer capable of a sustained performance of 50 Mflops. The industry in the meantime was moving towards providing systems with more than one processor. With these, operational centres requiring maximum power, such as those involved in numerical medium-range weather forecasting, could continue to meet real-time deadlines.

The present central computer, a CRAY Y-MP/C90 is some 100 times more powerful than the CRAY 1-A installed in 1978. The Centre is looking to increase its computing power to meet future demands of its operational and its research work and to this end it has issued an ITT for a computer at least 5 times more powerful than the C90, to be installed in 1996 and upgradeable in a further factor of 2 in 1998.

2. The model equations

The evolution equations for the atmospheric fields are the Navier-Stokes set of equations, cast in spherical coordinates in the horizontal and using a hybrid vertical coordinate h whose values at the interfaces between the vertical slabs are defined as

$$\eta_{k+\frac{1}{2}} = A_{k+\frac{1}{2}}/P_0 + B_{k+\frac{1}{2}} \tag{1}$$

where P_0 is a reference pressure. This vertical coordinate has a constant value of 0 at the top of the atmosphere and a value of 1 at the surface, the kinematic boundary condition is therefore simply $d\eta/dt = 0$ at both the top and the bottom of the model.

The momentum equation is

$$\frac{d\vec{V}}{dt} + 2\vec{\Omega} \times \vec{V} + RT\,\vec{\nabla}\,ln(p) + \vec{\nabla}\phi = \vec{P}_{\vec{v}} + \vec{\mathbf{K}}_{\vec{v}} \tag{2}$$

the thermodynamic equation

$$\frac{dT}{dt} - \mathrm{K}\,T\,\frac{\omega}{p} = P_T + K_T \tag{3}$$

the moisture equation

$$\frac{dQ}{dt} = P_Q + K_Q \tag{4}$$

and the continuity equation

$$\frac{d}{dt}\left(\frac{\partial p}{\partial \eta}\right) + \frac{\partial p}{\partial \eta}\left(D + \frac{\partial \eta}{\partial \eta}\right) = C \tag{5}$$

where
$\vec{V}$ = "horizontal" velocity

$\vec{\Omega}$ = angular velocity of the Earth's rotation

R = ideal gas constant
T = temperature
Φ = geopotential
ω = "vertical" velocity (dp/dt)
P = "physics" contribution
K = diffusion contribution

3. Main characteristics of the model

- Spectral representation in the "horizontal" dimensions using spherical harmonics at triangular truncation representing waves up to wavenumber 213. These basis functions are appropriate to represent scalar fields. The velocity components of the wind are discontinuous at the poles and therefore cannot be represented as a linear combination of spherical harmonics, therefore the fields represented in spectral space are: vorticity

$$\xi = \vec{k} \cdot \vec{\nabla} \times \vec{V_h} \tag{6}$$

  and divergence

$$D = \vec{\nabla} \cdot \vec{V_h} \tag{7}$$

  where ∇ is the bi-dimensional "nabla" operator.
- Finite differences in the vertical. The pressures at the "half levels" or interfaces between the vertical slabs are given by

$$P_{k+\frac{1}{2}} = A_{k+\frac{1}{2}} + B_{k+\frac{1}{2}}\, P_s \tag{8}$$

  where the A's and B's are the coefficients defining the vertical coordinate and P_s is the surface pressure at the corresponding horizontal position.
- Semi-Lagrangian treatment of the total time derivative
- Semi-implicit treatment of the terms of the equations governing the gravity waves (these are terms 3 and 4 of the momentum equation, term 2 of the thermodynamic equation and term 2 of the continuity equation)
- Reduced Gaussian grid for representation in physical space
- Revised Legendre transforms

4. The spectral representation

In the spectral technique, a scalar field is represented as a linear combination of basis functions. For spherical geometry the proper basis functions are the spherical harmonics

$$Y_n^m = P_n^m(\mu)\, e^{im\lambda} \tag{9}$$

where λ is the geographical longitude, μ is the sine of the latitude, m is the zonal wavenumber, n is the total wavenumber and $P_n^m(\mu)$ are the associated Legendre Functions of the first kind. Some properties of the associated Legendre functions are:

$$P_n^{-m}(\mu) = P_n^m(\mu) \tag{10}$$

and

$$\frac{1}{2}\int_{-1}^{1} P_n^m(\mu)\, d\mu = \delta_{n,s} \tag{11}$$

The representation of a field $X(\lambda, \mu, \eta, t)$ in terms of these basis functions is

$$X(\lambda, \mu, \eta, t) = \sum_{m=M}^{M} \sum_{n=m}^{N(m)} X_n^m(\eta, t)\, Y_n^m \tag{12}$$

where X_n^m are the spectral coefficients (functions of the vertical coordinate η and of time) and can be computed from the "physical space" representation of X by means of

$$X_n^m(\eta, t) = \frac{1}{4\pi}\int_{-1}^{1}\int_{0}^{2\pi} X(\lambda, \mu, \eta t)\, P_n^m(\mu)\, e^{-im\lambda}\, d\lambda\, d\mu \tag{13}$$

The fields X being real, the spectral coefficients have the property that

$$X_n^{-m} = (X_n^m)^* \tag{14}$$

where $()^*$ represents the complex conjugate, therefore only the coefficients corresponding to non-negative zonal wave-numbers m have to be computed.

The application of the spectral method to the solution of non-linear partial differential equations is based on the "transform method". In this method, the spectral representation of the produce of two functions is computed in the following way:

- transform each of the fields to be multiplied from spectral to physical (or grid-point) space
- multiply both fields at each of the grid points
- transform the resulting field to spectral space

Therefore, a series of transforms have to be performed, at each time step of the model, from spectral to grid-point space and vice versa. This is done through the Fourier coefficients, defined as

$$X_m(\mu, \eta, t) = \frac{1}{2\pi}\int_{0}^{2\pi} X(\lambda, \mu, \eta, t)\, e^{-im\lambda}\, d\lambda \tag{15}$$

or, from the spectral coefficients

$$X_m(\mu,\eta,t) = \sum_{n=m}^{N(m)} X_n^m\,(\eta,t)\;P_n^m(\mu) \tag{16}$$

whose inverse is:

$$X_n^m(\eta,t) = \frac{1}{2}\int_{-1}^{1} X_m\;(\mu,\eta,t)\;P_n^m\;(\mu)\;d\mu \tag{17}$$

because of the orthogonality property (11) of the Legendre functions.

Therefore

$$X(\lambda,\mu,\eta,t) = \sum_{m=-M}^{M} X_m(\mu,\eta,t) = e^{im\lambda} \tag{18}$$

One of the advantages of the spectral technique is the possibility of computing the derivatives of the fields analytically, thus avoiding the introduction of numerical dispersion due to the space discretization. The longitudinal derivative of field X is

$$\left(\frac{\partial X}{\partial}\lambda\right)_n^m = im\,X_n^m \tag{19}$$

that is to say, the spherical harmonics are eigenfunctions of the operator $\partial/\partial\lambda$.

And the latitudinal derivative is

$$\left(\frac{\partial X}{\partial\mu}\right)_n^m = X_n^m\,\frac{dP_n^m}{d\mu} \tag{20}$$

where

$$(1-\mu^2)\frac{dP_n^m}{d\mu} = -n\varepsilon_{n+1}^m\;P_{n+1}^m + (n+1)\;\varepsilon_n^m P_{n-1}^m \tag{21}$$

with

$$\varepsilon_n^m = \left(\frac{n^2-m^2}{4n^2-1}\right)^{\frac{1}{2}} \tag{22}$$

Another very important property of the spherical harmonics is that they are eigenfunctions of the Laplacian:

$$\left(\nabla^2 X\right)_n^m = -\frac{n(n-1)}{a^2}X_n^m \tag{23}$$

In order to apply the transform method, we have to perform, at every time-step, an inverse spectral transform of every field needed in physical space for computing the non-linear terms of our equations. These transforms are done through the Fourier components as follows:

- Perform an inverse Legendre transform (16) for a series of values of the latitude coordinate μ.
- Perform an inverse Fourier transform (18)

Once the non-linear terms are computed in grid-point space using the local values of the corresponding functions, the resulting field has to be transformed to spectral space by means of:

- A direct Fourier transform (15)
- A direct Legendre transform (17)

The direct and inverse Fourier transforms can be performed exactly by means of the "Fast Fourier Transform" algorithm, by using a number of points equally spaced in the coordinate λ, and this will furthermore be alias- free for quadratic terms if the number of points is at least $3M+1$, where M is the truncation limit for the zonal wavenumber.

The direct Legendre transform and its inverse are also alias-free for quadratic terms if the direct form is computed by means of a Gaussian quadrature formula, using the points for which

$$P_{N_G}^{0}(\mu) = 0 \tag{24}$$

(Gaussian latitudes), N_G being the number of Gaussian latitudes needed which should be at least $(3N+1)/2$, N being the truncation limit for total wavenumber (equal to M for triangular truncation).

The set of points at the Gaussian latitudes and equally spaced in the longitudinal direction is known as the Gaussian grid of the model. The points at high latitudes are therefore much closer in terms of geographical distance than the points at lower latitudes. Nevertheless, one of the properties of the triangular truncation in a spectral representation is that it is homogeneous over the globe, that is, the highest wavenumber representable does not depend on the geographical position. In order to get a Gaussian grid closer to this condition, we decreased the number of points in each row of latitude as we go from equator to pole so that the geographical distance between points remains approximately constant. This is the "reduced" Gaussian grid (Hortal and Simmons, 1991).

5. Advective treatment of the Coriolis term

In the momentum equation (2), the Coriolis term

$$2\vec{\Omega} \times \vec{V}_h \tag{25}$$

can be expressed as

$$2\vec{\Omega} \times \frac{d\vec{r}}{dt} \tag{26}$$

where $\vec{r}$ is the position vector of the parcel of air being advected. Then, the sum of the two first terms of the equation is

$$\frac{d\left(\vec{V}_h + 2\vec{\Omega} \times \vec{r}\right)}{dt} \tag{27}$$

The quantity being advected is therefore $\vec{V} + 2\vec{\Omega} \times \vec{r}$ rather than $\vec{V}$.

6. The semi-Lagrangian procedure

The material time derivative of a quantity dX/dt is the rate of change of the property X of a parcel of air, following the movement of that parcel. It can be discretized, in a two-time-level scheme, as

$$\left(X^{+} - X^{-}\right)/\Delta t \tag{28}$$

across a time interval Δt, where X^{+} is the value of property X at the future time step (time $t_0 + \Delta t$) at the arrival point of the parcel trajectory and X^{-} is the value at time t_0 at the departure point of the parcel trajectory.

If we start the integration for the time step with a set of air parcels situated at a series of regularly spaced points (such as the Gaussian grid), we end up with the values of the property X at a series of points (the arrival points of each trajectory) which are irregularly distributed in space. This is called the forward Lagrangian procedure.

Instead, we may assume that the arrival points of the trajectories are going to be regularly distributed at the future time step and trace back the departure points of the trajectories which fulfil this condition. This is the upwind or semi-Lagrangian procedure (Bates and McDonald, 1982). The values of the quantity X at time t_0 can be computed by means of interpolation from the known values at a set of regularly distributed points at that time. For a review of semi-Lagrangian schemes used in atmospheric models, the reader is referred to Staniforth and Coté (1991) and references therein.

In order to compute the parcel trajectories with second order accuracy in time, we need to know the velocities at the moment $t_0 + \Delta t/2$. We have therefore two possibilities:

- In a three-time-levels semi-Lagrangian discretization we use the velocities at the present time t_0 to compute trajectories spanning a time interval $2\Delta t$. The departure points then correspond to time $t_0 - \Delta t$ and the arrival points to time $t_0 + \Delta t$ and the discretization is

$$dX/dt \approx (X^+ - X^-)/2\Delta t \tag{29}$$

- In a two-time-levels semi-Lagrangian discretization, the arrival points also correspond to time $t_0 + \Delta t$ but the departure points correspond to time t_0. The velocities needed for the computation of the trajectories should then correspond to time $t_0 + \Delta t/2$ and can be estimated by means of extrapolation in time from the known values at time t_0 and at time $t_0 - \Delta t$.

The computation of trajectories is performed by means of:

$$(\vec{r})^- = (\vec{r})^+ - 2\,(\vec{V})^0 \Delta \mathbf{t} \tag{30}$$

where $\vec{V}^0$ is the velocity at the centre of the trajectory going from $\vec{r}^{\,-}$ to $\vec{r}^{\,+}$. This is therefore an implicit equation and needs an iterative procedure to solve it. We start with a first guess for $\vec{V}^0$ which is the velocity at the arrival point $\vec{r}^{\,+}$. We then compute an estimation of the departure point $\vec{r}^{\,-}$. With a new estimate of $\vec{V}^0$, interpolated to the medium point between $\vec{r}^{\,-}$ and $\vec{r}^{\,+}$, we make a new estimate of $\vec{r}^{\,-}$ and continue until we get convergence. In practice, the number of iterations needed has been found to be just one for an accurate estimation of the trajectories.

Once the trajectories have been computed, we need to find the values of X^-. This is done by interpolation to the departure points $\vec{r}^{\,-}$. This interpolation should be of high order, otherwise the advected fields get unduly smoothed. A cubic interpolation using either Lagrange polynomials or Hermite polynomials or cubic splines is found to be accurate enough to avoid smoothing of the fields. Nevertheless, three-dimensional cubic interpolation is quite expensive in terms of computer time and a cheaper procedure has been adopted operationally at ECMWF, called quasi-cubic interpolation. In this procedure, the interpolations at the vertical levels next-nearest neighbours to the departure point are done by bi-linear interpolation rather than bi-cubic. At the nearest-neighbour levels to the departure point, the interpolation is cubic in longitude at the Gaussian rows nearest to the departure point but linear at the two other rows. This procedure is accurate enough and is much cheaper than the standard three-dimensional cubic interpolation.

A problem produced by a high order interpolation, when applied to non smooth fields is over/under-shooting. This means that the interpolated value may be larger than any of the values used for the interpolation, which is physically meaningless in an advection process and produces gravity waves in the model which increase the eddy kinetic energy throughout the integration. This can be avoided by means of some form of shape-preserving interpolation. The form chosen at ECMWF is a modification of the quasi-monotone procedure proposed by Bermejo and Staniforth (1992). In this procedure, after each cubic interpolation the interpolated value is compared with the two nearest-neighbouring values used in the interpolation. If the interpolated value lies outside the interval covered by these two values, it is modified as little as possible to stay within the interval. When the quasi-monotone procedure is applied in the vertical dimension, due to the coarse resolution of our model in this direction, a smoothing is produced in the temperature field at the tropopause and likewise the maxima of zonal wind in the stratosphere are smoothed. Therefore the quasi-monotone procedure is only applied in the horizontal dimensions.

7. The semi-implicit treatment of gravity waves

The general form of the model equations is

$$\frac{dF}{dt} = R \tag{31}$$

which is discretized in the three-time-level version of the semi-Lagrangian scheme as

$$\frac{(F)^{+} - (F_f)^{-}}{2\mathbf{\Delta t}} = (R)^{0} \tag{32}$$

where $()^{+}$, $()^{-}$ and $()^{0}$ correspond to the evaluation of terms at the end of the trajectory time (time $t+\mathbf{\Delta t}$), the beginning of the trajectory (time $t-\mathbf{\Delta t}$) and the middle of it (time t) resp. The subindex f means a time-filtered value to filter-out the computational mode of a three-time-scheme.

An option in the model is to compute the value $()^{0}$ at time t as an average between the value at the end of the trajectory (no interpolation needed for that point) and its value at the departure point (trilinear interpolation used to that point). This is the option used operationally.

The semi-implicit treatment of a linearized gravity-wave term $\bar{X}$ is achieved by adding to the explicit r.h.s. of the corresponding equation the quantity.

$$\bar{X} \to \frac{\beta}{2}\left[(\bar{X}^{+}(\bar{X}_f)^{-} - 2(\bar{X}^{0}\right] \tag{33}$$

this is computed as

$$\bar{X} = \frac{\beta}{2}\left[\bar{X}^{+}(t+\Delta t) + \bar{X}_f^{-}(t-\Delta t) - \bar{X}^{+}(t) - \bar{X}^{-}(t)\right] \tag{34}$$

Currently, $\beta = 1$ for the operational run (time-step of 15 minutes) and $\beta = 1.5$ for a run made daily to 72 hours with a time-step of 22.5 minutes.

This larger value of the semi-implicit scheme but worsens its dispersion properties.

In the momentum equation, the linearized term:

$$-\vec{\nabla}_\eta(\bar{\phi} + R_d\, T_r\, \mathbf{In}\; p_s) \tag{35}$$

is treated semi-implicity

Here

$$\bar{\phi} = \phi_s - R_d \int_1^\eta T\, \frac{d}{d\eta}(ln\bar{p})\, d\eta = \phi_s + \mathbf{Y}\, T \tag{36}$$

T_r is a reference temperature used in the linearization (=300K)
$\bar{p}$ is the pressure corresponding to a reference constant surface pressure (80000 hPa)

γ is a matrix acting on the vertical vector of temperature values, which represents the finite differences discretization of the vertical integral in equation (36).

R_d is the ideal gas constant for dry air and
Φ_s is the geopotential at the Earth's surface (orography).

In the thermodynamic equation, the term

$$\mathbf{K}\, T_r\left(\frac{\bar{\omega}}{p}\right) = -\frac{K\, T_r}{\bar{p}} \int_0^\eta D\, \frac{d\bar{p}}{d\eta}\, d\eta = -T\, D \tag{37}$$

is treated semi-implicitly, here D is the divergence
τ is a matrix representing the vertical discretization of the integral

In the continuity equation, the term

$$-\frac{d\bar{p}}{d\eta}\, D \tag{38}$$

is also treated semi-implicitly.

Ignoring the diffusion terms which are treated independently (time splitting) in spectral space, the discretized equations read:

Momentum:

$$\left[\vec{V}_h + \Delta t\beta\vec{\nabla}_\eta(\mathbf{Y}(\mathbf{T}(\mathbf{t}+\Delta\mathbf{t}) - \mathbf{T}(\mathbf{t})) + \mathbf{R_d}\;\mathbf{T_r}(\mathbf{ln}\;\mathbf{p_s}(\mathbf{t}+\Delta\mathbf{t}) - \mathbf{lnp_s}(\mathbf{t})))\right]^+ \tag{39}$$

$$= \left[\vec{V}_{hf} - \Delta t\beta\vec{\nabla}_\eta(\mathbf{Y}(T(t-\Delta t)_f - T(t)) + R_dT_r(lnp_s(t-\Delta t)^f - lnp_s(t)))\right] + 2\Delta t\left[\vec{MT}^0\right] \tag{40}$$

where MT is the explicit r.h.s. of the momentum equation without diffusion.

Applying the operator $\vec{k} \cdot \vec{\nabla}_\eta x$ to this equation we get:

$$\xi^+ = L \tag{41}$$

and with the operator $\vec{\nabla}_\eta \cdot$

$$\Big[D + \mathbf{\Delta} t \beta \nabla^2 (\mathbf{Y} T + R_d T_r ln\, p_s)\Big]^+ = M \tag{42}$$

In order to compute the r.h.s. of these equations, one has to add together vectors referring to different points and therefore they have to be expressed in a common reference frame. This reference frame has been chosen to be the unit vectors lying on the tangential plane at the arrival of the trajectories (the Gaussian grid points) and pointing to the north pole and to the east.

The discretized thermodynamic equation is

$$[T + \mathbf{\Delta}\, t\, \beta\, \mathbf{T}\, D]^+ = T_1 \tag{43}$$

and the continuity equation

$$[ln\, p_s + \mathbf{\Delta}\, t\, \beta \mathbf{v}\, D]^+ = L_0 \tag{44}$$

Eliminating $(ln\, p_s)^+$ and T^+ we finally get

$$(1 - \lceil\, \nabla^2) D^+ = DT' \tag{45}$$

where the matrix **lceil** is a constant coefficient matrix with dimension the number of vertical levels. This Helmholtz equation is easily solved in spectral space as the spherical harmonics are eigenvectors of the Laplacian operator and matrix $\lceil$ is a small matrix.

The set of prognostic equations presented in this paper are complemented with a set of diagnostic equations to compute the geopotential $\mathbf{\Phi}$ at the model levels by integrating the hydrostatic equation, and the vertical velocities $d\eta/dt$ and dp/dt. Those last ones are derived integrating the continuity equation over a layer of the atmosphere.

8. Parallelization of the code

In order to use more efficiently the multitasking capability of the present CRAY central computer, the code has to be parallelized in such a way that the individual processors are idle as little as possible. On the other hand, the limited amount of central memory available precludes keeping in memory the values of all the variables over the whole globe, needed for the computation at a certain time step.

The points to be taken into account for an efficient parallelization of the code are:

- Fourier transforms need the values of the field to be transformed at all the longitude points at a certain Gaussian latitude.
- Legendre transforms need the values of the Fourier coefficients corresponding to a certain zonal wave-number at all the Gaussian latitudes.
- Semi-Lagrangian advection needs a number of points around the arrival point of the trajectory large enough to account for the length of the trajectory plus the need for cubic interpolation.

Taking all these constraints into consideration, the model uses a two-scan strategy as follows:

The values of the fields at the northernmost row of latitude (or of a set of rows if we are near the poles where the lines are shorter) are read into memory from a work-file. These values are given to one of the processors to compute the r.h.s. of the equations. In the meanwhile, the next row (or set of rows) of latitude southwards is read and supplied to a second processor, and so on until all the processors are busy. When one of the processors finishes computing a row of latitude, the next free row of latitude is read and supplied to that processor. Eventually, enough number of latitude rows will have been computed so that the semi-Lagrangian trajectories arriving at the points of the first row will have their departure points within the band of latitude rows present in central memory. At this stage, the first available processor is given the task of computing the semi-Lagrangian part of the solution for this first row, do the Fourier transforms on that row and write the Fourier coefficients to a work-file.

This scan of rows continues until we reach the south pole.

Next, the Fourier coefficients corresponding to zonal wavenumber $m = 0$ for all latitudes are read from the work-file and given to a processor to perform the Legendre transforms at that zonal wavenumber, do the spectral computations (including horizontal diffusion) and do the necessary inverse Legendre transforms. The coefficients for the next zonal wavenumber are in the meanwhile read and given to the next processor and so on until all the available processors are busy. Once a processor finishes its task and writes back to the work-file the set of computed Fourier coefficients, it is ready to handle the next zonal wavenumber waiting to be computed.

This completes the computations needed in a time-step.

In the first scan, rather than the values of the fields at the grid-points, values of Fourier coefficients are read and the inverse Fourier transforms performed at each row.

9. Discussion

In this article we have examined the implementation of the semi-Lagrangian semi-implicit method in a high-resolution version of the ECMWF forecast model. A fuller description, including the vertical discretization for conservation of angular momentum and energy for frictionless adiabatic flow can be found in Ritchie et al (1995). A novel feature of this model is the use of the semi- Lagrangian treatment of advection in a baroclinic spectral model in conjunction with the reduced Gaussian grid, as described by Hortal and Simmons (1991), in which the "fully reduced grid" gives a more homogeneous resolution in grid- point space and leads to savings of about 25% in the computational cost of the model.

The momentum equation formulation is considerably more economical than its vorticity-divergence counterpart and, by also incorporating economies in the Legendre transforms demonstrated by Temperton (1991) for the shallow-water equations, reduced the number of multilevel Legendre transforms to a total of 12 per time step.

The semi-implicit treatment of the terms describing the gravity waves is achieved by adding a semi-implicit correction to the explicit formulation of the equations. Quasi-cubic interpolation using a bi-dimensional modification of the quasi-monotonic procedure first suggested by Bermejo and Staniforth (1992) is applied to the fields being advected while an averaging along the semi-Lagrangian trajectory is used for the righ-hand-side terms of the equations, using in this case trilinear interpolation to the departure points. By interpolating first in the longitudinal direction, it is relatively simple to accommodate the irregular mesh resulting from the use of the reduced Gaussian grid.

Several important computational details related to the operational implementation of the model in a modestly parallel processor machine were presented in section 8. It has been found that the semi-Lagrangian version with a 15-min time step gives an accuracy equivalent to that of an Eulerian version with a 3-min time step, giving an efficiency improvement of about a factor of 4 after allowing for the 20% of time spent in the semi-Lagrangian computations. No special filters were required in order for the application of the semi-Lagrangian method in hybrid coordinates to work successfully in this three-time-level model.

References

Bates, J. R., and A. McDonald, (1982) Multiply-upstream, semi-Lagrangian advective schemes: Analysis and applications to a multilevel primitive equation model. *Mon. Wea. Rev.*, **110, pp. 1831–1842**

Bermejo, R., and A. Staniforth, (1992) The conversion of semi-Lagrangian advection schemes to quasi-monotone schemes. *Mon. Wea. Rev.*, **120, pp. 2622–2632**

Hortal, M., and A. J. Simmons, (1991) Use of reduced Gaussian grids in spectral models. *Mon. Wea. Rev.*, **119, pp. 1057–1074**

Ritchie, H., C. Temperton, A. J. Simmons, M. Hortal, T. Davies, D. Dent, and M. Hamrud, (1995) Implementation of the semi-Lagrangian method in a high- resolution version of the ECMWF forecast model. *Mon. Wea. Rev.*, **123, pp. 489–514**

Staniforth, A., and J. Coté (1991) Semi-Lagrangian integration schemes for atmospheric models - A review. *Mon. Wea. Rev.*, **119, pp. 2206–2223**

Temperton, C., (1991) On scalar and vector transform methods for global spectral models. *Mon. Wea. Rev.*, **119, pp. 1303–1307**

A LEVEL-SET CAPTURING SCHEME FOR COMPRESSIBLE INTERFACES.

SMADAR KARNI
Temple University, Philadelphia, PA 19122
and
Courant Institute, New York University, 251 Mercer Street, New York, NY 10012

1. Introduction

Extending single-component flow descriptions to two-component fluid flows represents only a small modelling generalization. For example, the standard conservation laws that described the dynamics of a single-component fluid flow may now be written for the two-component fluid mixture, augmented by an evolution equations that describe its composition. For two immiscible fluids, one may write an evolution equation for an indicator function whose given level-set tracks the material interface that separates them, thereby allowing to determine which volume in space is occupied by one fluid and which by the other. A major concern in numerical computations of multifluid dynamics is to ensure that fluid components remain in pressure equilibrium with each other across the material interface [1,2,5,6,7,15]. Within the class of shock-capturing schemes, such interfaces have non-zero thickness due to numerical diffusion, and pressure equilibrium need to be maintained across the interface transition.

A level-set approach is extremely attractive for computations of interface motion because of its potential of dealing with complicated geometries with great simplicity and straightforwardness. As a formulation, level-set has a front-tracking 'flavour'. Yet, it allows an overall front-capturing discretization algorithm, thus avoiding the complexities associated with front tracking. The last few years have seen a growing industry of level-set based schemes for interface propagation, quite a few of which for incompressible variable density flows ([3,13] and references therein). The incompressibility assumption allows the elimination of the pressure from the model, and so deviations from pressure equilibrium has no direct effect on the computed solution.

E.F. Toro and J.F. Clarke (eds.), Numerical Methods for Wave Propagation, 253–273.

Mulder et al. have proposed an attractive level-set formulation for compressible interface motion and used it to compute interfacial instabilities [9]. This flow model was analyzed in [5] where it was shown that, in fact, the model does not maintain pressure equilibrium between its components and induce erroneous pressure fluctuations across materials fronts.

A constructive observation was made in [5] that if pressure is computed using the Pressure Evolution Equation (PEE), rather than recovered in the standard fashion using the Equation of State (EOS), such erroneous fluctuations will not arise.

This has lead the author to develop a scheme based on discretizing the evolution equations for the so-called primitive variables, and including viscous correction terms to account for leading order conservation errors [4,5]. This primitive variable scheme has been successfully implemented in an Adaptive Mesh Refinement algorithm to study the complicated interaction of shock wave with a bubble inhomogeneity [10]. Two experiments of shock/bubble interaction [18] were reproduced to remarkable agreement, capturing all the intricate nonlinear mechanisms of the interaction and revealing flow details beyond those observed in the experiments. The excellent agreement lends credibility to several new observations that were made, which where previously left unexplained by the experimental results. The major limitation of the primitive variable approach is that conservation errors, although controlled, are inherent to the scheme. This makes the scheme not suitable for computations of strong shocks, yielding incorrect shock speed/strength.

In this paper we have used a level-set flow model for immiscible multifluids which is suitable for shocks ranging from weak, through moderate to *very* strong. The formulation automatically ensures pressure equilibrium between the fluid components, and amounts to a small, yet fundamental, modification of the level-set formulation of Mulder et al. [9]. It leads to an algorithm that is *essentially conservative*, in the following way: density and momentum are perfectly conserved, total energy is conserved *almost everywhere* and conservation error goes to zero under mesh refinement. One dimensional shock tube problems and shock/interface interactions were presented in [6], establishing that the energy conservation errors are *extremely* small on standard grids and converge to zero under mesh refinement. A flow model for miscible multifluids has also been described in [6]. The ultimate goal for all multifluid algorithms is to be able to handle very stiff problems such as gas/liquid interface dynamics. In this paper, we go some way toward this goal and demonstrate the suitability of the method for modelling stiff multicomponent gas dynamics. The method has been extended to 2D via operator splitting and we present results for computations of two physical interfacial instabilities: (i) Kelvin-Helmholtz and (ii) Rayleigh-Taylor instabilities. The former is a shear interface instability, where small perturbations induce a vortex sheet roll-up. The latter is an instability of a heavy fluid sitting on top of a light fluid, in equilibrium, in a gravitational field. Small perturbations to the interface grow unstably and develop familiar 'mushroom-like' fingers.

2. Flow Model

In one space dimension, the governing equations for inviscid two-component compressible flow are

$$\rho_t + (\rho u)_x = 0 \tag{1a}$$

$$(\rho u)_t + (\rho u^2 + p)_x = 0 \tag{1b}$$

$$E_t + (u(E+p))_x = 0 \tag{1c}$$

with ρ, u, p and E representing the density, velocity, pressure and total energy of the fluid mixture. A smooth function, ψ, is defined so that at time $t = 0$, its zero level-set coincides with the location of the material interface separating the fluid components. The material interface propagates with the fluid velocity, u, hence we evolve ψ according to

$$\psi_t + u\psi_x = 0$$

so that at any later time, t, the level set $\psi(x,t) = 0$ locates the material interface, and the sign of ψ determines which fluid component occupies the space volume. The above evolution equation for ψ may be combined with the continuity equation ($1a$) and recast as a conservation law

$$(\rho\psi)_t + (\rho\psi u)_x = 0 \tag{1d}$$

but this is a matter of computational convenience not a modelling necessity. The fluid components are assumed to be ideal gases with specific heat ratios $\gamma_{1,2}$ respectively, obeying the Equation of State (EOS)

$$p = (\gamma(\psi) - 1)(E - \frac{1}{2}\rho u^2). \tag{2}$$

with

$$\gamma(\psi) = \begin{cases} \gamma_1 & \psi > 0 \\ \gamma_2 & \psi < 0 \end{cases}$$

This model was proposed in [9] for computations of compressible interface motion. However, it has been shown in [5] that computing the pressure via the level-set dependent EOS (2) induces erroneous pressure fluctuations across material fronts, even if the data has perfectly uniform pressure and velocity. The fluctuations are $O(\gamma_2 - \gamma_1)$ and depending on the local CFL number, may be either positive or negative, i.e. they do not possess a definite sign. Once generated, these pressure fluctuations set off false dispersive mechanisms which contaminate the field solution of the other flow variables and may trigger false interface dynamics. A very useful observation made in [6] is that if the pressure field is evolved using the Pressure Evolution Equation (PEE),

$$p_t + up_x + \gamma pu_x = 0. \tag{1e}$$

then such oscillations cannot occur. Indeed, if the fluid components are in pressure equilibrium, both p and u are uniform (more generally, continuous) across the interface, and from (1e), pressure will remain uniform (continuous) in the solution. This equation, thus, automatically maintains pressure equilibrium between fluid components, and this property will be inherited by any consistent discretization scheme.

The two procedures for computing the pressure field, namely the Equation of State (2) and the Pressure Evolution Equation (1e), will, in general, give different answers near flow discontinuities, thus they provide us with a choice:

(i) The EOS (2) is suitable everywhere except near material interfaces where it might generate oscillations;

(ii) The PEE (1e) is suitable near interfaces but not near shocks where it might induce conservation errors;

We therefore propose the following flow model:

$$\begin{aligned} \rho_t &+ (\rho u)_x &= 0 \\ (\rho u)_t &+ (\rho u^2 + p)_x &= 0 \\ (\rho\psi)_t &+ (\rho\psi u)_x &= 0 \end{aligned} \tag{3}$$

Equations (3) are supplemented by one of the following:

(i) If $\psi = 0$ then

$$\begin{aligned} &p_t + up_x + \gamma pu_x = 0 \\ &E = \frac{p}{\gamma(\psi) - 1} + \frac{1}{2}u^2 \end{aligned} \tag{3a}$$

(ii) Otherwise

$$E_t + (u(E+p))_x = 0$$
$$p = (\gamma(\psi) - 1)(E - \tfrac{1}{2}u^2) \tag{3b}$$

The latter is a standard conservative update where the total energy is computed from the energy conservation law. The pressure is then found from the EOS. This procedure is used everywhere except across the material front, which in 1D amounts to only one (!) computational cell. In particular it is used near shocks where conservation is essential in order to ensure correct shock speed/strength.

The former solution update is only used across material fronts. There, the PEE becomes exact and ensures pressure equilibrium is maintained between the fluid components. In this update procedure, the pressure is computed using the PEE, then the EOS is used to compute the total energy. This step is not strictly conservative, but since the PEE is well-behaved near material fronts, the solution update is essentially conservative. This is supported by strong numerical evidence. The pressure solution update via the PEE may be regarded as a boundary condition imposed on the moving material front, ensuring that the pressure gradient is smooth in the direction normal to the interface.

3. Numerical Scheme

The common practice in numerical schemes for conservation laws is to update the cell average of the conserved flow variables by the net flux through the cell boundary. Equivalently, one may adopt a wave-based approach in the spirit of the [12,8], in which the solution is updated by signals (waves) that are generated by projecting the local flow gradient onto the characteristic fields. This framework is particularly attractive in the present context, since pressure is not a conserved variable and there is no flux function by which it gets updated. A wave-based approach allows both the conserved variables and the pressure to be updated by the same numerical scheme. The condition $\psi(x,t) = 0$ which identifies the interface, is replaced by the discrete statement $\psi_j \cdot \psi_{j+1} < 0$, which implies a sign change in ψ and indicates a material interface. If a material interface is identified between grid cells j and $j+1$, the wave signals across the cell interface are computed using equations ($3a$) for the pressure/energy fields. Otherwise, ($3b$) is used. We point out that wherever ($3b$) is used, the wave-based solution update is entirely equivalent to a flux-based solution update. The resulting numerical scheme perfectly conserves mass and momentum. Total energy is conserved everywhere except across the (one!) cell edge in question. Numerical results show that energy conservation errors are *extremely* small even for extremely strong shocks, and go to zero under mesh refinement. Furthermore, they are generated during shock/interface interaction [6] and appear as a slight phase shift (not an amplitude shift, which would be more harmful).

We have used a Roe-type scheme to solve the above model. The equations are locally linearized using a Roe linearization [11], the solution gradient is projected onto the linearized eigenvectors and the solution is updated by a limited Lax-Wendroff step on each of the characteristic fields [12,14]. On both physical and mathematical grounds, the eigenvalues of both the $(3,3a)$ model and $(3,3b)$ model are the same. The eigenvectors are, of course, different.

System $(3,3a)$ solves for the independent variables $\mathbf{W} = (\rho, \rho u, p, \rho\psi)^T$ (note change of order) and the corresponding eigenstructure is

$$\begin{array}{llll}
\mathbf{r}_1 = (1, & u-a, & a^2, & \psi)^T \qquad \lambda_1 = u-a; \\
\mathbf{r}_2 = (1, & u, & 0, & \psi)^T \qquad \lambda_2 = u; \\
\mathbf{r}_3 = (0, & 0, & 0, & 1)^T \qquad \lambda_3 = u; \\
\mathbf{r}_4 = (1, & u+a, & a^2, & \psi)^T \qquad \lambda_4 = u+a.
\end{array} \tag{4}$$

System $(3,3b)$ solves for the independent variables $\mathbf{W} = (\rho, \rho u, E, \rho\psi)^T$ (again note change of order) and the corresponding eigenstructure is

$$\begin{array}{llll}
\mathbf{r}_1 = (1, & u-a, & H-ua, & \psi)^T \qquad \lambda_1 = u-a; \\
\mathbf{r}_2 = (1, & u, & \frac{1}{2}u^2, & \psi)^T \qquad \lambda_2 = u; \\
\mathbf{r}_3 = (0, & 0, & \dfrac{-X}{\gamma-1}, & 1)^T \qquad \lambda_3 = u; \\
\mathbf{r}_4 = (1, & u+a, & H+ua, & \psi)^T \qquad \lambda_4 = u+a.
\end{array} \tag{5}$$

where we have used a to denote the speed of sound, H the total enthalpy and $X = \frac{p}{(\gamma-1)\rho}\gamma'(\psi)$, following the notation in [6,7,9]. We note that X is the contribution to changes in total energy due to changes in material properties. As discussed in [9], analytically X is a delta function. This fact is reflected by the discrete approximation in that the discrete Jacobian matrix becomes almost infinite (*i.e.* behaves like $= \frac{1}{\Delta x}$), which numerically might be a source of concern. The only place where $X \neq 0$, both analytically and discretely, is across the material interface. However in the algorithm proposed here, across the interface the total energy is updated by an altogether different procedure which does not use the energy conservation law (rather it uses the PEE $(1e)$

and the EOS (2)). And so X never really plays a role in the present solution procedure, and its singular character is avoided.

Both models can be conveniently represented in terms of *one* set of eigenvectors for all five variables $\mathbf{W} = (\rho, \rho u, E, p, \rho\psi)^T$

$$\begin{array}{lllllllllll}
\mathbf{r}_1 & = & (& 1, & u-a, & H-ua, & a^2, & \psi &)^T & \lambda_1 & = & u-a; \\
\mathbf{r}_2 & = & (& 1, & u, & \frac{1}{2}u^2, & 0, & \psi &)^T & \lambda_2 & = & u; \\
\mathbf{r}_3 & = & (& 0, & 0, & \frac{-X}{\gamma-1}, & 0, & 1 &)^T & \lambda_3 & = & u; \\
\mathbf{r}_4 & = & (& 1, & u+a, & H+ua, & a^2, & \psi &)^T & \lambda_4 & = & u+a.
\end{array} \tag{6}$$

with the understanding that the fourth components of the respective eigenvectors (corresponding to pressure changes) are used only when an interface is identified (*i.e.* $\psi_j \cdot \psi_{j+1} < 0$) and the third components (corresponding to energy changes) are used everywhere else (*i.e.* away from the interface). They are never used at the same time.

In either case, a small change in the solution $\delta\mathbf{W}$ can be projected onto the above set of eigenvectors

$$\delta\mathbf{W} = \sum_{k=1}^{4} \alpha_k \mathbf{r}_k \tag{7}$$

for the wave strengths

$$\begin{aligned}
\alpha_1 &= \frac{\delta p - \rho a \delta u}{2a^2} \\
\alpha_2 &= \frac{a^2 \delta\rho - \delta p}{a^2} \\
\alpha_3 &= \rho\delta\psi \\
\alpha_4 &= \frac{\delta p + \rho a \delta u}{2a^2}.
\end{aligned} \tag{8}$$

4. Numerical Results

4.1. ONE DIMENSIONAL SHOCK/INTERFACE INTERACTION

A shock wave of strength M_s is propagating from left to right and is hitting a gas interface. As a results of the shock impact, the interface sets into motion and the refracted shock wave propagates into the gas. The reflected acoustic wave is either an expansion or a shock wave, depending on the *Acoustic Impedance* of the receiving gas. A number of cases were presented in [6] both for (i) air/helium interface (reflected expansion) and (ii) air/freon (R22) interface (reflected shock). In these calculations, mass and momentum are perfectly conserved. Total energy is conserved everywhere except at the gas interface itself (one(!) grid cell interface). The error in total energy has been computed (against exact solution) and mesh convergence tests, summarized in Tables (1) and (2), are recalled from [6]. They correspond to the case of a $M_s = 9.2659$ and $M_s = 113.4$ shocks. Conservation error seems *not* to depend on shock strength. Moreover, it is *negligible* already on standard meshes, and further converges to zero under mesh refinement. The scheme thus qualifies as an *Essentially Conservative Scheme.* Figure (1) shows a $M_s = 10.017$ shock refraction at a 'stiff' interface separating two ideal gases with $\gamma_l/\gamma_r = 3.0/1.1$, and a density ratio $\rho_r/\rho_l = 3.0$. This is a more difficult test case. The compression ratio for a strong shock is $(\gamma + 1)/(\gamma - 1)$ which is 2 for the $\gamma_l = 3$ gas and 21 for the $\gamma_r = 1.1$ gas. The transmitted shock propagates at a speed which is only about 5% higher than that of the material interface which follows immediately after. In order for the computed interface and transmitted shock to separate by one cell, the interface must propagate through a minimum of 20 cells. The overall computed solution, therefore, takes longer to stabilize. Figure (1) shows excellent agreement between computed and exact solutions, on a 200 point mesh.

4.2. TWO-DIMENSIONAL INTERFACIAL INSTABILITIES

The Kelvin-Helmholtz Instability is a shear interface instability. Initial conditions are uniform $\rho = 1.0$ and $p = 1$, horizontal velocity $u = 0$ and vertical velocity $v = \pm 0.5$ on either sides of the interface. The computational domain is one by two units in the vertical/horizontal directions, with periodic boundary conditions on the top/bottom boundaries and solid wall conditions on the side boundaries. The solution is perturbed by a small sinusoidal single-mode perturbation [17] of amplitude $A = 0.05$. The initial perturbation begins to grow unstably and rolls up into a vortex. Results are shown first for a single gas air-air instability. In the single gas case, the level-set function is nothing more than a passive scalar. It is carried with the fluid particles but feeds no information back into the interface dynamics. The level-set formulation (1a-d) works fine for this case, and we can use it to test the performance of the nonstrictly conservative scheme based on model $(3 - 3ab)$. Figures $(2a)$ and $(2b)$ show results computed by the conservative model (1a-d) and the hybrid level-set model $(3 - 3ab)$ at later times (T=0.25, 0.5, 1.0, 1.5, 2.0, 2.5, 3.0, and 3.5). The agreement between the two computed solutions is excellent. The instability geometry and growth rate are practically identical in both calculations. This lends credibility to the nonstrictly conservative calculations $(3 - 3ab)$, and implies that conservation losses in the total energy have negligible effect on this highly unstable calculation. Figures (3*a*) and (3*b*) show the same instability for a two-fluid interface. In this test, $\rho_1 = 1$, $\rho_2 = 0.5$, $\gamma_1 = 1.4$ and $\gamma_2 = 1.667$. In this case, the level-set formulation is no longer a passive scalar. It feeds back information into the solution by indicating which grid cells are occupied by each of the gas components and serve to select the appropriate gas parameters. In this case, the scheme based on the conservative level-set model fails to maintain pressure equilibrium and generates strong oscillations near the interface. By comparison, the hybrid level-set scheme based on $(3 - 3ab)$ produces clean oscillation-free interface roll-up. Figure (4) shows an air/hel Kelvin-Helmholtz shear instability. In this case, $\rho_{air} / \rho_{helium} = 29/4$, $\gamma_{air} = 1.4$ and $\gamma_{helium} = 1.667$. Again, the nonconservative level-set scheme produces nice and tight vortex sheet roll-up, with no oscillations.

The Rayleigh-Taylor Instability occurs at the interface between a heavy fluid sitting on top of a light fluid. The fluids are in equilibrium in the gravitational field, so both density and pressure are nonuniform in the vertical direction. The solution is perturbed sinusoidally using initial conditions from linearized stability analysis [16]. The important nondimensional parameters in this problem are (i) $D = \rho_b/\rho_a$ the density ratio across the interface and (ii) M a parameter that defines the ratio of gravitational time scales to the sound speed time scale. Following [16,9] we took $M^2 = g\lambda/c_b^2$ where g is the gravitational constant, λ is the wavelength of the perturbation and c_b is the speed of sound (below the interrace). The computational domain is one by six units in the vertical/horizontal directions, with periodic boundary conditions on the top/bottom boundaries and solid side walls. Gravity is pointing rightwards. Figures (5*a*) and (5*b*) show the dynamics of an air/air instability. The initial perturbation of amplitude $A = 0.005$, $D = 2$ and $M^2 = 1.0$, and the solution is shown at later times (T=5,7,9,10,11,12,13 and 14). The initial small perturbation grows into elongated 'mushroom-like' fingers of heavy fluid that penetrate into the light fluid. In this case, the level-set function again is just a passive scalar. The excellent agreement between the conservative (5*a*) and the nonstriclty conservative (6*b*) calculations serves primarily to lend credibility to the latter. The two solutions are practically identical, both in interface geometry and growth rate. Figures (6*a*) and (6*b*) show the same instability in a two component case. The initial amplitude is again $A = 0.005$, $D = 2$, $M^2 = 1.0$ and the specific heat ratios are $\gamma_b = 1.4$ and $\gamma_a = 1.67$. The conservative level-set method produces very oscillatory solutions due to nonphysical fluctuations in the pressure field (6*a*). The hybrid level-set scheme produces completely oscillation-free solutions. Figure

(7) shows results for a air/helium interface. Here, $A = 0.005$, $D = 29/4$, $M^2 = 1.0$ and the specific heat ratios are $\gamma_b = 1.4$ and $\gamma_a = 1.667$. Again, computed solutions by the hybrid level-set scheme are completely oscillation-free.

5. Acknowledgement

This work was supported in part by an NSF Postdoctoral Fellowship, NSF grant #DMS 94 96155 and ONR grant #N00014-94-I-0525 and DOE contract #DEFG0288ER25053.

References

1. Abgrall, R. (1996) How to prevent pressure oscillations in multicomponent flow calculations: a quasi conservative approach. To appear in *J. Comp. Phys.*.
2. Colella, P., Glaz, H.M. & Ferguson, R.E. (1989) Multifluid algorithms for Eulerian finite difference methods. *unpublished manuscript.*
3. Chang Y.C., Hou T.Y., Merriman B and Osher S. (1994) A level-set formulation of Eulerian interface capturing methods for incompressible fluid flows. UCLA CAM Report 94-4.
4. Karni, S. (1992) Viscous shock profiles and primitive formulations. *SIAM J. Num. Anal.* **29**, 1592-1609.
5. Karni, S. (1994) Multi-component flow calculations by a consistent primitive algorithm. *J. Comp. Phys.*, **112**, 31-43.
6. Karni, S. (1995) Hybrid Multifluid Algorithms. To appear in *SIAM J. Sci. Comp.*.
7. Larrouturou, B. (1991) How to preserve the mass fraction positive when computing compressible multi-component flows. *J. Comp. Phys.* **95**, 59-84.
8. Leveque R.J. (1985) A large time step generalization of Godunov's method for systems of conservation laws. *SIAM J. Num. Anal.*, **22**, 1051-1073.
9. Mulder, W., Osher, S. & Sethian, J.A. (1992) Computing interface motion: The compressible Rayleigh-Taylor and Kelvin-Helmholtz instabilities. *J. Comp. Phys.*, **100**, 209.
10. Quirk, J.J. & Karni, S. (1994) On the dynamics of a shock-bubble interaction. ICASE Report # 94-75. To appear in *J. Fluid Mech.*.
11. Roe, P.L. (1981) Approximate Riemann solvers, parameter vectors and difference schemes. *J. Comp. Phys.*, **43**, 357-372.
12. Roe, P.L. (1982) Fluctuations and signals - A framework for numerical evolution problems. in *Numerical methods for fluid dynamics* (eds. Morton, K.W. and Baines, M.J.), 219-257, Academic Press, New York.
13. Sussman M., Smereka P., and Osher S. (1994) A Level-set approach for computing solutions to incompressible two-phase flow. UCLA CAM Report 94-5.
14. Sweby, P.K. (1984) High resolution schemes using flux limiters for hyperbolic conservation laws. *SIAM J. Num. Anal.*, **21**, 995-1011.
15. Ton, V.T. (1995) Improved Shock-Capturing Methods for Multicomponent and Reacting Flows. Submitted to *J. Comp. Phys.*.
16. Gardner C.L., Glimm G., McBryan O. and Zhang Q. (1988) The dynamics of bubble growth for Rayleigh-Taylor unstable interfaces. *Phys. Fluids*, **31** (3), 447-465.
17. Krasny R. (1986) Desingularization of Periodic Vortex Sheet Roll-up. *J. Comp. Phys.*, **65**, 292-313.
18. Haas J.-F. & Sturtevant B. (1987) *J. Fluid. Mech.*, **127**, 539-561.

TABLE 1. Energy conservation errors for a $M_s = 9.2659$ shock impinging on an interface

Air/Helium at t=4.81 $\times 10^{-2}$		Air/R22 at t=6.41 $\times 10^{-2}$	
No. of Points	Relative Error in %	No. of Points	Relative Error in %
400	0.060	400	0.084
800	0.035	800	0.048
1600	0.021	1600	0.030

TABLE 2. Energy conservation error for a $M_s = 113.4$ shock impinging on a air/helium interface.

No. of Points	**Relative Error in %**
200	0.1445
400	0.0788
800	0.0418
1600	0.0226

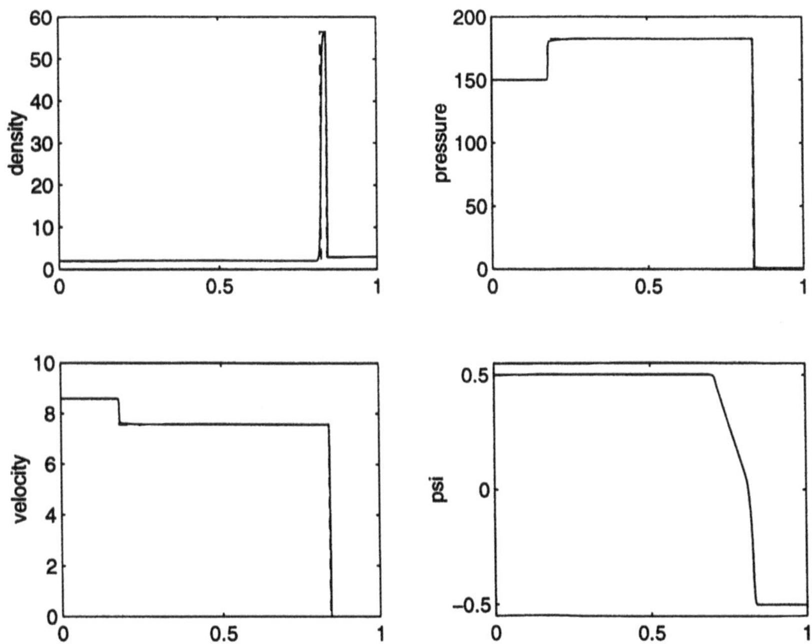

Figure 1. A strong shock wave, $M_s = 10.017$, refracting at a stiff gas interface ($\gamma_l/\gamma_r = 3.0/1.1$). Computed (solid) and exact (dash) solutions.

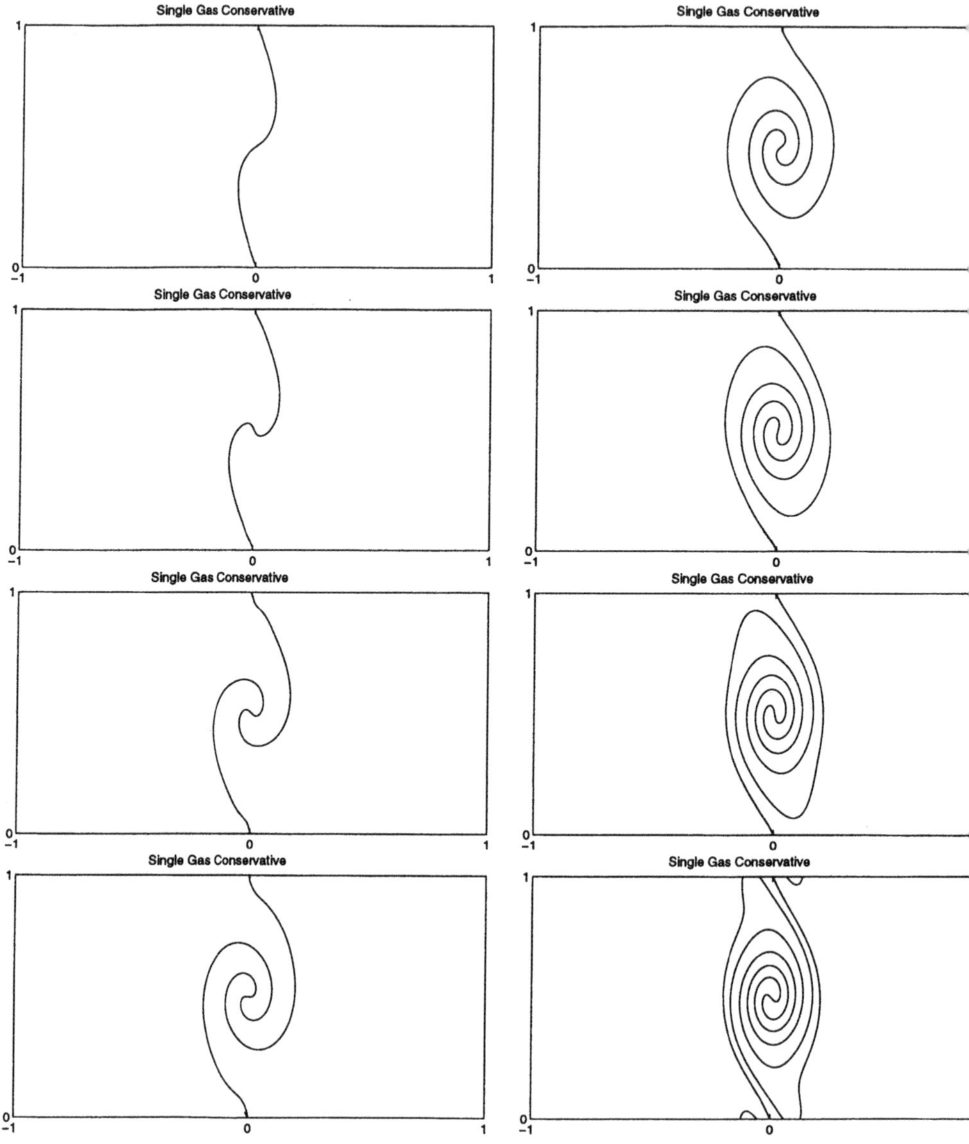

Figure 2a. Air-Air Kelvin-Helmholtz Instability by Conservative Model (1)

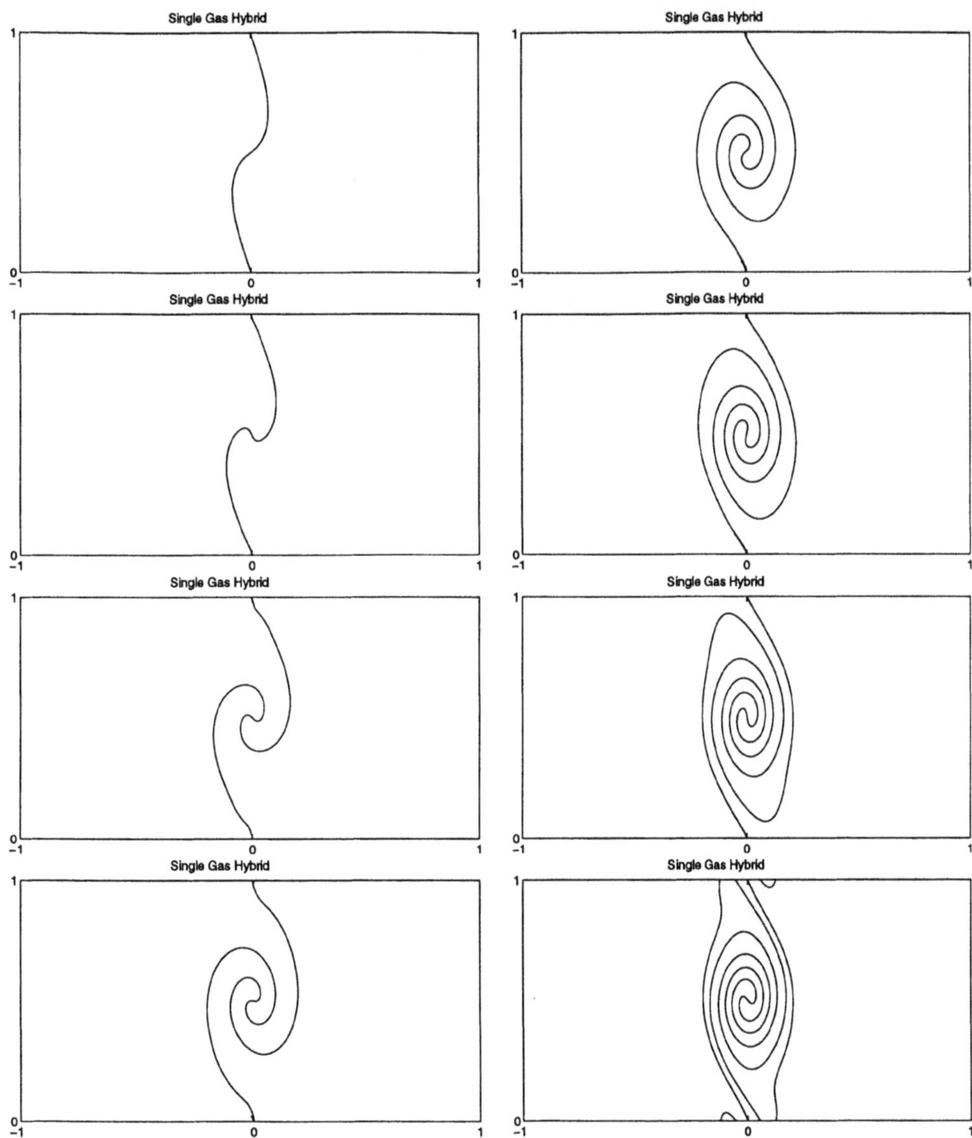

Figure 2b. Air-Air Kelvin-Helmholtz Instability by Hybrid Model (3-3ab)

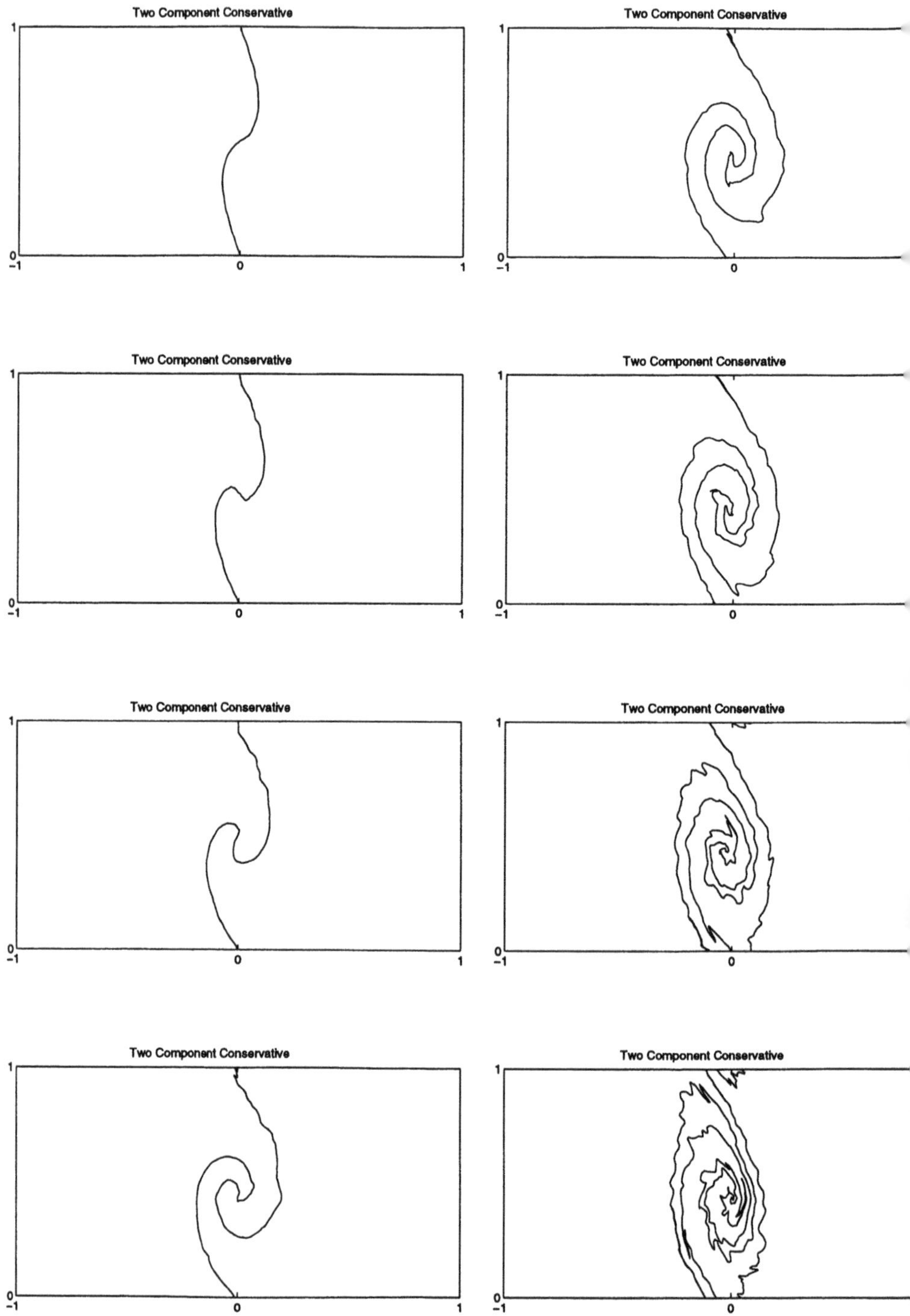

Figure 3a. Two-Gas Kelvin-Helmholtz Instability by Conservative Model (1)

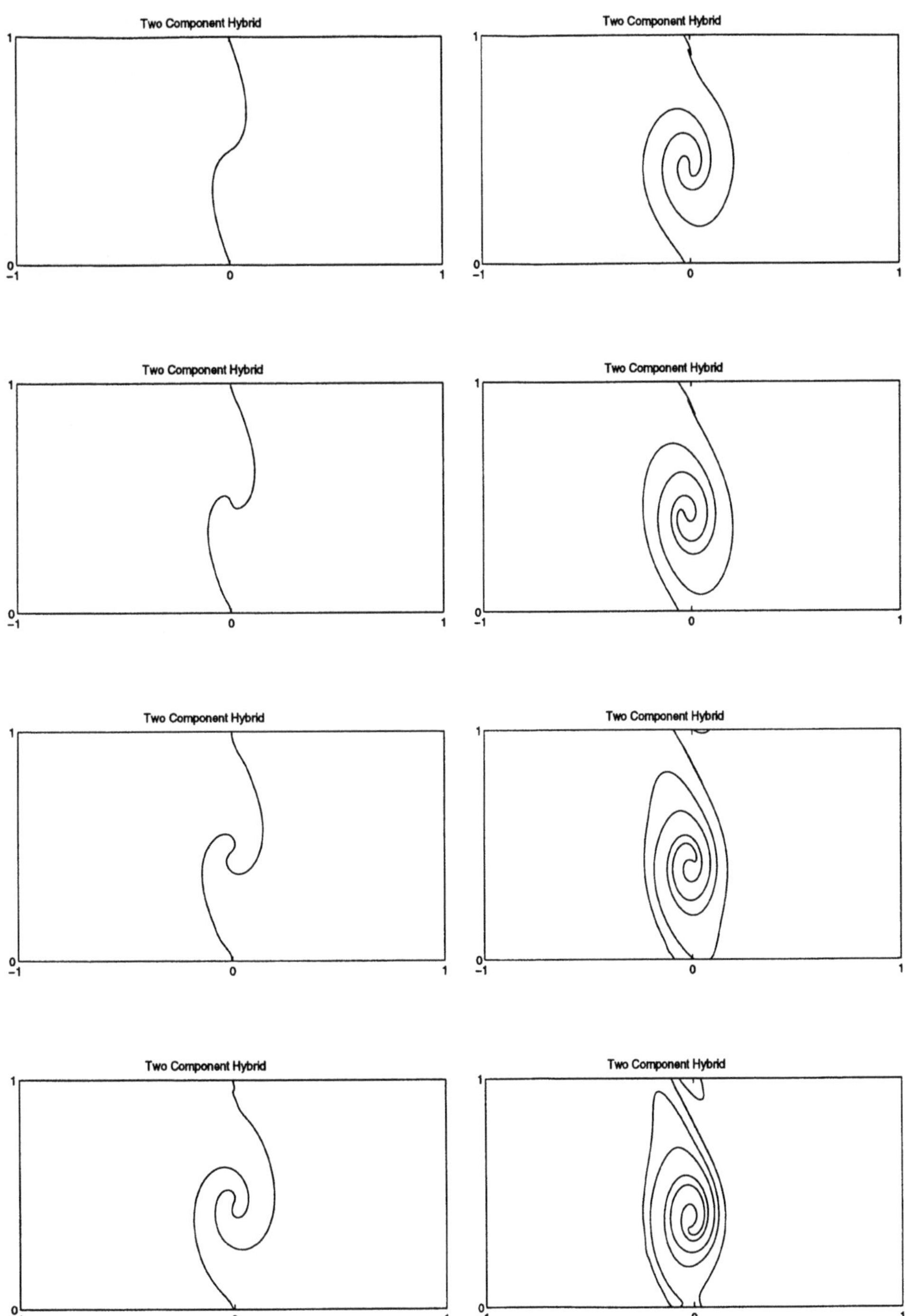

Figure 3b. Two-Gas Kelvin-Helmholtz Instability by Hybrid Model (3-3ab)

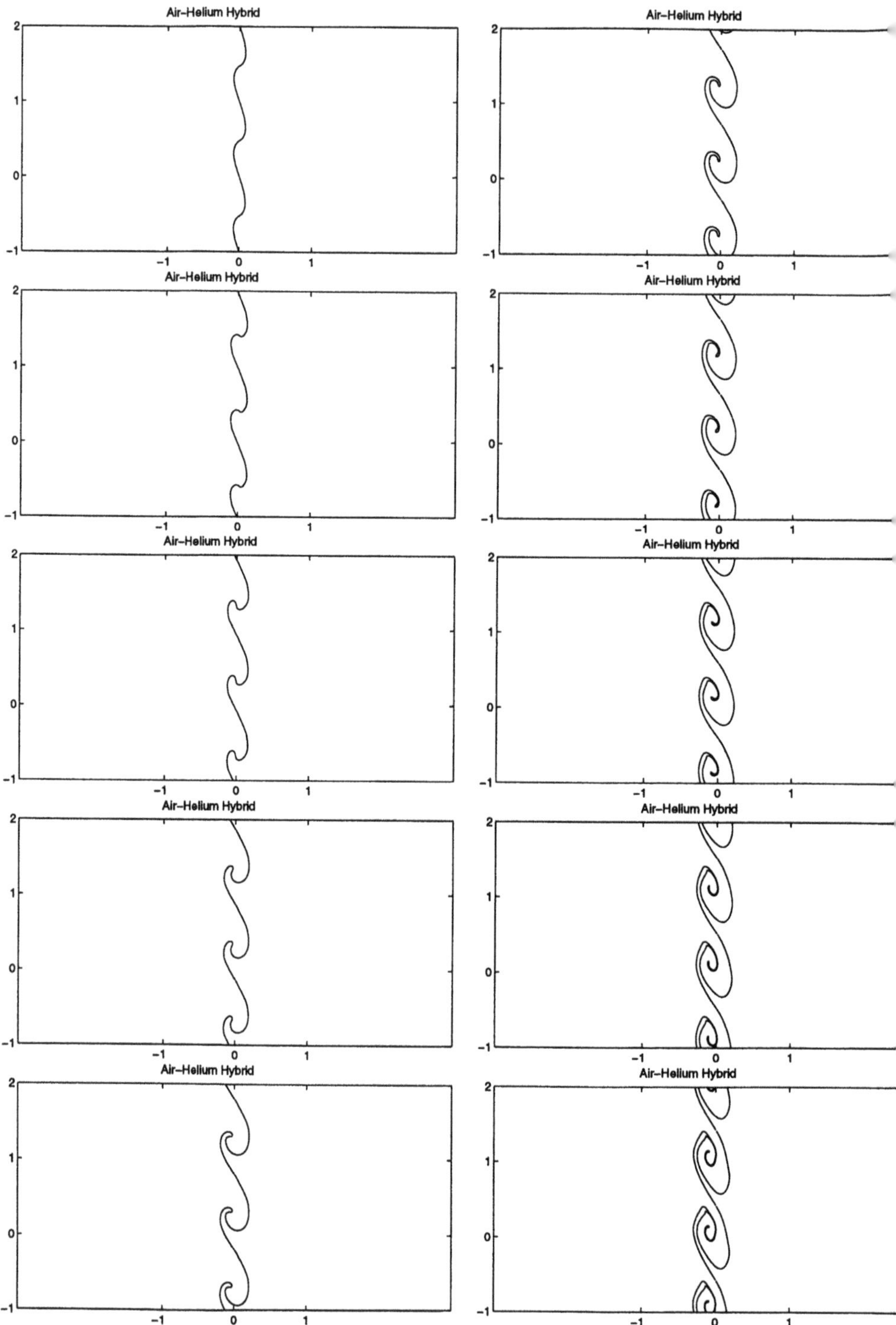

Figure 4. Air-Helium Kelvin-Helmholtz Instability by Hybrid Model (3-3ab)

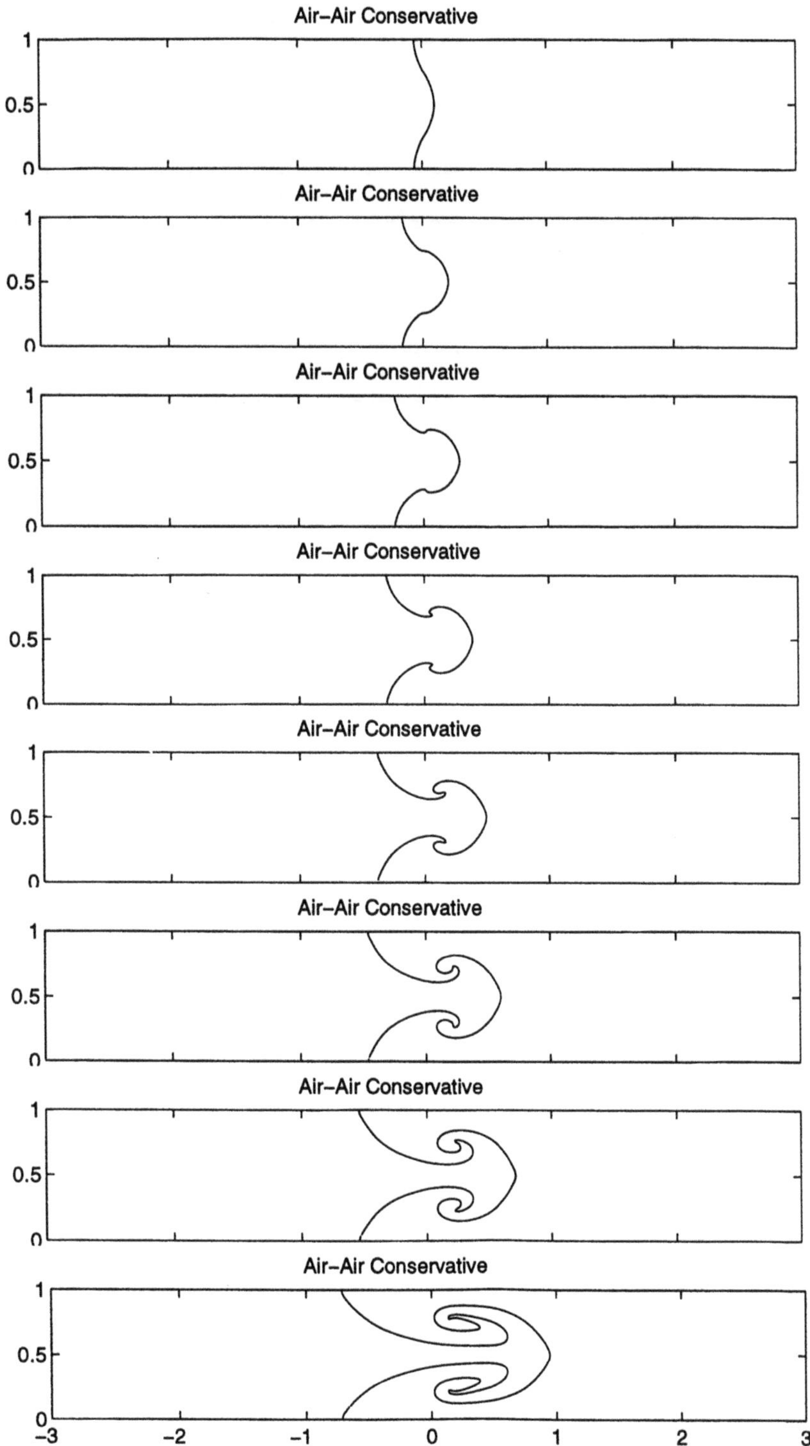

Figure 5a. Air-Air Rayleigh-Taylor Instability by Conservative Model (1)

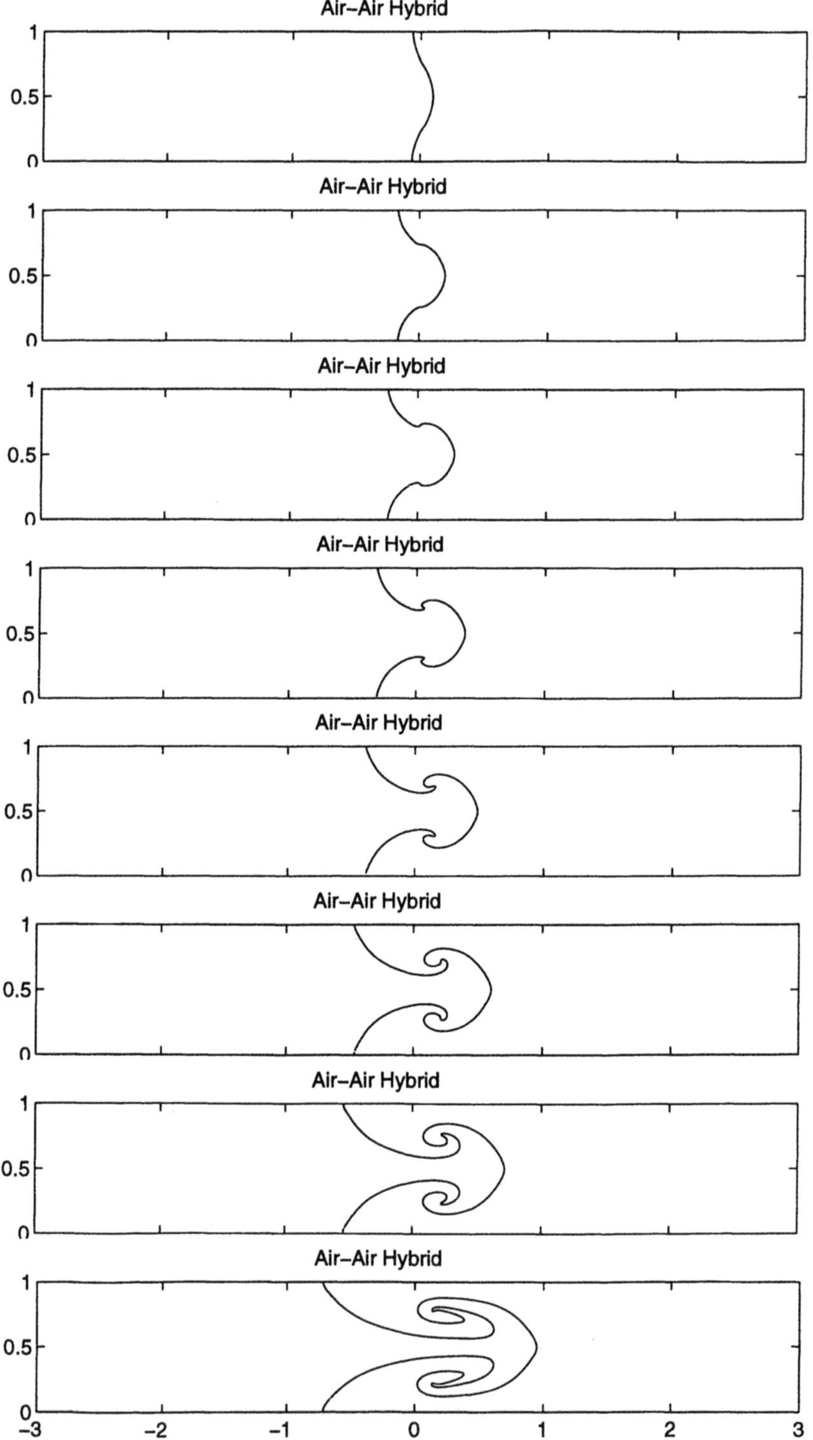

Figure 5b. Air-Air Rayleigh-Taylor Instability by Hybrid Model (3-3ab)

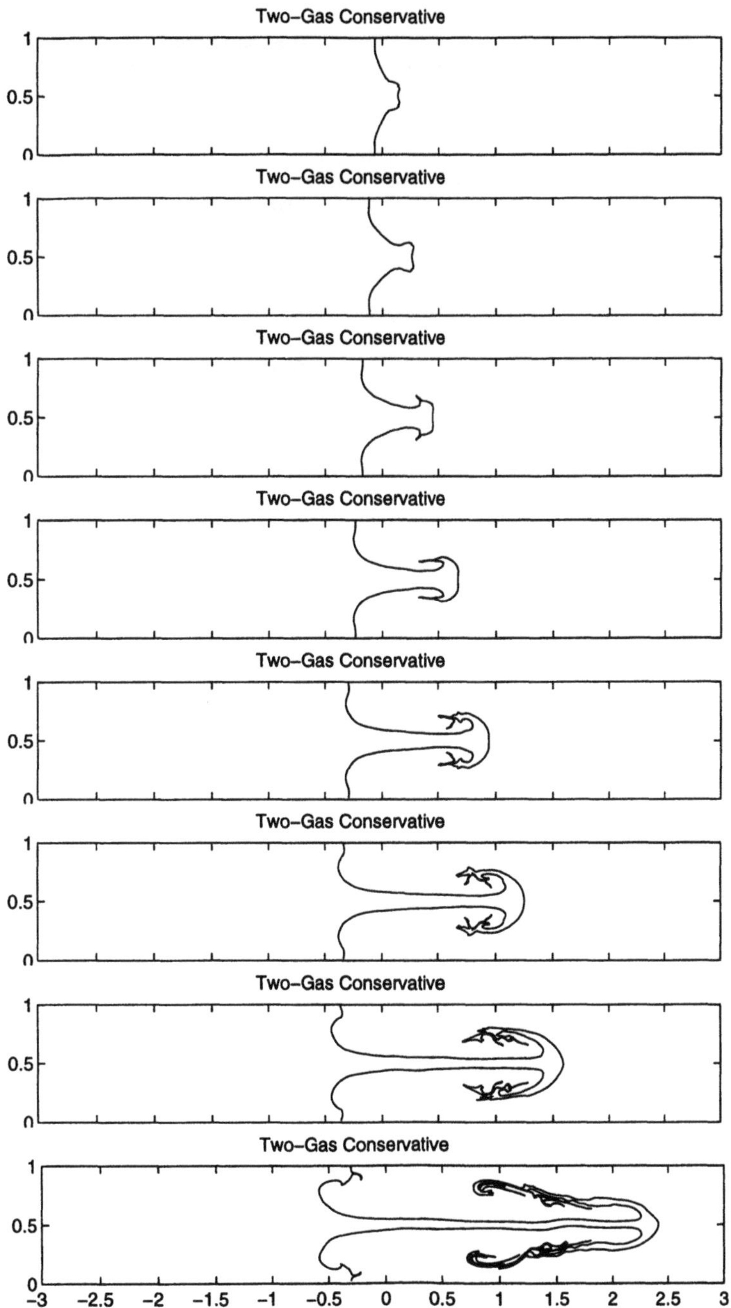

Figure 6a. Two-Gas Rayleigh-Taylor Instability by Conservative Model (1)

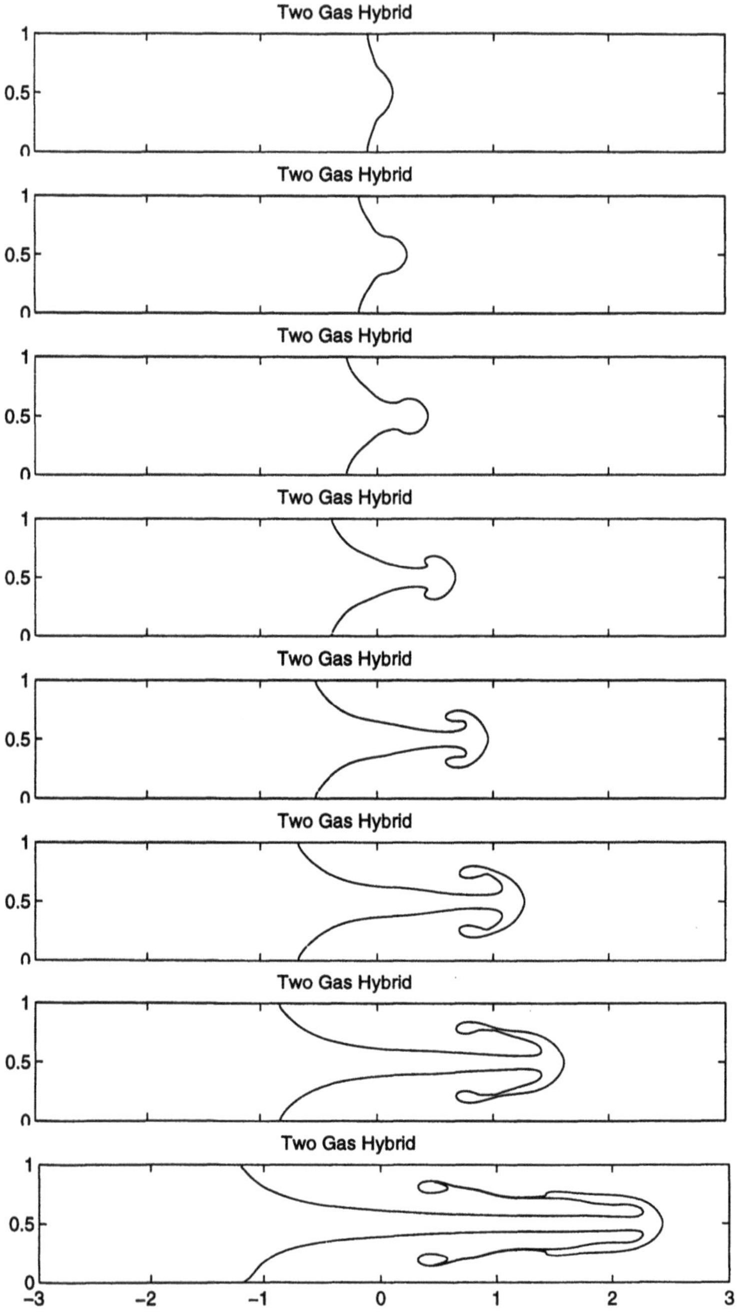

Figure 6b. Two-Gas Rayleigh-Taylor Instability by Hybrid Model (3-3ab)

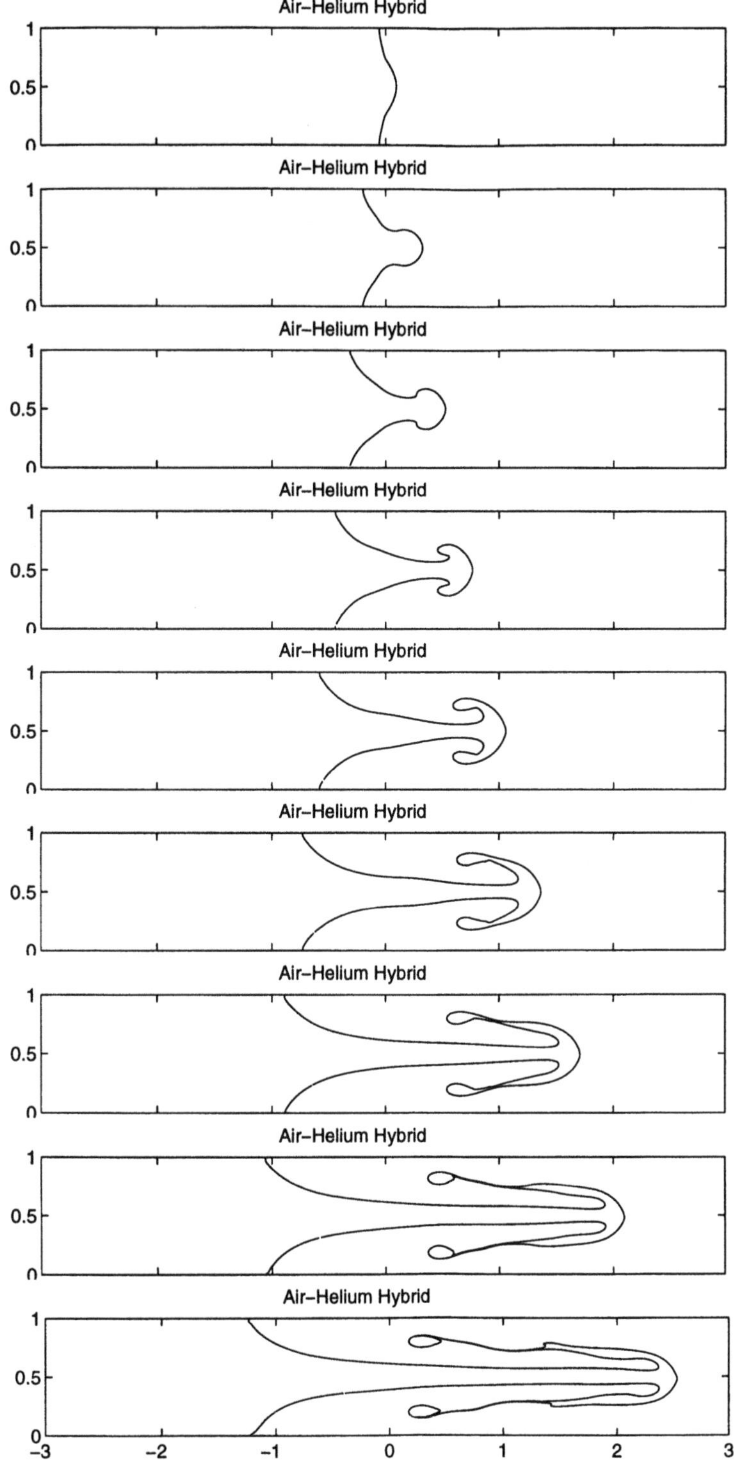

Figure 7. Air-Helium Rayleigh-Taylor Instability by Hybrid Model (3-3ab)

AN ENTROPY DIMINISHING CRITERION FOR HYPERBOLIC CONSERVATION LAWS

PHILIPPE G. LEFLOCH
Centre de Mathématiques Appliquées and Centre National de la Recherche Scientifique,
Ecole Polytechnique,
91128 Palaiseau Cedex, France,
and Department of Mathematics,
University of Southern California,
Los Angeles, California 90089–1113.

Abstract. This paper presents a new approach for the high order numerical approximation of hyperbolic systems of conservation laws. It is proposed to be used as a building principle an *entropy diminishing criterion* instead of the familiar total variation diminishing criterion introduced by Harten for scalar equations. Based on this criterion, *entropy diminishing projections* are obtained that ensure, both, the second order of accuracy and a complete set of discrete entropy inequalities. The resulting scheme is a nonlinear version of the classical Van Leer's MUSCL scheme. The strong convergence of this scheme is proved for systems of two equations.

1. Introduction

We report here on recent joint work with Frédéric Coquel (University of Paris VI, France) on the numerical approximation of the discontinuous solutions to an hyperbolic system of conservation laws

$$\partial_t u + \partial_x f(u) = 0, \qquad u(x,t) \in \mathcal{U}, \quad \S \in \boldsymbol{R}, \sqcup > \prime, \tag{1}$$

supplemented with an initial condition at time $t = 0$

$$u(x,0) = u_0(x), \qquad x \in \boldsymbol{R}, \tag{2}$$

E.F. Toro and J.F. Clarke (eds.), Numerical Methods for Wave Propagation, 275–295.

and the entropy inequality

$$\partial_t U(u) + \partial_x F(u) \leq 0. \tag{3}$$

In (1), the subset $\mathcal{U} \subset \mathbb{R}^{\mathcal{N}}$ is convex and open, the flux-function $f : \mathcal{U} \to \mathbb{R}^{\mathcal{N}}$ is a given smooth function, and the initial data $u_0 : \mathbf{R} \to \mathcal{U}$ belongs to $L^\infty(\mathbf{R})^N$. It is assumed that (1) admits at least one strictly convex entropy pair, i.e. a pair $(U, F) : \mathcal{U} \times \mathcal{U} \to \mathbb{R} \times \mathbb{R}$ of class $C^\in$ such that $\nabla F = \nabla U \cdot \nabla f$, and $\nabla^2 U \geq C Id$ for some $C > 0$. For background on hyperbolic systems, we refer to Lax (Lax P.D., 1973) and, for the theory of existence of entropy solutions, to Glimm (Glimm J., 1965) for initial data with small total variation, and DiPerna (Diperna R.J., 1983; Diperna R.J., 1983) for systems of two equations with L^∞ initial data.

Following previous works in the field by Goodman-LeVeque (Goodman J. and LeVeque R., 1988), Osher (Osher S.J., 1985), Tadmor (Tadmor E., 1987; Tadmor E., 1988) and others, the author recently analysed various approaches to deriving discrete versions of the entropy criterion: Chen-LeFloch (Chen G. -Q. and LeFloch P.G., 1995), Cockburn-Coquel-LeFloch (Cockburn B., Coquel F., and Lefloch P.G, 1995), and LeFloch-Liu (LeFloch P.G. and Liu J.-G, 1994). The present work is devoted to obtaining (3) for a class of high order schemes for systems.

Among the many efficient and robust numerical methods proposed those last fifteen years for systems of conservation laws, the second and higher order accurate versions of the (first order) Godunov scheme have been recognised as particularly attractive. The (second order) MUSCL scheme was introduced by Van Leer (Van Leer B., 1981) for gas dynamics computations, and extended to the third order of accuracy (PPM method) by Colella-Woodward (Colella P. and Woodward P.R., 1984). A Godunov-type scheme decomposes into two steps.

1. *Time marching*: The approximate solution is piecewise polynomial at every time level, and so one is left with a Cauchy problem for (1) with piecewise polynomial initial data. This sequence of non-interacting generalised Riemann problems can be solved up to any degree of approximation (Ben-Artzi and Falcovitz, 1984; Bourgeade A., LeFloch P.G., and Raviart P.-A., 1989; Harabetian E., 1988; LeFloch P.G. and

Raviart P.-A., 1988; Li T.-T. and W.-C. Yu, 1985). This allows one to march in time and obtain an approximate solution defined up to the next time level.

2. *Projection*: The above approximation is generally not piecewise polynomial and a projection step is necessary. Projecting is also needed to prevent spurious numerical oscillations.

Several strategies have been proposed in the literature that make the above method computationally attractive. The MUSCL scheme follows the above lines but uses classical Riemann solutions instead of generalised Riemann solutions, and the famous min-mod limiter for the projection step. In the MUSCL scheme, a piecewise constant approximation at a given time level is determined first by a standard L^2 projection; then an interpolation of the latter is built and nonlinearly corrected using the min-mod limiter. The min-mod function is primarily defined for scalar valued functions. Thus to apply the idea to systems, characteristic variables can be considered in order to formally reduce the system (1) to several scalar equations. MUSCL scheme is well recognised as a very successful tool for many applications especially in computational fluid dynamics.

The convergence of the MUSCL scheme was successfully established in recent years. The results concern *scalar* conservation laws only and are based on the observation that one can enforce high order Godunov-type methods to mimic, at the discrete level, many of the properties of the entropy solutions to (1). Such a scheme, by construction, is stable in the total variation norm or TVD (Harten (Harten A., 1974)), and therefore converges to a weak solution to (1). We recall that a high order scheme however cannot satisfy all of the discrete entropy inequalities (3) satisfied by first order schemes (Osher S.J. and E. Tadmor, 1988). (C.f. however (Coquel F. and LeFloch P.G., 1993; Coquel F. and LeFloch P.G., 1991; Vila J.P., 1988) for schemes based on a non-homogeneous correction.) The consistency with the entropy criterion (3) was established under some technical assumption in a pioneering work by Osher (Osher S.J., 1985), and recently proved in the general case by Lions-Sougadinis (Lions P.-L. and P. Sougadinis, 1990), Yang (Yang H., 1991), and LeFloch-Liu (LeFloch P.G. and Liu J.-G, 1994). The monotonicity property (i.e. the fact that the solution-operator for (1) preserves the regions of monotonicity of the solution) was an essential ingredient in the proofs. Another approach was investigated by Brenier-Osher (Brenier Y. and Osher S., 1988) and Goodman-LeVeque (Goodman J. and LeVeque R., 1988) based on a discrete version of Oleinik's entropy criterion. C.f. also, for related results, Nessyahu-Tadmor (Nessyahu H. and E. Tadmor, 1990) and Osher-Tadmor (Osher S.J. and E. Tadmor, 1988). More recently, Bouchut-Bourdarias-Perthame (Bouchut F., Bourdarias C., and Perthame B., 1993) have constructed a second-order TVD scheme for scalar conservation laws that satisfies a new discrete form of the entropy inequalities (3). The present work is motivated by this result and represents an extension to systems of (Bouchut F., Bourdarias C., and Perthame B., 1993). Cf. also Khobalatte-Perthame (Khobalatte B. and Perthame B., 1994) for an interesting version of the MUSCL scheme.

Van Leer's approach seems difficult to analyse theoretically in the case of *systems*. It is based on the so-called Total Variation Diminishing (TVD) criterion satisfied by scalar equations only. The total variation of a solution to (1) in general is not a diminishing function of time. Van Leer's scheme is defined for systems based on the intuitive idea that the system (1) can be formally diagonalised and the nonlinear coupling between the equations can be neglected. Each characteristic variable is classically controlled via the min-mod limiter. As we shall see, our theory establishes that the decoupling is in fact justified in a first approximation.

The principal objective is deriving a new version of the MUSCL scheme for systems, which will have a rigourous mathematical basis. A rigourous extension of Van Leer's approach to systems requires a new concept of nonlinear projection that would not create spurious numerical oscillations. Since the TVD criterion does not make sense for a system, one has to look for another way to control the oscillations: a criterion that at least will ensure the strong convergence of the scheme. Recent work on the convergence of schemes has pointed out the relevance of discrete entropy inequalities. For multidimensional scalar conservation laws, entropy dissipation estimates have been deduced from the latters and have successfully replaced the (stronger) TVD estimate, C.f. (Cockburn B., Coquel F., and LeFloch P.G., 1994; Cockburn B., Coquel F., and Lefloch P.G, 1995;

Coquel F. and LeFloch P.G., 1993; Coquel F. and LeFloch P.G., 1991). For first order schemes for one-dimensional systems, we refer to (Chen G. -Q, 1991; Chen G. -Q. and LeFloch P.G., 1995; Diperna R.J., 1983; Tadmor E., 1988). We also refer to Johnson-Szepessy (Johnson C., Szepessy A., and Hansbo P., 1990) for high-order finite element methods for systems.

Motivated by these results, we introduce a notion of *entropy diminishing projection* that ensures, both, the second order of accuracy and all of the discrete entropy inequalities. This notion is based on a property satisfied by (1): the mathematical entropy associated with an entropy solution is a diminishing function of the time variable. So we require that the projection should diminish the entropy. The existence of high order entropy diminishing projections for systems is established. Detailed proofs are given in LeFloch-Coquel (Coquel F. and LeFloch P.G., 1995; Coquel F. and LeFloch P.G., 1996).

The resulting scheme satisfies all discrete entropy inequalities (3), preserves the (available) invariant regions, and thus can only converge to an entropy weak solution. Strong convergence of this second order, entropy satisfying scheme is proved for systems of two equations by using the compensated compactness method. As a matter of fact, the scheme is a nonlinear version of the classical Van Leer's MUSCL scheme. The projection operator turns out to be a *nonlinear* version of the min-mod limiter. Finally our results show that the derivation of discrete entropy inequalities given in (3) carries over to the case of systems.

Numerical experiments have demonstrated the relevance of the scheme proposed here in computational fluid dynamics. The actual implementation uses the approximation of the generalised Riemann problem derived by Ben-Artzi-Falcovitz (Ben-Artzi and Falcovitz, 1984) and Bourgeade-LeFloch-Raviart (Bourgeade A., LeFloch P.G., and Raviart P.-A., 1989). At its current stage of development, the scheme gives satisfactory computational results but it should be considered as preliminary because of its relative complexity. Further developments of this approach should lead to an interesting scheme for CFD computations.

2. Entropy Diminishing Criterion

Let us specify some assumptions and notation that will be of constant use. We assume that for each $u \in \mathcal{U}$ the Jacobian matrix $\nabla f(u)$ has N real and distinct eigenvalues denoted by $\lambda_i(u)$, $i = 1, \cdots, N$. Let $r_i(u)$, $i = 1, \cdots, N$, be a basis of right eigenvectors. Each characteristic field is assumed to be either genuinely nonlinear, i.e. $\nabla\lambda_i \cdot r_i = 1$, or linearly degenerate, i.e. $\nabla\lambda_i \cdot r_i = 0$.

Assuming that a projection operator is given, we recall how a high order Godunov-type scheme is classically defined.

Let $\pi : L^\infty(0,1)^N \to \mathbf{P}(0,1)^N$ be a projection operator over a vector space $\mathbf{P}(0,1)^N$ of (smooth enough) vector-valued functions defined on the interval $(0,1)$. Denote by $\mathbf{P_q}(0,1)^N$ the space of all vector-valued polynomial functions of degree less or equal to a given integer q. The classical Godunov scheme will correspond to the simplest choice

$$\mathbf{P}(0,1)^N = \mathbf{P}_0(0,1)^N, \quad \pi\mathbf{v} = \int_0^1 \mathbf{v}(y)\,dy, \quad \mathbf{v} \in L^\infty(0,1)^N.$$

Our main example will be to take $\mathbf{P}(0,1)^N = \mathbf{P}_1(0,1)^N$. One can also consider the space of all functions of the form $T(P)$ with $P \in \mathbf{P_q}(0,1)^N$, where the integer q and the smooth mapping $T : \mathcal{U} \to \boldsymbol{R}^N$ are fixed. A special case is given by the characteristic variables of (1).

We introduce a uniform mesh for time and space discretisation: $t_n = n\tau$ $(n \geq 0)$ and $x_j = jh$ $(j \in \mathbf{Z})$. We also set $x_{j+1/2} = (j+1/2)h$ $(j \in \mathbf{Z})$. The increments τ and h will tend to zero with the ratio $\lambda = \tau/h$ kept constant and λ will have to satisfy a CFL stability restriction. For each $j \in \mathbf{Z}$, consider the operator $\sigma_{h,j} : L^\infty(x_{j-1/2}, x_{j+1/2})^N \to L^\infty(0,1)^N$ defined for $v \in L^\infty(x_{j-1/2}, x_{j+1/2})^N$ by

$$(\sigma_{h,j}v)(y) = v(x_{j-1/2} + hy), \quad y \in (0,1). \tag{4}$$

Abusing notation, we shall write $\sigma_{h,j}v$ for $\sigma_{h,j}(v_{|(x_{j-1/2},x_{j+1/2})})$. Let $\mathbf{P_h}(\boldsymbol{R})^N$ be the space of all vector-valued functions defined on $\boldsymbol{R}$ that are discontinuous at each "interface" $x_{j+1/2}$ and are polynomial functions in each cell $(x_{j-1/2}, x_{j+1/2})$:

$$\mathbf{P_h}(\boldsymbol{R})^N = \left\{ w : \boldsymbol{R} \to \boldsymbol{R}^N \,/\, \forall j, \exists v_j \in \mathbf{P}(0,1)^N, w_{|(x_{j-1/2}, x_{j+1/2})} = \sigma_{h,j} v_j \right\}.$$

Using the operator π, we define a projection operator $\pi_h : L^\infty(\boldsymbol{R}) \to \mathbf{P_h}(\boldsymbol{R})^N$ by the relation

$$\pi_h = \sigma_{h,j}^{-1} \circ \pi \circ \sigma_{h,j} \qquad j = \cdots, -1, 0, 1, \cdots.$$

A second ingredient for the construction of the scheme is the generalised Riemann problem (G.R.P.) and its approximation. The latter is a Cauchy problem (1)-(3) with the initial data

$$u_0(x) = \begin{cases} u_L(x) & for\ x < 0 \\ u_R(x) & for\ x > 0 \end{cases} \tag{5}$$

where $u_L(x)$ and $u_R(x)$ are two sufficiently smooth functions. In recent years, this problem has received a lot of attention from the point of view of existence and approximation. It is known (Harabetian E., 1988; Li T.-T. and W.-C. Yu, 1985) that the solution to the generalised Riemann problem has the same local structure as the one of the classical Riemann problem with initial data

$$u_0(x) = \begin{cases} u_L(0-) & for\ x < 0 \\ u_R(0+) & for\ x > 0 \end{cases} \tag{6}$$

In contrast with the classical problem, the solutions to the generalised Riemann problem are not given by a closed formula. The practical implementation of the schemes can be based on the general theory of approximation in (LeFloch P.G. and Raviart P.-A., 1988). For the gas dynamics systems, see (Ben-Artzi and Falcovitz, 1984; Bourgeade A., LeFloch P.G., and Raviart P.-A., 1989).

Equipped with a projection operator and an (approximate) generalised Riemann solver, we now define the approximate solutions $u_h(x,t)$ to the problem (1)–(3) such that $u_h(t_n+)$ belongs to $\mathbf{P_h}(\boldsymbol{R})^N$ for all $n \geq 0$. Denote by u_j^n the cell average of $u_h(t_n+)$ in $(x_{j-1/2}, x_{j+1/2})$. The scheme is initiated in time by setting

$$u_h(0+) = \pi_h(u_0) \in \mathbf{P_h}(\boldsymbol{R})^N. \tag{7}$$

Suppose that the approximate solution $u_h(t)$ has been computed up to $t \leq t_n$.

Step 1: Time Marching. Define $u_h(t)$ for $t \in (t_n, t_{n+1})$ as the entropy solution to (1)–(3) with the initial condition $u_h(t_n+)$ at time $t = t_n$. Under a suitable CFL condition, this Cauchy problem consists of a series of non-interacting generalised Riemann problems.

Step 2: Projection. Using the operator π_h, we compute the projection of $u_h(t_{n+1}-)$ and set

$$u_h(t_{n+1}+) = \pi_h(u_h(t_{n+1}-)) \in \mathbf{P_h}(\boldsymbol{R})^N. \tag{8}$$

In particular, the cell average values u_j^n satisfy the discrete conservation law

$$u_j^{n+1} = u_j^n - \lambda\,(g_{j+1/2}^n - g_{j-1/2}^n), \tag{9}$$

where the numerical flux $g_{j+1/2}^n$ is given by

$$g_{j+1/2}^n = \frac{1}{\tau} \int_{t_n}^{t_{n+1}} f(u_h(x_{j+1/2}, t))\, dt.$$

In the following, we refer to the scheme above as the *high order Godunov-type scheme* based on the projection π. Observe that exact solutions to the G.R.P. will always be considered in the convergence analysis. The analysis of *approximate* solver is out of the scope of this work. Note that the scheme is high order accurate in both space and time if π is a high order projection.

We now define the projection πv of any function $v \in L^\infty(0,1)^N$. Our objective is to obtain a scheme that simultaneously

1. is high-order accurate in the smooth regions of the solution,
2. does not create spurious oscillations, that might prevent the strong convergence to a weak solution.

In order to achieve both properties, it is known that the projection should be a nonlinear operator.

The treatment of the scalar conservation laws is based on the property that the total variation (in space) of an entropy solution is a diminishing function of time. The concept of Total Variation Diminishing (TVD) scheme has been introduced with great success for the purpose of achieving (1) and (2) above. Recall that a uniform bound for the total variation is a sufficient condition for the scheme to converge in a strong topology. On the other hand, Harten (Harten A., 1974) demonstrated that there exists *high order* accurate TVD schemes.

Considering next the numerical approximation of systems, we observe that the total variation of an entropy solution generally is not a diminishing function of time. The forthcoming analysis instead will be based on the entropy inequalities (3). As we shall see, the entropy inequalities provide natural stability properties for schemes of approximation for systems of conservation laws.

The distributional entropy inequality (3) implies that the entropy is a non-increasing function of time, i.e.

$$\int_{\mathbf{R}} U(u(x,t_2))\,dx \leq \int_{\mathbf{R}} U(u(x,t_1))\,dx \qquad t_2 > t_1. \tag{10}$$

Letting $t_1 \to t_2 = t$ and at least when u is piecewise smooth, we arrive at the following inequality

$$\int_{\mathbf{R}} U(u(x,t+))\,dx \leq \int_{\mathbf{R}} U(u(x,t-))\,dx, \qquad t > 0,$$

which can naturally be used as a building principle for numerical approximation. More generally, one can deduce from (3), at least formally,

$$\int_{y_1}^{y_2} U(u(x,t+))\,dx \leq \int_{y_1}^{y_2} U(u(x,t-))\,dx, \quad y_1 < y_2, \ \ t > 0. \tag{11}$$

We suggest here to use (10)-(11) as a criterion for nonlinear stability: one should require (11) for suitable values of y_1 and y_2 (typically the mesh points) and *all* entropy functions of the system (1).
Let $\mathcal{E}$ be the set of all convex entropy functions of (1).

Definition 1 *The operator π is called an entropy diminishing projection if*

$$\int_0^1 U(\pi v)\,dx \ \leq \int_0^1 U(v)\,dx \tag{12}$$

$\in \mathcal{E}$ *and* $v \in L^\infty(0,1)^N$.

Roughly speaking the condition (12) means that π is a non-increasing operator for the "semi-norm" $||v||_U = \int_0^1 U(v)\,dx$. For second order projections πv of the form

$$\pi v(x) = \bar{v} + \beta(x - 1/2) \quad \textit{for some value} \quad \beta \in \boldsymbol{R}^N, \tag{13}$$

where $\bar{v} = \int_0^1 v(x)\,dx$, it is possible to show the following:

Proposition 1 The set of all β in $\boldsymbol{R}^N$ satisfying (12) for all entropies $U \in \mathcal{E}$ is a non-empty, closed, and convex subset of $\boldsymbol{R}^N$.

Remark 1 If v is a constant, then the Jensen's inequality in the proof above is a strict inequality, $B_U \setminus \{0\}$ is not empty, and the point $\beta = 0$ belongs to the interior of B_U. On the other hand, it is difficult to claim that the set B is non empty. It is conjectured that the second order of accuracy is obtained if β is chosen to be an extremal point of the set B and achieve $\max_{\beta \in B} |\beta|$.

Our next result (Theorem 1) provides a set of inequalities, that are sufficient to imply the condition (12) in Definition 1, but is explicit, and so can be used to construct projections for practical purposes. Consider the averaging operator $M : L^\infty(0,1)^N \to L^\infty(0,1)^N$ defined by

$$M(v)(x) = \frac{1}{x} \int_0^x v(y)\,dy, \qquad x \in (0,1),\ v \in L^\infty(0,1)^N. \tag{14}$$

The same formulation defines $M(\pi v)$. The quantity $M(\pi v) - M(v)$ will be used to estimate the interpolation error. Such averages arise more naturally when the conservation laws are rewritten in the form of Hamilton-Jacobi equations; they play a role in the ENO schemes by Harten-Engquist-Osher-Chakravarthy (Harten A., Engquist B., Osher S.J. and Chalravarthy S., 1987) and in the scheme by Bouchut-Bourdarias-Perthame (Bouchut F., Bourdarias C., and Perthame B., 1993).

Note that $M(v) \in L^\infty(0,1)^N \cap W^{1,\infty}_{loc}(0,1)^N$ and $\bar{v} \equiv \int_0^1 v(x)\,dx = M(v)(1)$. If v is a piecewise continuous function, then $M(v) \in W^{1,\infty}(0,1)^N$ and

$$\lim_{x \to 0} M(v)(x) = v(0+).$$

Since U is a convex function, one has

$$\begin{aligned}\int_0^1 U(\pi v)\,dx &\le \int_0^1 U(v)\,dx + \int_0^1 \nabla U(\pi v)\cdot(\pi v - v)\,dx \\ &\le \int_0^1 U(v)\,dx + \nabla U(\pi v(1))\cdot(M(\pi v)(1)-\bar{v}) \\ &- \int_0^1 \nabla^2 U(\pi v)\cdot\Big(\frac{d\,\pi v}{dx}, M(\pi v) - M(v)\Big)x\,dx,\end{aligned} \tag{15}$$

Proposition 2 Assume that the operator π satisfies the condition

$$M(\pi v)(1) = \bar{v}, \tag{16}$$

and for all $U \in \mathcal{E}$ and $v \in L^\infty(0,1)^N$

$$\nabla^2 U(\pi v)\cdot(\frac{d\,\pi v}{dx}, M(\pi v) - M(v)) \ge 0 \quad on\,(0,1). \tag{17}$$

Then π is an entropy diminishing projection.

Proposition 2 holds for projections of any order of accuracy. We restrict now to second order and establish the existence of an entropy diminishing projection for arbitrary systems of conservation laws. Strict hyperbolicity of the system under consideration is assumed.

Theorem 1 Consider continuous functions $v : (0,1) \to \boldsymbol{R}^N$ taking their values in a small neighbourhood of a given state $v_* \in \mathcal{U}$. There exists an entropy diminishing projection π (consistent with all convex entropies of (1)) leading to a second-order Godunov-type scheme. For every function v, πv has the form (13) where the slope β is suitably chosen in the closed, bounded, and convex set B of all vectors $\beta \in \boldsymbol{R}^N$ satisfying

$$\beta_k(D_k - \beta_k) \ge 0, \quad k = 1, 2, \cdots, N \tag{18}$$

with

$$\beta = \sum_{1\le i\le N} \beta_i\, r_i(\pi v) \tag{19}$$

and

$$D(v) = 2\,\frac{M(v)-\bar{v}}{x-1} = \sum_{1\le i\le N} D_i\, r_i(\pi v). \tag{20}$$

The set of inequalities (18)–(20) form a nonlinear system of coupled algebraic inequalities for the characteristic variables β_i. In the scalar case, the set B is an interval of $\boldsymbol{R}$ (possibly reduced to the point $\beta = 0$).

Finally we arrive at the convergence of the high order Godunov-type scheme.

Theorem 2 Assume that π is an entropy diminishing projection in the sense of Definition 1 and consider the approximations u_h constructed by the corresponding Godunov-type scheme. Then, for any entropy pair (U, F), the discrete entropy inequality

$$\frac{1}{h}\int_{x_{j-1/2}}^{x_{j+1/2}} U(u_h(x, t_{n+1}+))\,dx \quad - \quad \frac{1}{h}\int_{x_{j-1/2}}^{x_{j+1/2}} U(u_h(x, t_n+))\,dx \\ + \quad \lambda\,(G_{j+1/2}^n - G_{j-1/2}^n) \le 0 \qquad (21)$$

holds true, where

$$G_{j+1/2}^n = \frac{1}{\tau}\int_{t_n}^{t_{n+1}} F(u_h(x_{j+1/2}, t))\,dt \qquad (22)$$

is a numerical entropy flux consistent with F. Suppose that the scheme is stable in the L^∞ norm.

1. Then any Young measure associated with the u_h's is an entropy measure valued solution to (1)–(3) in the sense of DiPerna.
2. When the compensated compactness method applies (e.g. $p = 2$), the high-order Godunov-type converges strongly.

No strong compactness property is a priori available. The proof of Theorem 2 is based on the following result of entropy stability.

Proposition 3 For every $U \in \mathcal{E}$, the scheme satisfies the estimate

$$\int_{\boldsymbol{R}} U(u_h(x, t_n+))\,dx + E_h(t_n+) \le \int_{\boldsymbol{R}} U(u_0)\,dx$$

with

$$E_h(t_n+) = \frac{1}{2}\sum_{0\le q\le n} \int_{\boldsymbol{R}} C_U^h \cdot (u_h(t_q+) - u_h(t_q-), u_h(t_q+) - u_h(t_q-))\,dx$$

and

$$C_U^h = \int_0^1 \nabla^2 U(u_h(t_n-) + s(u_h(t_n+) - u_h(t_n-)))\,(1-s)\,ds.$$

Using a uniformly convex entropy function, one deduces from Proposition 3:

$$\sum_{0\le n\le\infty} \int_{\boldsymbol{R}} |u_h(t_n+) - u_h(t_n-)|^2\,dx \le O(1). \tag{23}$$

Such an *entropy dissipation* estimate was first derived, for multidimensional scalar equations, in (Cockburn B., Coquel F., and Lefloch P.G, 1995; Coquel F. and LeFloch P.G., 1993; Coquel F. and LeFloch P.G., 1991).

Proof of Proposition 3 The scheme consists of two steps. *The first one is based on solving a generalised Riemann problem. Since entropy solutions are used, the distributional inequality (3) holds and after integration (C.f. 22) for the definition of the entropy-flux) one gets*

$$\begin{aligned}\frac{1}{h}\int_{x_{j-1/2}}^{x_{j+1/2}} U(u_h(t_{n+1}-,x))\,dx \quad &- \quad \frac{1}{h}\int_{x_{j-1/2}}^{x_{j+1/2}} U(u_h(t_n+,x))\,dx \\ &+ \quad \lambda(G^n_{j+1/2} - G^n_{j-1/2}) \le 0. \end{aligned}\tag{24}$$

The second step is based on an entropy diminishing projection operator π_h. Let us revisit the derivation of inequality (15). The following is indeed a sharper version of (15):

$$\begin{aligned} &\int_{x_{j-1/2}}^{x_{j+1/2}} U(u_h(t_{n+1}+,x))\,dx - \int_{x_{j-1/2}}^{x_{j+1/2}} U(u_h(t_{n+1}-,x))\,dx \\ = &-\int_{x_{j-1/2}}^{x_{j+1/2}} \nabla^2 U(u_h(t_{n+1}+,x))\cdot(u_h(t_{n+1}-,x) - u_h(t_{n+1}+,x))\,dx \\ &-\frac{1}{2}\int_{x_{j-1/2}}^{x_{j+1/2}} C_U^h\cdot(u_h(t_{n+1}+,x) - u_h(t_{n+1}-,x), u_h(t_{n+1}+,x) \\ &\qquad\qquad -u_h(t_{n+1}-,x))\,dx. \end{aligned}$$

Our requirement (17) reads here

$$-\int_{x_{j-1/2}}^{x_{j+1/2}} \nabla^2 U(u_h(t_{n+1}+,x))\cdot(u_h(t_{n+1}-,x) - u_h(t_{n+1}+,x))\,dx \le 0.$$

Therefore the discrete entropy inequality holds:

$$\frac{1}{h}\int_{x_{j-1/2}}^{x_{j+1/2}} U(u_h(t_{n+1}+,x)\,dx - \frac{1}{h}\int_{x_{j-1/2}}^{x_{j+1/2}} U(u_h(t_n+,x)\,dx$$
$$+\lambda(G^n_{j+1/2} - G^n_{j-1/2}) \leq -\frac{1}{2h}\int_{x_{j-1/2}}^{x_{j+1/2}} C^h_U\cdot(u_h(t_{n+1}+,x)$$
$$-u_h(t_{n+1}-,x), u_h(t_{n+1}+,x) - u_h(t_{n+1}-,x))\,dx.$$

Since U is a convex function, the right hand side of this inequality remains non-positive. The desired result follows by summation over space and time.
□

Proof of Theorem 2: Since π is taken to be an entropy diminishing projection, the inequality (24) implies the discrete entropy inequality (21) for any convex entropy pair (U, F).

Assuming the scheme to be stable in the L^∞ norm, we consider a Young measure associated with u_h, that is a probability measure $\nu_{x,t}$ such that

$$\iint g(u_h(x,t))\,\varphi(x,t)\,dtdx \to \iint <\nu_{x,t)}, g> \varphi(x,t)\,dtdx \qquad (25)$$

for all continuous function g and all test-function φ. Let φ be a non-negative test-function. Since u_h satisfies the inequality (3) in each strip $t_n < t < t_{n+1}$, one has

$$\int_{t_n}^{t_{n+1}}\int_{\boldsymbol{R}} U(u_h)\,\partial_t\varphi + F(u_h)\partial_x\varphi\,dtdx - \int_{\boldsymbol{R}} U(u_h(t_{n+1}-))\,\varphi(t_{n+1})\,dx$$
$$+\int_{\boldsymbol{R}} U(u_h(t_n+))\,\varphi(t_n)\,dx \geq 0.$$

After summation over time, we arrive at

$$\begin{aligned} 0 \;&\leq\; \iint U(u_h)\,\partial_t\varphi + F(u_h)\,\partial_x\varphi\,dtdx \qquad (26)\\ &+\; \sum_{n=0,1,2,\ldots}\int_{\boldsymbol{R}} (U(u_h(t_n+)) - U(u_h(t_n-)))\,\varphi(t_n)\,dx \equiv A^1_h + A^2_h. \end{aligned}$$

In view of (25), one has

$$A^1_h \to \iint <\nu, U> \partial_t\varphi + <\nu, F> \partial_x\varphi\,dtdx.$$

We write $A_h^2 = A_h^{2,1} + A_h^{2,2}$ with

$$A_h^{2,1} = \sum_{n,j} \int_{x_{j-1/2}}^{x_{j+1/2}} (U(u_h(t_n+)) - U(u_h(t_n-)))\,(\varphi(t_n) - \varphi(t_n, x_j))\,dx,$$

$$A_h^{2,2} = \sum_{n,j} \varphi(t_n, x_j) \int_{x_{j-1/2}}^{x_{j+1/2}} (U(u_h(t_n+)) - U(u_h(t_n-)))\,dx.$$

The diminishing entropy criterion (12) again yields

$$A_h^{2,2} \leq 0.$$

Using the estimate (23), one has

$$\begin{aligned} |A_h^{2,1}| &\leq \tau^{1/2}\,\|\nabla U\|_{L^\infty}\,\|\phi\|_{H_0^1}\,\Big(\sum_{0\leq q\leq\infty} \int_R |u_h(t_q+) - u_h(t_q-)|^2\,dx\Big)^{1/2} \\ &\leq O(h^{1/2}). \end{aligned}$$

This proves the claim (1) in Theorem 2. In view of the full set of discrete entropy inequalities (21)–(22) and the entropy dissipation estimate (23), it is not hard to apply the compensated compactness method along the lines of DiPerna's work (Diperna R.J., 1983). C.f. also Chen-Liu (Chen G. -Q. and Liu J. -G., 1993) and Coquel-LeFloch (Coquel F. and LeFloch P.G., 1993). This completes the proof of Theorem 2. □

To complete this section, a couple of observations are in order.

We observe that (18)-(19) is a set of nonlinear inequalities that can only be solved by an iterative method. In fact the numerical experiments performed in (Coquel F. and LeFloch P.G., 1996) are based on a linearisation of (19) in which $r_i(\pi v)$ is replaced by the constant function $r_i(\bar{v})$. With this linearisation, the inequalities in (18) are decoupled from each other and are solved explicitly as was done in the case $N = 1$.

This observation can be actually extended to any system provided a nonlinear approximation is sought. Suppose we have a vector-valued function $\tilde{u}$ such that

$$U(u) = |\tilde{u}(u)|^2 \tag{27}$$

and $u \to \tilde{u}$ is a one-to-one smooth mapping. Then rewrite the system of conservation laws (1) in terms of this new variable v as

$$\partial_t \tilde{g}(\tilde{u}) + \partial_x \tilde{f}(\tilde{u}) = 0, \tag{28}$$

while (3) now reads

$$\partial_t|\tilde{u}|^2 + \partial_x \tilde{F}(\tilde{u}) \leq 0 \tag{29}$$

with

$$u = \tilde{g}(\tilde{u}), \qquad f(u) = \tilde{f}(\tilde{u}), \qquad F(u) = \tilde{F}(\tilde{u}).$$

The Hessian matrix of the entropy function is now the identity matrix and, given a function $\tilde{v}$, the inequality (17) can be rigourously *decoupled* in the form

$$\tilde{\beta}_k(\tilde{D}_k - \tilde{\beta}_k) \geq 0, \quad k = 1, 2, \cdots, N \tag{30}$$

with

$$\tilde{\beta} = \sum_{1 \leq i \leq N} \tilde{\beta}_i \, \tilde{r}_i(\bar{\tilde{v}}) \tag{31}$$

and

$$\tilde{D}(\tilde{v}) = 2\,\frac{M(\tilde{v}) - \bar{\tilde{v}}}{x - 1} = \sum_{1 \leq i \leq N} D_i \, \tilde{r}_i(\bar{\tilde{v}}). \tag{32}$$

Here $\tilde{r}_i$ are the eigenvectors of the Jacobian matrix of $\tilde{f}$. Indeed one entropy inequality (17) is rigourously satisfied in that case, while the other ones hold in some approximate sense, which may be enough both to prove the convergence of the scheme and for practical purposes.

3. Systems of Two or Three Conservation Laws

Systems of two conservation laws admit infinitely many entropy functions. In the present subsection, we take advantage of this property. Next we treat the system of three equations of the gas dynamics.

Consider the following 2×2 system

$$\begin{aligned} \partial_t u_1 + \partial_x f_1(u_1, u_2) &= 0, \\ \partial_t u_2 + \partial_x f_2(u_1, u_2) &= 0, \end{aligned} \tag{33}$$

where $f = (f_1, f_2)$ satisfies $f_{1,2} f_{2,1} > 0$ (C.f. Smoller (Smoller J., 1983)). Here we have set $f_{i,j} = \partial f_i / \partial u_j$. This assumption implies that (33) is strictly hyperbolic. We consider a set of two Riemann invariants (Φ_1, Φ_2), which by definition diagonalises (33). Denote by $u(0)$ the initial condition at the time $t = 0$. Consider the following two assumptions

1. For $i = 1, 2$ the i-Riemann invariant

$$\Phi_i \text{ is either convex or concave} \tag{34}$$

with respect to the conservative variables.

2. For $i = 1, 2$, there exists one convex entropy function $U_{\star i}$ with the property

$$\begin{aligned} U_{\star i}(u) &\geq 0, \\ U_{\star i}(u(0)) &= 0 \\ U_{\star i}(u) &> 0 \quad for\, |\Phi_i(u)| > C_i \end{aligned} \tag{35}$$

for some constants $C_i > 0$.

Under classical assumptions on the equation of state, the p-system of gas dynamics satisfies (34), while the isentropic Euler system satisfies (35) (C.f. (Lions P.-L., Perthame B.,and Tadmor E., 1994) for a construction of the entropy and also (Serre D., 1987) for the nonlinear elasticity system).

Theorem 3 Consider a strictly hyperbolic system of two equations with two genuinely nonlinear characteristic fields, (33).

1. *The inequalities (17) and (18)* are equivalent, *and, for any function v in $L^\infty(0,1)^2$, read as*

$$\frac{d}{dx}\Phi_i(\pi v(x))\Big\{ < \nabla_u \Phi_i(\pi v(x)), 2\frac{\bar{v} - M(v)(x)}{1-x} > -\frac{d\Phi_i(\pi v)}{dx}\Big\} \geq 0. \tag{36}$$

for $x \in (0,1)$ and $i = 1, 2$.

2. *Suppose that the system is endowed with bounded and convex invariant domains and one of the two assumptions (34) or (35) holds. Then the corresponding entropy diminishing Godunov-type scheme preserves the invariant domains of (33).*

This theorem provides the L^∞ stability of the scheme. ¿From Theorem 2 and the results in (Chen G. -Q, 1991; Chen G. -Q. and Liu J. -G., 1993; Diperna R.J., 1983)*, we immediately deduce that the second order Godunov-type scheme converges strongly. It is not hard to extend the* *convergence analysis to the isentropic Euler system of gas dynamics for a polytropic perfect gas.*

We consider now the full system of gas dynamics in Lagrangian coordinates. Since $N \geq 3$, we do not have a large class of mathematical entropies in general (Bereux F., Bonnetier E., and LeFloch P.G.,). As a consequence, the equivalence between (17) and (18) is lost. Consider the following system

$$\begin{aligned} \partial_t \tau - \partial_x u &= 0, \\ \partial_t u + \partial_x p &= 0, \\ \partial_t e + \partial_x (pu) &= 0, \end{aligned} \tag{37}$$

where $\tau > 0$, u, $e > 0$ denote the specific volume, velocity and internal energy per unit mass, respectively. The pressure $p > 0$ is given by

$$p = \frac{\gamma - 1}{\tau}(e - \frac{u^2}{2}), \qquad \gamma > 1.$$

This system has the form (1): it admits three real and distinct eigenvalues, i.e. $-g$, 0, and g, where $g = \sqrt{\frac{\gamma p}{\tau}}$ is the Lagrangian sound speed. The notation (14), (20) will be of use here. The following statement concerns first order projections $\pi v = (\pi\tau, \pi u, \pi e)$ of an arbitrary function $v = (\tau, u, e)$: $[0,1] \to \boldsymbol{R}_+ \times \boldsymbol{R} \times \boldsymbol{R}_+$.

Theorem 4. Consider the gas dynamics system (37) and an arbitrary function v. The condition (17) is equivalent to the following two inequalities:

$$\begin{aligned} \frac{\pi p}{(\gamma-1)} \frac{d\pi\tau}{dx}\Big(D(\tau) - \frac{d\pi\tau}{dx}\Big) + \Big(\frac{d\pi e}{dx} - \pi u \frac{d\pi u}{dx}\Big)\Big(D(e) - \frac{d\pi e}{dx}\Big) \\ + \Big(\pi u (\pi u \frac{d\pi u}{dx} - \frac{d\pi e}{dx}) + \frac{(\pi\tau)^2 \pi p}{(\gamma-1)} \frac{d\pi u}{dx}\Big)\Big(D(u) - \frac{d\pi u}{dx}\Big) \geq 0, \end{aligned} \tag{38}$$

$$\begin{aligned} \Big(\pi p \frac{d\pi\tau}{dx} - \pi u \frac{d\pi u}{dx} + \frac{d\pi e}{dx}\Big)\Big(\pi p (D(\tau) - \frac{d\pi\tau}{dx}) - \pi u (D(u) - \frac{d\pi u}{dx}) \\ + (D(e) - \frac{d\pi e}{dx})\Big) \geq 0. \end{aligned} \tag{39}$$

The condition (18) is equivalent to the following three inequalities:

$$\Big(\frac{\pi p}{(\gamma-1)} \frac{d\pi\tau}{dx} + (\pi u \pm \frac{\pi g}{(\gamma-1)}) \frac{d\pi u}{dx} - \frac{d\pi e}{dx}\Big)$$

$$\left(\frac{\pi p}{(\gamma-1)}(D(\tau)-\frac{d\pi\tau}{dx})+(\pi u\pm\frac{\pi g}{(\gamma-1)})\right.$$
$$\left.(D(u)-\frac{d\pi u}{dx})-(D(e)-\frac{d\pi e}{dx})\right)\geq 0, \tag{40}$$
$$\left(\pi p\frac{d\pi\tau}{dx}-\pi u\frac{d\pi u}{dx}+\frac{d\pi e}{dx}\right)\left(\pi p(D(\tau)-\frac{d\pi\tau}{dx})\right.$$
$$\left.-\pi u(D(u)-\frac{d\pi u}{dx})+(D(e)-\frac{d\pi e}{dx})\right)\geq 0. \tag{41}$$

In particular, the conditions (17) and (18) are not equivalent.

4. Acknowledgements

This work has been partially supported by the National Science Foundation through NSF grants DMS 92-09326 and DMS 94-09400 and a Career faculty award.

References

Ben-Artzi M. and Falcovitz J. (1984). A Second Order Godunov-Type Scheme for Compressible Fluid Dynamics. *J. Comput. Phys.*, 55:1–32, 1984.

Bereux F., Bonnetier E., and LeFloch P.G. Gas dynamics equations: two special cases *in preparation.*

Bouchut F., Bourdarias C., and Perthame B. An example of MUSCL method satisfying all the entropy inequalities *C.R. Acad. Sc. Paris, Série I*, **317** 1993 619–624; and article to appear

Bourgeade A., LeFloch P.G., and Raviart P.-A. An asymptotic expansion for the solution of the generalized Riemann problem. Part II: application to the gas dynamics equations 1989 **6** *Ann. Inst. H. Poincaré, Nonlinear Analysis* 437–480

Brenier Y. and Osher S. The one-sided Lipschitz condition for convex scalar conservation laws *SIAM J. Numer. Anal.* **25** 1988 8–23

Chen G.-Q. The compensated compactness method and the system of isentropic gas dynamics *Preprint, Mathematical Sciences Research Institute, Berkeley* 1991

Chen G.-Q. and LeFloch P.G. Entropy flux-splittings for hyperbolic conservation laws. Part I: general framework *to appear in Comm. Pure Appl. Math.* 1995

Chen G.-Q. and Liu J.-G. Convergence of second-order schemes for isentropic gas dynamics *Math. of Comp.* 1993 **61** 607–627

Cockburn B., Coquel F., and LeFloch P.G. Error estimates for finite volume methods for multidimensional conservation laws 1994 **63** *Math. of Comp.* 77–103

Cockburn B., Coquel F., and LeFloch P.G. Convergence of finite volume methods for multidimensional conservation laws *SIAM J. Numer. Anal.* 1995

Colella P. and Woodward P. R. The piecewise parabolic method (PPM) for gas dynamical simulations *J. Comp. Phys.* **54** 1984 174–201

Coquel F. and LeFloch P.G. Convergence of finite difference schemes for conservation laws in several space variables: a general theory *SIAM J. Numer. Anal.* **30** 1993 675–700

Coquel F. and LeFloch P.G. Convergence of finite difference schemes for scalar conservation laws in several space dimensions: the corrected antidiffusive flux approach *Math. of Comp.* **57** 1991 169–210

Coquel F. and LeFloch P.G. A second order entropy satisfying scheme for systems of conservation laws *Note C.R. Acad. Sc., Paris, Série I,* 1995

Coquel F. and LeFloch P.G. A second order entropy satisfying scheme for systems of conservation laws *to appear* 1996

DiPerna R.J. Convergence of the viscosity method for isentropic gas dynamics *Comm. Math. Phys.* **91** 1983 1–30

DiPerna R.J. Convergence of approximate solutions to conservation laws *Arch. Rat. Mech. Anal.* **82** 1983 27–70

Glimm J. Solutions in the large for nonlinear hyperbolic systems of equations *Comm. Pure Appl. Math.* **18** 1965 697–715

Goodman J. and LeVeque R. A geometric approach to high resolution TVD schemes *SIAM J. Numer. Anal.* **25** 1988 268–284

Harabetian E. The generalized Riemann problem with analytic data *Trans. Amer. Math. Soc.* 1988

Harten A. On a class of high order resolution total-variation stable finite difference schemes *SIAM J. Numer. Anal.* **21** 1974 1–23

Harten A., Engquist B., Osher S.J., and Chakravarthy S. Uniformly high order accurate essentially non-oscillatory schemes *J. Comp. Phys.* **71** 1987 231–303

Johnson C., Szepessy A., and Hansbo P. On the convergence of shock capturing streamline diffusion finite element methods for hyperbolic conservation laws *Math. of comp.* **54** 107–129 1990

Khobalatte B. and Perthame B. Maximum principle on the entropy and second-order kinetic schemes *Math. of Comp.* **62** 1994 119–135

Lax P.D. Hyperbolic Systems of Conservation Laws and the Mathematical Theory of Shock Waves *Conf. Board Math. Sci. 11, SIAM, Philadelphia* 1973

Van Leer B. Towards the ultimate conservative difference schemes, V: A second order sequel to Godunov's method *J. Comp. Phys.* **43** 1981 357–372

LeFloch P.G. and Liu J.-G. Discrete entropy and monotonicity criterion for hyperbolic conservation laws 1994 **319** *C.R. Acad. Sc., Paris,* 881–886; and article in preparation

LeFloch P.G. and Raviart P.-A. An asymptotic expansion for the solution of the generalized Riemann problem. Part I: general theory *Ann. Inst. H. Poincaré, Nonlin. Anal.* **5** 1988 179–207

Lions P.-L., Perthame B., and Tadmor E. Kinetic formulations for the p-system and Euler system *Comm. Math. Phys.* **163** 1994 415–431

Lions P.-L. and P. Sougadinis Convergence of MUSCL-type methods for scalar conservation laws *C.R. Acad. Sc. Paris, Série I,* **311** 1990 259–264

Li T.-T. and W.-C. Yu Boundary Value Problem for Quasilinear Hyperbolic Systems *Duke Univ. Math. Series* 1985

Nessyahu H. and E. Tadmor Non-oscillatory central differencing for hyperbolic conservation laws *J. Comp. Phys.* **87** 1990 408–462

Osher S.J. Convergence of generalized MUSCL schemes *SIAM J. Numer. Anal.* **22** 1985 947–961

Osher S.J. and S. Chakravarthy High resolution schemes and entropy condition *SIAM J. Numer. Anal.* **22** 1985 947–961

Osher S.J. and E. Tadmor On the convergence of difference approximations to scalar conservation laws *Math. of Comp.* **50** 1988 19–51

Serre D. Domaines invariants pour les systèmes hyperboliques de lois de conservation *J. Diff. Eq.* **69** 1987 46–62

Smoller J. Shock Waves and Reaction Diffusion Equations *Springer-Verlag, New York* 1983

Tadmor E. The numerical viscosity of entropy stable schemes for systems of conservation laws *Math. of Comp.* **49** 1987 91–103

Tadmor E. Semi-discrete approximations to nonlinear systems of conservation laws; consistency and L^∞-stability imply convergence *ICASE Report 88-41* 1988

Vila J.P. High-order schemes and the entropy condition for nonlinear hyperbolic systems of conservation laws *Math. of Comp.* **50** 1988 53–73

Yang H. Nonlinear wave analysis and convergence of MUSCL schemes *IMA Preprint 697, Minneapolis* 1991

HIGH RESOLUTION METHODS FOR RELATIVISTIC FLUID DYNAMICS

JOSE M. MARTI
Max-Planck-Institut für Astrophysik
Karl-Schwarzschild-Str. 1, 85740 Garching, Germany

1. Introduction

Most of the matter in the universe can, in some approximation, be treated as a fluid and in several cases a description in terms of relativistic dynamics is the most suitable. Relativistic fluid dynamics (RFD) should be applied to flows in which velocities (of individual particles or of the fluid as a whole) approach that of light in vacuum, c ($\sim 3\,10^{10}$ cm s^{-1}) or, alternatively, in scenarios involving huge gravitational potentials (of the order of the rest-mass energy, $\sim 9\,10^{20}$ erg g^{-1}), where a description in terms of the Einstein theory of gravity becomes necessary.

Relativistic effects appear under extreme conditions. As an example, let us note that the speed of sound in air, commonly used in aerodynamics is only a millionth of c and that departures from classical gravity in terrestrial or even solar scenarios are remarkably small (in the case of the Sun, the surface gravitational potential is about 10^{15} erg g^{-1}, several orders of magnitude below the limit given in the previous paragraph). Nevertheless, relativity is a necessary ingredient in astronomy for describing scenarios involving compact objets. Among them are supernovae, X-ray binaries, active galactic nuclei and, interesting in terms of the production of gravitational radiation, coalescing neutron stars. On the other hand, present-day heavy-ion collision experiments taking place in large particle accelerators produce beams with velocities equal to a large fraction of c. The aim of these experiments, is to gain insight into the equation of state for hot dense matter. In an astrophysical context, most of the scenarios cited above also involve the presence of flows at relativistic speeds, the most compelling one being the commonly observed jets in extragalactic radio sources associated to active galactic nuclei. Nowadays, in the accepted standard model, flow velocities

E.F. Toro and J.F. Clarke (eds.), Numerical Methods for Wave Propagation, 297–322.

as large as 99% of c are required to explain the apparent superluminal motion observed in more than 40 of these sources.

Simulations based on the numerical integration of the hydrodynamical equations provide a useful tool to confront the theoretical models with the observations (as in the astrophysical scenarios cited above) or the experimental results. Therefore it is clear why the development of hydro codes which work accurately under the required extreme conditions is of interest. In the case of RFD, the first Eulerian code was developed by Wilson (1972, 1979) and collaborators (Centrella & Wilson 1984, Hawley, Smarr & Wilson 1984) on the basis of explicit finite-differencing techniques and monotonic transport. The code was stabilized accross shocks by means of artificial viscosity (von Neumann and Richtmyer 1950). This code has been widely used in cosmology, axisymmetric relativistic stellar collapse, accretion onto compact objects and, more recently, collisions of heavy ions. However, despite its popularity (almost all the codes for numerical relativistic hydrodynamics in the eighties were based on Wilson's procedure) it has turned out to be unable to describe accurately extremely relativistic flows. The more relativistic the flow, the less accurate the code becomes and being unacceptable for flows with bulk Lorentz factors larger than 2 (see, e.g., Centrella & Wilson 1984). Norman and Winkler (1986) analized the problem in depth and proposed a fully implicit treatment of the equations as the only way to increase the accuracy of artificial viscosity formulations in the ultrarelativistic limit.

However, in recent years, and parallel to the evolution of computational methods for classical fluid dynamics, several new methods for numerical RFD have been designed which exploit the hyperbolic and conservative character of the relativistic equations. Once the equations have been written in conservation form and the corresponding set of unknowns and fluxes have been identified, almost every high-resolution method devised to solve hyperbolic systems of conservation laws can be extended to RFD (see, e.g., Martí et al. 1991, Marquina et al. 1992, Schneider et al. 1993). More recently, Martí & Müller (1994) have obtained the exact solution of the Riemann problem for RFD in one spatial dimension (1D) which has been further used to simulate extremely relativistic flows with great success (Martí & Müller 1995). Balsara (1994) has developed an approximate Riemann solver based in the two-shock approximation and Eulderink & Mellema (1994) extended the Roe solver (Roe 1981) to general relativistic fluid dynamics.

Judging from the results of several test calculations shown in these references, it can be concluded that an accurate description of ultrarelativistic flows with strong shock waves can be accomplished by writing the system of RFD in conservation form and using Riemann solvers. This talk is devoted to review the recent developments in the application of high-resolution methods based on (exact or approximate) Riemann solvers to the integration of RFD equations.

2. The equations of relativistic fluid dynamics

Special relativity (SR) has followed Einstein's postulate which established the universality of the speed of light: the speed of light relative to any unaccelerated observer is c, regardless of the motion of the light's source relative to the observer. All the counter-intuitive predictions of SR (e.g, the time dilation and space contraction in moving frames of reference, the equivalence of mass and energy) come from this postulate.

General relativity (GR) came into being historically as an extension of the special theory to describe the motion of particles under the presence of gravitational fields. In order to consider the motion of particles evolving in a gravitational field as free, GR abandones the Eucledian geometry and uses a curved four-dimensional manifold (the spacetime) to represent the effects of gravity on particles' trajectories. The information about the curvature of spacetime is contained in the Einstein tensor, G, defined in terms of the metric components, $g_{\mu\nu}$ $(\mu, \nu = 0, \ldots, 3)$ describing the geometry of spacetime. It is linked to the distribution of matter through Einstein's equations

$$G^{\mu\nu} = 8\pi T^{\mu\nu} \tag{1}$$

(throughout this Chapter, units in which $G = c = 1$ are used) where T is the energy-momentum tensor describing the physical properties of matter (vanishing in the case of pure vacuum).

The evolution of matter through spacetime is established by the conservation equations of energy-momentum and particle number describing the state of the fluid as a function of the coordinates.

Sometimes one can assume to good approximation that the spacetime structure is determined by a part of the energy momentum-tensor and that the remainder no longer alters the curvature (this is true, for example, in accretion problems, where the accreting mass is much larger than the surrounding material). Here one speaks of external field approximation (or test relativistic fluid dynamics). If the gravitational field can be completely neglected (as, for example, in ion collisions reactions or in extragalactic jets), one can simply assume SR. In these cases, the conservation equations solved in a given spacetime provide the neccesary conditions for a whole description of the fluid. The stellar collapse is an example where none of these simplifications can be considered because the fluid itself acts as a dynamical source of the curvature.

As stated above, the equations describing the evolution of a relativistic fluid are local conservation laws: the local conservation of energy-momentum and the local conservation of rest mass,

$$T^{\mu\nu}_{\ ;\mu} = 0, \tag{2}$$

$$(\rho u^{\mu})_{;\mu} = 0, \tag{3}$$

where "$;\mu$" stands for the covariant derivate with respect to coordinate x^{μ} and summation is extended over repeated indices. In this last equation, ρ stands for the rest-mass density measured in the fluid's local rest frame (i.e., moving along with the fluid elements), and u^{μ} is a vector field, the fluid four-velocity, representing the fluid-elements' paths through the spacetime.

In the following we will consider only perfect fluids (fluids in which viscosity and thermal conduction effects can be neglected). In this case, the energy-momentum tensor appearing in Eq. (2) can be written as

$$T^{\mu\nu} = \rho h u^{\mu} u^{\nu} + p g^{\mu\nu}, \tag{4}$$

where p stands for the isotropic pressure and h is the specific enthalpy, defined by

$$h = 1 + \varepsilon + \frac{p}{\rho}, \tag{5}$$

where ε is the specific internal energy.

The above system of conservation laws is closed by means of the normalization condition for the four-velocity

$$g_{\mu\nu} u^{\mu} u^{\nu} = -1 \tag{6}$$

and an equation of state (EOS) that we shall assume as given in the form

$$p = p(\rho, \varepsilon). \tag{7}$$

In order to discuss the recent developments in numerical RFD, we shall focus on the SR case. In the absence of gravity, the metric tensor $g_{\mu\nu}$ reduces to the Minkowski flat metric, $\eta_{\mu\nu}$, which in Cartesian coordinates is given by $\eta_{\mu\nu} = \mathrm{diag}(-1, 1, 1, 1)$.

In Minkowski spacetime and with the following definitions,

$$D = \rho W \tag{8}$$

$$S^{j} = \rho h W^{2} v^{j} \tag{9}$$

$$v^{j} = \frac{u^{j}}{W} \tag{10}$$

$$\tau = \rho h W^2 - p - \rho W, \tag{11}$$

the conservation equations (2), (3) can be written in a more compact form as

$$\frac{\partial \mathbf{F}^\mu}{\partial x^\mu} = 0, \tag{12}$$

where each of the four $\mathbf{F}^\mu$ is a five-dimensional vector defined by

$$\mathbf{F}^0 = (D, S^j, \tau)^\mathrm{T} \tag{13}$$

$$\mathbf{F}^i = (Dv^i, S^j v^i + \delta^{ij} p, \tau v^i + p v^i)^\mathrm{T} \tag{14}$$

$(i, j = 1, 2, 3)$. In the above equations, quantities D, S^j, v^j and τ are, respectively, the rest-mass density, momentum density, fluid flow velocity and total-energy (excluding rest-mass) density measured in the so-called laboratory frame. Quantity W stands for the Lorentz factor and verifies

$$W = \frac{1}{\sqrt{1 - \mathrm{v}^2}}, \tag{15}$$

with $\mathrm{v}^2 = (v^1)^2 + (v^2)^2 + (v^3)^2$. System (12) displays the conservative character of the equations of RFD and is written in a suitable form for the application of high resolution methods.

Introducing the Jacobian matrices $\mathcal{A}^\mu(\mathbf{w})$ associated to the five-vectors $\mathbf{F}^\mu(\mathbf{w})$ $(\mathbf{w} = (\rho, v^i, \varepsilon))$

$$\mathcal{A}^\mu = \frac{\partial \mathbf{F}^\mu}{\partial \mathbf{w}}, \tag{16}$$

system (12) can be written as a quasilinear system of first order partial differential equations for the unknown field $\mathbf{w}$

$$\mathcal{A}^\mu \frac{\partial \mathbf{w}}{\partial x^\mu} = 0 \tag{17}$$

The explicit expressions of $\mathcal{A}^\mu$ for the case of three-dimensional special relativistic hydrodynamics in Cartesian coordinates can be found in Font et al. (1994).

It can be demonstrated (see, e.g., Anile 1989) that the system of equations (2)–(3) is hyperbolic for causal EOS, i.e., for EOS verifying the condition $c_s < 1$, where c_s is the local sound speed. The spectral decomposition of the three 5×5 Jacobians associated to system (12)

$$\mathcal{B}^i = (\partial \mathbf{F}^i / \partial \mathbf{F}^0) = (\partial \mathbf{F}^i / \partial \mathbf{w})(\partial \mathbf{F}^0 / \partial \mathbf{w})^{-1} = \mathcal{A}^i (\mathcal{A}^0)^{-1}$$

is the following (case of $\mathcal{B}^x$):

Eigenvalues:

$$\lambda_0 = v^x \quad \text{(triple)} \tag{18}$$

$$\lambda_{\pm} = \frac{1}{1 - \mathrm{v}^2 c_s^2} \Big\{ v^x (1 - c_s^2) \pm \\ \pm c_s \sqrt{(1 - \mathrm{v}^2)[(1 - \mathrm{v}^2 c_s^2) - v^x v^x (1 - c_s^2)]} \Big\} . \quad (19)$$

Right-eigenvectors:

$$\mathbf{r}_{0,1} = \left(\frac{\tilde{\kappa}}{hW(\tilde{\kappa} - c_s^2)}, v^x, v^y, v^z, 1 - \frac{\tilde{\kappa}}{hW(\tilde{\kappa} - c_s^2)} \right)^{\mathrm{T}}$$

$$\mathbf{r}_{0,2} = \left(Wv^y, 2hW^2 v^x v^y, h(1 + 2W^2 v^y v^y), 2hW^2 v^z v^y, 2hW^2 v^y - Wv^y \right)^{\mathrm{T}}$$

$$\mathbf{r}_{0,3} = \left(Wv^z, 2hW^2 v^x v^z, 2hW^2 v^y v^z, h(1 + 2W^2 v^z v^z), 2hW^2 v^z - Wv^z \right)^{\mathrm{T}}$$

$$\mathbf{r}_{\pm} = \left(1, hW \left(v^x - \frac{v^x - \lambda_{\pm}}{1 - v^x \lambda_{\pm}} \right), hWv^y, hWv^z, hW \left(\frac{1 - v^x v^x}{1 - v^x \lambda_{\pm}} \right) - 1 \right)^{\mathrm{T}}$$

with $\tilde{\kappa} \equiv \kappa/\rho$.

This spectral decomposition provides the technical implements needed to develop approximate Riemann solvers based on local linearizations of the relativistic system of equations in a way which is identical to the classical fluid dynamics case. However, let us note that the characteristic wavespeeds in the relativistic case not only depend on the fluid velocity components in the wave propagation direction but also on the normal velocity components (see Eq. 19). This coupling adds new numerical difficulties which are specific to RFD.

3. The Riemann problem for the RFD equations

Given that it is of theoretical interest, let us start by considering the exact solution of the Riemann problem associated to the RFD equations. In classical fluid dynamics, the Riemann problem has played a very important role providing analytical solutions which can be used to test hydrodynamical codes (see, e.g., Sod 1978) as well as in the development of the codes themselves. In fact, most high-resolution methods are based on the pioneering work of Godunov (1959) who first used the exact solution of Riemann problems to construct a hydrodynamical code.

We are concerned with the breakup of an initial discontinuity which separates two constant states L (left) and R (right) of an ideal gas in arbitrary (1D) conditions, in the absence of any gravitational field (SR). This problem was considered by Martí & Müller (1994) who derived an exact solution. Prior to this work, the building blocks of this analytical solution, i.e., the elementary nonlinear special relativistic waves (Taub 1948) had been extensively treated in literature (for a review see, e.g., Anile 1989). However, none of these investigations aimed at solving the general Riemann problem. Thompson (1986) considered the analytical solution of the special relativistic Riemann problem for the particular case of zero initial flow velocities (the so-called shock-tube problem).

Both in relativistic and Newtonian hydrodynamics the discontinuity between the two constant initial states $\mathbf{V}_L$ and $\mathbf{V}_R$ ($\mathbf{V} = (p, \rho, v)$) decays into two elementary nonlinear waves (shocks or rarefactions), one moving towards the initial left state and the other towards the initial right state. Between the waves two new states, namely $\mathbf{V}_{L*}$ and $\mathbf{V}_{R*}$, appear which are separated from each other through a contact discontinuity moving along with the fluid. Across the contact discontinuity, pressure and velocity are continuous, while the density exhibits a jump. As in classical hydrodynamics (see, e.g., Courant & Friedrichs 1976) the self-similar character of the flow through rarefaction waves and the Rankine-Hugoniot conditions across shocks provide the conditions for linking the intermediate states $\mathbf{V}_{S*}$ ($S = L, R$) with their corresponding initial state $\mathbf{V}_S$. In particular, one can express the velocity of the intermediate states v_{S*} as a function of the pressure p_{S*} of these states. The smoothness of the velocity across the contact discontinuity then gives

$$v_{L*}(p_*) = v_{R*}(p_*) \,, \tag{20}$$

where $p_* = p_{L*} = p_{R*}$.

The functions $v_{S*}(p)$ are defined by

$$v_{S*}(p) = \begin{cases} \mathcal{R}^S(p) & \text{if } p \leq p_S \\ \mathcal{S}^S(p) & \text{if } p > p_S, \end{cases} \tag{21}$$

where $\mathcal{R}^S(p)$ ($\mathcal{S}^S(p)$) denotes the family of all states which can be connected through a rarefaction (shock) with a given state S ahead of the wave.

The fact that one Riemann invariant is constant through any rarefaction wave provides the relation needed to derive the function $\mathcal{R}^S$

$$\mathcal{R}^S(p) = \frac{(1+v_S)A_\pm(p) - (1-v_S)}{(1+v_S)A_\pm(p) + (1-v_S)} \tag{22}$$

with

$$A_\pm(p) = \left(\frac{\sqrt{\gamma-1} - c(p)}{\sqrt{\gamma-1} + c(p)} \frac{\sqrt{\gamma-1} + c_S}{\sqrt{\gamma-1} - c_S} \right)^{\pm \frac{2}{\sqrt{\gamma-1}}} \tag{23}$$

the + (−) sign of $A_{\pm}$ corresponding to $S = L$ ($S = R$). In the above equation, c_S is the sound speed of the state S, and $c(p)$ is given by

$$c(p) = \left(\frac{\gamma(\gamma-1)p}{(\gamma-1)\rho_S(p/p_S)^{1/\gamma} + \gamma p}\right)^{1/2} . \quad (24)$$

The family of all states $\mathcal{S}^S(p)$, which can be connected through a shock with a given state S ahead of the wave, is determined by the shock jump conditions. One obtains

$$\mathcal{S}^S(p) = \left(h_S W_S v_S \pm \frac{p - p_S}{j(p)\sqrt{1 - V_{\pm}(p)^2}}\right) \left[h_S W_S + (p - p_S)\left(\frac{1}{\rho_S W_S} \pm \frac{v_S}{j(p)\sqrt{1 - V_{\pm}(p)^2}}\right)\right]^{-1} , \quad (25)$$

where the + (−) sign corresponds to $S = R$ ($S = L$). $V_{\pm}(p)$ and $j(p)$ denote the shock velocity and the modulus of the mass flux across the shock front, respectively. They are given by

$$V_{\pm}(p) = \frac{\rho_S^2 W_S^2 v_S \pm j(p)^2 \sqrt{1 + (\rho_S/j(p))^2}}{\rho_S^2 W_S^2 + j(p)^2} \quad (26)$$

and

$$j(p) = \sqrt{\frac{p_S - p}{\dfrac{h_S^2 - h(p)^2}{p_S - p} - \dfrac{2h_S}{\rho_S}}} , \quad (27)$$

where the enthalpy $h(p)$ of the state behind the shock is the (unique) positive root of the quadratic equation

$$\left(1 + \frac{(\gamma-1)(p_S - p)}{\gamma p}\right) h^2 - \frac{(\gamma-1)(p_S - p)}{\gamma p} h + \frac{h_S(p_S - p)}{\rho_S} - h_S^2 = 0 , \quad (28)$$

which is obtained from the Taub adiabat (the relativistic version of the Hugoniot adiabat) for an ideal gas equation of state.

In Fig. 1, the functions $v_{L*}(p)$ and $v_{R*}(p)$ are displayed in a p-v diagram for a particular set of Riemann problems. Once p_* has been obtained, the remaining state quantities can easily be derived.

Figures 2-4 show the flow pattern produced in the three different types of Riemann problems.

4. An application of the exact relativistic Riemann solution: the relativistic PPM

Martí & Müller (1995) have used the procedure sketched above to construct an exact Riemann solver to be implemented in an extension of the popular

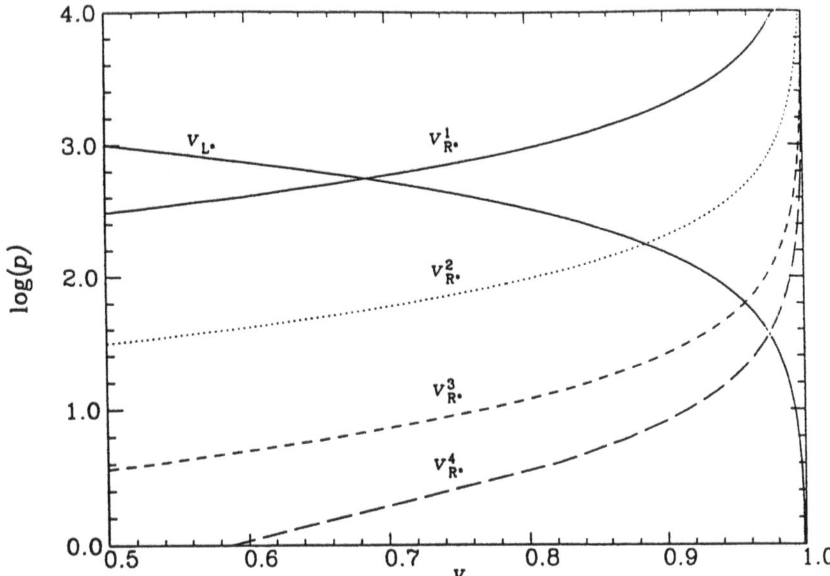

Figure 1. Graphical solution in the $p - v$ plane of the Riemann problems defined by the initial states $\mathbf{V}_L = (p_L = 10^3, \rho_L = 1, v_L = 0.5)$ and $\mathbf{V}^i_R = (p^i_R, \rho_R = 1, v_R = 0)$, $(i = 1, \ldots, 4)$ with $p^1_R = 10^2$, $p^2_R = 10$, $p^3_R = 1$, $p^4_R = 10^{-1}$. The adiabatic index of the equation of state is 5/3 in all cases. Note the asymptotic behaviour of the functions when they approach $v = 1$.

PPM method of Colella & Woodward (1984) for 1D RFD. In Martí & Müller (1995), detailed calculations show the accuracy and convergence rates of the code in the computation of extremely relativistic flows with strong shocks. Although the extension to multidimensional problems is difficult (see below) and quick progress is being made in the development of several kinds of approximate relativistic Riemann solvers, the results obtained in 1D problems with the exact Riemann solver and the relativistic extension of PPM set the level of accuracy demanded of present–day high–resolution relativistic methods.

The equations of (special) RFD in 1D and Cartesian coordinates can be written as

$$\frac{\partial \mathbf{u}}{\partial t} + \frac{\partial \mathbf{F}(\mathbf{u})}{\partial x} = 0 \tag{29}$$

with

$$\mathbf{u} = (D, S, \tau)^{\mathrm{T}} \tag{30}$$

and

$$\mathbf{F} = (Dv,\, Sv + p,\, S - Dv)^{\mathrm{T}}. \tag{31}$$

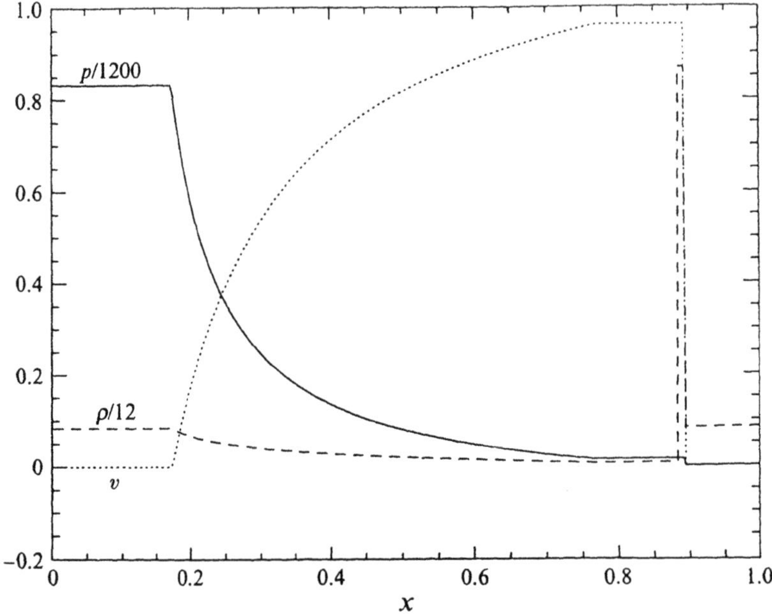

Figure 2. Analytical pressure, density and flow velocity profiles for the relativistic Riemann problem with initial data ($p_L = 10^3$, $\rho_L = 1$, $v_L = 0$), ($p_R = 10^{-2}$, $\rho_R = 1$, $v_R = 0$). The adiabatic index of the equation of state, γ, is 5/3 for the left and right states. The initial discontinuity is placed at $x = 0.5$. The solution is shown at $t = 0.4$.

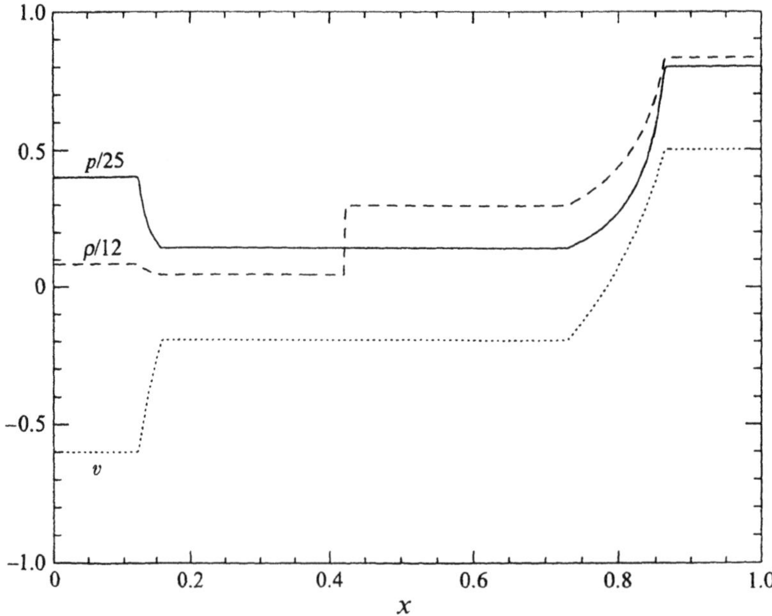

Figure 3. Same as Fig. 2 but for initial data ($p_L = 10$, $\rho_L = 1$, $v_L = -0.6$), ($p_R = 20$, $\rho_R = 10$, $v_R = 0.5$), $\gamma = 5/3$.

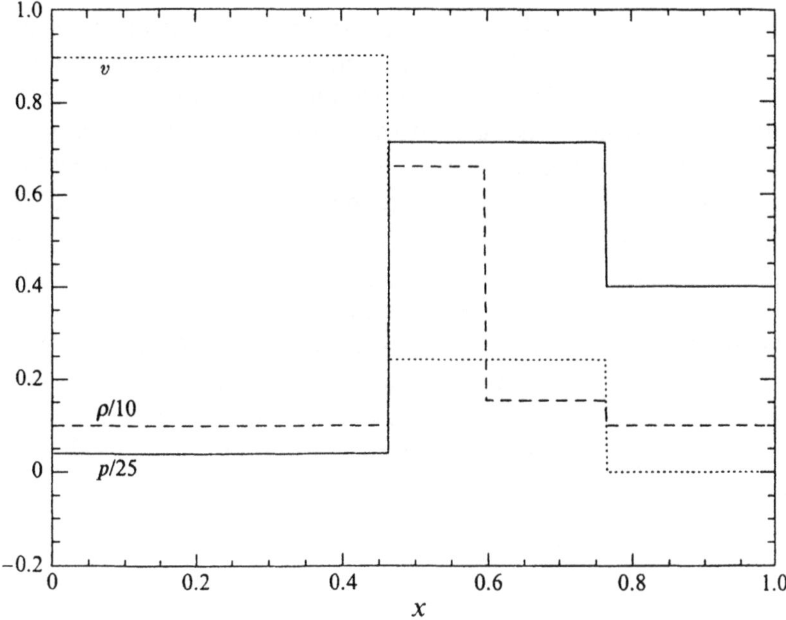

Figure 4. Same as Fig. 2 but for initial data ($p_L = 1$, $\rho_L = 1$, $v_L = 0.9$), ($p_R = 10$, $\rho_R = 1$, $v_R = 0$), $\gamma = 4/3$.

In order to solve system (29) we consider the conservative difference scheme

$$\mathbf{u}_j^{n+1} = \mathbf{u}_j^n + \frac{\Delta t}{\Delta x} \{\hat{\mathbf{F}}_{j-\frac{1}{2}} - \hat{\mathbf{F}}_{j+\frac{1}{2}}\} \tag{32}$$

where $\mathbf{u}_j^n$ and $\mathbf{u}_j^{n+1}$ are the zone-averaged values of the state vector $\mathbf{u}$ of zone j at times $t = t^n$ and $t = t^{n+1} = t^n + \Delta t$, respectively. $\hat{\mathbf{F}}_{j\pm\frac{1}{2}}$ are the time averaged numerical fluxes at the right (+) and left (−) interface of zone j. In a Godunov-type difference scheme appropriate left and right states are constructed from the zone-averaged values, which are then used to calculate the numerical fluxes $\hat{\mathbf{F}}_{j+\frac{1}{2}}$ by solving the corresponding Riemann problem.

In our relativistic version of PPM the interpolation algorithm described in the original paper by Colella & Woodward (1984) giving monotonic conservative parabolic profiles of variables within the numerical zones, is applied to zone averaged values of $\mathbf{V} = (p, \rho, v)$, which are obtained from zone averaged values of the conserved quantities $\mathbf{u}$. As in the Newtonian version of PPM, we determine for each zone j the quartic polynomial which has zone-averaged values a_{j-2}, a_{j-1}, a_j, a_{j+1}, a_{j+2} to interpolate the structure inside the zone, where a is one of the quantities p, ρ or v. Using this quartic polynomial values of a at the left and right interface of the zone,

$a_{L,j}$ and $a_{R,j}$, are obtained. These reconstructed values are then modified such that the parabolic profile, which is uniquely determined by $a_{L,j}$, $a_{R,j}$ and a_j, is monotonic inside the zone. Finally, the interpolation procedure is slightly modified near discontinuities to produce narrower jumps.

To obtain time-averaged fluxes at an interface $j + 1/2$ separating zones j and $j+1$, in PPM two spatially averaged states, $\mathbf{V}_{j+\frac{1}{2},S}$ ($S = L, R$ where L and R denote the left and right side of the interface, respectively), are constructed, taking into account the characteristic information reaching the interface from both sides during the time step. In our relativistic version of PPM, we have followed the same procedure as in Colella & Woorward (1984), but have considered the characteristic speeds and Riemann invariants of the equations of relativistic instead of Newtonian hydrodynamics.

In Godunov's approach the numerical fluxes $\hat{\mathbf{F}}_{j+\frac{1}{2}}$ are calculated according to

$$\hat{\mathbf{F}}_{j+\frac{1}{2}} = \mathbf{F}(\bar{\mathbf{u}}_{j+\frac{1}{2}}) \tag{33}$$

where $\bar{\mathbf{u}}_{j+\frac{1}{2}}$ is an approximation to $(1/\Delta t) \int \mathbf{u}(x_{j+\frac{1}{2}}, t)\, dt$, i.e. the time-averaged value of the solution at $x_{j+\frac{1}{2}}$, which is obtained solving the Riemann problem at $x_{j+\frac{1}{2}}$ with left and right states $\mathbf{V}_{j+\frac{1}{2},L}$ and $\mathbf{V}_{j+\frac{1}{2},R}$, respectively.

Once the intermediate state created in the breakup of the initial discontinuity has been obtained, we calculate the Riemann solution at the interfaces following the procedure developed by Colella & Glaz (1985). We omit the details and address the reader to Martí & Müller (1995).

4.1. FIRST EXAMPLE: SHOCK HEATING OF A COLD FLUID

The initial setup consists of an inflowing cold (i.e., ε=0) gas with coordinate velocity v_1 and Lorentz factor W_1, which fills the computational grid and hits a wall placed at the right edge of the grid. As the gas hits the wall, it is compressed and heated up converting its momentum into internal energy and giving rise to a shock, which starts to propagate off the wall. Behind the shock, the gas is at rest (v_2=0) and has a specific internal energy

$$\varepsilon_2 = W_1 - 1\,. \tag{34}$$

The compression ratio between shocked and unshocked gas, $\eta = \rho_2/\rho_1$, follows from

$$\eta = \frac{\gamma + 1}{\gamma - 1} + \frac{\gamma}{\gamma - 1}\varepsilon_2\,, \tag{35}$$

where γ is the adiabatic index of the equation of state.

This test problem has been widely used to check the accuracy of RFD codes: Centrella & Wilson (1984), Norman & Winkler (1986), Martí et

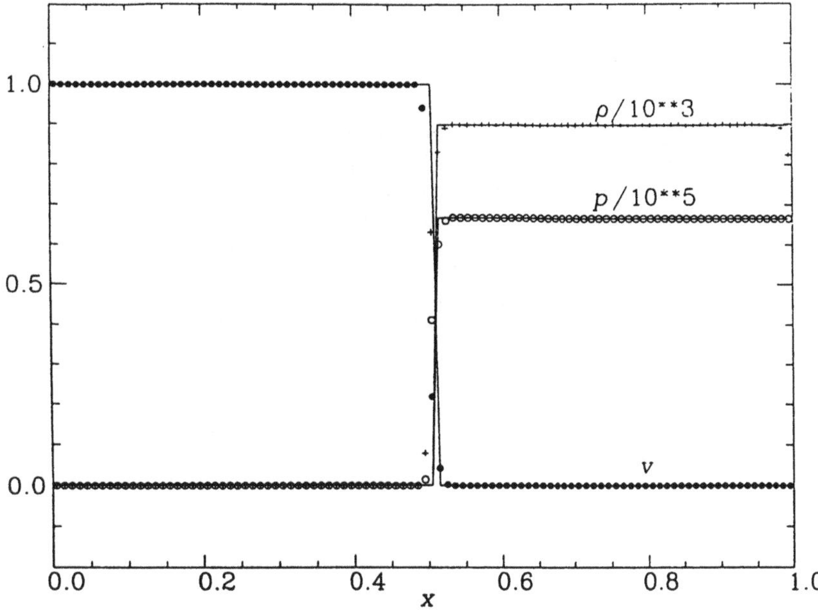

Figure 5. Exact (solid line) and numerical profiles of pressure, density and flow velocity for the shock heating problem with an inflow velocity $v_1 = 0.99999$, when the shock has propagated 50 zones off the wall (at $x = 1$). The computations have been performed on an equidistant grid of 100 zones.

al. (1991), Marquina et al. (1992), Schneider et al. (1993), Eulderink & Mellema (1995). Concerning explicit schemes, the numerical results improved significantly for this test problem, when numerical methods based on Riemann solvers were introduced.

In our test calculations we have used a gas with an adiabatic index $\gamma = 4/3$ and inflow velocities ranging from nearly Newtonian to ultrarelativistic values. The computational grid consisted of 100 equidistant zones covering the interval $x \in [0, 1]$. The wall was placed at $x = 1$. For numerical reasons, the specific internal energy of the inflowing gas was set to a small finite value ($\varepsilon_1 = 10^{-7} W_1$). Figure 5 shows the profiles of pressure, rest-mass density and flow velocity for a run with a gas inflow velocity $v_1 = 0.99999$ ($W_1 \approx 70$) after the shock has propagated 50 zones off the wall. The profiles obtained for other inflow velocities are qualitatively similar.

In our sample of calculations the mean relative error $\bar{\varepsilon}_r(\eta)$ is always smaller than 10^{-3} and, in accordance with other codes based on a Riemann solver, the accuracy of our results does not exhibit any significant dependence on the Lorentz factor of the inflowing gas. This can be seen in Table 1, which shows the relative errors in the compression ratio obtained with several numerical algorithms (including our relativistic version of PPM).

TABLE 1. Relative errors in the compression ratio in the problem of shock heating of a cold gas. v_1 is the inflow velocity of the gas in units of c and W_1 the corresponding Lorentz factor. AV stands for the results obtained with the artificial-viscosity method of McAbee et al. (1989); RHLLE corresponds to the Godunov-type method of Schneider et al. (1993) discussed in Section 5.3; RPPM is the relativistic PPM method.

W_1	v_1	$\bar{\epsilon}(\eta)$		
		AV	RHLLE	RPPM
1.15	0.5			<1.3E-04
1.50	0.75	1.9E-03		
2.00	0.866	7.4E-03	5.9E-04	
2.29	0.900			6.4E-04
4.00	0.968	3.5E-03		
5.00	0.980		8.9E-04	
7.09	0.990			3.9E-04
8.00	0.992	2.2E-02		
10.0	0.995	2.6E-02	1.5E-03	
22.4	0.999			8.6E-04
50.0	0.9998		2.1E-03	
70.7	0.9999			5.9E-04
223.6	0.99999			3.4E-04

4.2. SECOND EXAMPLE: A RELATIVISTIC SHOCK TUBE

Shock tubes represent a special class of Riemann problems in which the initial state on both sides of the discontinuity is at rest. They have become a useful tool in testing numerical codes, because their evolution involves shock waves and rarefactions. Here we present results in a mildly relativistic test (the results obtained in a more extreme problem can be found in Martí & Müller 1995). The initial state is as follows:

$$\begin{array}{ll} \rho_L = 10.0 & \rho_R = 1.0 \\ p_L = 13.3 & p_R = 0 \\ v_L = 0 & v_R = 0 \end{array}$$

The adiabatic index was set to $\gamma = 5/3$, and the initial discontinuity was placed at $x = 0.5$ (this problem has been considered by several authors,

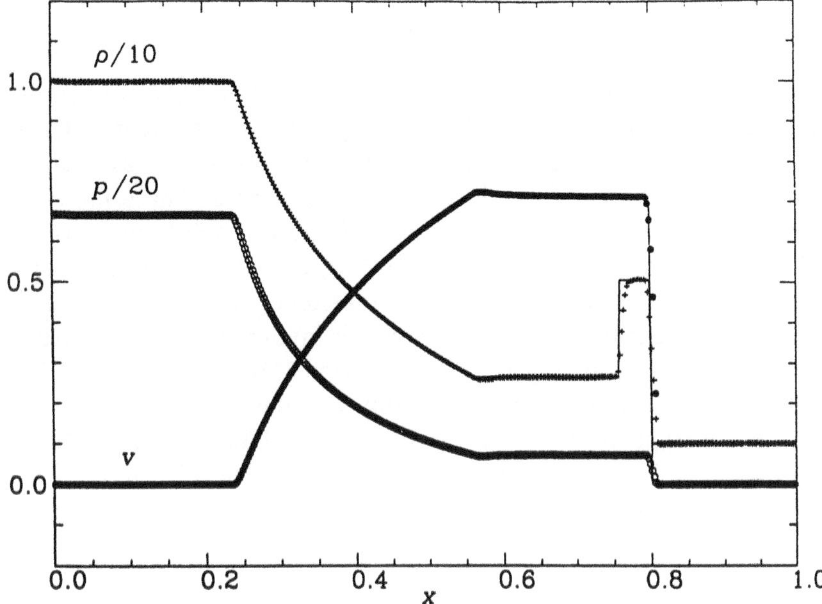

Figure 6. **Exact (solid line) and numerical profiles of pressure, density and flow velocity for the relativistic shock tube problem defined in the text. The computations have been performed on an equidistant grid of 400 zones.**

e.g., Centrella & Wilson 1984, Hawley et al. 1984, Schneider et al. 1993, Eulderink & Mellema 1995). For numerical reasons the pressure of the right state was set to a small finite value ($p_R = 0.66\ 10^{-6}$). The decay of the initial discontinuity gives rise to an intermediate state located between a shock wave and a rarefaction propagating to the right (i.e. positive x-direction) and left, respectively. The fluid in the intermediate state moves to the right with a velocity $v_{\rm shell} = 0.72$. Figure 6 shows the results for a grid of 400 equidistant zones. One recognizes that the shock is smeared across 4–5 zones, and that the largest errors occur for the post-shock density. Studying the convergence properties of the method under grid refinement, Martí & Müller (1995) found a convergence rate corresponding to an order of accuracy of roughly 1, which is expected for problems with discontinuities, and which is closely consistent with the Newtonian version of PPM (Woodward & Colella 1984). The corresponding convergence rate for the nominally first-order accurate Godunov variant of our method is only 0.66 (note that the accuracy obtained on the finest grid with the relativistic Godunov variant of our method is then achieved by the relativistic PPM with a grid which is four times coarser).

5. Approximate relativistic Riemann solvers

To solve Riemann problems exactly usually involves a time-consuming calculation. On the other hand, in Godunov-type methods, most of the structure of the Riemann solution is lost during the process of averaging over numerical cells that introduces large numerical errors. These facts suggest as an alternative the use of approximate Riemann solutions obtained by some less expensive means. Besides these arguments, in the case of RFD, there are other reasons concerning the extension of the exact Riemann solutions to multidimensional calculations which favour the use of approximate Riemann solvers. Let us explain this assertion. Most modern multidimensional (explicit) finite-difference methods are based on the technique of directional splitting in which the problem is split into (two or three) sets of 1D problems. Thus, a timestep consists of solving sets of 1D Riemann problems in the presence of a tangential velocity v^t. As it is well known (see, e.g., Landau & Lifshitz 1987) in classical fluid dynamics, the decay of an initial discontinuity does not depend on the tangential velocity and, furthermore, this velocity is constant accross shock waves and rarefactions. These facts reduce the problem to the solution of the 1D problem plus the determination of the correct value of the tangential velocity in the intermadiate states, the latter given by $v^t_{I*} = v^t_I$ $(I = L, R)$.

In relativistic calculations, however, one has to deal with the solution of Riemann problems in which the two components (normal and tangential) of the flow velocity are coupled due to the presence of the Lorentz factor in the equations. For shock waves, this coupling just results in an increase of the number of algebraic jump conditions to be solved. For rarefactions, however, the coupling implies the solution of a system of ordinary differential equations resulting in a really expensive calculation. An interesting way to overcome this difficulty without invoking linearized approximate Riemann solvers (which, however, are the most frequently used up to now in the scant number of multidimensional relativistic calculations) is the two-shock approximation introduced recently by Balsara (1994).

Linearized Riemann solvers are based in the exact solution of Riemann problems corresponding to a new system of equations obtained by a suitable linearization of the original one. The local linearization of the Jacobian matrices of the original system and its subsequent spectral decomposition is on the basis of all these solvers. Once the system of RFD has been written in conservation form and the corresponding sets of unknowns and fluxes have been identified, almost every high-resolution method devised to solve hyperbolic systems of conservation laws can be extended to RFD. Examples are the methods of Roe (1981) or Harten et al. (1983) which have been extended to relativistic fluid dynamics in recent years.

5.1. TWO-SHOCK APPROXIMATION FOR RELATIVISTIC HYDRODYNAMICS

This solver is obtained as a relativistic extension of Colella (1982) method for classical fluid dynamics. In this last case, it has turned out to be capable of computing shocks of arbitrary strength (Colella 1982; Woodward & Colella 1984). In order to construct Riemann solutions, in the two-shock approximation instead of using a rarefaction wave (when it is necessary) one uses the analytical continuation of shock waves onto the rarefaction side. In doing that, the coupling of the tangential components of the flow velocity reduces only to the algebraic complication of the Rankine-Hugoniot conditions accross oblique shocks. This complication can be further overcome by making successive Lorentz transformations between the shocks' rest frames (where the fluxes only depend on the normal components of the flow velocity) and the laboratory (i.e., Eulerian) frame.

The results obtained by Balsara (1994) in the resolution of mildly relativistic 1D test problems are very promising. Table 2 contains the converged solution for the intermediate states using Balsara's procedure corresponding to the Riemann problem defined in Section 4.2, together with the exact solution. Despite the fact that the true solution involves a strong rarefaction wave (see Fig. 6), the approximate solution is pretty accurate, with a maximum relative error in the density value of the intermediate state besides the rarefaction wave in the order of 10%.

TABLE 2. Two-shock approximation vs. exact solution for the Riemann problem defined in Section 2.

Method	p_*	v_*	ρ_{L*}	ρ_{R*}
Two-shock app.	1.440E+00	7.131E-1	2.990E+00	5.069E+00
Exact	1.445E+00	7.137E-1	2.640E+00	5.062E+00

5.2. GENERAL RELATIVISTIC ROE SOLVER

Eulderink (1993) and Eulderink & Mellema (1995) have extended Roe's Riemann solver (Roe 1981) to the general relativistic system of equations in arbitrary spacetimes. Eulderink looks for a local linearization of the Jacobian matrices of the system fulfilling the properties demanded by Roe in his original paper that we shall repit here. Let $\mathcal{A}$ be the Jacobian matrix associated to one of the fluxes $\mathbf{F}$ of the original system, and $\mathbf{u}$ the vector of unknowns. Then, the locally constant matrix $\tilde{\mathcal{A}}$, depending on $\mathbf{u}_L$ and $\mathbf{u}_R$

(the left and right set of values defining the local Riemann problem), must verify:

1. It constitutes a linear mapping from the vector space $\mathbf{u}$ to the vector space $\mathbf{F}$.
2. As $\mathbf{u}_\mathrm{L} \rightarrow \mathbf{u}_\mathrm{R} \rightarrow \mathbf{u}$, $\tilde{\mathcal{A}}(\mathbf{u}_\mathrm{L}, \mathbf{u}_\mathrm{R}) \rightarrow \mathcal{A}(\mathbf{u})$.
3. For any $\mathbf{u}_\mathrm{L}$, $\mathbf{u}_\mathrm{R}$, $\tilde{\mathcal{A}}(\mathbf{u}_\mathrm{L}, \mathbf{u}_\mathrm{R})(\mathbf{u}_\mathrm{R} - \mathbf{u}_\mathrm{L}) = \mathbf{F}(\mathbf{u}_\mathrm{R}) - \mathbf{F}(\mathbf{u}_\mathrm{L})$.
4. The eigenvectors of $\tilde{\mathcal{A}}$ are linearly independent.

Conditions 1 and 2 are necessary to recover the linearized algorithm from the non-linear one smoothly. Condition 3 (supposing 4 is fulfilled) ensures that if a single discontinuity is located at the interface, then the solution of the linearized problem gives the exact solution to the Riemann problem.

Once the averaged matrix satisfying Roe's conditions, $\tilde{\mathcal{A}}$, has been obtained for every numerical interface, the numerical fluxes are computed by solving the locally linear system. Roe's numerical flux is as follows

$$\hat{\mathbf{F}}^{\mathrm{ROE}} = \frac{1}{2}\left[\mathbf{F}(\mathbf{u}_\mathrm{L}) + \mathbf{F}(\mathbf{u}_\mathrm{R}) - \sum_p |\tilde{\lambda}^{(p)}| \tilde{\alpha}^{(p)} \tilde{\mathbf{r}}^{(p)}\right] \tag{36}$$

$$\tilde{\alpha}^{(p)} = \tilde{\mathbf{l}}^{(p)}(\mathbf{u}_\mathrm{R} - \mathbf{u}_\mathrm{L}) \tag{37}$$

$\tilde{\lambda}^{(p)}$, $\tilde{\mathbf{r}}^{(p)}$, $\tilde{\mathbf{l}}^{(p)}$ being the eigenvalues and right and left eigenvectors of $\tilde{\mathcal{A}}$, respectively (p extends from 1 to the number of equations of the system).

Eulderink has obtained the Roe linearization for the relativistic system of equations in a general spacetime in terms of the average state

$$\tilde{\mathbf{w}} = \frac{\mathbf{w}_\mathrm{L} + \mathbf{w}_\mathrm{R}}{k_\mathrm{L} + k_\mathrm{R}}, \tag{38}$$

$$\mathbf{w} = (ku^0, ku^1, ku^2, ku^3, k\frac{p}{\rho h}), \tag{39}$$

$$k^2 = \sqrt{-g}\rho h, \tag{40}$$

where g stands for the determinant of the metric tensor, $g_{\mu\nu}$. Note that the role played by ρ in the Cartesian non relativistic Roe solver as weight for averaging, is now played by k, which apart from geometrical factors, tends to ρ in the non relativistic limit. A Riemann solver for special relativistic flows and the generalization of the Roe solver for the Euler equations in arbitrary coordinate systems are easily deduced from Eulderink's work.

The results obtained by Eulderink & Mellema (1995) in 1D test problems for ultrarelativistic flows (up to Lorentz factors 625) in the presence of strong discontinuities and large gravitational background fields demonstrate the high performance of the Eulderink-Roe solver.

Finally, let us note that on relaxing condition 3 above, the solver is no longer exact for shocks but still produces accurate solutions. The advantage is that the reduced set of conditions is fulfilled by a large number of averages. A typical choice is the arithmetic mean. The 1D general relativistic hydro code developed by Romero et al. (1995) uses the flux formula (36) with such an average. It has succesfully passed a long series of tests including the spherical version of the relativistic shock reflection shown in Section 4.1.

5.3. RELATIVISTIC HLL METHOD

Schneider et al. (1993) have proposed the use of the method of Harten, Lax & van Leer (1983) (HLL here on) to integrate the equations of special RFD. This method avoids the explicit calculation of the eigenvalues and eigenvectors of the Jacobian matrices but is based on an over–simplified approximate solution of the original Riemann problems. In Roe's method, the exact solution of the Riemann problem, which consists of four constant states separated by elementary waves, is replaced by four constant states separated by linear discontinuities. The HLL Riemannn solver contains only three constant states and the approximate Riemann solution is

$$\mathbf{u}^{\mathrm{HLL}}(x/t; \mathbf{u}_{\mathrm{L}}, \mathbf{u}_{\mathrm{R}}) = \begin{cases} \mathbf{u}_{\mathrm{L}} & \text{for } x < a_{\mathrm{L}}t \\ \mathbf{u}_{*} & \text{for } a_{\mathrm{L}}t \leq x \leq a_{\mathrm{R}}t \\ \mathbf{u}_{\mathrm{R}} & \text{for } x > a_{\mathrm{R}}t \end{cases} \tag{41}$$

where a_{L} and a_{R} are, respectively, lower an upper bounds for the smallest and largest signal velocities. The intermediate state $\mathbf{u}_{*}$ is determined by requiring consistency of the approximate Riemann solution with the integral conservation equations over a grid zone. This gives

$$\mathbf{u}_{*} = \frac{a_{\mathrm{R}}\mathbf{u}_{\mathrm{R}} - a_{\mathrm{L}}\mathbf{u}_{\mathrm{L}} - \mathbf{F}(\mathbf{u}_{\mathrm{R}}) + \mathbf{F}(\mathbf{u}_{\mathrm{L}})}{a_{\mathrm{R}} - a_{\mathrm{L}}} \tag{42}$$

and for the numerical flux

$$\widehat{\mathbf{F}}^{\mathrm{HLL}} = \frac{a_{\mathrm{R}}^{+}\mathbf{F}(\mathbf{u}_{\mathrm{L}}) - a_{\mathrm{L}}^{-}\mathbf{F}(\mathbf{u}_{\mathrm{R}}) + a_{\mathrm{R}}^{+}a_{\mathrm{L}}^{-}(\mathbf{u}_{\mathrm{R}} - \mathbf{u}_{\mathrm{L}})}{a_{\mathrm{R}}^{+} - a_{\mathrm{L}}^{-}} \tag{43}$$

where

$$a_{\mathrm{L}}^{-} = \min\{0, a_{\mathrm{L}}\}, \quad a_{\mathrm{R}}^{+} = \max\{0, a_{\mathrm{R}}\}. \tag{44}$$

An essential ingredient of the scheme are good estimates for the smallest and largest signal velocities. In the non-relativistic case, Einfeldt (1988) proposed calculating them based on the smallest and largest eigenvalues of the Roe matrix. This HLL scheme with Einfeldt's recipe is a very robust upwind scheme for the Euler equations and possesses the property of being positively conservative. The method is exact for single shocks but its main drawback is, probably, that it is very dissipative, specially on contact discontinuities.

Schneider et al. (1993) have presented results in 1D ultrarelativistic hydrodynamics (see also Table 1) using a version of the HLL method with the signal velocities calculated according by

$$a_{\rm R} = (\bar{v} + \bar{c}_s)/(1 + \bar{v}\bar{c}_s) \tag{45}$$

$$a_{\rm L} = (\bar{v} - \bar{c}_s)/(1 - \bar{v}\bar{c}_s) \tag{46}$$

where c_s is the relativistic sound speed, and where the bar denotes the arithmetic mean between the initial left and right states. Duncan & Hughes (1994) have further generalized this method to 2D special RFD and applied it to the simulation of relativistic extragalactic jets, but they have not considered the coupling of the tangential flow velocities in the estimation of the signal speeds.

5.4. MARQUINA'S FLUX FORMULA

In a recent paper, Donat & Marquina (1995) have extended a numerical flux formula to systems which was first proposed by Shu & Osher (1989) for scalar equations. Donat & Marquina's work is motivated by the search for a robust and accurate approximate Riemann solver that avoids the common failures suffered by many Riemann solvers (see Quirk 1994 for a catalogue of these failures).

In the scalar case and for characteristic wavespeeds which do not change sign at the given numerical interface, Marquina's flux formula is identical to Roe's flux. Otherwise, the scheme switches to the more viscous, entropy satisfying local-Lax-Friedrichs scheme (see Shu & Osher 1989 for details), in order to avoid some relevant numerical pathologies. In the case of conservation law systems, the combination of Roe and local-Lax-Friedrichs solvers is carried out in each characteristic field after the local linearization and decoupling of the system of equations (see Donat & Marquina 1995). However, contrary to Roe's and other linearized methods, the extension of Marquina's method to systems is not based in any averaged intermediate state.

Martí et al. (1995a,b) have used this method in their simulations of relativistic jets. The resulting numerical code has been successfully used to describe ultrarelativistic flows in both one and two spatial dimensions with great accuracy (a large set of test calculations using Marquina's Riemann solver can be found in Appendix II of Martí et al. 1995b). Besides, the numerical experimentation in two dimensions confirms that the dissipation of the scheme is sufficient to eliminate the carbuncle phenomenon (see Quirk 1994), appearing in high Mach number relativistic jet simulations when using other standard solvers.

6. An SR application: Simulation of extragalactic relativistic jets

It is generally assumed that the jets in extragalactic radio sources originate from the vicinity of a large (10^7 to 10^9 solar masses), rotating black hole. Such compact objects are the most efficient engines known for converting gravitational potential energy from the surrounding accretion disk into bulk outflow and radiation. The outflow, collimated in the first parsecs to form a pair of twin supersonic beams of relativistic electron-positron (maybe electron-proton) plasma, propagate through the interstellar medium and, eventually, the intergalactic medium, up to distances of hundreds of kiloparsecs, being observable at radio frequencies due to the emission of synchrotron radiation from the charged particles spiralling in the embedded magnetic field. The interaction of beams with the surrounding medium produces the hot spots (sites of enhanced emission) observed at the end of the beams in the most powerful radio galaxies and quasars.

Numerical simulations of jets within the fluid approximation have become an important tool in the understanding of the morphology and dynamics of extragalactic jets since Norman et al. (1982) verified the jet model by Blandford & Rees (1974) for double radio sources, using hydrodynamical simulations of axisymmetric supersonic flows. Nowadays, these numerical simulations are succesfully used to explain the structures observed in many large-scale radio jets. However, despite the fact that within the currently accepted standard model (Blandford & Königl 1979) the observation of apparent superluminal motions in parsec scale extragalactic jets and the asymmetries in the radio flux emitted by the twin jets of powerful radio sources are explained by assuming relativistic bulk flow velocities within the beams, the numerical simulations have been traditionaly performed in the framework of classical fluid dynamics and only with the advent of hydrodynamical codes based on high-resolution shock-capturing techniques the simulation of unsteady relativistic jets has become a feasible task.

In Martí et al. (1995a,b) we have developed and used a code to solve the special RFD equations for the conserved relativistic densities of rest-mass, D, momentum, $\mathbf{S}$, and energy, τ, with the purpose of studying the morphology, dynamics and propagation properties of extragalactic relativistic jets. In our calculations, the flow is supposed to be axisymmetric and hence we use two-dimensional cylindrical coordinates (r,z). Expressed in this coordinate system and neglecting viscosity and thermal conduction, the relativistic hydrodynamical equations are

$$\frac{\partial D}{\partial t} + \frac{1}{r}\frac{\partial r D v^r}{\partial r} + \frac{\partial D v^z}{\partial z} = 0 \tag{47}$$

$$\frac{\partial S^r}{\partial t} + \frac{1}{r}\frac{\partial r(S^r v^r + p)}{\partial r} + \frac{\partial S^r v^z}{\partial z} = \frac{p}{r} \tag{48}$$

$$\frac{\partial S^z}{\partial t} + \frac{1}{r}\frac{\partial r S^z v^r}{\partial r} + \frac{\partial (S^z v^z + p)}{\partial z} = 0 \tag{49}$$

$$\frac{\partial \tau}{\partial t} + \frac{1}{r}\frac{\partial r(S^r - D v^r)}{\partial r} + \frac{\partial (S^z - D v^z)}{\partial z} = 0 \tag{50}$$

or, in vector notation,

$$\frac{\partial \mathbf{u}}{\partial t} + \frac{1}{r}\frac{\partial r \mathbf{F}^r}{\partial r} + \frac{\partial \mathbf{F}^z}{\partial z} = \mathbf{S} \tag{51}$$

with the definitions of the vector of unknowns, $\mathbf{u}$,

$$\mathbf{u} = (D, S^r, S^z, \tau)^{\mathrm{T}} \tag{52}$$

fluxes,

$$\mathbf{F}^r = (D v^r, S^r v^r + p, S^z v^r, S^r - D v^r)^{\mathrm{T}} \tag{53}$$

$$\mathbf{F}^z = (D v^z, S^r v^z, S^z v^z + p, S^z - D v^z)^{\mathrm{T}} \tag{54}$$

and sources

$$\mathbf{S} = (0, \frac{p}{r}, 0, 0)^{\mathrm{T}}. \tag{55}$$

An equation of state has to be supplied to close the system of equations (51). In our calculations we have assumed an ideal gas equation of state

$$p = (\gamma - 1)\rho\varepsilon \tag{56}$$

where $4/3 \leq \gamma \leq 5/3$ is the (constant) adiabatic index, for both the jet and the external medium fluids.

In our calculations, the system of equations (12) is solved on a discrete numerical grid and the time variation of quantities $\mathbf{u}$ within the numerical cells is calculated by considering the fluxes across the numerical interfaces. In the form of a method of lines, our discretization reads

$$\frac{d\mathbf{u}_{i,j}}{dt} = -\frac{1}{r_i \Delta r}[r_{i+\frac{1}{2}}\hat{\mathbf{F}}^r_{i+\frac{1}{2},j} - r_{i-\frac{1}{2}}\hat{\mathbf{F}}^r_{i-\frac{1}{2},j}] - \frac{1}{\Delta z}[\hat{\mathbf{F}}^z_{i,j+\frac{1}{2}} - \hat{\mathbf{F}}^z_{i,j-\frac{1}{2}}] + \mathbf{S}_{i,j} \tag{57}$$

where subscripts i, j are related respectively to the r- and z-discretizations and refer to mean values in the corresponding two-dimensional cell. Quantities $\hat{\mathbf{F}}^r_{i+\frac{1}{2},j}$, $\hat{\mathbf{F}}^z_{i,j+\frac{1}{2}}$ are the numerical fluxes at the cell interfaces and are calculated using Marquina's flux formula on the basis of the spectral decomposition of the Jacobian matrices of the RFD system of equations shown in Section 2. The spatial accuracy of our algorithm is improved up to second order by means of the PPM reconstruction technique described in Section 4. Integration in time is done simultaneously in both spatial directions using a total-variation-diminishing Runge-Kutta scheme of high order (Shu & Osher 1988). Finally, at each time step a one-dimensional Newton-Raphson iteration is used to compute the primitive variables $\{\rho, \varepsilon, p, v^r, v^z\}$ from the conserved ones (see Martí & Müller 1995 for details).

Numerical studies of classical jets have been performed by varying two fundamental parameters: the internal beam Mach number, M_b, and the beam to external medium density ratio, η. In the relativistic case the important parameters are the internal to rest-mass beam energy density ratio (i.e., the specific internal energy of the beam gas, ε_b) and the beam flow velocity, v_b, which control the strength of the relativistic effects. The initial set of parameters is completed with η, the beam to external medium pressure ratio, K, and the adiabatic exponent of the ideal gas law.

Figure 7 shows a snapshot in the evolution of a relativistic jet characterised by the following set of parameters: $\eta = 0.01$, $K = 1$, $\gamma = 5/3$, $v_b = 0.999c$, $\varepsilon_b = 0.61c^2$. The gross morphological elements typical of non-relativistic calculations (see, e.g., Norman et al. 1982) are clearly identified. A central beam (the radio jet) powered by the continuous injection of gas from the left boundary is found to propagate supersonically through the external medium. The beam plasma is thermalized and decelerated in a terminal shock at the end of the beam forming a high pressure region (the hot spot). The deflected beam material forms an extended cocoon (the radio lobe) in a process of mixing with the (colder and denser) ambient material. Finally, a bow shock in front of the jet appears due to the supersonic propagation of the jet. Beam and ambient materials are separated by a contact discontinuity clearly seen in the rest-mass density and specific internal energy frames. The motion of the fluid elements can be followed in the last frame of Fig. 7 showing the modulus and direction of the fluid velocity vectors. The beam is characterized by an almost constant and parallel flow at the initial speed, only perturbed by the presence of conical shocks (see, e.g., the thermal pressure frame) associated with the radio knots observed in the jets of the most powerful radio galaxies and quasars. The beam flow velocity is drastically reduced after the terminal shock where a fraction of the beam plasma is redirected towards the source.

In Martí et al. (1995b) we have considered a large sample of relativistic

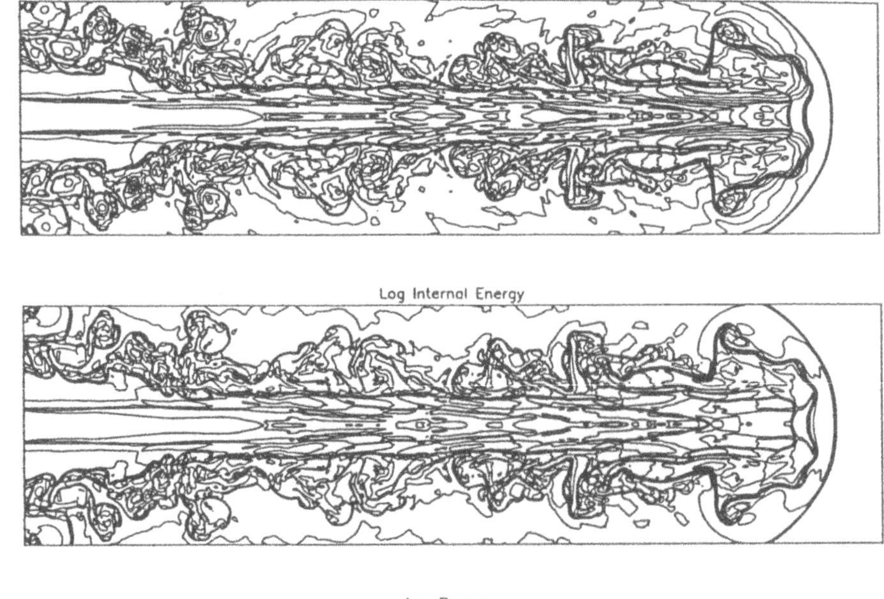

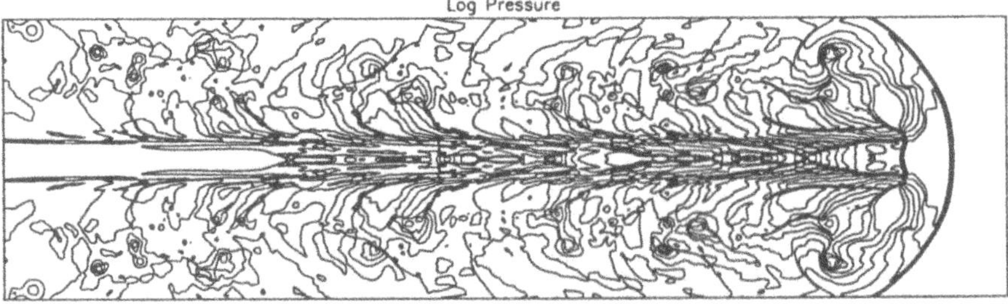

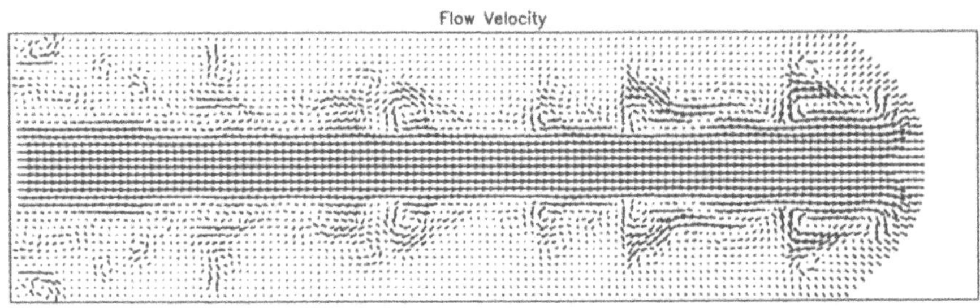

Figure 7. Contour plots of the logarithm of the proper rest-mass density (top panel), specific internal energy (second from top) and thermal pressure (third) of the relativistic jet model discussed in the text. Thirty contours span linearly the interval between minimum and maximum values: rest-mass density: $[-3.17, 0.68]$, specific internal energy: $[-2.19, 1.15]$, thermal pressure: $[-3.48, -0.20]$. The corresponding flow pattern is shown at the bottom panel.

jet models covering a wide range of the parameter space. Our results show that relativistic jets are significantly more stable than Newtonian ones due to the enhancement of the effective inertial mass of the beam (through the presence of the specific enthalpy and the Lorentz factor in the momentum density). Relativistic jets are found to propagate very efficiently at speeds which are consistent with the estimate based on the one–dimensional momentum balance. The propagation speed for the model considered here is $0.76c$, a 98% of the expected velocity from the 1D estimate. Moreover, the propagation efficiency of relativistic jets increases with the beam flow velocity and the internal beam energy and, as a consequence, the cocoons of relativistic jets are in general less prominent than those of Newtonian jets. Models with higher internal beam energies tend to have more stable beams without internal shocks and reduced cocoons. Colder models (i.e., $\varepsilon_b \ll c^2$) have extended cocoons and beams with prominent internal structure. The model considered in Fig. 7 is intermediate between hot and cold jets.

7. Concluding remarks

The numerical experience accumulated in recent years allows us to assert that the use of high–resolution methods based on Riemann solvers is the most accurate way to describe ultrarelativistic flows. However, more work is needed to attain the same level of development in relativistic codes as that achieved in classical computational fluid dynamics. For example, relativistic Riemann solvers not based in local linearizations (e.g., Martí & Müller 1994, Balsara 1994) have only been tested in 1D problems. On the other hand, a major part of the (theoretical and numerical) work has been performed in special RFD, whereas most problems of interest involve non–negligible, dynamical (i.e., time–dependent) gravitational fields. In this respect, although the works of Eulderink (1993) and Banyuls et al. (1995) on the spectral decomposition of the general RFD system of equations set the theoretical background to extend high–resolution methods to general relativistic problems (see, e.g., Romero et al. 1995), the performance of the numerical procedures has again only been tested in flows with particular symmetries.

The author acknowledges the careful reading of parts of this manuscript by J.Mª Ibáñez, A. Marquina and J.A. Miralles and also wants to thank his collaborators E. Müller, A. Marquina, J.Mª Ibáñez, J.A. Font, J.V. Romero, F. Banyuls and J.A. Miralles, with whom most of the theoretical and computional work used for this talk was done. The author is also very grateful to Prof. E.F. Toro for economical support. This work was carried out under the HCM Program of the European Union (contract ERBCH-BICT930496).

References

Anile, A.M. (1989) *Relativistic Fluids and Magnetofluids*. Cambridge University Press, Cambridge.

Balsara, D.S. (1994), *J. Comput. Phys.*, **114**, 284

Banyuls, F., Font, J.A., Ibáñez, J.Mª, Martí, J.Mª and Miralles, J.A. (1995), submitted

Blandford, R.D. and Königl, A. (1979), *Astrophys. J.*, **232**, 34

Blandford, R.D. and Rees, M.J. (1974), *Mo. Not. Royal Astron. Soc.*, **169**, 395

Centrella, J. and Wilson, J.R. (1984), *Astrophys. J. Suppl. Ser.*, **54**, 229

Colella, P. (1982), *SIAM J. Sci. Stat. Comput.*, **3**, 76

Colella, P. and Glaz, H.M. (1985), *J. Comput. Phys.*, **59**, 264

Colella, P. and Woodward, P.R. (1984), *J. Comput. Phys.*, **54**, 174

Courant, R. and Friedrichs, K.O. (1976) *Supersonic Flows and Shock Waves*. Springer, Berlin.

Donat, R. and Marquina, A. (1995), *J. Comput. Phys.*, in press

Duncan, G.C. and Hughes, P.A. (1994), *Astrophys. J.*, **436**, L119

Einfeldt, B. (1988), *SIAM J. Numer. Anal.*, **25**, 294

Eulderink, F. (1993), Ph.D. thesis, Rijksuniverteit te Leiden.

Eulderink, F. and Mellema, G. (1995), *Astron. Astrophys.*, in press

Font, J.A., Ibáñez, J.Mª, Marquina, A. and Martí, J.Mª (1994), *Astron. Astrophys.*, **282** 304

Godunov, S.K. (1959), *Mat. Sb.*, **47**, 271

Harten, A., Lax, P.D. and van Leer, B. (1983), *SIAM Rev.*, **25**, 35

Hawley, J.F., Smarr, L.L. and Wilson, J.R. (1984), *Astrophys. J. Suppl. Ser.*, **55**, 211

Landau, L.D. and Lifshitz, E.M. (1987) *Fluid Mechanics*. Pergamon, New York.

Marquina, A., Martí, J.Mª, Ibáñez, J.Mª, Miralles, J.A. and Donat, R. (1992), *Astron. Astrophys.*, **258**, 566

Martí, J.Mª, Ibáñez, J.Mª and Miralles, J.A. (1991), *Phys. Rev.*, **D43**, 3794

Martí, J.Mª and Müller, E. (1994), *J. Fluid Mech.*, **258**, 317

Martí, J.Mª and Müller, E. (1995), *J. Comput. Phys.*, in press

Martí, J.Mª, Müller, E., Font, J.A. and Ibáñez, J.Mª (1995a), *Astrophys. J.*, **448**, L105

Martí, J.Mª, Müller, E., Font, J.A., Ibáñez, J.Mª and Marquina A. (1995b), *Astrophys. J.*, submitted

McAbee, T.L., Wilson, J.R., Zingman, J.A. and Alonso, C.T. (1989), *Mod. Phys. Lett. A*, **4**, 983

Norman, M.L., Smarr, L., Winkler, K.-H.A. and Smith, M.D. (1982), *Astron. Astrophys.*, **113**, 285

Norman, M.L. and Winkler, K.-H.A. (1986), in *Astrophysical Radiation Hydrodynamics*, Norman, M.L. and Winkler, K.-H.A., eds. Reidel, Dordrecht.

Quirk, J. (1994), *Intl. J. Meth. Fluids*, **18**, 555

Roe, P. (1981), *J. Comput. Phys.*, **43**, 357

Romero, J.V., Ibáñez, J.Mª, Martí, J.Mª and Miralles, J.A. (1995), *Astrophys. J.*, in press

Schneider, V., Katscher, U., Rischke, D.H., Waldhauser, B., Maruhn, J.A. and Munz, C.-D. (1993), *J. Comput. Phys.*, **105**, 92

Shu, C.W. and Osher, S.J. (1988), *J. Comput. Phys.*, **77**, 439

Shu, C.W. and Osher, S.J. (1989), *J. Comput. Phys.*, **83**, 32

Sod, G.A. (1978), *J. Comput. Phys.*, **27**, 1

Taub, A.H. (1948), *Phys. Rev.*, **74**, 328

Thompson, K.W. (1986), *J. Fluid Mech.*, **171**, 365

von Neumann, J. and Richtmyer, R.D. (1950), *J. Appl. Phys.*, **21**, 232

Wilson, J.R. (1972), *Astrophys. J.*, **173**, 431

Wilson, J.R. (1979) in *Sources of Gravitational Radiation*, Smarr, L.L., ed. Cambridge University Press, Cambridge.

Woodward, P.R. and Colella, P. (1984), *J. Comput. Phys.*, **54**, 115

PRIMITIVE, CONSERVATIVE AND ADAPTIVE SCHEMES FOR HYPERBOLIC CONSERVATION LAWS

E. F. TORO
Department of Computing and Mathematics,
Manchester Metropolitan University,
Chester Street, Manchester, M1 5GD, U.K.
Emails: E.F.Toro@doc.mmu.ac.uk (work)
Lostoros@aol.com (home)
Webpage: www.doc.mmu.ac.uk/STAFF/E.F.Toro/index.html

1. Introduction

For the last two or three decades, it has become an accepted practice to utilise conservative methods when solving numerically hyperbolic conservation laws. Shock waves are the solution features that demand the utilisation of conservative methods. Practical computational experience shows that the use of a non–conservative method results in the wrong shock strength and thus the wrong propagation speed. To some extent an exception is the Random Choice Method (RCM) of Glimm (Glimm, 1965). This method is non–conservative and yet it gives the correct shock strength and, on average the correct propagation speed. Theoretically, Lax and Wendroff proved in 1960 (Lax and Wendroff, 1960) that if a conservative method converges, it does so to a weak solution of the conservation laws. Today it is known, (Harten, 1983), that if the scheme also satisfies an entropy condition, then the converged solution is the physical weak solution. Hence, there are very good reasons for utilising conservative methods.

Some very recent work of Hou and Le Floch (Hou and LeFloch, 1994) concerns the analysis of some non–conservative methods for non–linear scalar conservation laws. Two of their results stand out. First, they proved

E.F. Toro and J.F. Clarke (eds.), Numerical Methods for Wave Propagation, 323–385.

that their non–conservative methods converge and that they converge to a solution that corresponds to a new conservation law with a source term. This source term is generally non–zero in the presence of shock waves but vanishes for smooth flow, in which case the non–conservative schemes do converge to the correct solution. The second result that is relevant to the themes of this paper is this, if their non–conservative schemes are locally corrected by utilising a conservative method, then the hybrid scheme is proved to converge to the correct solution. The local correction is applied in the vicinity of shock waves.

For a few years now, it has been known that conservative methods produce erroneous solutions when applied to problems involving material interfaces. Some initial observations on this phenomenon were reported by Clarke and colleagues (Clarke *et. al.*, 1993), who were concerned with interfaces between reactive and non–reactive high–energy solid materials. Later, Karni (Karni, 1994) showed that the Godunov first–order upwind method with the Roe Riemann solver produced a spurious pressure perturbation in the vicinity of the interface. She also showed that her non–conservative method (Karni, 1992) avoided the difficulty. An extension of this work was reported in (Toro, 1994), (Toro, 1995d). In these works it is shown that there are two classes of initial value problems (IVPs) for the compressible Euler equations for which a range of conservative methods give erroneous solutions. One class of problems is that of material interfaces, which also includes the case of multicomponent flows. The other class of IVPs concerns shear waves in the two and three dimensional Euler equations. These results are quite general and are reported in full detail in (Toro, 1997a). In (Toro, 1994) and (Toro, 1995c) it is shown that a range of non–conservative methods avoid these difficulties. There is a third class of problems for which conservative methods also give incorrect solutions. These are symmetric data Riemann problems and occur natural at solid boundaries in one and multidimensional computations. Here it is observed that conservative schemes give a spurious density dip in the vicinity of the trivial contact. Primitive schemes, while not avoiding the problem altogether, give significantly better results.

There are therefore compelling reasons for seriously considering the task of constructing good non–conservative schemes, which we shall also call primitive–variable schemes or simply primitive schemes. Of course there is still the issue of shock waves, but we shall assume that these will be treated by switching to a conservative scheme, locally, in an adaptive fashion. For systems in which there are no shock waves or these are very weak, then primitive methods may be adequate. Examples are Acoustics, Potential Flow and the Artificial Compressibility Equations arising from the steady incompressible Navier–Stokes equations (Chorin, 1967). As a matter of fact, in this last example, the equations have traditionally been expressed in conservative form, although such formulation is physically meaningless, as the *conserved* variables are actually the primitive variables pressure and velocity components. There are non–linear hyperbolic systems for which the conservative form of the equations is unknown. In this case there is no option but to apply primitive methods. For linear systems with constant coefficients, primitive schemes are perfectly adequate; in fact for this example, the primitive schemes presented in this paper are entirely equivalent to conservative methods.

This paper is, in the main, about the construction of numerical methods based on non–conservative formulations of hyperbolic conservation laws. The methods are explicit and Riemann–problem based upwind. Schemes of first, second and third order of accuracy are constructed. Non–linear version of the schemes are derived by imposing Total Variation Diminishing (TVD) criteria. A finite difference formula that is analogous to the well–known explicit conservative formula is first put forward. This requires the specification of a matrix point value and intermediate states that are akin to intercell fluxes in explicit conservative methods. Four methodologies are then presented for computing these two items of information. These methodologies were first presented for constructing conservative methods. The first approach presented is the Weighted Average Flux (WAF) method of Toro (Toro, 1989). The other three methods are based on the MUSCL data reconstruction approach, first proposed by van Leer (van Leer, 1976) for constructing high–order extensions of the (conservative) Godunov method. The three approaches are the MUSCL–Hancock (van Leer, 1984); the Generalised Riemann Problem (GRP) approach of Ben–Artzi and Falcovitz (Ben-Artzi and Falcovitz, 1984) and the Piece–wise Linear (PLM) method of Colella

(Colella, 1985). The four approaches presented here contain a first–order monotone scheme that is the non–conservative analogue of the conservative Godunov first–order upwind method. Conservative versions of these four classes of schemes are constructed in a straightforward manner, namely by evaluating the physical flux vector at the intermediate states of the respective primitive scheme. Hybrid, adaptive methods are then constructed by switching from the primitive scheme to the corresponding conservative scheme in the cells crossed by a shock wave. Numerical experiments suggest that the hybrid schemes converge to the correct solution. It is therefore desirable to theoretically study the convergence properties of the proposed schemes; this is the subject of current research.

The rest of this paper is organised as follows. In Sect. 2 we recall some well–known hyperbolic systems of conservation laws and state their primitive form along with their eigenstructure. In Sect. 3 we recall the well–known explicit conservative formula and state a non–conservative analogue, which is the basis for constructing our primitive schemes. Both formulae become identical in the case of a linear hyperbolic system with constant coefficients. Riemann solvers that are suitable for primitive methods are reviewed in Sect. 4. In Sect. 5 we briefly review primitive methods based on the WAF approach. In Sect. 6 we construct MUSCL–Hancock type primitive schemes. In Sect. 7 we construct GRP–type primitive schemes while in Sect. 8 we present primitive schemes of the PLM type. In Sect. 9 we discuss TVD versions of the MUSCL schemes. Adaptive primitive–conservative schemes are presented in Sect. 10 while in Sect. 11 we present numerical results. Conclusions are given in Sect. 12.

2. Equations and their Eigenstructure

The equations of interest here may be written in differential conservation law form as

$$\mathbf{U}_t + \mathbf{F}(\mathbf{U})_x + \mathbf{G}(\mathbf{U})_y + \mathbf{H}(\mathbf{U})_z = \mathbf{0} \,, \tag{1}$$

where $\mathbf{U}$ is the vector of conserved variables and $\mathbf{F}(\mathbf{U})$, $\mathbf{G}(\mathbf{U})$ and $\mathbf{H}(\mathbf{U})$ are vectors of fluxes. Subscripts denote partial differentiation.

2.1. SCALAR EXAMPLES

A simple example is the Partial Differential Equation (PDE)

$$q_t + (aq)_x + (bq)_y + (cq)_z = 0 \,, \tag{2}$$

where $q = q(x, y, z, t)$ is a scalar quantity, such as the concentration of a chemical species, and $\mathbf{V} = \mathbf{V}(x, y, z, t) \equiv [a, b, c]$ is a velocity vector. Assuming differentiability in (2) we expand derivatives to obtain the non–conservative form of the PDE

$$q_t + aq_x + bq_y + cq_z = -q \, \nabla \cdot \mathbf{V} \;, \tag{3}$$

where

$$\nabla \cdot \mathbf{V} \equiv \operatorname{div} \mathbf{V} \equiv \frac{\partial a}{\partial x} + \frac{\partial b}{\partial y} + \frac{\partial c}{\partial z} \tag{4}$$

is the divergence of the velocity vector $\mathbf{V}$. There are applications in which the flow is divergence free, that is $\operatorname{div} \mathbf{V} = 0$. In this case (3) becomes a homogeneous linear advection equation in non–conservative form. The simplest case of all is the one–dimensional PDE

$$q_t + aq_x = 0 \,, \tag{5}$$

in which the velocity component a is constant. This is the linear advection equation with constant coefficient and is the most popular model equation for constructing numerical methods for hyperbolic PDEs.

Another scalar example is the inviscid, one–dimensional Burgers equation, which written in conservation form reads

$$q_t + f(q)_x = 0 \,, \quad f(q) = (\frac{1}{2}q^2) \,. \tag{6}$$

Under the assumption of differentiability one obtains

$$q_t + qq_x = 0 \,, \tag{7}$$

which is the non–conservative, or primitive–variable, form of the inviscid Burgers equation. The next two examples are non–linear systems.

2.2. SHALLOW FLUIDS

An example of considerable practical interest are the time–dependent two–dimensional shallow water equations. With the appropriate interpretation, these equations apply to a variety of physical situations, flow of shallow water being the most obvious. When written in the form (1) the vectors of conserved variables and fluxes are

$$\left.\begin{array}{lcl} \mathbf{U} & = & [\phi, \phi u, \phi v]^T \;, \\ \mathbf{F}(\mathbf{U}) & = & [\phi u, \phi u^2 + \frac{1}{2}\phi^2, \phi uv]^T \;, \\ \mathbf{G}(\mathbf{U}) & = & [\phi v, \phi uv, \phi v^2 + \frac{1}{2}\phi^2]^T \;, \end{array}\right\} \tag{8}$$

where $[.]^T$ denotes the transpose of $[.]$. Here ϕ is the geopotential

$$\phi = gh \;, \tag{9}$$

where g is the acceleration due to gravity and h is the total depth of fluid. The velocity vector is $\mathbf{V} = [u, v]$. Equations (1), (8) represent the physical laws of conservation of mass and momentum in the x and y directions. For details on the derivation of the shallow water equations see (Stoker, 1992) and (Toro, 1997b).

A non–conservative formulation of equations (1), (8) requires, first of all, a choice of variables $\mathbf{W}$. The most immediate choice is the conserved variables themselves, i.e. $\mathbf{W} \equiv \mathbf{U}$. Assuming differentiability one then obtains

$$\mathbf{W}_t + \mathbf{A}(\mathbf{W})\mathbf{W}_x + \mathbf{B}(\mathbf{W})\mathbf{W}_y = \mathbf{0} \;, \tag{10}$$

where the coefficient matrices are, for this choice of variables, the Jacobian matrices

$$\mathbf{A}(\mathbf{W}) = \frac{\partial \mathbf{F}}{\partial \mathbf{U}} \;; \;\; \mathbf{B}(\mathbf{W}) = \frac{\partial \mathbf{G}}{\partial \mathbf{U}}$$

corresponding to the fluxes $\mathbf{F}(\mathbf{U})$ and $\mathbf{G}(\mathbf{U})$ in (8) respectively. They are given as

$$\mathbf{A}(\mathbf{W}) = \begin{bmatrix} 0 & 1 & 0 \\ a^2 - u^2 & 2u & 0 \\ -uv & v & u \end{bmatrix} \;; \;\; \mathbf{B}(\mathbf{W}) = \begin{bmatrix} 0 & 0 & 1 \\ -uv & v & u \\ a^2 - v^2 & 0 & 2v \end{bmatrix} . \tag{11}$$

The eigenvalues of $\mathbf{A}(\mathbf{W})$ are

$$\lambda_1 = u - a \ , \qquad \lambda_2 = u \ , \qquad \lambda_3 = u + a \ , \tag{12}$$

where a is the *celerity* given by

$$a = \sqrt{\phi} \ . \tag{13}$$

The corresponding right eigenvectors are

$$\mathbf{K}^{(1)} = \begin{bmatrix} 1 \\ u - a \\ v \end{bmatrix} , \quad \mathbf{K}^{(2)} = \begin{bmatrix} 0 \\ 0 \\ 1 \end{bmatrix} , \quad \mathbf{K}^{(3)} = \begin{bmatrix} 1 \\ u + a \\ v \end{bmatrix} , \tag{14}$$

for an appropriate choice of scaling factors.

A popular choice of non–conservative variables (or primitive variables) is

$$\mathbf{W} \equiv [\phi, u, v]^T \ . \tag{15}$$

Then, the non–conservative form of the equations reads as (10), with the coefficient matrices $\mathbf{A}(\mathbf{W})$ and $\mathbf{B}(\mathbf{W})$ given by

$$\mathbf{A}(\mathbf{W}) = \begin{bmatrix} u & \phi & 0 \\ 0 & u & 0 \\ 1 & 0 & u \end{bmatrix} , \quad \mathbf{B}(\mathbf{W}) = \begin{bmatrix} v & \phi & 0 \\ 0 & v & 0 \\ 1 & 0 & v \end{bmatrix} . \tag{16}$$

The eigenvalues of $\mathbf{A}(\mathbf{W})$ are

$$\lambda_1 = u - a \ , \qquad \lambda_2 = u \ , \qquad \lambda_3 = u + a \tag{17}$$

and the corresponding right eigenvectors are

$$\mathbf{K}^{(1)} = \begin{bmatrix} -a \\ 1 \\ 0 \end{bmatrix} , \quad \mathbf{K}^{(2)} = \begin{bmatrix} 0 \\ 0 \\ 1 \end{bmatrix} , \quad \mathbf{K}^{(3)} = \begin{bmatrix} a \\ 1 \\ 0 \end{bmatrix} , \tag{18}$$

for an appropriate choice of scaling factors.

2.3. THE EULER EQUATIONS

The Euler equations may be written in conservative form as (1) with

$$\left.\begin{array}{lcl} \mathbf{U} & = & [\rho, \rho u, \rho v, \rho w, E]^T \ , \\ \mathbf{F}(\mathbf{U}) & = & [\rho u, \rho u^2 + p, \rho uv, \rho uw, u(E+p)]^T \ , \\ \mathbf{G}(\mathbf{U}) & = & [\rho v, \rho uv, \rho v^2 + p, \rho vw, v(E+p)]^T \ , \\ \mathbf{H}(\mathbf{U}) & = & [\rho w, \rho uw, \rho vw, \rho w^2 + p, w(E+p)]^T \ . \end{array}\right\} \tag{19}$$

The choice $\mathbf{W} \equiv \mathbf{U}$, and assuming differentiability, leads to the three–dimensional non–conservative system

$$\mathbf{W}_t + \mathbf{A}(\mathbf{W})\mathbf{W}_x + \mathbf{B}(\mathbf{W})\mathbf{W}_y + \mathbf{C}(\mathbf{W})\mathbf{W}_z = \mathbf{0} \,. \tag{20}$$

The first coefficient matrix $\mathbf{A}(\mathbf{W})$ is, for this choice of variables, the Jacobian matrix corresponding to the flux $\mathbf{F}(\mathbf{U})$ in (19). Similarly, $\mathbf{B}(\mathbf{W})$ and $\mathbf{C}(\mathbf{W})$ are the Jacobian matrices corresponding to $\mathbf{G}(\mathbf{U})$ and $\mathbf{H}(\mathbf{U})$ in (19), respectively. The matrix $\mathbf{A}(\mathbf{W})$ is given by

$$\mathbf{A}(\mathbf{W}) = \frac{\partial \mathbf{F}}{\partial \mathbf{U}} = \begin{bmatrix} 0 & 1 & 0 & 0 & 0 \\ \hat{\gamma}H - u^2 - a^2 & (3-\gamma)u & -\hat{\gamma}v & -\hat{\gamma}w & \hat{\gamma} \\ -uv & v & u & 0 & 0 \\ -uw & w & 0 & u & 0 \\ \frac{1}{2}u[(\gamma-3)H - a^2] & H - \hat{\gamma}u^2 & -\hat{\gamma}uv & -\hat{\gamma}uw & \gamma u \end{bmatrix} . \tag{21}$$

Here H is the total specific enthalpy

$$H = (E+p)/\rho = \frac{1}{2}\mathbf{V}^2 + \frac{a^2}{(\gamma-1)} \,, \mathbf{V}^2 = u^2 + v^2 + w^2 \,, \; \hat{\gamma} = \gamma - 1 \,; \tag{22}$$

γ is the ratio of specific heats; a is the sound speed and for ideal gases is given by

$$a = \sqrt{\frac{\gamma p}{\rho}} \,; \tag{23}$$

E is the total energy per unit volume

$$E = \rho(\frac{1}{2}\mathbf{V}^2 + e) \,, \tag{24}$$

where e is the specific internal energy given by an equation of state $e = e(p, \rho)$. For ideal gases

$$e = e(p,\rho) = \frac{p}{(\gamma-1)\rho} \,. \tag{25}$$

The eigenvalues of $\mathbf{A}(\mathbf{W})$ given by (21) are

$$\lambda_1 = u - a \,, \; \lambda_2 = \lambda_3 = \lambda_4 = u \,, \; \lambda_5 = u + a \,. \tag{26}$$

The matrix $\mathbf{K}$ of corresponding right eigenvectors is

$$\mathbf{K} = \begin{bmatrix} 1 & 1 & 0 & 0 & 1 \\ u-a & u & 0 & 0 & u+a \\ v & v & 1 & 0 & v \\ w & w & 0 & 1 & w \\ H - ua & \frac{1}{2}\mathbf{V}^2 & v & w & H + ua \end{bmatrix} . \tag{27}$$

With the choice of non–conservative variables (or primitive variables)

$$\mathbf{W} = [\rho, u, v, w, p]^T \ , \tag{28}$$

the three–dimensional compressible Euler equations read

$$\left.\begin{array}{c} \rho_t + u\rho_x + v\rho_y + w\rho_z + \rho(u_x + v_y + w_z) = 0 \ , \\ u_t + uu_x + vu_y + wu_z + \frac{1}{\rho}p_x = 0 \ , \\ v_t + uv_x + vv_y + wv_z + \frac{1}{\rho}p_y = 0 \ , \\ w_t + uw_x + vw_y + ww_z + \frac{1}{\rho}p_z = 0 \ , \\ p_t + up_x + vp_y + wp_z + \rho a^2(u_x + v_y + w_z) = 0 \ . \end{array}\right\} \tag{29}$$

In quasi–linear form, these equations may be written as (20), with the coefficient matrices $\mathbf{A}(\mathbf{W})$, $\mathbf{B}(\mathbf{W})$ and $\mathbf{C}(\mathbf{W})$ given by

$$\mathbf{A}(\mathbf{W}) = \begin{bmatrix} u & \rho & 0 & 0 & 0 \\ 0 & u & 0 & 0 & 1/\rho \\ 0 & 0 & u & 0 & 0 \\ 0 & 0 & 0 & u & 0 \\ 0 & \rho a^2 & 0 & 0 & u \end{bmatrix} , \tag{30}$$

$$\mathbf{B}(\mathbf{W}) = \begin{bmatrix} v & \rho & 0 & 0 & 0 \\ 0 & v & 0 & 0 & 0 \\ 0 & 0 & v & 0 & 1/\rho \\ 0 & 0 & 0 & v & 0 \\ 0 & 0 & \rho a^2 & 0 & v \end{bmatrix} , \tag{31}$$

$$\mathbf{C}(\mathbf{W}) = \begin{bmatrix} w & \rho & 0 & 0 & 0 \\ 0 & w & 0 & 0 & 0 \\ 0 & 0 & w & 0 & 0 \\ 0 & 0 & 0 & w & 1/\rho \\ 0 & 0 & 0 & \rho a^2 & w \end{bmatrix} . \tag{32}$$

The eigenvalues of the coefficient matrix $\mathbf{A}(\mathbf{W})$ are given by

$$\lambda_1 = u - a \ , \ \lambda_2 = \lambda_3 = \lambda_4 = u \ , \ \lambda_5 = u + a \ . \tag{33}$$

The matrix $\mathbf{K}$ of corresponding right eigenvectors (as columns) is given by

$$\mathbf{K} = \begin{bmatrix} \rho & 1 & \rho & \rho & \rho \\ -a & 0 & 0 & 0 & a \\ 0 & v & 1 & v & 0 \\ 0 & w & w & 1 & 0 \\ \rho a^2 & 0 & 0 & 0 & \rho a^2 \end{bmatrix} . \tag{34}$$

3. Numerical Methods

The numerical methods of interest in this paper are based on dimensional splitting, or predictor corrector procedures, to deal with the multidimensional version of the equations. We therefore study the methods in the context of *augmented* one–dimensional conservation laws

$$\mathbf{U}_t + \mathbf{F}(\mathbf{U})_x = \mathbf{0} . \tag{35}$$

This augmented system consists of the usual conservation laws (e.g.. mass, momentum and energy) plus the conservative form of one or more advection equations. In non–conservative form we write equations (35) as

$$\mathbf{W}_t + \mathbf{A}(\mathbf{W})\mathbf{W}_x = \mathbf{0} . \tag{36}$$

For the purpose of this paper we shall select the one–dimensional Euler equations along with a single advection equation. In the conservative form (35) the vectors $\mathbf{U}$ and $\mathbf{F}(\mathbf{U})$ are

$$\left.\begin{array}{lcl} \mathbf{U} & = & [\rho, \rho u, E, \rho q]^T , \\ \mathbf{F}(\mathbf{U}) & = & [\rho u, \rho u^2 + p, u(E+p), \rho u q]^T , \end{array}\right\} \tag{37}$$

where q is a passive scalar advected with the fluid of speed u. In particular, q can be taken to be the y component of velocity in two–dimensional flow, in which case it enters the definition of total energy E in (24). An obvious choice of non–conservative variables is

$$\mathbf{W} = [\rho, u, p, q]^T . \tag{38}$$

There are several other possibilities; see (Toro, 1997b). For the choice (38) the coefficient matrix is

$$\mathbf{A}(\mathbf{W}) = \begin{bmatrix} u & \rho & 0 & 0 \\ 0 & u & \frac{1}{\rho} & 0 \\ 0 & \rho a^2 & u & 0 \\ 0 & 0 & 0 & u \end{bmatrix} . \tag{39}$$

Note the appropriate ordering of the corresponding eigenvalues.

Explicit conservative schemes to solve (35) read

$$\mathbf{U}_i^{n+1} = \mathbf{U}_i^n + \frac{\Delta t}{\Delta x}\left[\mathbf{F}_{i-\frac{1}{2}}^{n+\frac{1}{2}} - \mathbf{F}_{i+\frac{1}{2}}^{n+\frac{1}{2}}\right] , \tag{40}$$

where $\mathbf{U}_i^n$ is an integral average over the computing cell (or finite volume) $I_i \equiv [x_{i-\frac{1}{2}}, x_{i+\frac{1}{2}}]$ of length $\Delta x = x_{i+\frac{1}{2}} - x_{i-\frac{1}{2}}$, see Fig. 1. Given initial data $\mathbf{U}_i^n$ at time $t = t^n$ (or time level n) and having chosen the mesh dimensions Δx and Δt, in order to update the solution to a time $t^{n+1} = t^n + \Delta t$ (time level $n+1$) via (40) one needs to define the intercell numerical fluxes $\mathbf{F}_{i+\frac{1}{2}}^{n+\frac{1}{2}}$. A variety of ways of doing this exist, see for example (Hirsch, 1990) and (Toro, 1997b).

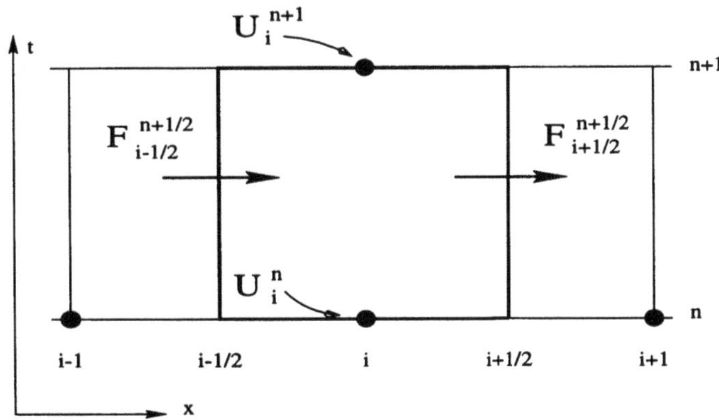

Figure 1. Stencil for conservative schemes. Average values $\mathbf{U}_i^n$ are updated to $\mathbf{U}_i^{n+1}$ using numerical fluxes $\mathbf{F}_{i+\frac{1}{2}}^{n+\frac{1}{2}}$

The main concern of this paper is the construction of non–conservative numerical schemes. To this end we propose explicit *finite difference* schemes

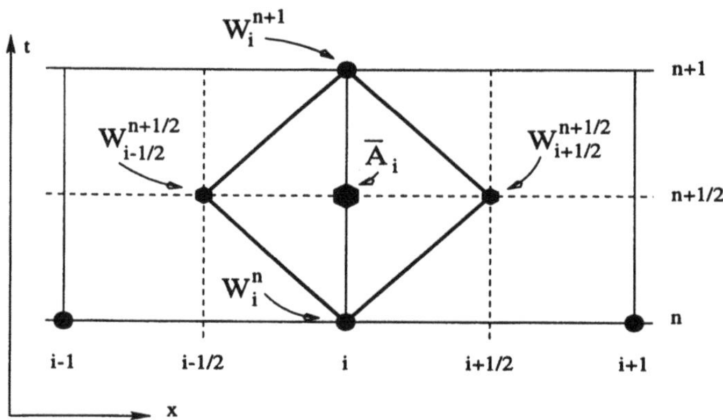

Figure 2. Stencil for primitive–variable schemes on a staggered grid. Point values $\mathbf{W}_i^n$ are updated to $\mathbf{W}_i^{n+1}$ using intermediate values $\mathbf{W}_{i-\frac{1}{2}}^{n+\frac{1}{2}}$, $\mathbf{W}_{i+\frac{1}{2}}^{n+\frac{1}{2}}$ and the average matrix $\overline{\mathbf{A}}_i$

for (36), which are of the form

$$\mathbf{W}_i^{n+1} = \mathbf{W}_i^n + \frac{\Delta t}{\Delta x}\overline{\mathbf{A}}_i \left[\mathbf{W}_{i-\frac{1}{2}}^{n+\frac{1}{2}} - \mathbf{W}_{i+\frac{1}{2}}^{n+\frac{1}{2}}\right] . \tag{41}$$

Here $\mathbf{W}_i^n$ is a point value at grid point i at time level n; $\overline{\mathbf{A}}_i$ in (41) is a value of the matrix $\mathbf{A}$ in (36) at the grid point i, and is yet to be defined; $\mathbf{W}_{i+\frac{1}{2}}^{n+\frac{1}{2}}$ is an intermediate value of the vector $\mathbf{W}$ at the half–time level $n+\frac{1}{2}$ at the staggered grid position $i+\frac{1}{2}$. See Fig. 2. One may view scheme (41) as a *staggered grid* scheme. The solution is sought at grid points i at time levels n and at intermediate values in space and time at $i+\frac{1}{2}, n+\frac{1}{2}$. We also require values for the coefficient matrix $\mathbf{A}(\mathbf{W})$, and these may be seen as defined at points $i, n+\frac{1}{2}$.

Note the analogy between the finite difference (non–conservative) scheme (41) and the finite volume (conservative) scheme (40), see Figs 1 and 2. There is a close relationship between the two classes of schemes. As a matter of fact, for the case of a linear system with constant coefficients, they are identical, provided the proper interpretation for the solution values is adopted. We shall exploit this similarity of the schemes for the purpose of converting a primitive scheme into a conservative scheme, which is an important step for constructing hybrid schemes, whereby the scheme to be used is adapted to local flow features, such as shocks.

The implementation of a primitive scheme (41) requires the definition of two items of information, namely, the matrix $\overline{\mathbf{A}}_i$ and the intermediate states $\mathbf{W}_{i+\frac{1}{2}}^{n+\frac{1}{2}}$. Various approaches are studied in subsequent sections. Three types of primitive variable schemes are of interest here, namely (i) monotone first–order upwind schemes; (ii) high–order extensions of these and (iii) TVD versions that combine the previous schemes. We shall utilise two approaches for construction the high–order schemes, namely the Weighted Average Flux (WAF) approach and the MUSCL approach. Both of these were first conceived for constructing conservative methods. Their extension for constructing non–conservative methods was initiated by the author and partial results were reported in (Toro, 1994), (Toro, 1995b) and (Toro, 1995c). Ivings *et. al.* (Ivings, 1996) have successfully applied some of these schemes to problems involving multi fluids. All the schemes of this paper utilise the solution of the Riemann problem, which is the subject of the next section.

4. Riemann Solvers

Both classes of schemes (40) and (41) of interest in this paper, may be represented as in Fig. 3, where at positions $x_{i-\frac{1}{2}}$ and $x_{i+\frac{1}{2}}$ there emerge systems of waves due to the interaction of the neighbouring pairs of states. For example, these pairs could be $(\mathbf{W}_{i-1}^n, \mathbf{W}_i^n)$ and $(\mathbf{W}_i^n, \mathbf{W}_{i+1}^n)$ respectively; in this case, the wave system at $x_{i+\frac{1}{2}}$ results from solving the Riemann problem for the pair $(\mathbf{W}_i^n, \mathbf{W}_{i+1}^n)$. We denote such Riemann problem by $RP(\mathbf{W}_i^n, \mathbf{W}_{i+1}^n)$.

4.1. THE RIEMANN PROBLEM

The Riemann problem for a system of PDEs is the initial value problem (IVP) for the original PDEs, such as (35), subject to the initial condition

$$\mathbf{W}(x,0) = \begin{cases} \mathbf{W}_i^n & \equiv \mathbf{W}_L \quad \text{if} \quad x < 0\,, \\ \mathbf{W}_{i+1}^n & \equiv \mathbf{W}_R \quad \text{if} \quad x > 0\,. \end{cases} \tag{42}$$

The solution is a similarity solution and is denoted by $\mathbf{W}_{i+\frac{1}{2}}^*(x/t)$; for system (35), (37) it has structure as depicted in Fig. 4. The variables x and t here are understood as local variables arising from the translation

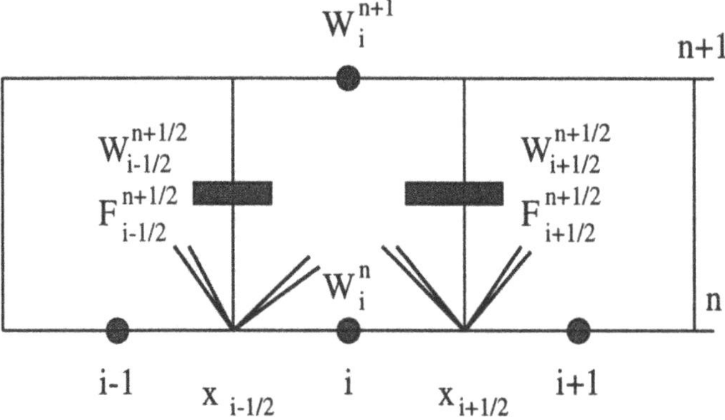

Figure 3. **Illustration of conservative scheme (40) and primitive scheme (41)**

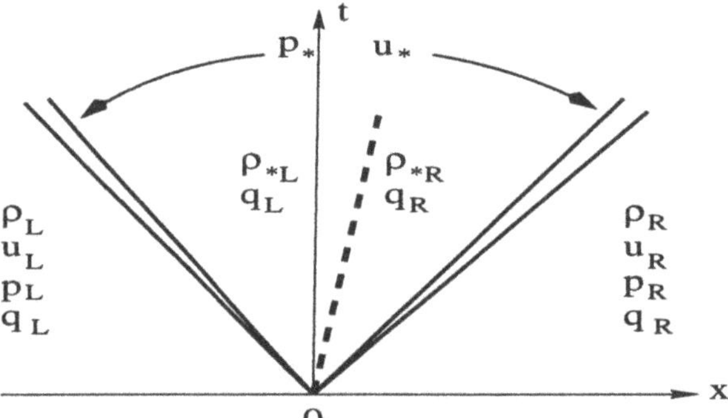

Figure 4. **Structure of the solution of the Riemann problem for the augmented one–dimensional time dependent Euler equations**

$x - x_{i+\frac{1}{2}} \rightarrow x$; $t - t^n \rightarrow t$. For an upto date and comprehensive study of Riemann solvers see (Toro, 1997b).

The solution of the Riemann problem for the augmented Euler equations (35), (37), as depicted in Fig. 4, consists of four wave families separating four constant states. There are two families of non–linear waves, shown in double lines. These are *associated* with the eigenvalues $\lambda_1 = u - a$ and $\lambda_3 = u + a$. The middle wave families correspond to the repeated eigenvalue $\lambda_2 = \lambda_4 = u$; these waves are linear in character, they are always discontinuous and are associated with contact discontinuities, shear waves and material interfaces. The pressure p_* and velocity u_* are constant in the region between the left and right waves, the *Star Region*. In general, p_* and velocity u_* change across the non–linear waves. Within the *Star Region*

ρ changes discontinuously from ρ_{*L} to ρ_{*R} and q changes discontinuously from q_L to q_R. The advected quantity q does not change across the non-linear waves; it only changes across the middle wave. This highlights how important it is to utilise the correct Riemann solver, if advected quantities are to be resolved sharply by the associated numerical method.

The Riemann solvers to be deployed with primitive variable schemes will not necessitate of the sophistication built into those solvers for strong shock waves. These can be expensive, they are unnecessary and moreover, they may be defective for some delicate flow features such as shear waves and interfaces. In general, Riemann solvers provide a state $\mathbf{W}^*_{i+\frac{1}{2}}$ or a flux $\mathbf{F}^*_{i+\frac{1}{2}}$. For primitive variable schemes we require an intermediate state $\mathbf{W}^{n+\frac{1}{2}}_{i+\frac{1}{2}}$. This rules out the use of popular Riemann solvers such as those of the HLL type (Harten *et. al.*, 1983) and Osher's Riemann Solver (Osher and Solomon, 1982), unless some *decoding* of fluxes is used to obtain corresponding values for the primitive variables.

In the next two sections we briefly review two approximate Riemann solvers that are suitable for the numerical methods developed in this paper.

4.2. THE PRIMITIVE–VARIABLE RIEMANN SOLVER

A very simple Riemann solver in terms of primitive variables, called PVRS, was proposed in (Toro, 1991). This results from linearising the equations in primitive variable form about an arithmetic mean of the initial data. The solution values in the *Star Region* are

$$\left.\begin{aligned} p_\star &= \tfrac{1}{2}(p_L + p_R) - \tfrac{1}{2}(u_R - u_L)C_1 \,, \\ u_\star &= \tfrac{1}{2}(u_L + u_R) - \tfrac{1}{2}(p_R - p_L)/C_1 \,, \\ \rho_{\star L} &= \rho_L + (u_L - u_\star)C_2 \,, \\ \rho_{\star R} &= \rho_R + (u_\star - u_R)C_2 \,, \\ q_{\star L} &= q_L \,, \\ q_{\star R} &= q_R \,, \end{aligned}\right\} \tag{43}$$

where the data dependent constants C_1 and C_2 are

$$C_1 = \frac{1}{4}(\rho_L + \rho_R)(a_L + a_R)\,, \quad C_2 = \frac{\rho_L + \rho_R}{a_L + a_R}\,. \tag{44}$$

For an isolated contact, PVRS gives the exact solution. Also the solution for $q^{\star}_{i+\frac{1}{2}}(x,t)$ in the general case, is exact, up to the approximation for $u_\star$. Some sophisticated Riemann solvers give erroneous solutions for this quantity.

4.3. THE TWO–SHOCK RIEMANN SOLVER

The Two–Shock Riemann Solver, called TSRS, was proposed by Toro (Toro, 1995a). This results from assuming a two–shock structure in the solution of the Riemann problem and assuming further that some of the functions involved may be evaluated at estimated values of the solution. The solution values in the *Star Region* are

$$\left.\begin{array}{lcl} p_\star & = & \frac{g_L(p_0)p_L+g_R(p_0)p_R-(u_R-u_L)}{g_L(p_0)+g_R(p_0)} , \\ u_\star & = & \frac{1}{2}(u_L+u_R)+\frac{1}{2}[(p_\star-p_R)g_R(p_0)-(p_\star-p_L)g_L(p_0)] , \\ \rho_{\star L} & = & \rho_L[\frac{p_\star/p_L+\Gamma}{\Gamma p_\star/p_L+1}] , \\ \rho_{\star R} & = & \rho_R[\frac{p_\star/p_R+\Gamma}{\Gamma p_\star/p_R+1}] , \\ q_{\star L} & = & q_L , \\ q_{\star R} & = & q_R . \end{array}\right\} \quad (45)$$

Here the functions $g_s(p)$, for $s = L, R$, are

$$g_s(p) = \sqrt{\frac{A_s}{p+B_s}} \quad (46)$$

and A_s, B_s, Γ are data dependent constants, namely

$$A_s = \frac{2}{(\gamma+1)\rho_s} \; ; \; B_s = (\frac{\gamma-1}{\gamma+1})p_s \; ; \; \Gamma = \frac{\gamma-1}{\gamma+1} . \quad (47)$$

As to p_0 in (45), this is an estimate for the solution for pressure and we use

$$p_0 = \max\ (0, p_{pvrs}) , \quad (48)$$

where p_{pvrs} is the solution for pressure in (43) given by the PVRS solver. This two shock Riemann solver is very efficient and exceedingly robust when compared with other popular Riemann solvers.

The previous Riemann solvers do not handle sonic flow conditions automatically. As proposed in (Toro, 1991), we recommend the use of the exact Riemann solver in such cases, as the solution is direct (no iteration) and is simple. The idea of using various Riemann solvers adaptively was also proposed and assessed in (Toro, 1991). Such scheme is utilised in (NAG, 1996) for providing wave speed estimates for HLL–type Riemann solvers. For details see Sect. 9.5 of Chapt. 9 in (Toro, 1997b). When the purpose is to implement adaptive primitive–conservative schemes, then the suggestion of utilising adaptive Riemann solvers is even more attractive. Broadly speaking, the requirements of primitive schemes on the Riemann solver are different from those of conservative schemes. See Sect. 10.

5. WAF–Type Primitive Variable Schemes

A WAF–type primitive variable scheme is reported in (Toro, 1997b). Here, we first briefly review the main features of that scheme as applied to systems. We then illustrate the WAF scheme for a general non–linear scalar PDE in non–conservative form.

5.1. REVIEW OF WAF SCHEMES FOR SYSTEMS

The schemes are based on the formula (41). We need to specify a procedure to compute the intermediate states $\mathbf{W}_{i+\frac{1}{2}}^{n+\frac{1}{2}}$ and the coefficient matrix $\overline{\mathbf{A}}_i$. First we define the intermediate states as weighted averages

$$\mathbf{W}_{i+\frac{1}{2}}^{n+\frac{1}{2}} = \frac{1}{\Delta x}\int_{-\frac{1}{2}\Delta x}^{\frac{1}{2}\Delta x} \mathbf{W}_{i+\frac{1}{2}}^{*}(x, \frac{1}{2}\Delta t)dx \; . \tag{49}$$

Here $\mathbf{W}_{i+\frac{1}{2}}^{*}(x,t)$ is the solution of the conventional Riemann problem with data $(\mathbf{W}_i^n, \mathbf{W}_{i+1}^n)$. An approximation to the integral is

$$\mathbf{W}_{i+\frac{1}{2}}^{n+\frac{1}{2}} = \sum_{k=1}^{N} \beta_k \mathbf{W}_{i+\frac{1}{2}}^{(k)} \; , \tag{50}$$

where N is the total number of waves in the solution of the Riemann problem ($N = 4$ in the present case). The weights or coefficients are given by

$$\left.\begin{aligned} &\beta_k = \tfrac{1}{2}(c_{i+\frac{1}{2}}^{(k)} - c_{i+\frac{1}{2}}^{(k-1)}) \; , \\ &c_{i+\frac{1}{2}}^{(k)} = \frac{\Delta t S_k}{\Delta x} \; , \quad c_{i+\frac{1}{2}}^{(0)} = -1 \; , \quad c_{i+\frac{1}{2}}^{(N+1)} = 1 \; , \end{aligned}\right\} \tag{51}$$

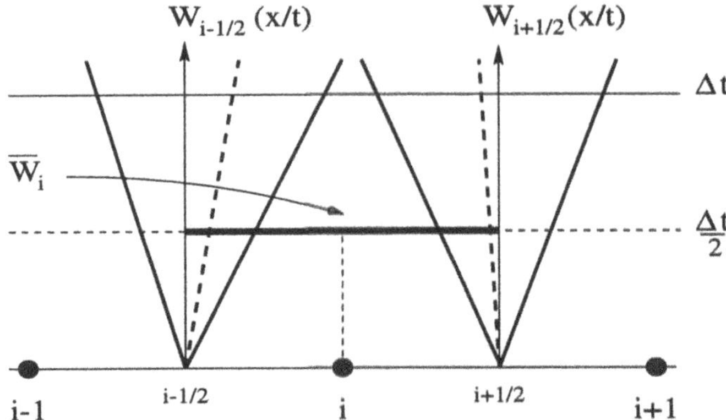

Figure 5. State $\overline{\mathbf{W}}_i$ to compute the coefficient matrix in primitive schemes

where $c^{(k)}_{i+\frac{1}{2}}$ is the Courant number for wave k of speed S_k. Manipulation of (50)–(51) leads to

$$\mathbf{W}^{n+\frac{1}{2}}_{i+\frac{1}{2}} = \frac{1}{2}(\mathbf{W}^n_i + \mathbf{W}^n_{i+1}) - \frac{1}{2}\sum_{k=1}^{N} c^{(k)}_{i+\frac{1}{2}}[\mathbf{W}^{(k+1)}_{i+\frac{1}{2}} - \mathbf{W}^{(k)}_{i+\frac{1}{2}}] \,. \tag{52}$$

To compute the coefficient matrix $\overline{\mathbf{A}}_i$ in (41) we first define a state $\overline{\mathbf{W}}_i$ and then set

$$\overline{\mathbf{A}}_i = \mathbf{A}(\overline{\mathbf{W}}_i) \,. \tag{53}$$

The simplest choice is obviously $\overline{\mathbf{W}}_i = \mathbf{W}^n_i$. This choice however, may not be sufficiently robust to be used with confidence. A better choice for $\overline{\mathbf{A}}_i$ is

$$\overline{\mathbf{W}}_i = \frac{1}{\Delta x}\int_0^{\frac{1}{2}\Delta x} \mathbf{W}^*_{i-\frac{1}{2}}(x, \frac{1}{2}\Delta t)dx + \frac{1}{\Delta x}\int_{-\frac{1}{2}\Delta x}^{0} \mathbf{W}^*_{i+\frac{1}{2}}(x, \frac{1}{2}\Delta t)dx \,, \tag{54}$$

which is an integral average of the solutions $\mathbf{W}^*_{i-\frac{1}{2}}(x/t)$ and $\mathbf{W}^*_{i+\frac{1}{2}}(x/t)$ of the two Riemann problems *affecting* mesh point i, at the half–time level. See Fig. 5. An approximation to this integral is

$$\overline{\mathbf{W}}_i = \sum_{j=j_L}^{N+1} \beta^{(j)}_{i-\frac{1}{2}} \mathbf{W}^{*(j)}_{i-\frac{1}{2}} + \sum_{j=1}^{j_R} \beta^{(j)}_{i+\frac{1}{2}} \mathbf{W}^{*(j)}_{i+\frac{1}{2}} \,, \tag{55}$$

with the coefficients defined as

$$\left.\begin{aligned} \beta^{(j)}_{i-\frac{1}{2}} &= \tfrac{1}{2}(c^{(j)}_{i-\frac{1}{2}} - c^{(j-1)}_{i-\frac{1}{2}}) \text{ for } j > j_L \,;\ \beta^{(j_L)}_{i-\frac{1}{2}} = \tfrac{1}{2}c^{(j_L)}_{i+\frac{1}{2}} \,, \\ \beta^{(j)}_{i+\frac{1}{2}} &= \tfrac{1}{2}(c^{(j)}_{i+\frac{1}{2}} - c^{(j-1)}_{i+\frac{1}{2}}) \text{ for } j < j_R \,;\ \beta^{(j_R)}_{i+\frac{1}{2}} = -\tfrac{1}{2}c^{(j_R)}_{i+\frac{1}{2}} \,. \end{aligned}\right\} \tag{56}$$

The superscript j_L refers to the value of k in the solution of the Riemann problem with data $(\mathbf{W}_{i-1}^n, \mathbf{W}_i^n)$ such that the speed S_{k-1} is negative and S_k is positive. A similar interpretation holds for j_R. Another approximation to (54), that is reported to work well (Ivings, 1996), is

$$\overline{\mathbf{W}}_i = \frac{1}{2}(\mathbf{W}_{i-\frac{1}{2}}^{n+\frac{1}{2}} + \mathbf{W}_{i+\frac{1}{2}}^{n+\frac{1}{2}}) , \tag{57}$$

which makes use of information made available by (49) or (52). The coefficient matrix is computed according to (53).

In summary, a WAF–type primitive variable scheme of the form (41) is completely determined once the intermediate states $\mathbf{W}_{i+\frac{1}{2}}^{n+\frac{1}{2}}$ are found according to (52) and the coefficient matrix $\overline{\mathbf{A}}_i$ is computed according to (53), with $\overline{\mathbf{W}}_i$ given by (55) or (57). The resulting scheme is second–order accurate in space and time. A first–order method is obtained if the intermediate states are obtained by evaluating the integral (49) by the mid–point rule in space. We shall expand on this point in the next section, where we apply the approach to a general scalar PDE. In the vicinity of large gradients, the second order schemes will produce spurious oscillations. These are eliminated, at least for the scalar case, by constructing Total Variation Diminishing (TVD) versions of the schemes. See Sect. 5.2.3 and Sect. 5.3.

5.2. WAF SCHEMES FOR A GENERAL SCALAR PDE

Consider the general scalar PDE in non–conservative form

$$q_t + a(q)q_x = 0 , \tag{58}$$

where $a = a(q)$ is the characteristic speed. For $a = constant$ we obtain the linear advection equation (5) and for $a = q$ we obtain Burgers equation (7). The scalar version of the primitive scheme (41), as applied to (58), is

$$q_i^{n+1} = q_i^n + \frac{\Delta t}{\Delta x}\overline{a}_i \left[q_{i-\frac{1}{2}}^{n+\frac{1}{2}} - q_{i+\frac{1}{2}}^{n+\frac{1}{2}}\right] . \tag{59}$$

We need a procedure to find $\overline{a}_i$ and the intermediate values $q_{i+\frac{1}{2}}^{n+\frac{1}{2}}$.

5.2.1. *The Riemann Problem*

In the spirit of linearised Riemann solvers, such as the PVRS Riemann solver of Sect. 4.2, we find the exact solution of the Riemann problem for a PDE that approximates (58), namely the PDE

$$q_t + a_{i+\frac{1}{2}} q_x = 0 , \tag{60}$$

where $a_{i+\frac{1}{2}}$ is the characteristic speed at the interface $i + \frac{1}{2}$ and is assumed constant. Various choices for $a_{i+\frac{1}{2}}$ are possible. Assuming a spatial distribution of $a(q)$ at time level n, with grid point values a_i^n we may take a linear interpolation, in which case we obtain

$$a_{i+\frac{1}{2}} = \frac{1}{2}[a_i^n + a_{i+1}^n] . \tag{61}$$

High–order interpolation schemes may also be used. The solution of the Riemann problem is now

$$q^*_{i+\frac{1}{2}}(x/t) = \begin{cases} q_i^n & \text{if} \quad x/t < a_{i+\frac{1}{2}} , \\ q_{i+1}^n & \text{if} \quad x/t > a_{i+\frac{1}{2}} . \end{cases} \tag{62}$$

5.2.2. *Schemes*

First we evaluate the coefficient $\overline{a}_i$ in (59) using (54). There are at least two ways of doing this. First we can evaluate an average state $\overline{q}_i$ via (54) and then set $\overline{a}_i = a(\overline{q}_i)$ in (53). A second possibility is to interpret (54) as an average of $a(q)$ directly. Following the first option we obtain

$$\overline{q}_i = \frac{1}{2} c^+_{i-\frac{1}{2}} q^n_{i-1} + [1 - \frac{1}{2}(c^+_{i-\frac{1}{2}} - c^-_{i+\frac{1}{2}})] q^n_i - \frac{1}{2} c^-_{i+\frac{1}{2}} q^n_{i+1} , \tag{63}$$

where

$$c^+_{i+\frac{1}{2}} = max(0, c_{i+\frac{1}{2}}) , \qquad c^-_{i+\frac{1}{2}} = min(0, c_{i+\frac{1}{2}}) , \tag{64}$$

with

$$c_{i+\frac{1}{2}} = \frac{\Delta t}{\Delta x} a_{i+\frac{1}{2}} \tag{65}$$

the interface Courant number. Then we can take $\overline{a}_i = a(\overline{q}_i)$. From the second option we obtain directly

$$\overline{a}_i = \frac{1}{2} c^+_{i-\frac{1}{2}} a^n_{i-1} + [1 - \frac{1}{2}(c^+_{i-\frac{1}{2}} - c^-_{i+\frac{1}{2}})] a^n_i - \frac{1}{2} c^-_{i+\frac{1}{2}} a^n_{i+1} . \tag{66}$$

Next we evaluate the intermediate states $q_{i+\frac{1}{2}}^{n+\frac{1}{2}}$ in (59). These are obtained from (49), that is

$$q_{i+\frac{1}{2}}^{n+\frac{1}{2}} = \frac{1}{\Delta x} \int_{-\frac{1}{2}\Delta x}^{\frac{1}{2}\Delta x} q_{i+\frac{1}{2}}^{*}(x, \frac{1}{2}\Delta t)dx \ , \tag{67}$$

where $q_{i+\frac{1}{2}}^{*}(x/t)$ is the solution of the Riemann problem, given by (62). A first order scheme is obtained from evaluating (67) by the mid–point rule in space, the result being

$$q_{i+\frac{1}{2}}^{n+\frac{1}{2}} = q_{i+\frac{1}{2}}^{*}(0) \ , \tag{68}$$

where $q_{i+\frac{1}{2}}^{*}(0)$ is the value of the solution of the Riemann problem at $x/t = 0$, i.e. along the t–axis. As a matter of fact, this is identical to the value used in the Godunov first–order upwind method (conservative) to evaluate intercell numerical fluxes. The resulting non–conservative method reads

$$q_i^{n+1} = q_i^n + \frac{\Delta t}{\Delta x}\bar{a}_i \left[q_{i-\frac{1}{2}}^{*}(0) - q_{i+\frac{1}{2}}^{*}(0) \right] \ . \tag{69}$$

Note that for $a(q) = constant$, scheme (69) reduces to the Courant, Isaacson and Rees (CIR) scheme (Courant *et. al.*, 1952). In this case it is also identical to the (conservative) Godunov first order upwind method. Scheme (69) is the non–conservative upwind analogue of the conservative first–order upwind method of Godunov.

Second–order schemes are obtained by evaluating (67) exactly. The result is

$$q_{i+\frac{1}{2}}^{n+\frac{1}{2}} = \frac{1}{2}(1 + c_{i+\frac{1}{2}})q_i^n + \frac{1}{2}(1 + c_{i+\frac{1}{2}})q_{i+1}^n \ . \tag{70}$$

Note that for $a(q) = constant$, scheme (59) with (66) and (70) reduces to the Lax–Wendroff method, which is second–order accurate in space and time. We thus expect the scheme resulting from (70) to produce spurious oscillations in the vicinity of large gradients.

5.2.3. TVD *Schemes*

A Total Variation Diminishing (TVD) version of the scheme will eliminate the oscillations of the linear second–order methods (59), with intermediate states as given by (70). This is achieved by replacing $c_{i+\frac{1}{2}}$ in (70) by a WAF limiter function $\phi_{i+\frac{1}{2}}$, to obtain

$$q_{i+\frac{1}{2}}^{n+\frac{1}{2}} = \frac{1}{2}(1+\phi_{i+\frac{1}{2}})q_i^n + \frac{1}{2}(1+\phi_{i+\frac{1}{2}})q_{i+1}^n \; . \tag{71}$$

Here $\phi_{i+\frac{1}{2}} \equiv \phi_{i+\frac{1}{2}}(r_{i+\frac{1}{2}}, |c_{i+\frac{1}{2}}|)$, where $r_{i+\frac{1}{2}}$ is a *flow parameter*, given in terms of upwind changes Δ_{upwind} and local changes Δ_{local} in q. In fact

$$r_{i+\frac{1}{2}} = \frac{\Delta_{upwind}}{\Delta_{local}} \; , \tag{72}$$

where

$$\left.\begin{array}{l} \Delta_{upwind} = \left\{ \begin{array}{lll} q_i^n - q_{i-1}^n & \text{if} & c_{i+\frac{1}{2}} > 0 \\ q_{i+2}^n - q_{i+1}^n & \text{if} & c_{i+\frac{1}{2}} < 0 \end{array} \right. \\ \\ \Delta_{local} = q_{i+1}^n - q_i^n \end{array}\right\} . \tag{73}$$

The WAF limiter functions $\phi_{i+\frac{1}{2}}$ are related, and are entirely equivalent, to conventional flux limiters functions $B_{i+\frac{1}{2}}$ (Sweby, 1982), (Sweby, 1984), (Roe, 1983). In fact they are related via

$$\phi_{i+\frac{1}{2}} = 1 - (1 - |c_{i+\frac{1}{2}}|)B_{i+\frac{1}{2}} . \tag{74}$$

Thus any conventional flux limiter function $B_{i+\frac{1}{2}}$ can be used to construct WAF limiters $\phi_{i+\frac{1}{2}}$ for use in (71). Two well–known examples of flux limiters are SUPERBEE

$$B_{sup} = max[0, min(2r/|c|, 1), min(r, 2/(1-|c|))] \tag{75}$$

and MINMOD

$$B_{min} = \text{minmod}(r, 1) \; , \tag{76}$$

where subscripts have been omitted. The main difference between these limiters is that B_{sup} is *compressive* and will resolve discontinuities very sharply, but at the expense of *squaring* smooth waves, while B_{min} tends to be too diffusive, which results in smeared sharp fronts and *clipping of extrema*. A good compromise is the more recent limiter of Arora and Roe (Arora and Roe, 1997)

$$\left.\begin{array}{l} B_{aroe} = max[0, min\{b_1 r, 1 + b_2(r-1), b_{max}\}] \\ \\ b_1 = 2/|c| \; , \; \; b_2 = (1+|c|)/3 \; , \; \; b_{max} = 2/(1-|c|) \end{array}\right\} . \tag{77}$$

This limiter eliminates the squaring effect of B_{sup} and reduces the clipping of extrema of B_{min}, but at the same time it will resolve discontinuities over more cells, not as sharply as B_{sup}.

5.2.4. *Numerical Experiments*

To illustrate the performance of some of the schemes, we solve here the inviscid Burgers equation (7) in the domain $x \in [0, 4]$ with initial conditions

$$q(x,0) = \begin{cases} 1 & \text{if} \quad x < x_0 \,, \\ 0 & \text{if} \quad x > x_0 \,, \end{cases} \tag{78}$$

where $x_0 = \frac{1}{2}$. We discretise the domain with $M = 50$ grid points and use a CFL number of 0.8. Solution profiles are compared with the exact solution at the output times $t = 4.0$ and $t = 6.0$. The exact solution consists of a shock propagating to the right with speed $s = \frac{1}{2}$.

Fig. 6 shows results for the first–order primitive scheme (69) with the choice of speed $\overline{a}_i$ as given by (66). The numerical results (symbols) are compared with the exact solution (line) at times $t = 4$ (left) and $t = 6$ (right). The results are interesting on two counts. First, the position of the shock wave is incorrect, as expected from a non–conservative method; second, the shock is oscillation free. Fig. 7 shows the corresponding results for the first–order conservative scheme (40), the Godunov method, at the same times. As expected, the Godunov method gives oscillation free shocks and the correct shock propagation speed.

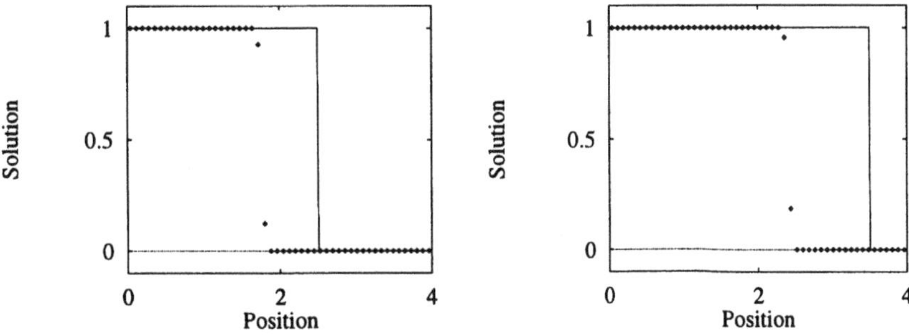

Figure 6. First order primitive scheme (69) (symbol) and exact solution (line) at times $t = 4.0$ (left) and $t = 6.0$ (right), using the average speed (66)

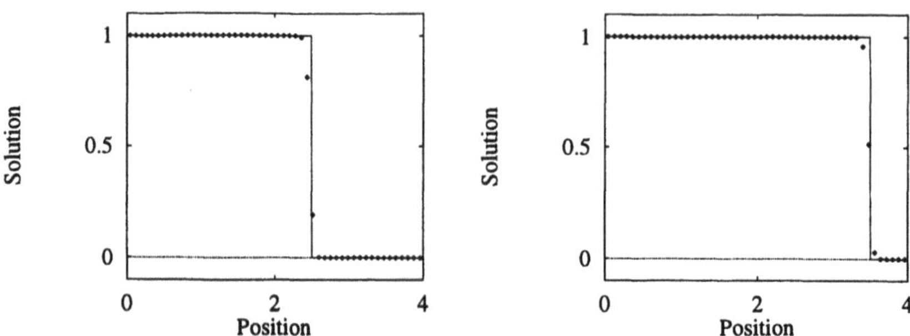

Figure 7. First order conservative (Godunov's method) (40) (symbol) and exact solution (line) at times $t = 4.0$ (left) and $t = 6.0$ (right)

5.3. FIRST–ORDER MONOTONE AND TVD SCHEMES

A first–order non–conservative upwind scheme that is the analogue of the conservative Godunov method is

$$\mathbf{W}_i^{n+1} = \mathbf{W}_i^n + \frac{\Delta t}{\Delta x}\overline{\mathbf{A}}_i \left[\mathbf{W}^*_{i-\frac{1}{2}}(0) - \mathbf{W}^*_{i+\frac{1}{2}}(0)\right] . \tag{79}$$

Here $\mathbf{W}^*_{i+\frac{1}{2}}(0)$ is the value of the solution of the appropriate Riemann problem, evaluated along the t–axis, just as in the Godunov method when evaluating the intercell fluxes. This scheme is the generalisation to non–linear systems of the scalar scheme (69). For the linear advection equation (5) this scheme reduces to the first–order upwind scheme of Courant Isaacson and Rees, the CIR scheme (Courant *et. al.*, 1952).

Scheme (79) will be be used as the basis for constructing high–order Total Variation Diminishing (TVD) versions of the second–order primitive method. This is most easily accomplished by replacing (52) by

$$\mathbf{W}_{i+\frac{1}{2}}^{n+\frac{1}{2}} = \frac{1}{2}(\mathbf{W}_i^n + \mathbf{W}_{i+1}^n) - \frac{1}{2}\sum_{k=1}^{N} \mathrm{sign}(c_{i+\frac{1}{2}}^{(k)})\phi_{i+\frac{1}{2}}^{(k)}[\mathbf{W}_{i+\frac{1}{2}}^{(k+1)} - \mathbf{W}_{i+\frac{1}{2}}^{(k)}] , \tag{80}$$

where $\phi_{i+\frac{1}{2}}^{(k)}$ is the limiter at the intermediate position $i + \frac{1}{2}$ corresponding to wave k and $c_{i+\frac{1}{2}}^{(k)}$ is the Courant number for wave k. The limiter can be taken as any of the WAF limiter functions discussed in Sect. 5.2.3. The underlying monotone scheme (79) is obtained from the TVD intermediate state (80) by setting all the flux limiters $\phi_{i+\frac{1}{2}}^{(k)} = sign(c_{i+\frac{1}{2}}^{(k)})$.

We summarise the application of the primitive WAF scheme based on (41) as applied to (36): one first solves the Riemann problems with data $(\mathbf{W}_i^n, \mathbf{W}_{i+1}^n)$ to find $\mathbf{W}_{i+\frac{1}{2}}^*(x/t)$; one then computes the intermediate state vectors $\mathbf{W}_{i+\frac{1}{2}}^{n+\frac{1}{2}}$ according to (80) and the coefficient matrix $\overline{\mathbf{A}}_i$ according to (53), with $\overline{\mathbf{W}}$ obtained from (55) or (57); the primitive scheme (41) is thus completely determined. The linearised stability condition of the scheme as applied to the linear advection equation (5) is $|c| \leq 1$, where c is the Courant number.

For the purpose of implementing adaptive primitive–conservative schemes in Sect. 10, we remark that a conservative version of the primitive method (41) is most easily obtained by computing an intercell flux as

$$\mathbf{F}_{i+\frac{1}{2}} = \mathbf{F}(\mathbf{W}_{i+\frac{1}{2}}^{n+\frac{1}{2}}) \tag{81}$$

and advancing the solution via the conservative formula (40).

In the next three sections we shall construct non–conservative methods of the type (41) following the MUSCL approach, first introduced for constructing conservative methods. Partial results were reported by the author in (Toro, 1995b). We shall extend three MUSCL approaches to the construction of non–conservative methods, namely the MUSCL–Hancock method (MHM) (van Leer, 1984), the Generalised Riemann Problem (GRP) method of Ben–Artzi and Falcovitz (Ben-Artzi and Falcovitz, 1984) and the Piece–wise Linear Method (PLM) of Colella (Colella, 1985).

6. MUSCL–Hancock Type Primitive Schemes

The purpose of this section is to construct non–conservative schemes following the MUSCL Hancock method (MHM) (van Leer, 1984), described in Sect. 6.1. Schemes of first, second and third order accuracy are derived, as well as TVD versions of these. Before proceeding to the details of MHM we recall the general MUSCL approach, which stands for Monotone Upstream Scheme for Conservation Laws. This was first introduced by van Leer (van Leer, 1976) for constructing high–order conservative upwind methods.

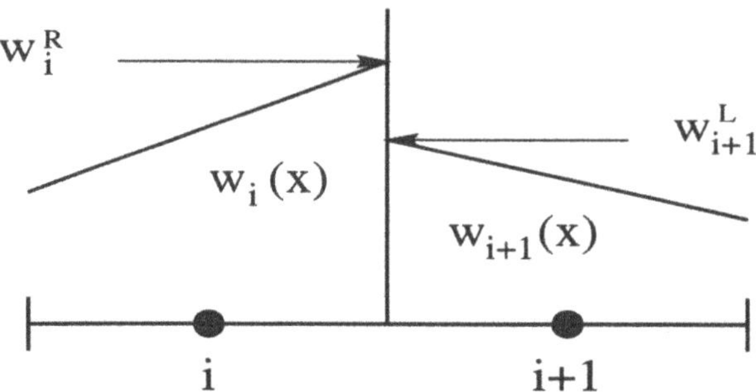

Figure 8. MUSCL reconstruction of a single component w of the vector **W** and corresponding boundary extrapolated values

6.1. THE MUSCL APPROACH

The MUSCL approach provides a way of extending first–order accurate methods to higher order of accuracy. Here, the underlying first–order scheme is the upwind non–conservative scheme (79), which is the upwind non–conservative analogue of the Godunov method. A first step common to all MUSCL schemes is the data reconstruction procedure, whereby the grid point values (cell averages in conservative methods) are extended to an interval $[x_{i-\frac{1}{2}}, x_{i+\frac{1}{2}}]$ around the grid point x_i as

$$\mathbf{W}_i(x) = \mathbf{W}_i^n + \frac{(x - x_i)}{\Delta x}\Delta_i \ , \tag{82}$$

where $\frac{\Delta_i}{\Delta x}$ is an estimate for the slope of **W** at the grid point i. For generality, we write the vector *difference* Δ_i as

$$\left.\begin{array}{lll} \Delta_i & = & \alpha\Delta_{i-\frac{1}{2}} + (1-\alpha)\Delta_{i+\frac{1}{2}} \ , \\ \Delta_{i-\frac{1}{2}} & = & \mathbf{W}_i^n - \mathbf{W}_{i-1}^n \ , \\ \Delta_{i+\frac{1}{2}} & = & \mathbf{W}_{i+1}^n - \mathbf{W}_i^n \ , \end{array}\right\} \tag{83}$$

where α is a parameter in the range $[0, 1]$ and is open to choice. Fig. 8 illustrates the reconstruction procedure in cells i and $i + 1$, applied to a single component w of the vector **W**. Also shown there are the extrapolated values to the boundary $i + \frac{1}{2}$, namely w_i^R and w_{i+1}^L. The complete vectors are denoted by $\mathbf{W}_i^R$ and $\mathbf{W}_{i+1}^L$.

In order to use the non–conservative scheme (41) we need a procedure for computing the coefficient matrix $\overline{\mathbf{A}}_i$ and the intermediate states $\mathbf{W}_{i+\frac{1}{2}}^{n+\frac{1}{2}}$.

6.2. INTERMEDIATE STATES

Here we propose to compute the intermediate states in essentially the same way as for computing intercell fluxes for the MUSCL–Hancock conservative scheme. After the reconstruction procedure (82)–(83) there are three distinct steps, namely

1. **Extrapolation**
   Using the extended distribution of the reconstructed solution in (82) we evaluate the vectors $\mathbf{W}_i(x)$ and $\mathbf{W}_{i+1}(x)$ at the intermediate position $x_{i+\frac{1}{2}}$ to obtain, see Fig. 8, the two extrapolated values

$$\mathbf{W}_i^R = \mathbf{W}_i^n + \frac{1}{2}\Delta_i \, ; \quad \mathbf{W}_{i+1}^L = \mathbf{W}_{i+1}^n - \frac{1}{2}\Delta_{i+1} \, . \tag{84}$$

2. **Evolution**
   We then propose to evolve the values $\mathbf{W}_i^{L,R}$ by a time $\frac{1}{2}\Delta t$ according to

$$\hat{\mathbf{W}}_i^{L,R} = \mathbf{W}_i^{L,R} + \frac{1}{2}\frac{\Delta t}{\Delta x}\mathbf{A}_i^n(\mathbf{W}_i^L - \mathbf{W}_i^R) \, . \tag{85}$$

3. **The Riemann Problem**
   In order to compute $\mathbf{W}_{i+\frac{1}{2}}^{n+\frac{1}{2}}$ we propose to find the solution $\mathbf{W}_{i+\frac{1}{2}}^*(x/t)$ of the Riemann problem with piece–wise constant data $(\hat{\mathbf{W}}_i^R, \hat{\mathbf{W}}_{i+1}^L)$ and set

$$\mathbf{W}_{i+\frac{1}{2}}^{n+\frac{1}{2}} = \mathbf{W}_{i+\frac{1}{2}}^*(0) \, , \tag{86}$$

   i.e. we take the value of $\mathbf{W}_{i+\frac{1}{2}}^*(x/t)$ along the t–axis.

6.3. THE COEFFICIENT MATRIX

There are several possible choices for the matrix $\overline{\mathbf{A}}_i$ in (41). As for the WAF schemes of Sect. 5, we first compute a state $\overline{\mathbf{W}}_i$ and then evaluate the matrix according to (53). Here we propose two choices for $\overline{\mathbf{W}}_i$, namely

$$\overline{\mathbf{W}}_i = \frac{1}{\Delta x}\int_0^{\frac{1}{2}\Delta x} \mathbf{W}_{i-\frac{1}{2}}^*(x, \frac{1}{2}\Delta t)dx + \frac{1}{\Delta x}\int_{-\frac{1}{2}\Delta x}^0 \mathbf{W}_{i+\frac{1}{2}}^*(x, \frac{1}{2}\Delta t)dx \tag{87}$$

and

$$\overline{\mathbf{W}}_i = \frac{1}{2}(\hat{\mathbf{W}}_i^R + \hat{\mathbf{W}}_i^L) \, . \tag{88}$$

We note that expressions (87) and (88) are not identical to (54) and (57) respectively. After some algebra expression (88) yields

$$\overline{\mathbf{W}}_i = \mathbf{W}_i^n - \frac{1}{2}\frac{\Delta t}{\Delta x}\mathbf{A}(\mathbf{W}_i^n)\Delta_i \ , \tag{89}$$

which says that $\mathbf{W}_i^n$ has been evolved by a time $\frac{1}{2}\Delta t$ to obtain $\overline{\mathbf{W}}_i$. Expression (87) is effectively a time–space integral average of the combined solutions $\mathbf{W}^*_{i-\frac{1}{2}}(x/t)$ and $\mathbf{W}^*_{i+\frac{1}{2}}(x/t)$ of the neighbouring Riemann problems with data $(\hat{\mathbf{W}}^R_{i-1}, \hat{\mathbf{W}}^L_i)$ and $(\hat{\mathbf{W}}^R_i, \hat{\mathbf{W}}^L_{i+1})$. Using a trapezium–rule approximation to (87) we obtain

$$\overline{\mathbf{W}}_i = \frac{1}{2}[\mathbf{W}^*_{i-\frac{1}{2}}(0) + \mathbf{W}^*_{i+\frac{1}{2}}(0)] = \frac{1}{2}[\mathbf{W}^{n+\frac{1}{2}}_{i-\frac{1}{2}} + \mathbf{W}^{n+\frac{1}{2}}_{i+\frac{1}{2}}] \ . \tag{90}$$

The last equality follows from (86). A better approximation to the integral (87) gives a summation of the form (55), (56). We have thus provided expressions for the coefficient matrix and the intermediate states following the MUSCL–Hancock approach, so that the non–conservative scheme (41) is now completely determined.

We note that the MHM approach reproduces the non–conservative first–order upwind scheme (79) as a special case, namely when the slopes in (82) are identically zero. For other choices of the slopes we produce second and third order accurate schemes. This is most easily seen by applying the MHM to the linear advection equation (5). Using (83) we write

$$\Delta_i = \alpha(q_i^n - q_{i-1}^n) + (1-\alpha)(q_{i+1}^n - q_i^n) \ . \tag{91}$$

The extrapolated values at the intermediate position $x_{i+\frac{1}{2}}$ are found using (84), namely

$$\left.\begin{array}{rcl} q_i^R & = & q_i^n + \frac{1}{2}\Delta_i \ , \\ q_{i+1}^L & = & q_{i+1}^n - \frac{1}{2}\Delta_{i+1} \ . \end{array}\right\} \tag{92}$$

The evolved extrapolated states are, according to (85),

$$\left.\begin{array}{rcl} \hat{q}_i^R & = & q_i^n + \frac{1}{2}(1-c)\Delta_i \ , \\ \hat{q}_{i+1}^L & = & q_{i+1}^n - \frac{1}{2}(1+c)\Delta_{i+1} \ , \end{array}\right\} \tag{93}$$

where c is the Courant number $c = \frac{\Delta t}{\Delta x}a$. The solution of the Riemann problem for (5) with data $(\hat{q}_i^R, \hat{q}_{i+1}^L)$ is

$$q^*_{i+\frac{1}{2}}(x/t) = \left\{ \begin{array}{ll} \hat{q}_i^R & if \ \ \frac{x}{t} < a \ , \\ \hat{q}_{i+1}^L & if \ \ \frac{x}{t} > a \ . \end{array}\right. \tag{94}$$

We finally write the intermediate value $q_{i+\frac{1}{2}}^{n+\frac{1}{2}}$ as

$$q_{i+\frac{1}{2}}^{n+\frac{1}{2}} = \delta_1 q_{i-1}^n + \delta_2 q_i^n + \delta_3 q_{i+1}^n \,, \tag{95}$$

with coefficients δ_k as given by Table 1. Substitution of (95) into (41) reproduces the schemes of Table 1.

Note that depending on the choice of α we reproduce three well known schemes that are second–order accurate in space and time, namely the schemes of Lax–Wendroff (LW), Fromm (FR) and Warming–Beam (WB). For any α such that $0 \leq \alpha \leq 1$ one can easily see that the resulting scheme (GSO) is second–order accurate in space and time; for the particular choice

$$\alpha = \frac{1}{3}(1 + c)$$

the resulting scheme (THO) is third–order accurate in space and time.

A conservative version of the MHM scheme is constructed by evaluating an intercell flux $\mathbf{F}_{i+\frac{1}{2}}^{n+\frac{1}{2}}$ according to (81), with the intermediate state $\mathbf{W}_{i+\frac{1}{2}}^{n+\frac{1}{2}}$ obtained from (86), and advancing the solution according to the conservative formula (40).

7. GRP–Type Primitive Schemes

Here we present another way of constructing fully discrete non–conservative schemes of the form (41). We follow the Generalised Riemann Problem (GRP) approach of Ben–Artzi and Falcovitz (Ben-Artzi and Falcovitz, 1984), first put forward for constructing conservative methods. The essence of the approach is the solution of the Riemann problem with *piece–wise linear data* obtained by a MUSCL reconstruction as in (82)–(83). Finding the GRP solution is a formidable task, particularly for complex sets of PDEs. Here we first modify the GRP approach by replacing the solution of the Generalised Riemann Problem by the solution of two (conventional) piece–wise constant data Riemann problems. In fact this modified GRP (or MGRP) scheme lends itself easily to higher–order extensions e.g. third or fourth order of accuracy. We then apply the MGRP method to construct primitive–variable schemes

TABLE 1. Coefficients for (95) and associated numerical methods obtained from the non-conservative MUSCL Hancock approach as applied to the linear advection equation.

	δ_1	δ_2	δ_3	
α	$-\frac{1}{2}(1-c)\alpha$	$1+\frac{1}{2}(1-c)(2\alpha-1)$	$\frac{1}{2}(1-c)(1-\alpha)$	GSO
0	0	$\frac{1}{2}(1+c)$	$\frac{1}{2}(1-c)$	LW
$\frac{1}{2}$	$-\frac{1}{4}(1-c)$	1	$\frac{1}{4}(1-c)$	FR
1	$\frac{1}{2}(c-1)$	$\frac{1}{2}(3-c)$	0	WB
$\frac{1}{3}(1+c)$	$-\frac{1}{6}(1-c^2)$	$1+\frac{1}{6}(1-c)(2c-1)$	$\frac{1}{6}(1-c)(2c-1)$	THO

of the form (41). One requires definitions for the coefficient matrix $\overline{\mathbf{A}}_i$ and the intermediate states $\mathbf{W}_{i+\frac{1}{2}}^{n+\frac{1}{2}}$.

7.1. INTERMEDIATE STATES

Starting from the reconstruction step (82)–(83), one obtains the generalised Riemann problem

$$\left.\begin{array}{l} \mathbf{W}_t + \mathbf{A}(\mathbf{W})\mathbf{W}_x = \mathbf{0}\,, \\ \mathbf{W}(x,0) = \left\{ \begin{array}{ll} \mathbf{W}_i(x) & \text{if } x < 0\,, \\ \mathbf{W}_{i+1}(x) & \text{if } x > 0\,. \end{array}\right. \end{array}\right\} \qquad (96)$$

We denote the GRP solution by $\mathbf{W}_{i+\frac{1}{2}}(x,t)$. The intermediate state $\mathbf{W}_{i+\frac{1}{2}}^{n+\frac{1}{2}}$ is obtained from a time Taylor expansion about $t = 0$, at $x_{i+\frac{1}{2}}$ ($x = 0$ in local coordinates), namely

$$\mathbf{W}_{i+\frac{1}{2}}^{n+\frac{1}{2}} = \mathbf{W}_{i+\frac{1}{2}}^{(0)} + \frac{\Delta t}{2}\frac{\partial}{\partial t}\mathbf{W}_{i+\frac{1}{2}} + O(\Delta t^2) . \tag{97}$$

Here $\mathbf{W}_{i+\frac{1}{2}}^{(0)}(x/t)$ represents the interaction of data states in (96) immediately after the initial time. It is the solution of the piece–wise constant data (conventional) Riemann problem for the PDEs in (96) with left and right data

$$\mathbf{W}(x,0) = \begin{cases} \mathbf{W}_i^R & = & \mathbf{W}_i^n + \frac{1}{2}\Delta_i & \text{if} \quad x < 0 \, , \\ \mathbf{W}_{i+1}^L & = & \mathbf{W}_{i+1}^n - \frac{1}{2}\Delta_{i+1} & \text{if} \quad x > 0 \, . \end{cases} \tag{98}$$

The data states are obtained by evaluating the functions $\mathbf{W}_i(x)$ and $\mathbf{W}_{i+1}(x)$ at position $i+\frac{1}{2}$. The second term in (97) involves a time derivative of the solution $\mathbf{W}_{i+\frac{1}{2}}$ at $t = 0$. This is the most difficult part of the original GRP approach. Our modification starts from a linearisation of the PDEs in (96) about the state $\mathbf{W}_{i+\frac{1}{2}}^{(0)}$ to obtain an estimate for the term $\frac{\partial}{\partial t}\mathbf{W}_{i+\frac{1}{2}}$. We thus have the linear system

$$\mathbf{W}_t + \mathbf{A}_{i+\frac{1}{2}}^{(0)}\mathbf{W}_x = \mathbf{0} \, , \tag{99}$$

where the constant coefficient matrix is given by

$$\mathbf{A}_{i+\frac{1}{2}}^{(0)} = \mathbf{A}(\mathbf{W}_{i+\frac{1}{2}}^{(0)}) . \tag{100}$$

Then we replace the time derivative in (97) by a space derivative using (99) to obtain

$$\mathbf{W}_{i+\frac{1}{2}}^{n+\frac{1}{2}} = \mathbf{W}_{i+\frac{1}{2}}^{(0)} - \frac{\Delta t}{2}\mathbf{A}_{i+\frac{1}{2}}^{(0)}\mathbf{W}_{i+\frac{1}{2}}^{(1)} \, , \tag{101}$$

where second and higher order terms have been neglected and

$$\mathbf{W}_{i+\frac{1}{2}}^{(1)} = \frac{\partial}{\partial x}\mathbf{W}_{i+\frac{1}{2}} \tag{102}$$

is a space–derivative term. We find this term from the following result: For any vector

$$\mathbf{Z} = \frac{\partial^m}{\partial x^m}\mathbf{W} \tag{103}$$

we have

$$\frac{\partial \mathbf{Z}}{\partial t} + \mathbf{A}_{i+\frac{1}{2}}^{(0)}\frac{\partial \mathbf{Z}}{\partial x} = \mathbf{0} \, , \tag{104}$$

where $\mathbf{A}^{(0)}_{i+\frac{1}{2}}$ is the constant matrix in (99) and m is a non–negative integer. The proof is straightforward. The case $m = 1$ follows from differentiating (99) with respect to x, assuming sufficient smoothness. This result says that the gradients of $\mathbf{W}$ obey the same system of PDEs as $\mathbf{W}$ itself. One can therefore define the *Riemann problem for the gradients*, namely the solution of the PDEs

$$\frac{\partial}{\partial t}(\mathbf{W}_x) + \mathbf{A}^{(0)}_{i+\frac{1}{2}}\frac{\partial}{\partial x}(\mathbf{W}_x) = \mathbf{0}\,, \tag{105}$$

with initial conditions

$$\mathbf{W}_x(x,0) = \begin{cases} \frac{\Delta_i}{\Delta x} & = & \mathbf{W}^{(1)}_L \quad \text{if} \quad x < 0\,, \\ \\ \frac{\Delta_{i+1}}{\Delta x} & = & \mathbf{W}^{(1)}_R \quad \text{if} \quad x > 0\,. \end{cases} \tag{106}$$

The equations are linear with constant coefficients and thus the solution of the Riemann problem can be found from standard theory; see for example the textbook of Toro (Toro, 1997b). By denoting $\mathbf{W}^{(0)}_{i+\frac{1}{2}} = (\overline{\rho}, \overline{u}, \overline{p}, \overline{q})^T$ we write the matrix $\mathbf{A}^{(0)}_{i+\frac{1}{2}}$ as

$$\mathbf{A}^{(0)}_{i+\frac{1}{2}} = \begin{bmatrix} \overline{u} & \overline{\rho} & 0 & 0 \\ 0 & \overline{u} & 1/\overline{\rho} & 0 \\ 0 & \overline{\rho}\,\overline{a}^2 & \overline{u} & 0 \\ 0 & 0 & 0 & \overline{u} \end{bmatrix}. \tag{107}$$

The eigenvalues are

$$\overline{\lambda}_1 = \overline{u} - \overline{a} \quad ; \quad \overline{\lambda}_2 = \overline{u} = \overline{\lambda}_4 \quad ; \quad \overline{\lambda}_3 = \overline{u} + \overline{a} \tag{108}$$

and the corresponding right eigenvectors are

$$\mathbf{K}^{(1)} = \begin{bmatrix} 1 \\ -\overline{a}/\overline{\rho} \\ \overline{a}^2 \\ 0 \end{bmatrix}; \quad \mathbf{K}^{(2)} = \begin{bmatrix} 1 \\ 0 \\ 0 \\ 0 \end{bmatrix}; \quad \mathbf{K}^{(3)} = \begin{bmatrix} 1 \\ \overline{a}/\overline{\rho} \\ \overline{a}^2 \\ 0 \end{bmatrix}; \quad \mathbf{K}^{(4)} = \begin{bmatrix} 0 \\ 0 \\ 0 \\ 1 \end{bmatrix}. \tag{109}$$

In order to completely determine the intermediate state $\mathbf{W}^{n+\frac{1}{2}}_{i+\frac{1}{2}}$ in (101) we require the solution of the Riemann problem for (105)–(106) at $x_{i+\frac{1}{2}}$ i.e.

along the t–axis in local coordinates. This is given by

$$\mathbf{W}_{i+\frac{1}{2}}^{(1)} = \mathbf{W}_L^{(1)} + \sum_{\lambda_j \leq 0} \alpha_j \mathbf{K}^{(j)} = \mathbf{W}_R^{(1)} - \sum_{\lambda_j \geq 0} \alpha_j \mathbf{K}^{(j)} \quad (110)$$

where the coefficients are given by

$$\left.\begin{array}{l} \alpha_1 = \frac{1}{2\overline{\rho a}^2}\left[\Delta p^{(1)} - \overline{\rho a}\Delta u^{(1)}\right] , \\ \alpha_2 = \Delta\rho^{(1)} - \frac{\Delta p^{(1)}}{\overline{a}^2} , \\ \alpha_3 = \frac{1}{2\overline{\rho a}^2}\left[\Delta p^{(1)} + \overline{\rho a}\Delta u^{(1)}\right] , \\ \alpha_4 = \Delta q^{(1)} . \end{array}\right\} \quad (111)$$

and $\Delta r^{(1)} = r_R^{(1)} - r_L^{(1)}$ for $r^{(1)}$ any component of the gradient vectors $\mathbf{W}_L^{(1)}$, $\mathbf{W}_R^{(1)}$.

7.2. THE COEFFICIENT MATRIX

As for the previous primitive schemes, the coefficient matrix $\overline{\mathbf{A}}_i$ in (41) is computed from (53), where the state $\overline{\mathbf{W}}_i$ is now to be found following the GRP approach. A possible choice is

$$\overline{\mathbf{W}}_i = \frac{1}{\Delta x}\int_0^{\frac{1}{2}\Delta x} \mathbf{W}_{i-\frac{1}{2}}^{(0)}(x, \frac{1}{2}\Delta t)dx + \frac{1}{\Delta x}\int_{-\frac{1}{2}\Delta x}^{0} \mathbf{W}_{i+\frac{1}{2}}^{(0)}(x, \frac{1}{2}\Delta t)dx , \quad (112)$$

where $\mathbf{W}_{i+\frac{1}{2}}^{(0)}(x,t)$ is the solution of the conventional Riemann problem for the state $\mathbf{W}$ with data (98). Recall that $\mathbf{W}_{i+\frac{1}{2}}^{(0)}(x,t)$ is the first term in the expansion for $\mathbf{W}_{i+\frac{1}{2}}^{n+\frac{1}{2}}$, see (97). Note that (112) is different from (87), the average state for the MUSCL–Hancock scheme, and is also different from (54), the average state for the WAF scheme. A trapezium rule approximation to (112) gives

$$\overline{\mathbf{W}}_i = \frac{1}{2}[\mathbf{W}_{i-\frac{1}{2}}^{(0)}(0) + \mathbf{W}_{i+\frac{1}{2}}^{(0)}(0)] . \quad (113)$$

A more accurate choice is given by summation (55) with coefficients as given by (56) and with the appropriate interpretation for the states j.

As described, it is easily checked that the primitive–variable GRP scheme reproduces the first–order non–conservative scheme (79), when the slopes in the data reconstruction stage are identically zero. Also, it is easy to verify that for the linear advection equation (5) the GRP approach of this section reproduces the second and third order accurate schemes listed in Table 1.

A conservative version of the GRP scheme is constructed by evaluating an intercell flux $\mathbf{F}_{i+\frac{1}{2}}^{n+\frac{1}{2}}$ according to (81), with the intermediate state $\mathbf{W}_{i+\frac{1}{2}}^{n+\frac{1}{2}}$ obtained from (101), and advancing the solution according to the conservative formula (40).

8. PLM–Type Primitive Schemes

The Piece–wise Linear Method, PLM, was first put forward by Colella (Colella, 1985) to construct conservative second–order upwind methods. Here we apply the PLM idea to the construction of primitive variable schemes of the form (41), where the coefficient matrix $\overline{\mathbf{A}}_i$ and the intermediate states $\mathbf{W}_{i+\frac{1}{2}}^{n+\frac{1}{2}}$ are yet to be defined.

8.1. INTERMEDIATE STATES

A direct application of the PLM philosophy to non–conservative systems (36) provides $\mathbf{W}_{i+\frac{1}{2}}^{n+\frac{1}{2}}$ from the solution of the generalised Riemann problem (96) used for the GRP non–conservative method of the previous section, where the piece–wise linear data $\mathbf{W}_i(x)$, $\mathbf{W}_{i+1}(x)$ is obtained from a MUSCL reconstruction step (82)–(83) in the usual way. Then we set

$$\mathbf{W}_{i+\frac{1}{2}}^{n+\frac{1}{2}} = \frac{1}{2}[\mathbf{W}_{i+\frac{1}{2}}^{(0)} + \mathbf{W}_{i+\frac{1}{2}}^{(\Delta t)}] \,, \tag{114}$$

where $\mathbf{W}_{i+\frac{1}{2}}^{(0)}$ is the solution of the piece–wise constant Riemann problem for (36) with initial data

$$\mathbf{W}(x,0) = \begin{cases} \mathbf{W}_i^R & = & \mathbf{W}_i^n + \frac{1}{2}\Delta i & \text{if} \;\; x < 0 \,, \\ \mathbf{W}_{i+1}^L & = & \mathbf{W}_{i+1}^n - \frac{1}{2}\Delta_{i+1} & \text{if} \;\; x > 0 \,. \end{cases} \tag{115}$$

Note that $\mathbf{W}_{i+\frac{1}{2}}^{(0)}$ is identical to the first term in the expansion (97) for the GRP scheme.

The second term $\mathbf{W}^{(\Delta t)}_{i+\frac{1}{2}}$ in (114) is computed by using the characteristic equations. These are obtained from

$$\ell^{(k)} . \begin{pmatrix} d\rho \\ du \\ dp \\ dq \end{pmatrix} = \mathbf{0} \tag{116}$$

where $\ell^{(k)}, k = 1, ..., 4$ are the left eigenvectors of $\mathbf{A}$ in (36) given by

$$\left.\begin{array}{l} \ell^{(1)} = [0, 1, -1/\rho a, 0] \ , \\ \ell^{(2)} = [\rho, 0, -\rho/a^2, 0] \ , \\ \ell^{(3)} = [0, 1, 1/\rho a, 0] \ , \\ \ell^{(4)} = [0, 0, 0, q] \ . \end{array}\right\} \tag{117}$$

The characteristic equations are

$$\left.\begin{array}{llll} dp - \rho a du & = 0 & \text{along} \ \frac{dx}{dt} = & \lambda_1 = u - a \ , \\ dp - a^2 d\rho & = 0 & \text{along} \ \frac{dx}{dt} = & \lambda_2 = u \ , \\ dq & = 0 & \text{along} \ \frac{dx}{dt} = & \lambda_4 = u \ , \\ dp + \rho a du & = 0 & \text{along} \ \frac{dx}{dt} = & \lambda_3 = u + a \ . \end{array}\right\} \tag{118}$$

Now we propose a slight modification to the PLM approach, namely we linearise equations (118) about the state $\mathbf{W}^{(0)}_{i+\frac{1}{2}}$ already available. Using the notation

$$\mathbf{W}^{(0)}_{i+\frac{1}{2}} = [\overline{\rho}, \overline{u}, \overline{p}, \overline{q}]^T$$

we have

$$\left.\begin{array}{llll} dp - \overline{\rho}\,\overline{a} du & = & 0 \ \text{along} & \frac{dx}{dt} = \overline{u} - \overline{a} \ , \\ dp - \overline{a}^2 d\rho & = & 0 \ \text{along} & \frac{dx}{dt} = \overline{u} \ , \\ dq & = & 0 \ \text{along} & \frac{dx}{dt} = \overline{u} \ , \\ dp + \overline{\rho}\,\overline{a} du & = & 0 \ \text{along} & \frac{dx}{dt} = \overline{u} + \overline{a} \ . \end{array}\right\} \tag{119}$$

The objective is to use (119) to trace characteristics through the point $(x_{i+\frac{1}{2}}, \Delta t)$ back to the x–axis, see Fig. 9, to find intersections and thus values to construct $\mathbf{W}^{(\Delta t)}_{i+\frac{1}{2}}$ in (114). Define Courant numbers

$$c_1 = \frac{\Delta t}{\Delta x}(\overline{u} - \overline{a}) \ , \ \ c_2 = \frac{\Delta t}{\Delta x}\overline{u} \ , \ \ c_3 = \frac{\Delta t}{\Delta x}(\overline{u} + \overline{a}) \ . \tag{120}$$

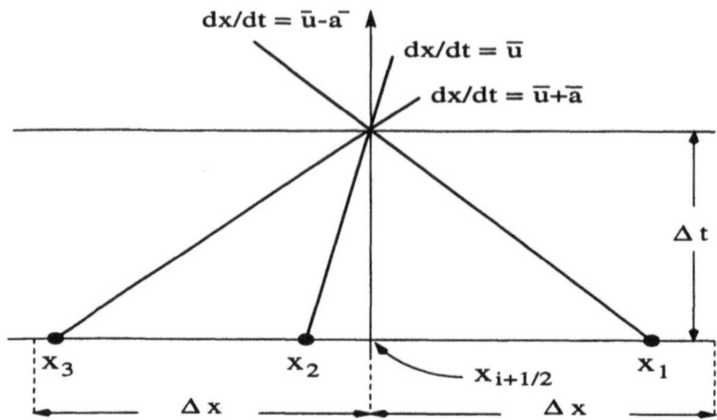

Figure 9. State $\mathbf{W}^{(\Delta t)}_{i+\frac{1}{2}}$ is obtained by tracing characteristics back to the reconstructed data. Case illustrated is that of subsonic flow with positive particle velocity

The characteristic curves of speeds $\overline{\lambda}_1 = \overline{u} - \overline{a}$, $\overline{\lambda}_2 = \overline{u}$ and $\overline{\lambda}_3 = \overline{u} + \overline{a}$ through the point $(x_{i+\frac{1}{2}}, \Delta t)$ intersect the x–axis at three points x_1, x_2 and x_3 respectively. Fig. 9 illustrates the situation for the subsonic case: $\overline{\lambda}_1 = \overline{u} - \overline{a} < 0$, $\overline{\lambda}_2 = \overline{u} > 0$, $\overline{\lambda}_3 = \overline{u} + \overline{a} > 0$, for which we have

$$x_1 = c_1 \Delta x, \quad x_2 = (1 - c_2)\Delta x, \quad x_3 = (1 - c_3)\Delta x\ . \tag{121}$$

In general

$$x_k = \begin{cases} c_k \Delta x & \text{if} \quad c_k \leq 0\,, \\ (1 - c_k)\Delta x & \text{if} \quad c_k \geq 0\,. \end{cases} \tag{122}$$

Recall that the reconstructed data has the form

$$\left.\begin{array}{lcll} \mathbf{W}_i(x) & = & \mathbf{W}_i^n + \frac{(x - x_i)}{\Delta x}\Delta_i & \text{if} \quad x \in [0, \Delta x] \\ \mathbf{W}_{i+1}(x) & = & \mathbf{W}_{i+1}^n + \frac{(x - x_{i+1})}{\Delta x}\Delta_i & \text{if} \quad x \in [0, \Delta x] \end{array}\right\}. \tag{123}$$

Substitution of the points x_k in (122) into the appropriate function in (123) will give corresponding values for the vector $\mathbf{W}$ at the foot of the characteristics. It is easy to see that

$$\mathbf{W}(x_k) = \begin{cases} \mathbf{W}_i^n + (\frac{1}{2} - c_k)\Delta_i & if \quad c_k \geq 0\,, \\ \mathbf{W}_{i+1}^n - (\frac{1}{2} + c_k)\Delta_{i+1} & if \quad c_k \leq 0\,. \end{cases} \tag{124}$$

Let us denote the values at $(x_{i+\frac{1}{2}}, \Delta t)$ by $\mathbf{W}^{(\Delta t)}_{i+\frac{1}{2}} = [\rho^\star, u^\star, p^\star, q^\star]^T$. Then, integration of the linearised equations (119) along characteristics produces

$$\left.\begin{array}{lcl} p^\star - \overline{\rho}\,\overline{a}u^\star & = & p_1 - \overline{\rho}\,\overline{a}u_1 \\ p^\star - \overline{a}^2\rho^\star & = & p_2 - \overline{a}^2\rho_2 \\ q^\star & = & q_2 \\ p^\star + \overline{\rho}\,\overline{a}u^\star & = & p_3 + \overline{\rho}\,\overline{a}u_3 \end{array}\right\} \tag{125}$$

The sought solution is given by

$$\left.\begin{array}{rcl} p^{\star} & = & \frac{1}{2}(p_1+p_3)+\frac{1}{2}(u_3-u_1)\overline{\rho}\,\overline{a}\,, \\ u^{\star} & = & \frac{1}{2}(u_1+u_3)+\frac{1}{2}(p_3-p_1)/\overline{\rho}\,\overline{a}\,, \\ \rho^{\star} & = & \rho_2+\frac{p^{\star}-p_2}{\overline{a}^2}\,, \\ q^{\star} & = & q_2\,. \end{array}\right\} \qquad (126)$$

This completes the construction of the intermediate state in (114) following the PLM approach.

8.2. THE COEFFICIENT MATRIX

As for previous schemes we first construct an average vector $\overline{\mathbf{W}}_i$ and then set

$$\overline{\mathbf{A}}_i = \mathbf{A}(\overline{\mathbf{W}}_i)\,.$$

As to the choices for $\overline{\mathbf{W}}_i$ we propose the same expressions as for the GRP schemes, namely (112) and their approximations (113) and (55)–(56).

As described, it is easily checked that the primitive–variable PLM scheme reproduces the first–order non–conservative scheme (79), when the slopes in the data reconstruction stage (82)– (83) are identically zero. Also, it is straightforward to verify that for the linear advection equation (5) the PLM approach of this section reproduces the second and third order accurate schemes listed in Table 1.

A conservative version of the PLM scheme is constructed by evaluating an intercell flux $\mathbf{F}_{i+\frac{1}{2}}^{n+\frac{1}{2}}$ according to (81), with the intermediate state $\mathbf{W}_{i+\frac{1}{2}}^{n+\frac{1}{2}}$ obtained from (114), and advancing the solution according to the conservative formula (40).

9. TVD Version of the MUSCL–Type Schemes

As stated so far, all MUSCL schemes of Sects. 6 to 8 will produce spurious oscillations in the vicinity of large gradients. A Total Variation Diminishing (TVD) version of these schemes is produced by utilising standard theory first developed for conservative methods. There are two possible approaches. The most well–known of the two replaces the slopes Δ_i by limited slopes

$$\overline{\Delta}_i = \begin{cases} max[0, min(\beta\Delta_{i-\frac{1}{2}}, \Delta_{i+\frac{1}{2}}), min(\Delta_{i-\frac{1}{2}}, \beta\Delta_{i+\frac{1}{2}})] \,, & \Delta_{i+\frac{1}{2}} > 0 \,, \\ min[0, max(\beta\Delta_{i-\frac{1}{2}}, \Delta_{i+\frac{1}{2}}), max(\Delta_{i-\frac{1}{2}}, \beta\Delta_{i+\frac{1}{2}})] \,, & \Delta_{i+\frac{1}{2}} < 0 \end{cases} \tag{127}$$

for particular values of the parameter β. The value $\beta = 1$ reproduces the MINBEE (or MINMOD) flux limiter and $\beta = 2$ reproduces the SUPERBEE flux limiter, see Sect. 5.2.3. Here $\overline{\Delta}_i$ is a vector and thus the limiting process is applied to each component of the system being solved.

Another approach is to replace the slope vector Δ_i in the reconstruction stage, see equations (82)–(83), by a limited slope $\overline{\Delta}_i$,

$$\overline{\Delta}_i = \xi(r)\Delta_i \,. \tag{128}$$

where $\xi(r)$ is a slope limiter that lies in the TVD region

$$\xi(r) = 0 \text{ for } r \le 0 \,, \;\; 0 \le \xi(r) \le \min\{\xi_L(r), \xi_R(r)\} \text{ for } r > 0 \,, \tag{129}$$

with

$$\left.\begin{aligned} \xi_L(r) &= \frac{2\beta_{i-\frac{1}{2}} r}{1-\omega+(1+\omega)r} \,, \\ \xi_R(r) &= \frac{2\beta_{i+\frac{1}{2}}}{1-\omega+(1+\omega)r} \,, \\ r &= \frac{\Delta_{i-\frac{1}{2}}}{\Delta_{i+\frac{1}{2}}} \,, \end{aligned}\right\} \tag{130}$$

and

$$\beta_{i-\frac{1}{2}} = \frac{2}{1+c} \,, \quad \beta_{i+\frac{1}{2}} = \frac{2}{1-c} \,. \tag{131}$$

Possible slope limiters are now given. A slope limiter that is analogous to SUPERBEE is

$$\xi_{sb}(r) = \begin{cases} 0 \,, & \text{if } r \le 0 \,, \\ 2r \,, & \text{if } 0 \le r \le \frac{1}{2} \,, \\ 1 \,, & \text{if } \frac{1}{2} \le r \le 1 \,, \\ \min\{r, \xi_R(r), 2\} \,, & \text{if } r \ge 1 \,. \end{cases} \tag{132}$$

A van Leer–type slope limiter is

$$\xi_{vl}(r) = \begin{cases} 0 \,, & r \le 0 \,, \\ \min\{\frac{2r}{1+r}, \xi_R(r)\} \,, & r \ge 0 \,. \end{cases} \tag{133}$$

A van Albada–type slope limiter is

$$\xi_{va}(r) = \begin{cases} 0\,, & r \leq 0\,, \\ \min\{\dfrac{r(1+r)}{1+r^2}, \xi_R(r)\}\,, & r \geq 0\,. \end{cases} \tag{134}$$

A MINBEE–type slope limiter is

$$\xi_{mb}(r) = \begin{cases} 0\,, & r \leq 0\,, \\ r\,, & 0 \leq r \leq 1\,, \\ \min\{1, \xi_R(r)\}\,, & r \geq 1\,. \end{cases} \tag{135}$$

For general background and full details on slope limiters see the textbook (Toro, 1997b).

10. Adaptive Primitive—Conservative Schemes

Computational experience and recent theoretical results by Hou and LeFloch (Hou and LeFloch, 1994) show that in the presence of shock waves, primitive schemes will compute a solution propagating at the wrong speed. See computational results of Fig. 6 and compare these with their conservative counterparts of Fig. 7. If a local correction to the primitive scheme in the vicinity of the shock is applied, then the correct solution may be obtained. In (Toro, 1994), see also (Toro, 1995c) and (Ivings, 1996), it was proposed to apply such correction by switching to a conservative method in a cell i, only if the two neighbouring Riemann problems contain a shock travelling in the direction of the mesh point i. Details of this method are given in (Toro, 1997b), but for completeness we briefly review the scheme here.

Consider a mesh point i and the two neighbouring Riemann problem solutions $\mathbf{W}^*_{i-\frac{1}{2}}(x/t)$, $\mathbf{W}^*_{i+\frac{1}{2}}(x/t)$. These local solutions contain all the information we require to switch scheme. Denote by $p^*_{i-\frac{1}{2}}$ and $p^*_{i+\frac{1}{2}}$ the solutions for pressure in the star regions of $\mathbf{W}^*_{i-\frac{1}{2}}(x/t)$ and $\mathbf{W}^*_{i+\frac{1}{2}}(x/t)$ respectively. Denote by $S_{i-\frac{1}{2}}$ and $S_{i+\frac{1}{2}}$ the speeds of the fastest shocks in the solutions $\mathbf{W}^*_{i-\frac{1}{2}}(x/t)$ and $\mathbf{W}^*_{i+\frac{1}{2}}(x/t)$ respectively, travelling towards the mesh point i. Now define

$$S_{switch} = 1 + \epsilon\,, \tag{136}$$

where ϵ is a small positive quantity yet to be defined. If

$$\left.\begin{array}{l} p^*_{i-\frac{1}{2}}/p^n_i > S_{switch} \quad \text{and } S_{i-\frac{1}{2}} > 0\,, \\ \text{or if} \\ p^*_{i+\frac{1}{2}}/p^n_i > S_{switch} \quad \text{and } S_{i+\frac{1}{2}} < 0\,, \end{array}\right\} \tag{137}$$

then mesh point i is advanced via a conservative method (40); otherwise one advances the solution via a primitive scheme (41).

First we note the physical character of the switching mechanism. Strictly speaking, for any $S_{switch} > 1$, that is for any ϵ with $\epsilon > 0$, a shock wave exists. In practice it is sufficient to select a small positive value for ϵ. The choice of ϵ in (136) is not too critical and any value in the range $(0, 0.1)$ gives satisfactory results. If ϵ is too large one risks computing shocks of *moderate* strength with the primitive scheme, which will lead to shock waves being in the wrong position. Smaller values of ϵ simply mean that the conservative method is used at more mesh points, perhaps unnecessarily.

Alternative hybrid schemes have been presented by Karni (Karni, 1995) and independently by Abgrall (Abgrall, 1996).

11. Numerical Results

Here we select three test problems to illustrate the performance of some of the methods presented in this paper. Tests 1 and 2 pertain to the one-dimensional time dependent Euler equations for a gamma–law gas with $\gamma = 1.4$. Test 3 concerns the Euler equations augmented by an extra conservation law to model the interaction of two gamma–law gases, see equations (35), (37). Tests 1 and 3 have exact solution, which we use to compare with the numerical results. In all test problems we use the methods with the exact Riemann solver and with a CFL coefficient $C_{cfl} = 0.9$.

Test 1 has domain [0,4] with shock–tube like data for $\mathbf{W} = [\rho, u, p]^T$; the initial discontinuity is positioned at $x_0 = 2.0$. The data states are $\mathbf{W}_L = [1.0, 0.0, 1.0]^T$ for $x \leq x_0$ and $\mathbf{W}_R = [0.125, 0.0, 0.1]^T$ for $x > x_0$; the chosen output time is $t = 1.0$ units. Test 2 has domain [0,1] with double shock–tube like data for $\mathbf{W} = [\rho, u, p]^T$. This is the Woodward and Colella blast wave problem (Woodward and Colella, 1984). The initial discontinuities are positioned at $x_0 = 0.1$ and $x_1 = 0.9$. The data states are $\mathbf{W}_L = [1.0, 0.0, 1.0]^T$ for $x \leq x_0$, $\mathbf{W}_M = [1.0, 0.0, 1000.0]^T$ for $x_0 \leq x \leq x_1$ and $\mathbf{W}_R = [1.0, 0.0, 100.0]^T$ for $x > x_1$. The chosen output time is $t = 0.038$. Test 3 has domain [0,4] with shock–tube like data for $\mathbf{W} = [\rho, u, p, \gamma]^T$, with the initial discontinuity at $x_0 = 2$. The data states are $\mathbf{W}_L = [1.0, 1.0, 1.0, 1.4]^T$ for $x \leq x_0$ and $\mathbf{W}_R = [0.1, 1.0, 1.0, 1.2]^T$ for $x > x_0$. The chosen output time is $t = 1.0$.

The results for Test 1 are shown in Figs. 10 to 17. Fig. 10 shows results for the first–order primitive upwind scheme. As anticipated the shock wave has the wrong strength and the wrong position. The position error grows with time. The rarefaction wave and the position of the contact appear to be correct. Fig. 11 shows the corresponding result for the second–order primitive WAF method without the TVD condition. The position of the shock is still incorrect and there are now spurious oscillations, not just in the vicinity of discontinuities. Fig. 12 shows the corresponding results from the WAF TVD primitive scheme. The rarefaction and contact position are very accurate by the shock is still in the wrong place. To assess the convergence trend of this method we computed the solution with a mesh of 500 points; the result of Fig. 13 clearly indicates that the primitive TVD scheme converges but does so to the wrong solution. Fig. 14 shows the result obtained from the adaptive primitive/conservative scheme as described in Sect. 10; here both the primitive and conservative schemes are of the WAF type. It appears as if now the solution is correct. Fig. 15 shows the corresponding result from the conservative WAF method. This confirms that the adaptive scheme is correct, compare with Fig. 14. In the adaptive scheme the primitive method was used in 97% of all the computations; the switching pressure ratio parameter was $\epsilon = 0.05$. Fig. 16 shows the adaptive scheme with a mesh of 500 points. It appears as if the method does converge and does so to the correct solution. In this case the primitive method acted in 99% of the computations. Compare this result with that of Fig. 17 for the conservative method.

Test 2 does not have an exact solution and hence we use a well–tested conservative method with a fine mesh to assess the new methods. The main feature of this test is the presence of very strong shock waves and complex wave interaction; the left and right boundaries are stationary reflective walls. The shock wave emerging from the left *diaphragm* has shock Mach number 198. Obviously the ideal gas assumption is wholly inadequate here, but the challenging nature of the test problem for numerical methods remains unquestionable. Results for Test 2 are shown in Figs. 18 and 19. In Fig. 18 we compare the primitive second–order WAF TVD scheme (symbols) and the conservative WAF TVD scheme (line). By regarding the conservative solution as the correct solution, it is seen that the primitive scheme solution is completely wrong. Fig. 19 show a comparison between the adaptive primitive/conservative WAF TVD scheme (symbol) and the conservative WAF TVD scheme (line). To plotting accuracy the results are indistinguishable. In the adaptive scheme the primitive algorithm acted in 99% of the computations.

Results for Test 3 are shown in Figs. 20 to 26. This problem has exact solution and involves a contact discontinuity and an interface between two gamma–law gases. Both pressure and velocity are constant. Fig. 20 shows the result for the Godunov (conservative) method. Note the spurious pressure and velocity perturbations. The density profile is not monotone and the gamma profile is in the wrong position. Fig. 21 shows the result for a fine mesh of 500 cells. A slight improvement is observed, but the question is, does the numerical solution converge to the correct solution ? Note that the origin of the spurious oscillations is to be found in the first–order method and has nothing to do with the limiter, the Riemann solver or even the method, although these are factors that will influence the particular shape of the oscillations. Their origin is in the conservative character of the method (Toro, 1997a). In fact the perturbations are also present for CENTRED methods, such as Lax–Friedrichs, which is the most diffusive of all stable methods. It is found theoretically (Toro, 1997a) that for a range of Courant numbers the pressure perturbations of the Lax–Friedrichs method are actually larger than those of the Godunov method. Fig. 22 shows the corresponding result for the WAF TVD scheme (conservative). A slight improvement is observed, relative to the Godunov method of Fig. 20. The fine mesh solution of Fig. 23 shows a small improvement relative to Fig. 22. Fig. 24 shows the result from the primitive WAF TVD scheme; the solution looks very accurate, there are no spurious pressure and velocity perturbations.

The fine mesh results of Fig. 25 suggest that the method converges and does so to the correct solution. Fig. 26 shows the result from the adaptive primitive/conservative WAF scheme. The result is correct and is indistinguishable from the primitive result of Fig. 24. Fig. 27 shows the corresponding fine mesh solution.

12. Conclusions

A collection of numerical methods for non–linear systems of hyperbolic conservation laws have been presented. New primitive variable second–order and TVD schemes have been constructed following the WAF, MUSCL–HANCOCK, GRP and PLM approaches. These contain a first–order upwind non–conservative scheme that is the analogue of the Godunov method. For scalar conservation laws third accurate schemes are also derived. Each primitive scheme has an associated conservative method that is most easily constructed from the information made available by the primitive scheme. In the presence of shock waves, the primitive schemes can be locally corrected by switching to the associated conservative method. Partial results suggest that the adaptive primitive/conservative schemes compute the correct solution, indistinguisable from those of conservative methods, even in the presence of very strong shock waves and complex wave interaction, and they give better results than conservative methods for certain problems, see Test 3.

It would be desirable to systematically assess all the methods presented here, and for a wide range of flow conditions. Such task is beyond the scope of the present paper. Having established that the schemes are useful for solving non–linear systems of conservation laws, it would then be highly desirable to study in more detail some of the theoretical aspects of the methods, such as converge and error estimates for the primitive and adaptive schemes. This is the subject of current investigations.

Some of the numerical experiments, see results for Test 3, raise a number of issues. First, they *explain* difficulties observed, but not widely reported, in the computation of problems involving material interfaces (Clarke *et. al.*, 1993), (Karni, 1994). It is tempting to blame such oscillations on the Riemann solver or the flux limiter, when using high–resolution upwind methods. As indicated in (Toro, 1994), the same difficulty is observed when computing shear waves in the two and three dimensional compressible Euler equations. A problem of current interest is the so called *carbuncle phenomenon*. The author conjectures that the *cause* of such problem is to be found in the conservative character of the methods for the two classes of problems above, and not on the Riemann solver or the flux limiter, although these may influence the evolution of the problem.

References

Abgrall, R. (1996). How to Prevent Pressure Oscillations in Multicomponent Flow Calculations: A Quasiconservative Approach. *J. Comput. Phys.*, 125:150–160, 1996.

Arora, M. and Roe, P. L. (1997). A Well–Behaved TVD Limiter for High–Resolution Calculations of Unsteady Flow. *J. Comput. Phys.*, 132:3–11, 1997.

Ben-Artzi M. and Falcovitz J. (1984). A Second Order Godunov-Type Scheme for Compressible Fluid Dynamics. *J. Comput. Phys.*, 55:1–32, 1984.

Chorin, A. J. (1967). A Numerical Method for Solving Viscous Flow Problems. *J. Comput. Phys.*, 2:12–26, 1967.

Clarke, J. F., Karni, S., Quirk, J. J., Simmons, L. G., Roe, P. L. and Toro, E. F. (1993). Numerical Computation of Two–Dimensional, Unsteady Detonation Waves in High Energy Solids. *J. Comput. Phys.*, 106:215–233, 1993.

Colella, P. (1985). A Direct Eulerian MUSCL Scheme for Gas Dynamics. *SIAM J. Sci. Stat. Comput.*, 6:104–117, 1985.

Courant, R., Isaacson, E. and Rees, M. (1952). On the Solution of Nonlinear Hyperbolic Differential Equations by Finite Differences. *Comm. Pure. Appl. Math.*, 5:243–255, 1952.

Glimm, J. (1965). Solution in the Large for Nonlinear Hyperbolic Systems of Equations. *Comm. Pure. Appl. Math.*, 18:697–715, 1965.

Harten, A. (1983). High Resolution Schemes for Hyperbolic Conservation Laws. *J. Comput. Phys.*, 49:357–393, 1983.

Harten, A., Lax, P. D. and van Leer, B. (1983). On Upstream Differencing and Godunov–Type Schemes for Hyperbolic Conservation Laws. *SIAM Review*, 25(1):35–61, 1983.

Hirsch, C. (1990). *Numerical Computation of Internal and External Flows, Vol. II: Computational Methods for Inviscid and Viscous Flows.* Wiley, 1990.

Hou, T. Y. and LeFloch, P. (1994). Why Non–Conservative Schemes Converge to the Wrong Solutions: Error Analysis. *Math. of Comput.*, 62:497–530, 1994.

Ivings, M. J., Causon, D. M. and Toro, E. F. (1996). On Hybrid High–Resolution Upwind Methods for Multicomponent Flows. *ZAMM*, 77, Issue 9:645–668, 1997.

Karni, S. (1992). Viscous Shock Profiles and Primitive Formulations. *SIAM J. Numer. Anal.*, 29(6):1592–1609, 1992.

Karni, S. (1994). Multicomponent Flow Calculations Using a Consistent Primitive Algorithm. *J. Comput. Phys:*, 112(1):31–43, 1994.

Karni, S. (1995). Hybrid Multifluid Algorithms. Technical Report 95-001, Courant Mathematics and Computing Laboratory, 1995.

Lax P. D. and Wendroff, B. (1960). Systems of Conservation Laws. *Comm. Pure Appl. Math.*, 13:217–237, 1960.

NAG Library (1996). NAG Library Routines, Mark 18, Routines D03PWF and D03PXF, 1996.

Osher, S. and Solomon, F. (1982). Upwind Difference Schemes for Hyperbolic Conservation Laws. *Math. Comp.*, 38,158:339–374, 1982.

Roe, P. L. (1983). Some Contributions to the Modelling of Discontinuous Flows. In *Proceedings of the SIAM/AMS Seminar*, San Diego, 1983.

Stoker, J. J. (1992). *Water Waves. The Mathematical Theory with Applications.* John Wiley and Sons, 1992.

Sweby, P. K. (1982). *Shock Capturing Schemes.* PhD thesis, Department of Mathematics, University of Reading, UK, 1982.

Sweby, P. K. (1984). High Resolution Schemes Using Flux Limiters for Hyperbolic Conservation Laws. *SIAM J. Numer. Anal.*, 21:995–1011, 1984.

Toro, E. F. (1989). A Weighted Average Flux Method for Hyperbolic Conservation Laws. *Proc. Roy. Soc. London*, A423:401–418, 1989.

Toro, E. F. (1991). A Linearised Riemann Solver for the Time–Dependent Euler Equations of Gas Dynamics. *Proc. Roy. Soc. London*, A434:683–693, 1991.

Toro, E. F. (1994). Defects of Conservative Approaches and Adaptive Primitive-Conservative Schemes for Computing Solutions to Hyperbolic Conservation Laws. Technical Report MMU 9401, Department of Mathematics and Physics, Manchester Metropolitan University, UK, 1994.

Toro, E. F. (1995a). Direct Riemann Solvers for the Time–Dependent Euler Equations. *Shock Waves*, 5:75–80, 1995.

Toro, E. F. (1995b). MUSCL–Type Primitive Variable Schemes. Technical Report MMU–9501, Department of Mathematics and Physics, Manchester Metropolitan University, UK, 1995.

Toro, E. F. (1995c). On Adaptive Primitive-Conservative Schemes for Conservation Laws. In M. M. Hafez, editor, *Sixth International Symposium on Computational Fluid Dynamics: A Collection of Technical Papers*, volume 3, pages 1288–1293, Lake Tahoe, Nevada, USA, September 4-8, 1995.

Toro, E. F. (1995d). Some IVPs for Which Conservative Methods Fail Miserably. In M. M. Hafez, editor, *Sixth International Symposium on Computational Fluid Dynamics: A Collection of Technical Papers*, volume 3, pages 1294–1299, Lake Tahoe, Nevada, USA, September 4-8, 1995.

Toro, E. F. (1997a). Anomalies of Conservative Methods: Analysis and Numerical Evidence. *Submitted*, 1997.

Toro, E. F. (1997b). *Riemann Solvers and Numerical Methods for Fluid Dynamics.* Springer–Verlag, 1997. Berlin, Heidelberg.

Toro, E. F. and Roe, P. L. (1987). A Hybridised High-Order Random Choice Method for Quasi-Linear Hyperbolic Systems. In Grönig, editor, *Proc. 16th Intern. Symp. on Shock Tubes and Waves*, pages 701–708, Aachen, Germany, July 1987.

van Leer, B. (1976). MUSCL, A New Approach to Numerical Gas Dynamics. In *Computing in Plasma Physics and Astrophysics*, Max–Planck–Institut für Plasma Physik, Garchung, Germany, April 1976.

van Leer, B. (1984). On the Relation Between the Upwind-Differencing Schemes of Godunov, Enguist-Osher and Roe. *SIAM J. Sci. Stat. Comput.*, 5(1):1–20, 1985.

Woodward, P. and Colella, P. (1984). The Numerical Simulation of Two–Dimensional Fluid Flow with Strong Shocks. *J. Comput. Phys.*, 54:115–173, 1984.

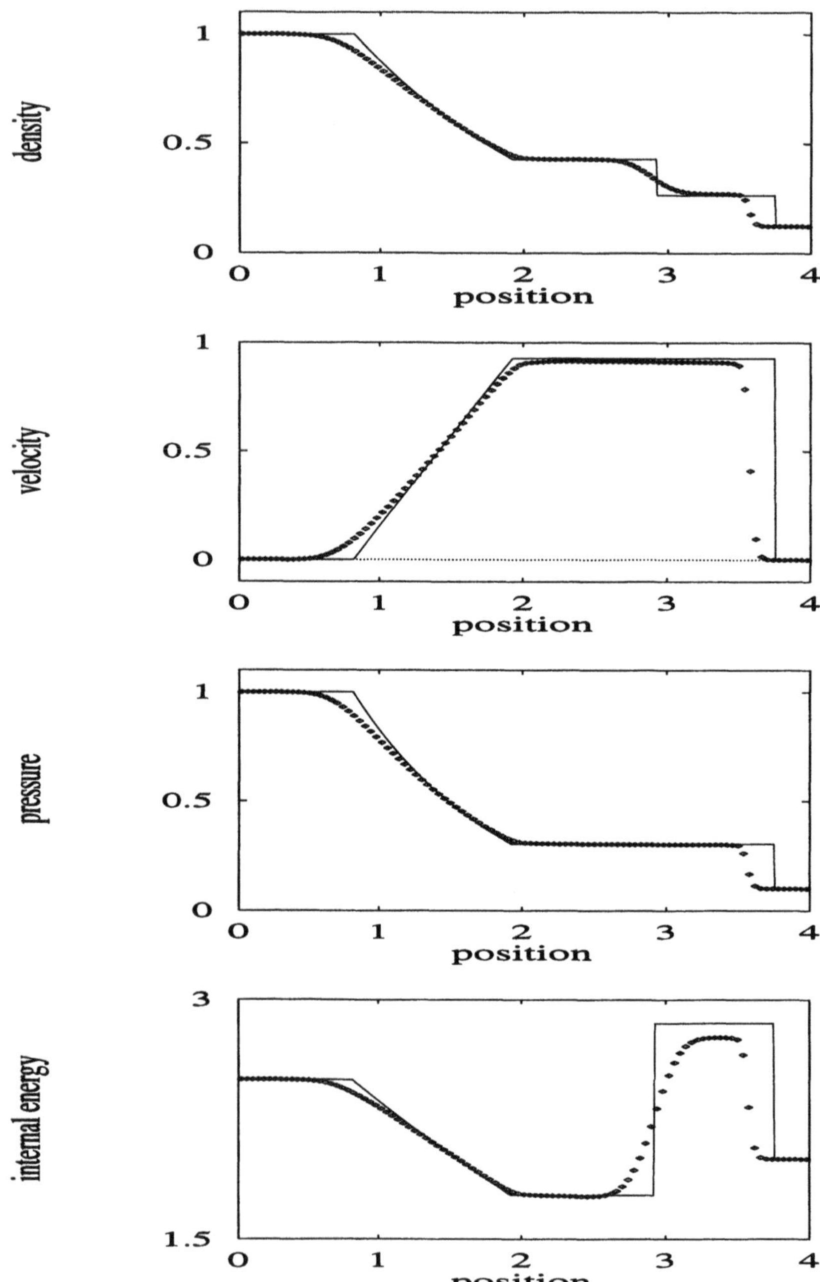

Figure 10. TEST 1. Primitive first order upwind scheme with mesh $M = 100$. Numerical (symbol) and exact (line) solutions are compared at time 1.0

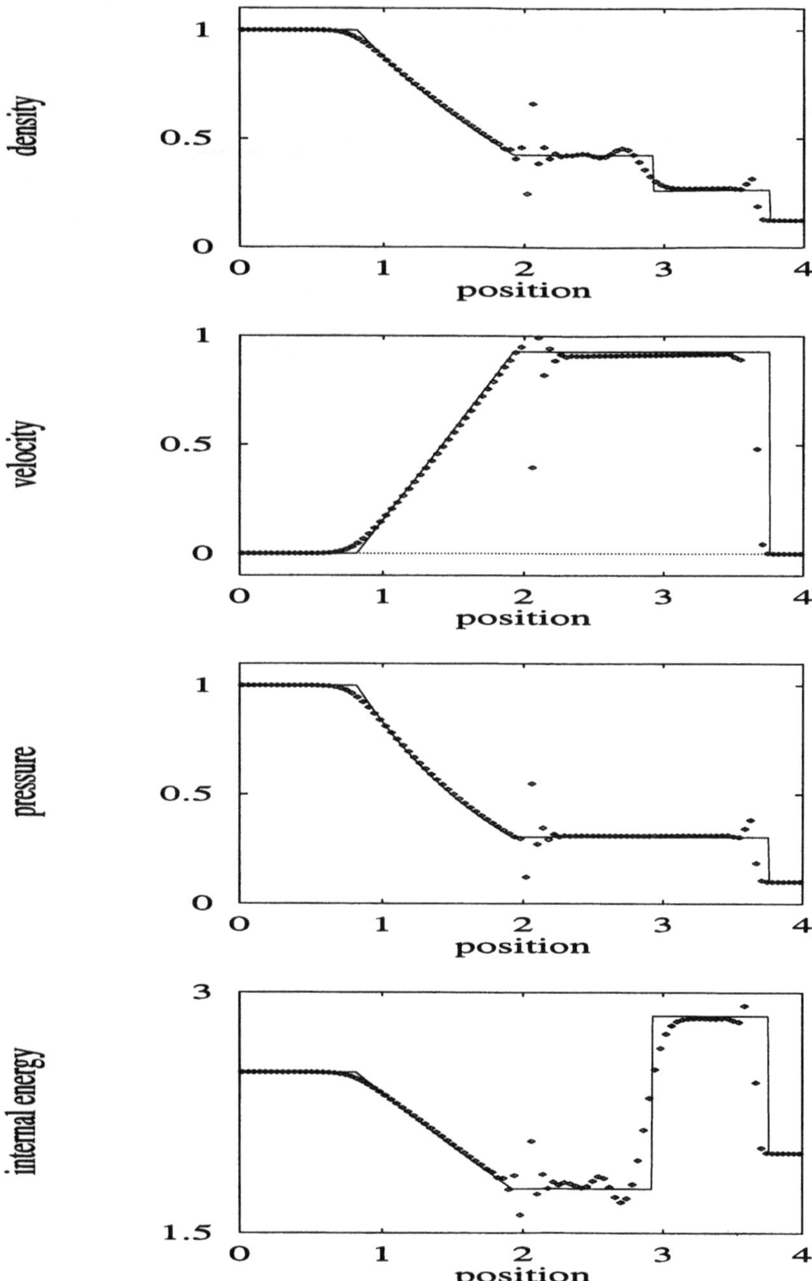

Figure 11. TEST 1. Primitive second order upwind scheme with mesh $M = 100$. Numerical (symbol) and exact (line) solutions are compared at time 1.0

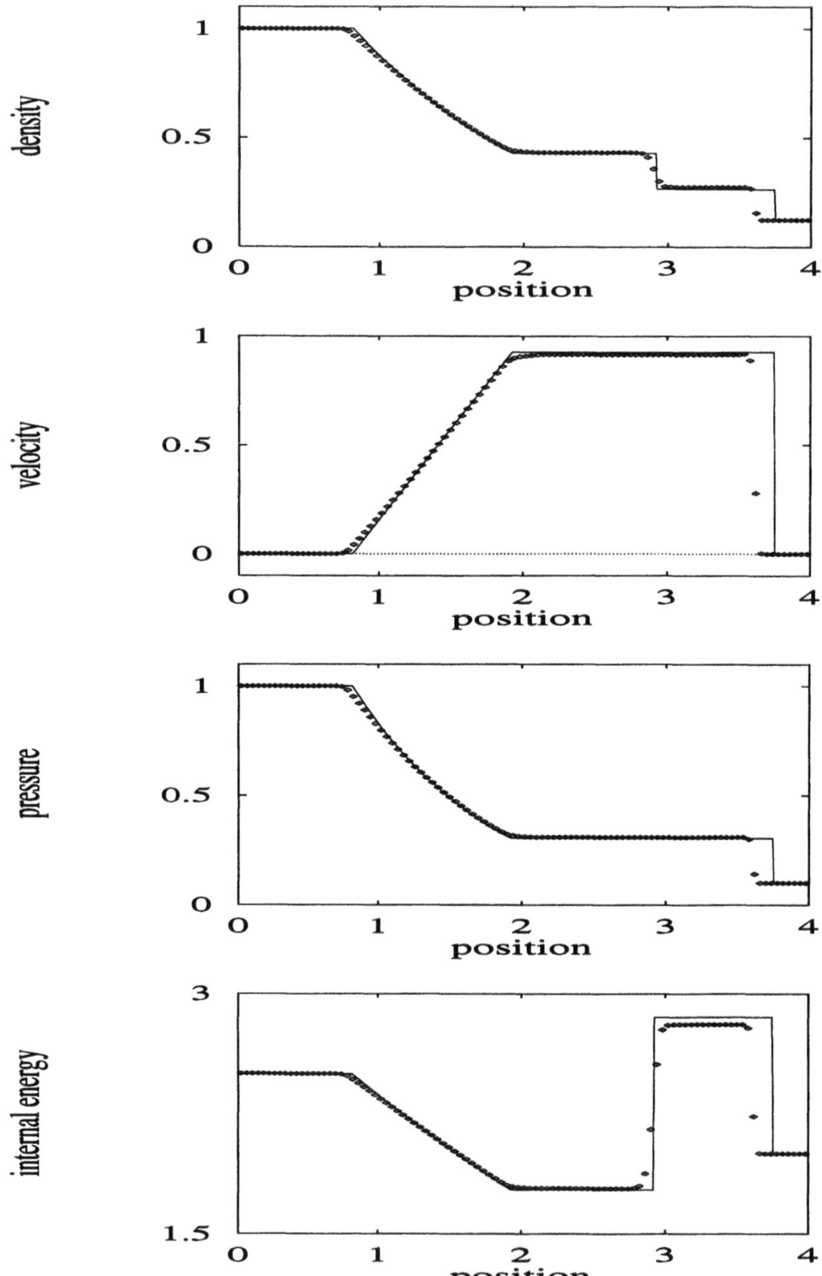

Figure 12. TEST 1. Primitive WAF TVD scheme with SUPERBEE limiter and mesh $M = 100$. Numerical (symbol) and exact (line) solutions are compared at time 1.0

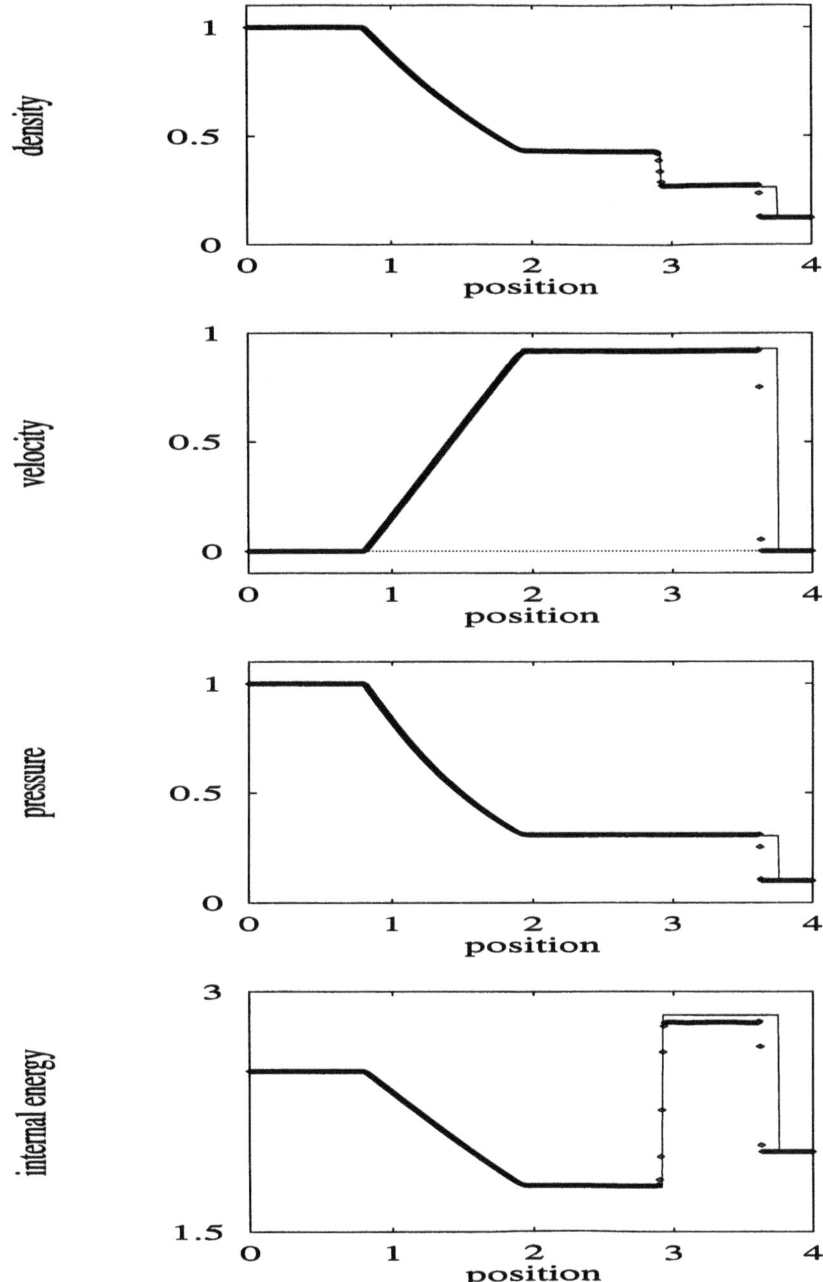

Figure 13. TEST 1. Primitive WAF TVD scheme with SUPERBEE limiter and mesh $M = 500$. Numerical (symbol) and exact (line) solutions are compared at time 1.0

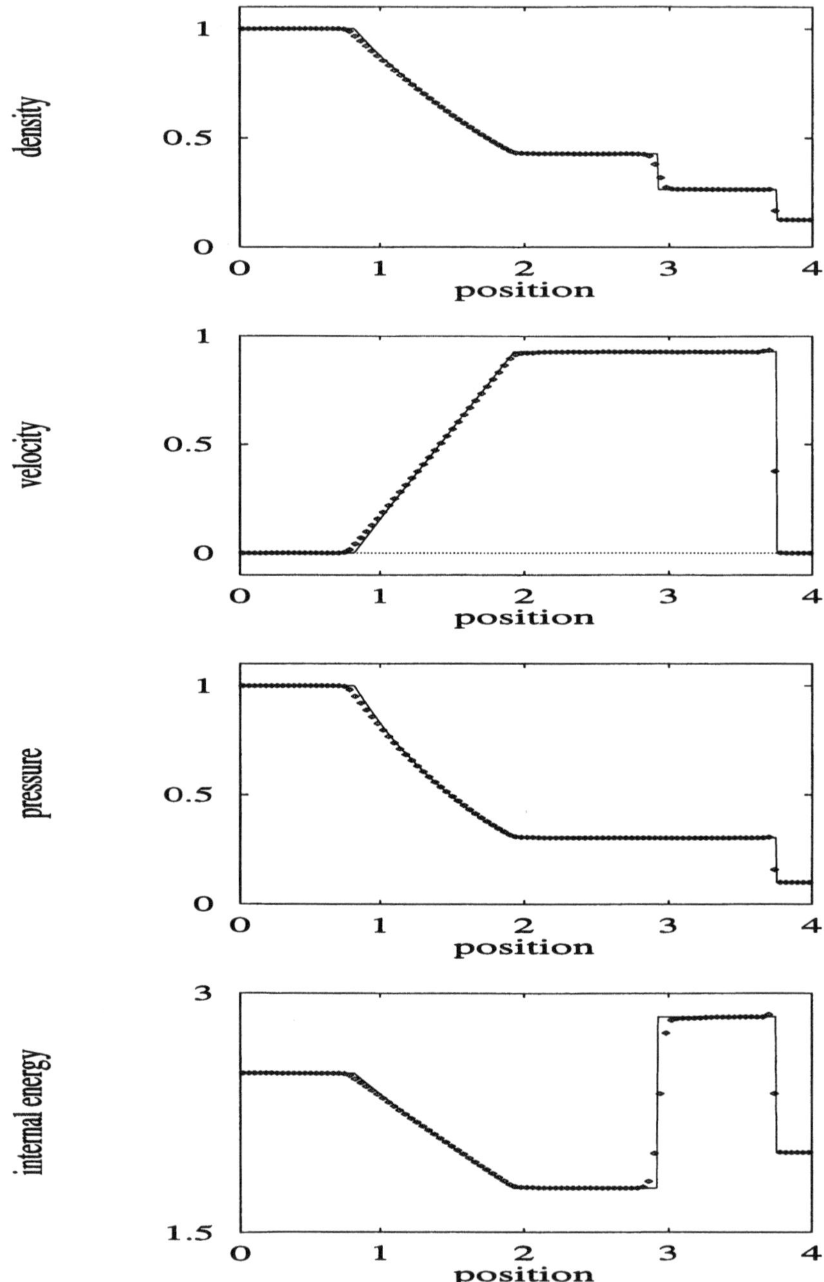

Figure 14. TEST 1. Adaptive primitive/conservative WAF TVD scheme with SUPERBEE limiter and mesh $M = 100$. Numerical (symbol) and exact (line) solutions are compared at time 1.0

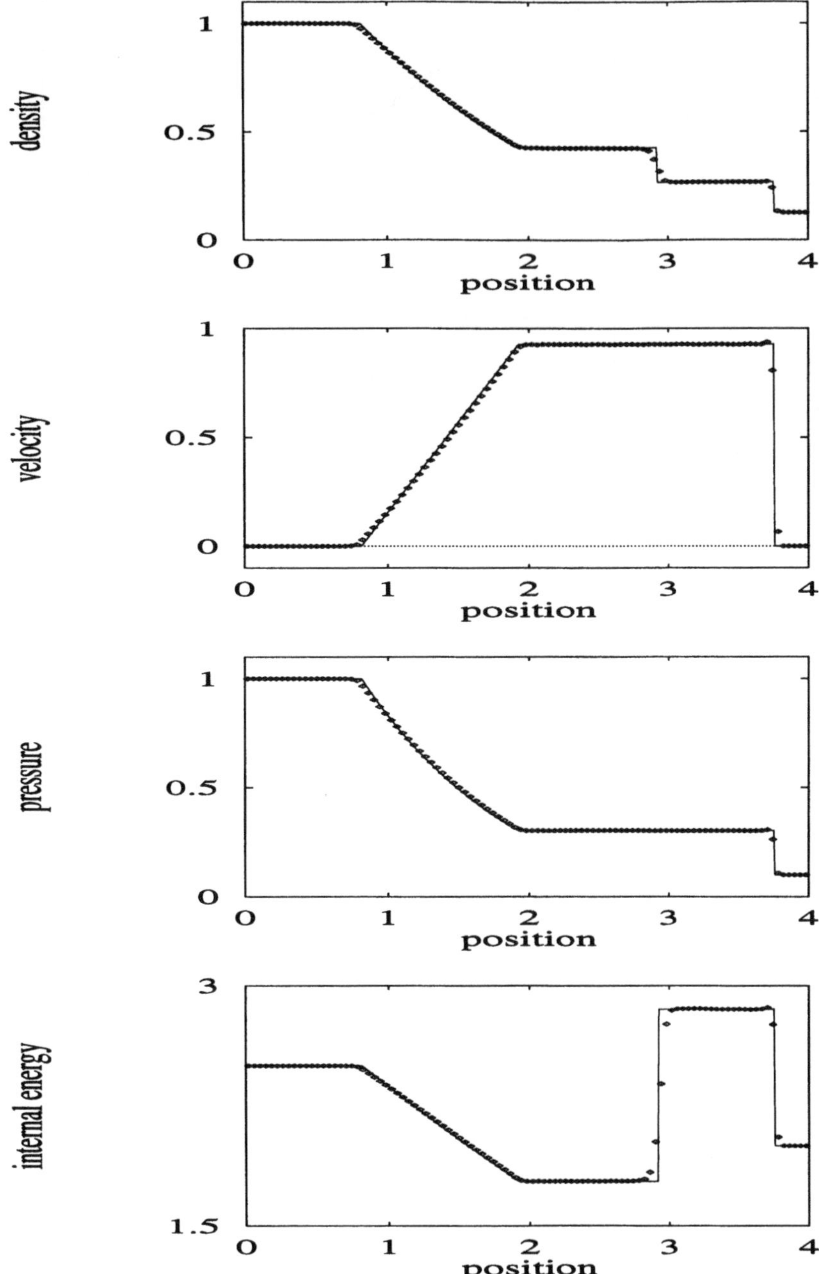

Figure 15. TEST 1. Conservative WAF TVD scheme with SUPERBEE limiter and mesh $M = 100$. Numerical (symbol) and exact (line) solutions are compared at time 1.0

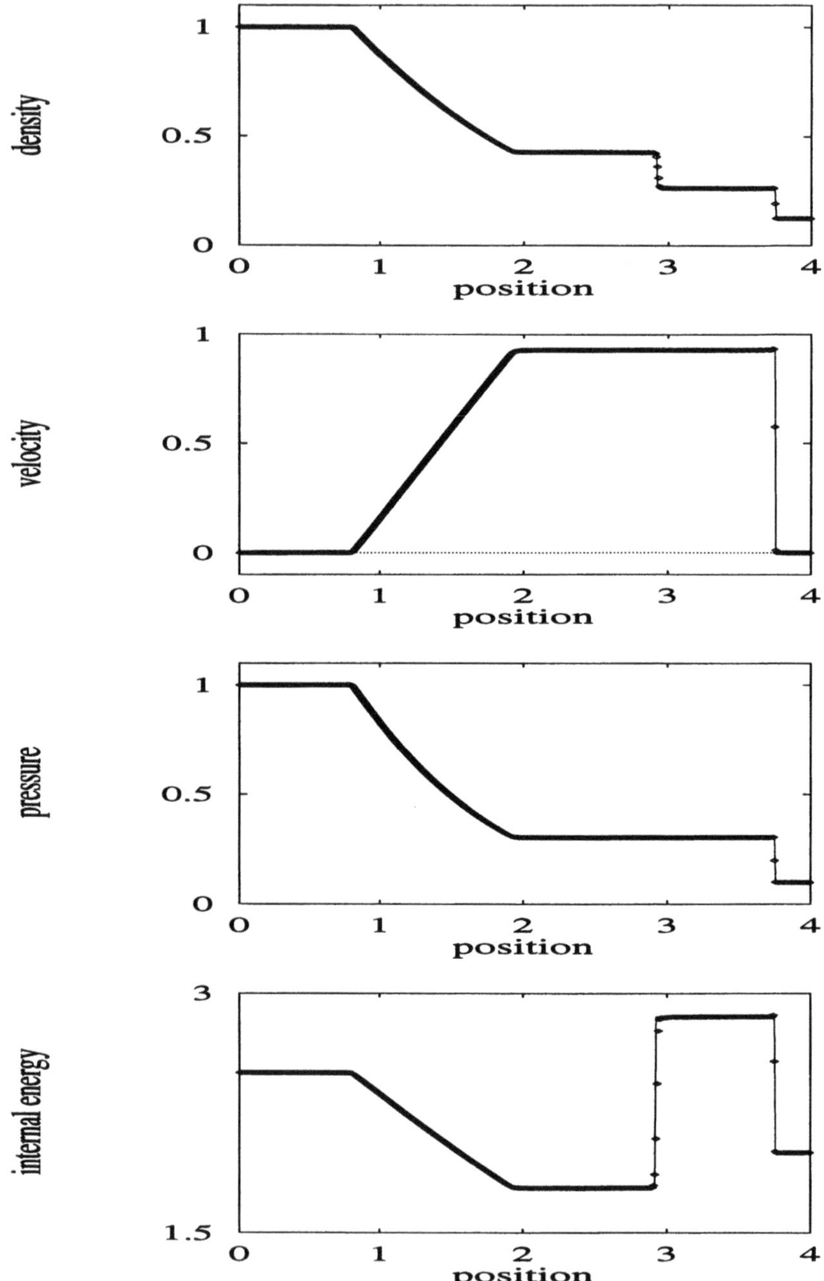

Figure 16. TEST 1. Adaptive primitive/conservative WAF TVD scheme with SUPERBEE limiter and mesh $M = 500$. Numerical (symbol) and exact (line) solutions are compared at time 1.0

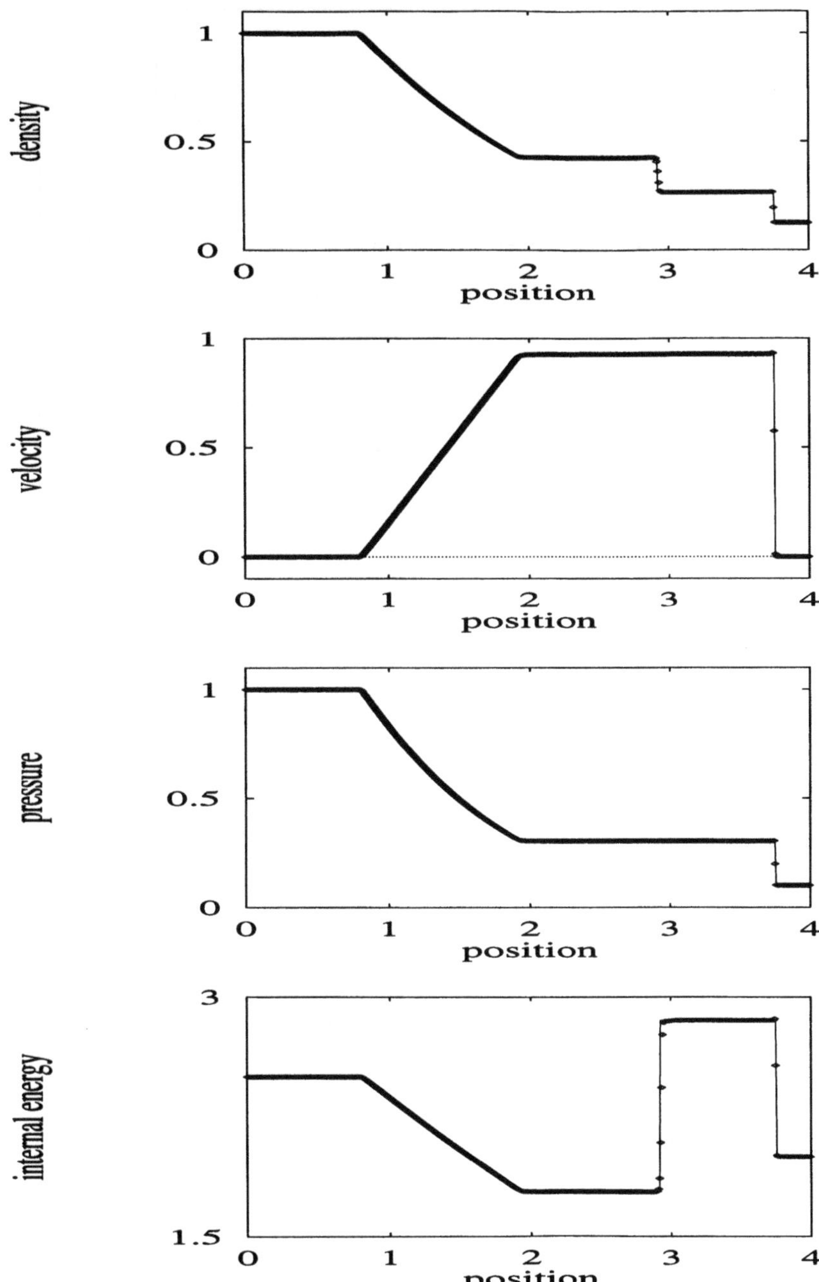

Figure 17. TEST 1. Conservative WAF TVD scheme with SUPERBEE limiter and mesh $M = 500$. Numerical (symbol) and exact (line) solutions are compared at time 1.0

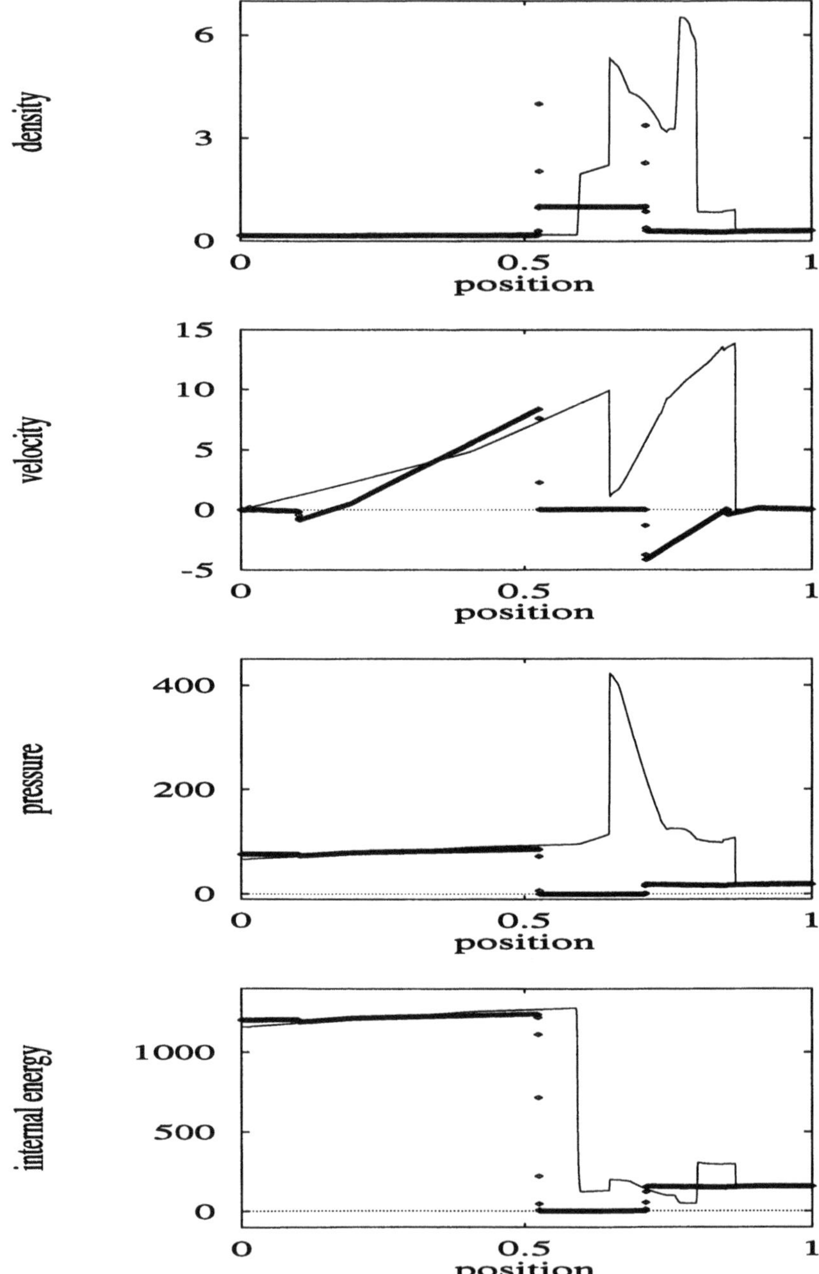

Figure 18. TEST 2. Comparison between the primitive TVD WAF scheme (symbol) and the conservative TVD WAF scheme (line), using mesh $M = 2000$, at time 0.038

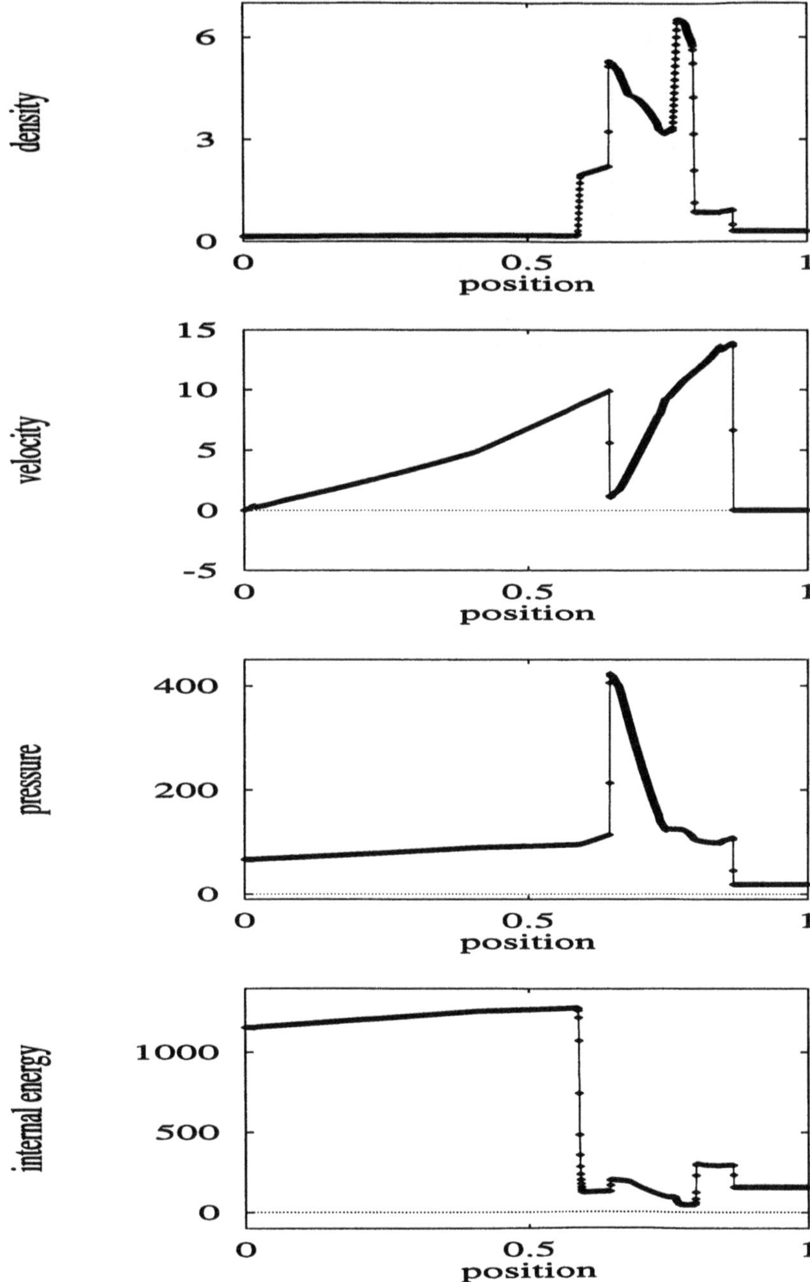

Figure 19. TEST 2. Comparison between the adaptive primitive/conservative TVD WAF scheme (symbol) and the conservative TVD WAF scheme (line), using mesh $M = 2000$, at time 0.038

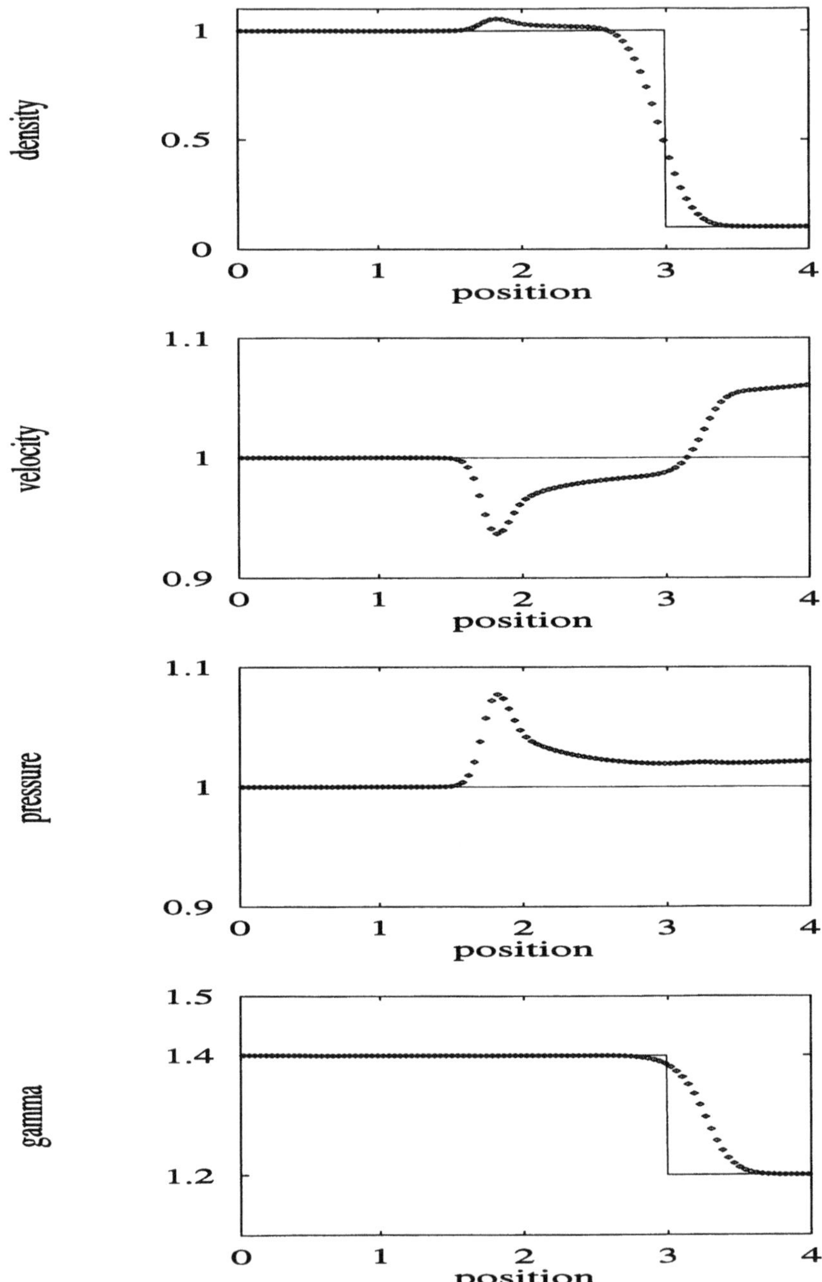

Figure 20. TEST 3. Comparison between the first–order Godunov (conservative) scheme (symbol) using mesh $M = 100$ and the exact solution (line), at time 1.0

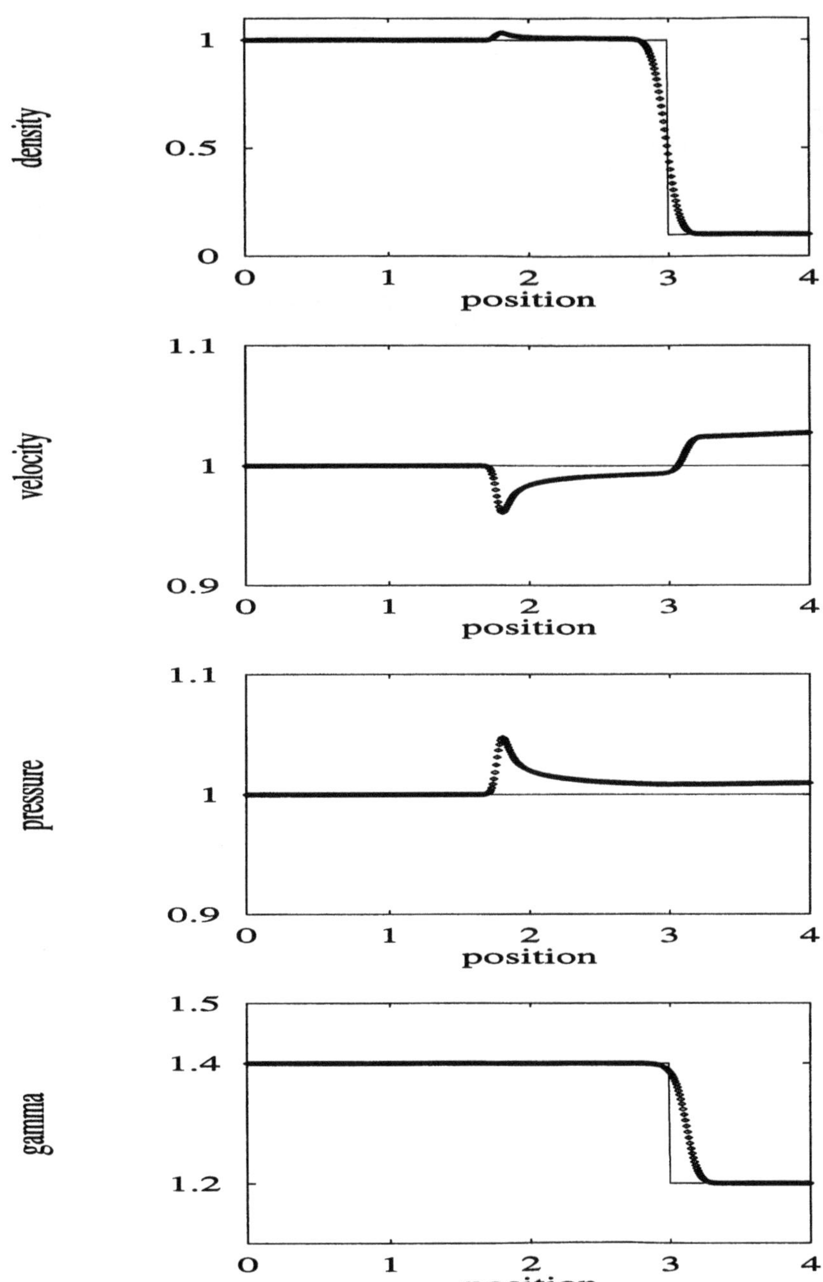

Figure 21. TEST 3. Comparison between the first–order Godunov (conservative) scheme (symbol) using mesh $M = 500$ and the exact solution (line), at time 1.0

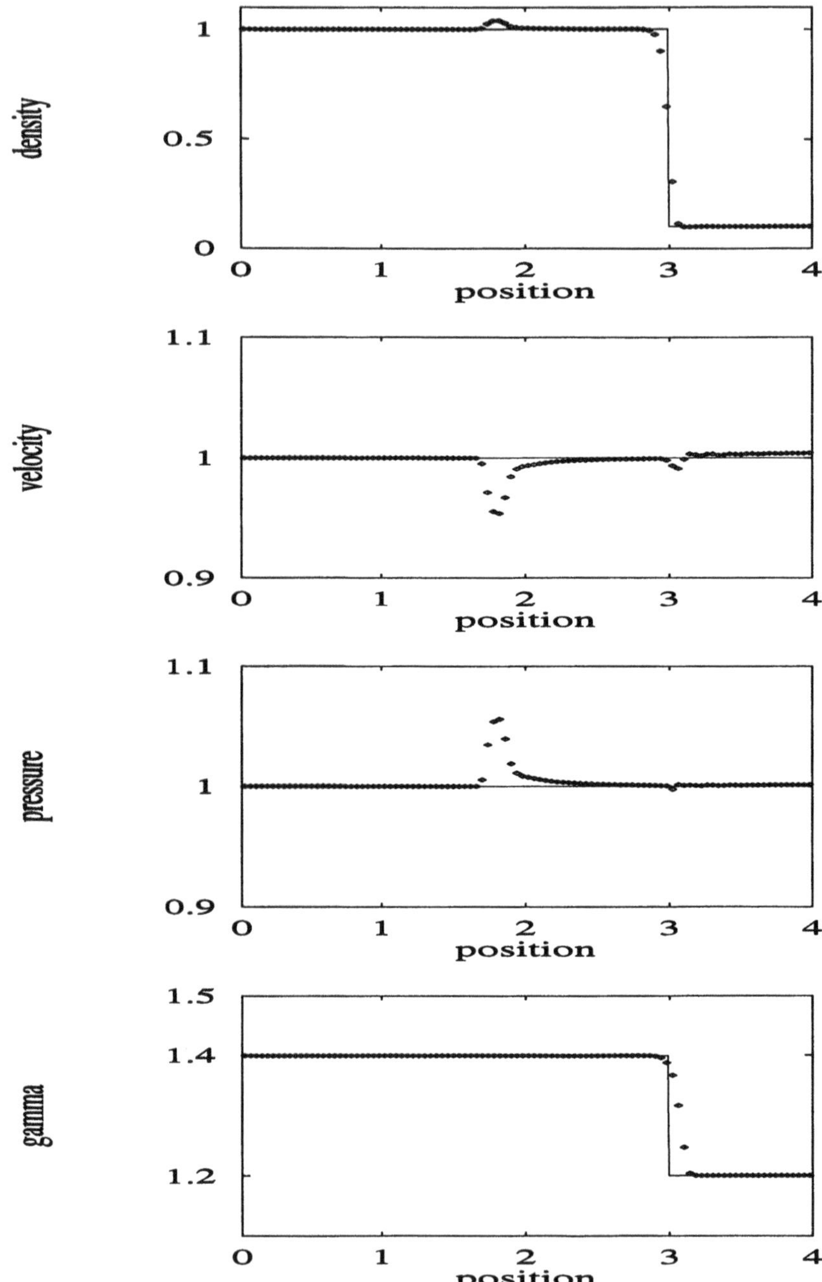

Figure 22. TEST 3. Comparison between the WAF TVD (conservative) scheme (symbol) using mesh $M = 100$ and the exact solution (line), at time 1.0

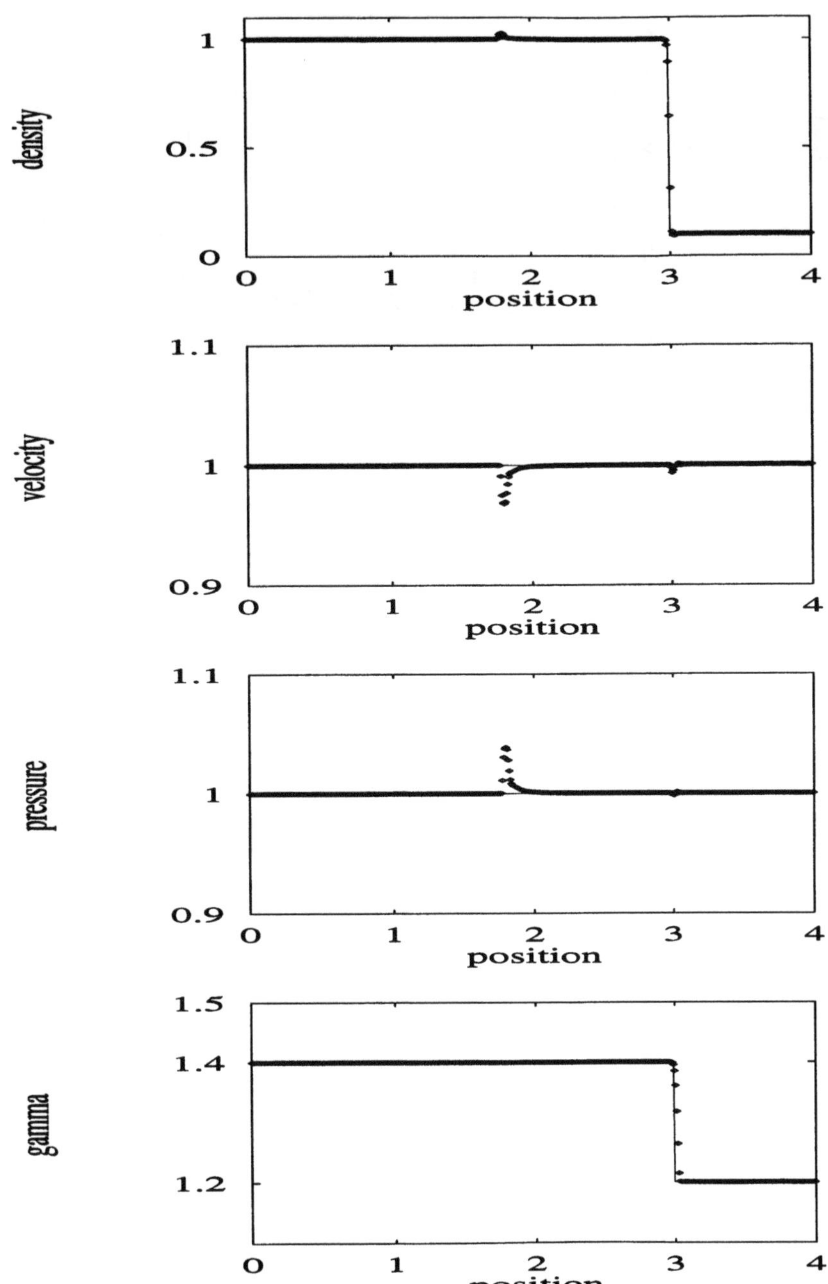

Figure 23. TEST 3. Comparison between the WAF TVD (conservative) scheme (symbol) using mesh $M = 500$ and the exact solution (line), at time 1.0

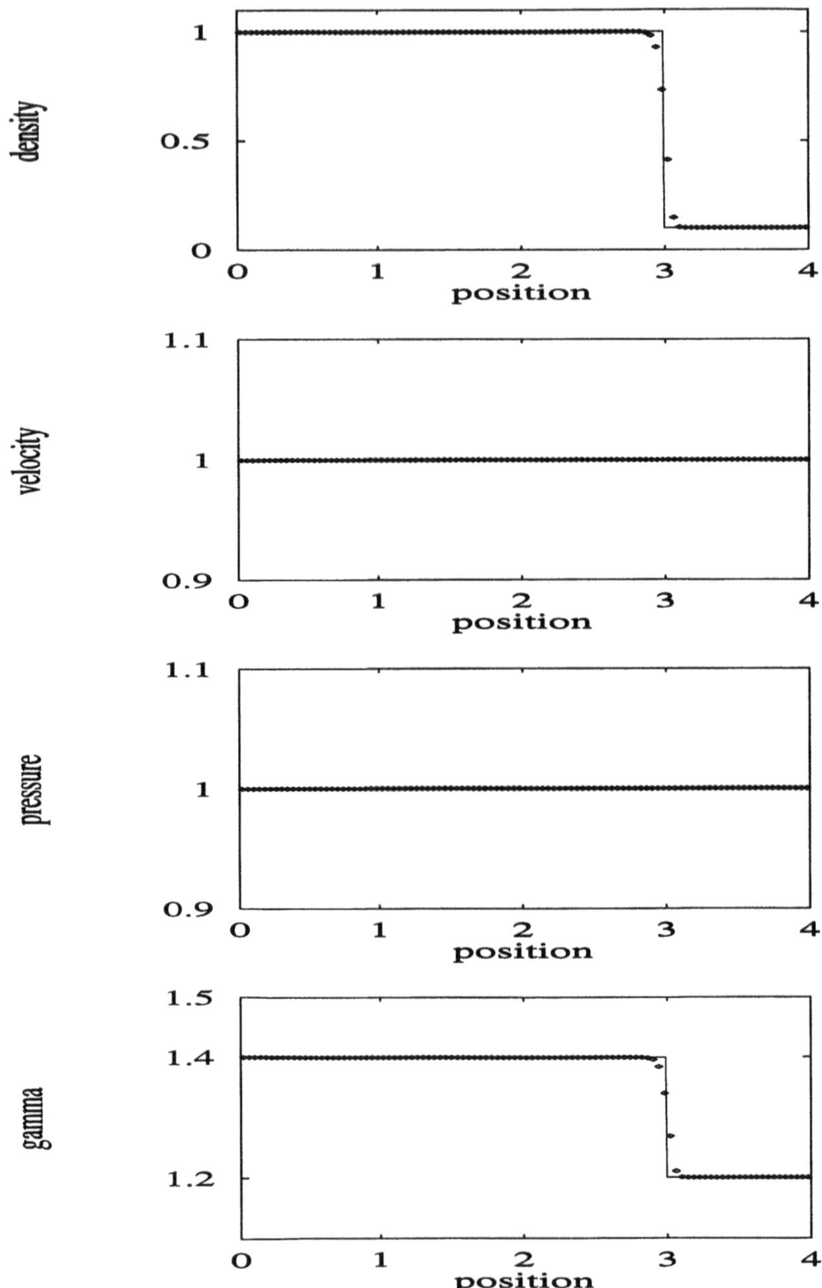

Figure 24. TEST 3. Comparison between primitive WAF TVD scheme (symbol) using mesh $M = 100$ and the exact solution (line), at time 1.0

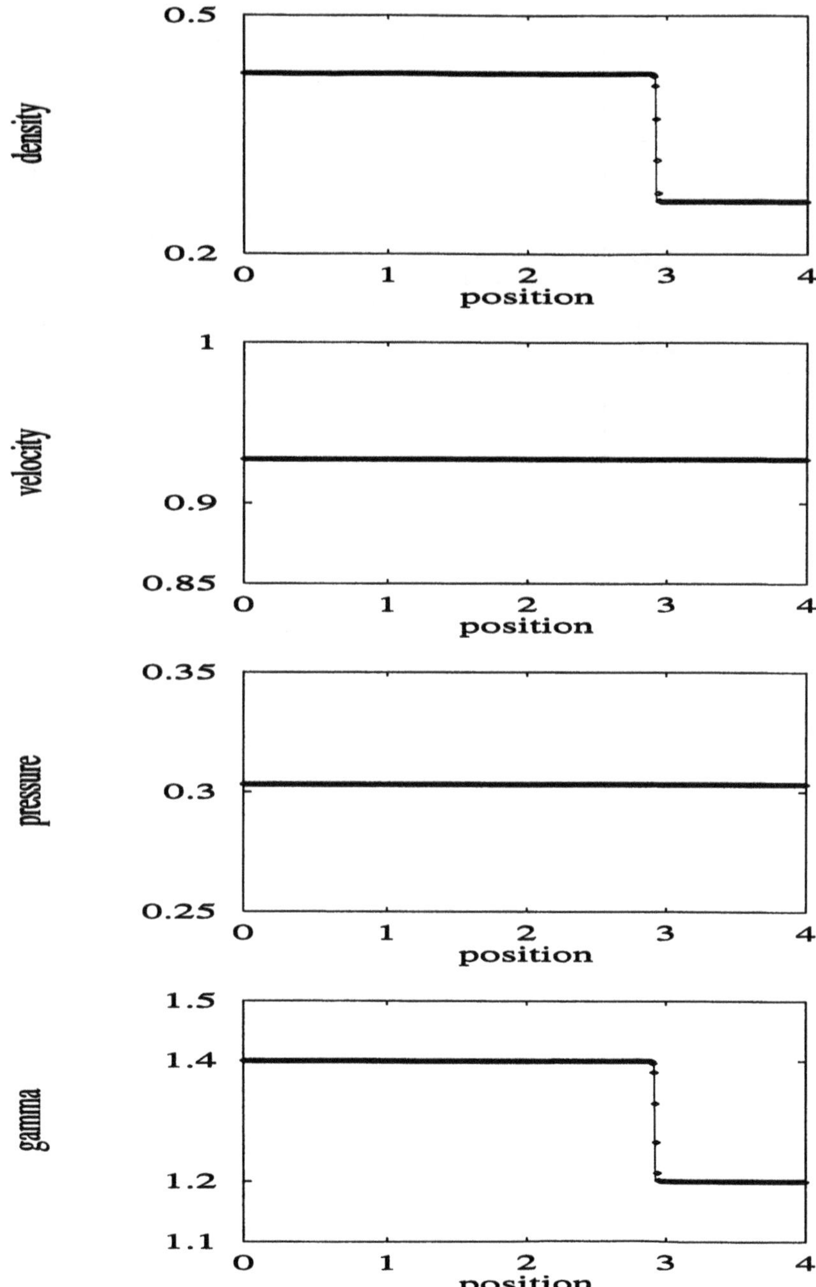

Figure 25. TEST 3. Comparison between primitive WAF TVD scheme (symbol) using mesh $M = 500$ and the exact solution (line), at time 1.0

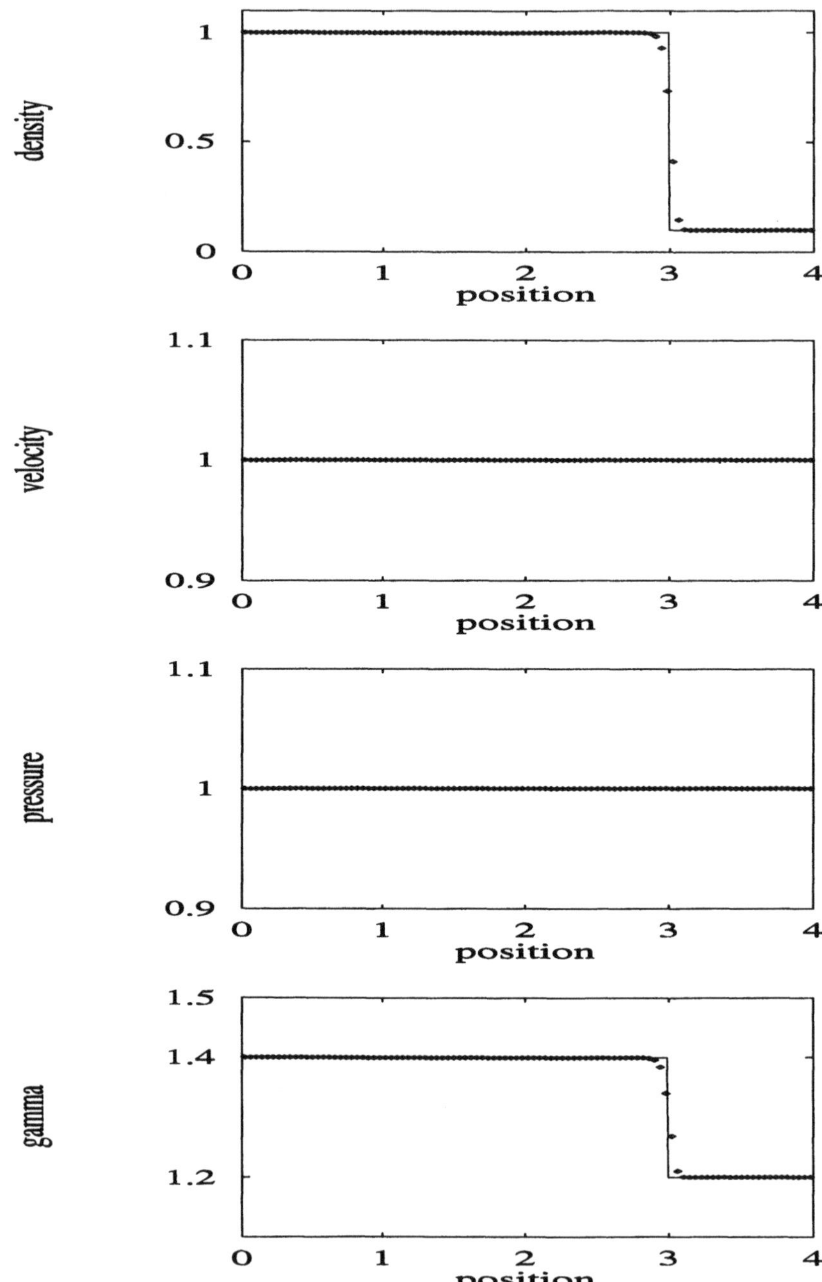

Figure 26. TEST 3. Comparison between adaptive primitive/conservative WAF TVD scheme (symbol) using mesh $M = 100$ and the exact solution (line), at time 1.0

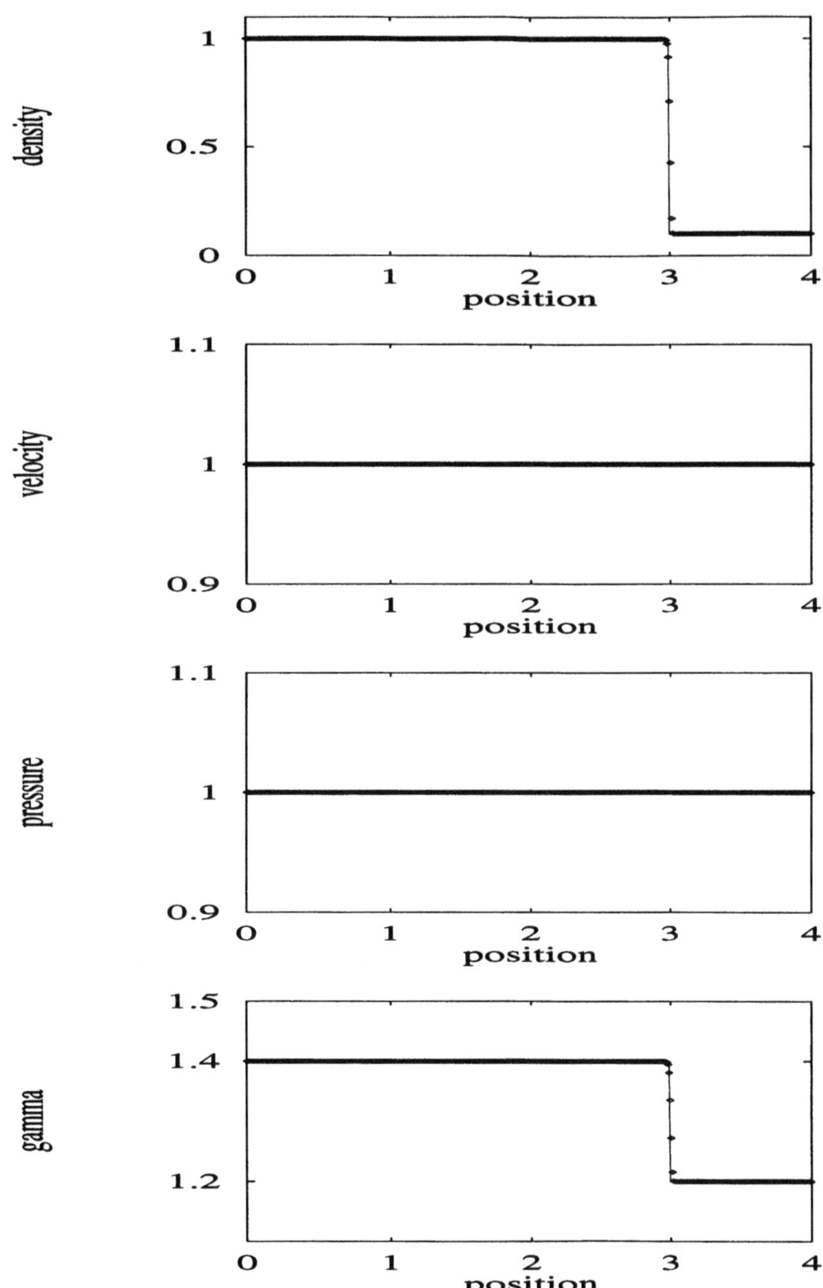

Figure 27. TEST 3. Comparison between adaptive primitive/conservative WAF TVD scheme (symbol) using mesh $M = 500$ and the exact solution (line), at time 1.0

Mechanics

FLUID MECHANICS AND ITS APPLICATIONS

Series Editor: R. Moreau

Aims and Scope of the Series

The purpose of this series is to focus on subjects in which fluid mechanics plays a fundamental role. As well as the more traditional applications of aeronautics, hydraulics, heat and mass transfer etc., books will be published dealing with topics which are currently in a state of rapid development, such as turbulence, suspensions and multiphase fluids, super and hypersonic flows and numerical modelling techniques. It is a widely held view that it is the interdisciplinary subjects that will receive intense scientific attention, bringing them to the forefront of technological advancement. Fluids have the ability to transport matter and its properties as well as transmit force, therefore fluid mechanics is a subject that is particularly open to cross fertilisation with other sciences and disciplines of engineering. The subject of fluid mechanics will be highly relevant in domains such as chemical, metallurgical, biological and ecological engineering. This series is particularly open to such new multidisciplinary domains.

1. M. Lesieur: *Turbulence in Fluids.* 2nd rev. ed., 1990 ISBN 0-7923-0645-7
2. O. Métais and M. Lesieur (eds.): *Turbulence and Coherent Structures.* 1991 ISBN 0-7923-0646-5
3. R. Moreau: *Magnetohydrodynamics.* 1990 ISBN 0-7923-0937-5
4. E. Coustols (ed.): *Turbulence Control by Passive Means.* 1990 ISBN 0-7923-1020-9
5. A.A. Borissov (ed.): *Dynamic Structure of Detonation in Gaseous and Dispersed Media.* 1991 ISBN 0-7923-1340-2
6. K.-S. Choi (ed.): *Recent Developments in Turbulence Management.* 1991 ISBN 0-7923-1477-8
7. E.P. Evans and B. Coulbeck (eds.): *Pipeline Systems.* 1992 ISBN 0-7923-1668-1
8. B. Nau (ed.): *Fluid Sealing.* 1992 ISBN 0-7923-1669-X
9. T.K.S. Murthy (ed.): *Computational Methods in Hypersonic Aerodynamics.* 1992 ISBN 0-7923-1673-8
10. R. King (ed.): *Fluid Mechanics of Mixing.* Modelling, Operations and Experimental Techniques. 1992 ISBN 0-7923-1720-3
11. Z. Han and X. Yin: *Shock Dynamics.* 1993 ISBN 0-7923-1746-7
12. L. Svarovsky and M.T. Thew (eds.): *Hydroclones.* Analysis and Applications. 1992 ISBN 0-7923-1876-5
13. A. Lichtarowicz (ed.): *Jet Cutting Technology.* 1992 ISBN 0-7923-1979-6
14. F.T.M. Nieuwstadt (ed.): *Flow Visualization and Image Analysis.* 1993 ISBN 0-7923-1994-X
15. A.J. Saul (ed.): *Floods and Flood Management.* 1992 ISBN 0-7923-2078-6
16. D.E. Ashpis, T.B. Gatski and R. Hirsh (eds.): *Instabilities and Turbulence in Engineering Flows.* 1993 ISBN 0-7923-2161-8
17. R.S. Azad: *The Atmospheric Boundary Layer for Engineers.* 1993 ISBN 0-7923-2187-1
18. F.T.M. Nieuwstadt (ed.): *Advances in Turbulence IV.* 1993 ISBN 0-7923-2282-7
19. K.K. Prasad (ed.): *Further Developments in Turbulence Management.* 1993 ISBN 0-7923-2291-6
20. Y.A. Tatarchenko: *Shaped Crystal Growth.* 1993 ISBN 0-7923-2419-6

Kluwer Academic Publishers – Dordrecht / Boston / London

Mechanics

FLUID MECHANICS AND ITS APPLICATIONS

Series Editor: R. Moreau

21. J.P. Bonnet and M.N. Glauser (eds.): *Eddy Structure Identification in Free Turbulent Shear Flows.* 1993 ISBN 0-7923-2449-8
22. R.S. Srivastava: *Interaction of Shock Waves.* 1994 ISBN 0-7923-2920-1
23. J.R. Blake, J.M. Boulton-Stone and N.H. Thomas (eds.): *Bubble Dynamics and Interface Phenomena.* 1994 ISBN 0-7923-3008-0
24. R. Benzi (ed.): *Advances in Turbulence V.* 1995 ISBN 0-7923-3032-3
25. B.I. Rabinovich, V.G. Lebedev and A.I. Mytarev: *Vortex Processes and Solid Body Dynamics.* The Dynamic Problems of Spacecrafts and Magnetic Levitation Systems. 1994 ISBN 0-7923-3092-7
26. P.R. Voke, L. Kleiser and J.-P. Chollet (eds.): *Direct and Large-Eddy Simulation I.* Selected papers from the First ERCOFTAC Workshop on Direct and Large-Eddy Simulation. 1994 ISBN 0-7923-3106-0
27. J.A. Sparenberg: *Hydrodynamic Propulsion and its Optimization.* Analytic Theory. 1995 ISBN 0-7923-3201-6
28. J.F. Dijksman and G.D.C. Kuiken (eds.): *IUTAM Symposium on Numerical Simulation of Non-Isothermal Flow of Viscoelastic Liquids.* Proceedings of an IUTAM Symposium held in Kerkrade, The Netherlands. 1995 ISBN 0-7923-3262-8
29. B.M. Boubnov and G.S. Golitsyn: *Convection in Rotating Fluids.* 1995 ISBN 0-7923-3371-3
30. S.I. Green (ed.): *Fluid Vortices.* 1995 ISBN 0-7923-3376-4
31. S. Morioka and L. van Wijngaarden (eds.): *IUTAM Symposium on Waves in Liquid/Gas and Liquid/Vapour Two-Phase Systems.* 1995 ISBN 0-7923-3424-8
32. A. Gyr and H.-W. Bewersdorff: *Drag Reduction of Turbulent Flows by Additives.* 1995 ISBN 0-7923-3485-X
33. Y.P. Golovachov: *Numerical Simulation of Viscous Shock Layer Flows.* 1995 ISBN 0-7923-3626-7
34. J. Grue, B. Gjevik and J.E. Weber (eds.): *Waves and Nonlinear Processes in Hydrodynamics.* 1996 ISBN 0-7923-4031-0
35. P.W. Duck and P. Hall (eds.): *IUTAM Symposium on Nonlinear Instability and Transition in Three-Dimensional Boundary Layers.* 1996 ISBN 0-7923-4079-5
36. S. Gavrilakis, L. Machiels and P.A. Monkewitz (eds.): *Advances in Turbulence VI.* Proceedings of the 6th European Turbulence Conference. 1996 ISBN 0-7923-4132-5
37. K. Gersten (ed.): *IUTAM Symposium on Asymptotic Methods for Turbulent Shear Flows at High Reynolds Numbers.* Proceedings of the IUTAM Symposium held in Bochum, Germany. 1996 ISBN 0-7923-4138-4
38. J. Verhás: *Thermodynamics and Rheology.* 1997 ISBN 0-7923-4251-8
39. M. Champion and B. Deshaies (eds.): *IUTAM Symposium on Combustion in Supersonic Flows.* Proceedings of the IUTAM Symposium held in Poitiers, France. 1997 ISBN 0-7923-4313-1
40. M. Lesieur: *Turbulence in Fluids.* Third Revised and Enlarged Edition. 1997 ISBN 0-7923-4415-4; Pb: 0-7923-4416-2

Kluwer Academic Publishers – Dordrecht / Boston / London

Mechanics

FLUID MECHANICS AND ITS APPLICATIONS

Series Editor: R. Moreau

41. L. Fulachier, J.L. Lumley and F. Anselmet (eds.): *IUTAM Symposium on Variable Density Low-Speed Turbulent Flows.* Proceedings of the IUTAM Symposium held in Marseille, France. 1997 ISBN 0-7923-4602-5
42. B.K. Shivamoggi: *Nonlinear Dynamics and Chaotic Phenomena.* An Introduction. 1997 ISBN 0-7923-4772-2
43. H. Ramkissoon, *IUTAM Symposium on Lubricated Transport of Viscous Materials.* Proceedings of the IUTAM Symposium held in Tobago, West Indies. 1998 ISBN 0-7923-4897-4
44. E. Krause and K. Gersten, *IUTAM Symposium on Dynamics of Slender Vortices.* Proceedings of the IUTAM Symposium held in Aachen, Germany. 1998 ISBN 0-7923-5041-3
46. U. Frisch (ed.): *Advances in Turbulence VII.* Proceedings of the Seventh European Turbulence Conference, held in Saint-Jean Cap Ferrat, 30 June -- 3 July 1998. 1998. ISBN 0-7923-5115-0
47. E.F. Toro and J.F. Clarke, *Numerical Methods for Wave Propagation.* Selected Contributions from the Workshop held in Manchester, UK. 1998 ISBN 0-7923-5125-8

Kluwer Academic Publishers – Dordrecht / Boston / London

ICASE/LaRC Interdisciplinary Series in Science and Engineering

1. J. Buckmaster, T.L. Jackson and A. Kumar (eds.): *Combustion in High-Speed Flows*. 1994 ISBN 0-7923-2086-X
2. M.Y. Hussaini, T.B. Gatski and T.L. Jackson (eds.): *Transition, Turbulence and Combustion*. Volume I: Transition. 1994 ISBN 0-7923-3084-6; set 0-7923-3086-2
3. M.Y. Hussaini, T.B. Gatski and T.L. Jackson (eds.): *Transition, Turbulence and Combustion*. Volume II: Turbulence and Combustion. 1994 ISBN 0-7923-3085-4; set 0-7923-3086-2
4. D.E. Keyes, A. Sameh and V. Venkatakrishnan (eds): *Parallel Numerical Algorithms*. 1997 ISBN 0-7923-4282-8
5. T.G. Campbell, R.A. Nicolaides and M.D. Salas (eds.): *Computational Electromagnetics and Its Applications*. 1997 ISBN 0-7923-4733-1
6. V. Venkatakrishnan, M.D. Salas and S.R. Chakravarthy (eds.): *Barriers and Challenges in Computational Fluid Dynamics*. 1998 ISBN 0-7923-4855-9

KLUWER ACADEMIC PUBLISHERS – DORDRECHT / BOSTON / LONDON

GPSR Compliance
The European Union's (EU) General Product Safety Regulation (GPSR) is a set of rules that requires consumer products to be safe and our obligations to ensure this.

If you have any concerns about our products, you can contact us on

ProductSafety@springernature.com

In case Publisher is established outside the EU, the EU authorized representative is:

Springer Nature Customer Service Center GmbH
Europaplatz 3
69115 Heidelberg, Germany

www.ingramcontent.com/pod-product-compliance
Ingram Content Group UK Ltd.
Pitfield, Milton Keynes, MK11 3LW, UK
UKHW021858190726
13853UKWH00003B/1325

* 9 7 8 9 4 0 1 5 9 1 3 8 6 *